WITHDRAWN
W9-ADY-826
LIBRARY

SUPERACIDS

SUPERACIDS

GEORGE A. OLAH
G. K. SURYA PRAKASH

Donald P. and Katherine B. Loker Hydrocarbon Research Institute
and Department of Chemistry University of Southern California
Los Angeles, California

JEAN SOMMER

Department of Chemistry
Louis Pasteur University
Strasbourg, France

A WILEY-INTERSCIENCE PUBLICATION

JOHN WILEY & SONS

New York • Chichester • Brisbane • Toronto • Singapore

Library of Congress Cataloging in Publication Data:

Olah, George A. (George Andrew), 1927–
Superacids.

"A Wiley-Interscience publication."
Includes bibliographies and index.
1. Superacids. I. Prakash, G. K. Surya. II. Sommer, Jean. III. Title.

QD499.043 1985 546'.24 84-20980
ISBN 0-471-88469-3

Printed in the United States of America

10 9 8 7 6 5 4 3 2 1

In memory of
James Bryant Conant,

who first recognized
superacids

and named them
accordingly

PREFACE

The chemistry of superacids, that is, of acid systems stronger than conventional strong mineral Brønsted acids such as sulfuric acid or Lewis acids like aluminum trichloride, has developed in the last two decades into a field of growing interest and importance. It was J. B. Conant who in 1927 gave the name "superacids" to acids that were capable of protonating certain weak bases such as carbonyl compounds and called attention to acid systems stronger than conventional mineral acids. The realization that Friedel-Crafts reactions are, in general, acid catalyzed with conjugate Lewis-Brønsted acid systems frequently acting as the de facto catalysts extended the scope of acid-catalyzed reactions. Friedel-Crafts acid systems, however, are usually only 10^3 to 10^6 times stronger than 100% sulfuric acid. The development in the early 1960s of Magic Acid, fluoroantimonic acid, and related conjugate superacids, 10^7 to 10^{19} times stronger than sulfuric acid added a new dimension to and revival of interest in superacids and their chemistry. The initial impetus was given by the discovery that stable, long-lived, electron-deficient cations, such as carbocations, acidic oxonium ions, halonium ions, and halogen cations can be obtained in these highly acidic systems. Subsequent work opened up new vistas of chemistry and a fascinating, broad field of chemistry is developing at superacidities. Because acidity is a term related to a reference base, superacidity allows extension of acid-catalyzed reactions to very weak bases and thus extends, for example, hydrocarbon chemistry to saturated systems including methane.

Some years ago in two review articles (*Science 206,* 13, 1979; *La Recherche 10,* 624, 1979), we briefly reviewed some of the emerging novel aspects of superacids. However, we soon realized that the field was growing so fast that to be able to provide a more detailed survey for the interested chemist a more comprehensive review was required. Hence, we welcomed the suggestion of our publisher and Dr. Theodore P. Hoffman, chemistry editor of Wiley-Interscience, that we write a monograph on superacids.

We are unable to thank all of our friends and colleagues who directly or indirectly contributed to the development of the chemistry of superacids. The main credit goes to all researchers in the field whose work created and continues to enrich this fascinating area of chemistry. Professor R. J. Gillespie's pioneering work on the inorganic chemistry of superacids was of immense value and inspiration to the development of the whole field. Our specific thanks are due to Drs. David Meidar and Khosrow Laali, who helped with the review of solid superacid systems and their reactions. Professor E. M. Arnett is thanked for reading part of our manuscript and for his thoughtful comments.

Finally we would like to thank Mrs. R. Choy, who tirelessly and always cheerfully typed the manuscript.

GEORGE A. OLAH
G. K. SURYA PRAKASH
JEAN SOMMER

Los Angeles, California
January 1985

CONTENTS

CHAPTER 2 SUPERACID SYSTEMS 33

CHAPTER 3 CARBOCATIONS IN SUPERACIDS 65

CHAPTER 4 HETEROCATIONS IN SUPERACIDS 177

CHAPTER 5 SUPERACID CATALYZED REACTIONS 243

SUPERACIDS

chapter 1

GENERAL ASPECTS

1.1 DEFINING ACIDITY

1.1.1 Acids and Bases

The concept of acidity was born in ancient times to describe the physiological property such as taste of food or beverage (in Latin *acidus:* sour; *acetum:* vinegar). Later during the development of experimental chemistry it was soon realized that mineral acids such as sulfuric, nitric, and hydrochloric acids played a key role in chemical transformations. Our present understanding of acid-induced or catalyzed reactions covers an extremely broad field ranging from large-scale industrial processes in hydrocarbon chemistry to enzyme-controlled reactions in the living cell.

The chemical species that plays a unique and privileged role in acidity is the hydrogen nucleus, i.e., the proton: H^+. Since its $1s$ orbital is empty, the proton is not prone to electronic repulsion and by itself has a powerful polarizing effect. Due to its very strong electron affinity, it cannot be found as a free "naked" species in the condensed state but is always associated with one or more molecules of the acid itself or of the solvent. Free protons exist only in the gas phase (such as in mass spectrometric studies). Regardless, as a shorthand notation, one generally depicts the proton in solution chemistry as "H^+." Due to its very small size (10^5 times smaller than any other cation) and the fact that only the $1s$ orbital is used in bonding by hydrogen, proton transfer is a very facile chemical reaction and does not necessitate important reorganization of the electronic valence shells. Understanding the nature of the proton is important while generalizing quantitative relationships in acidity measurements.[1]

The first clear definition of acidity can be attributed to Arrhenius, who between 1880 and 1890 elaborated the theory of ionic dissociation in water to explain the variation in strength of different acids.[2] Based on electrolytic experiments such as conductance measurement, he defined acids as substances that dissociate in water and yield the hydrogen ion whereas bases dissociate to yield hydroxide ion. In 1923 J. N. Brønsted generalized this concept to other solvents.[3] He defined an acid as a species that can donate a proton and base as a species that can accept it. This definition is generally known as the Brønsted-Lowry concept. The dissociation of an acid HA in a solvent S can be written as an acid-base equilibrium.

$$HA + S \rightleftharpoons A^- + SH^+ \tag{1}$$

The ionization of the acid HA in solvent S leads to a new acid HS^+ and a base A^-. Equation (1) has a very wide scope and can be very well applied to neutral and positively and negatively charged acid systems. The acid-base pair that differs only by a proton is referred to as the conjugate acid-base pair. Thus, H_2O is the conjugate base of the acid H_3O^+. An obvious consequence of the concept is that the extent to which an acid ionizes depends on the basicity of the solvent in which the ionization takes place. This shows the difficulty in establishing an absolute acidity scale. Acidity scales are energy scales and thus they are arbitrary both with respect to the reference point and to the magnitude of units chosen.

Fortunately, many of the common solvents by themselves are capable of acting as acids and bases. These amphoteric or amphiprotic solvents undergo self-ionization, for example,

$$2H_2O \rightleftharpoons H_3O^+ + OH^- \tag{2}$$

$$2HF \rightleftharpoons H_2F^+ + F^- \tag{3}$$

and, in a general way,

$$2HA \rightleftharpoons H_2A^+ + A^- \tag{4}$$

This equilibrium is characterized by the autoprotolysis constant K_{ap}, which under the usual high dilution conditions can be written as the following:

$$K_{ap} = [H_2A^+]\,[A^-] \tag{5}$$

Indeed the extent of dissociation of the solvent is very small (in HF, $K_{ap} \sim 10^{-11}$; in H_2O, $K_{ap} = 10^{-14}$). The pK_{ap} value that gives the acidity range will be discussed later.

It was G. N. Lewis who extended and generalized the acid-base concept to nonprotonic systems.[4] He defined an acid as a substance that can accept electrons and a base as a substance that can donate electrons. Lewis acids are electron-deficient molecules or ions such as BF_3 or carbocations, whereas Lewis bases are molecules that contain readily available nonbonded electron pairs (as in ethers, amines, etc.)

$$BF_3 + :O(CH_3)_2 \rightleftharpoons \overset{-}{B}F_3{:}\overset{+}{O}(CH_3)_2$$

Of course, in a generalized way, the proton H^+ is also a Lewis acid and the Brønsted acids and bases also fall into the Lewis categories.

Considering the general Equation (4) for the autoionization of solvent HA, one can define an acid as any substance that will increase $[H_2A^+]$ and a base as any substance that will increase $[A^-]$ and thus decrease $[H_2A^+]$. This definition, which includes both Lewis' and Brønsted's concepts, is used in practice while measuring the acidity of a solution by pH.

A number of strategies have been developed for acidity measurements of both aqueous and nonaqueous solutions. We will briefly review the most important ones and their use in establishing acidity scales.

1.1.2 The pH Scale

The concentration of the acid itself is of little significance other than analytical, with the exception of strong acids in dilute aqueous solutions. The concentration of H^+ itself is not satisfactory because it is solvated diversely and the ability of transferring a proton to another base depends on the nature of the medium. The real physical quantity describing the acidity of a medium is the activity of the proton $a_H{}^+$. The experimental determination of the activity of the proton requires the measurement of the potential of a hydrogen electrode or a glass electrode in equilibrium with the solution to be tested. The equation is of the following type, wherein C is a constant.

$$E = C - \frac{RT}{F} \log_{10} (a_{H^+}) \tag{6}$$

It was Sørensen's[5] idea to use this relationship, which can be considered as a basis to the modern definition of the pH scale of acidity for aqueous solutions. The pH of a dilute solution of acid is related to the concentration of the solvated proton from equation: $\text{pH} = -\log [HS^+]$. Depending on the dilution, the proton can be further solvated by two or more solvent molecules.

When the acid solution is highly diluted in water, the pH measurement is convenient, but it becomes critical when the acid concentration increases and, even more so, if nonaqueous media are employed. Since a reference cell is used with a liquid junction, the potential at the liquid junction also has to be known. The hydrogen ion activity cannot be measured independently and for this reason the equality $\text{pH} = -\log_{10}(a_{H^+})$ cannot be definitely established for any solution. Under the best experimental conditions, the National Bureau of Standard has set up a series of standard solutions of pH from which the pH of any other aqueous solution can be extrapolated as long as the ionic strength of the solution is not higher than 0.1 M. For more concentrated solutions, the pH scale will no longer have any real significance. In extending the limit to 1 M solutions, it is apparent that the available range of acidity is directly related to the autoprotolysis constant (Equation 5), as the minimum value of pH in a solution is zero and the maximum value is pK_{ap}; $pK_{ap} = p(HA_2{}^+) + p(A^-)$. Thus, the range of pH (ΔpH) is pK_{ap} (for water, 14 pH units). These limiting conditions are rather unfortunate because many chemical transformations are achieved beyond this range and under much less ideal conditions.

1.1.3 The Acidity Functions

Considering the limited application of the pH scale, a quantitative scale is needed to express the acidity of more concentrated or nonaqueous solutions.

A knowledge of the acidity parameter should permit one to estimate the degree of transformation of a given base (to its protonated form) in its conjugate acid. This should allow one to relate these data to the rate of acid-catalyzed reactions. Hammett and Deyrup[6a] in 1932 were the first to suggest a method for measuring the degree of protonation of weakly basic indicators in acid solution. The proton transfer equilibrium in the acid solution between an electro-neutral weak base B and the solvated proton can be written as follows:

$$\mathrm{B} + \mathrm{AH_2}^+ \rightleftharpoons \mathrm{BH}^+ + \mathrm{AH} \qquad (7)$$

Bearing in mind that the proton is solvated (AH_2^+) and that AH is the solvent, the equilibrium can be written

$$\mathrm{B} + \mathrm{H}^+ \rightleftharpoons \mathrm{BH}^+ \qquad (8)$$

The corresponding thermodynamic equilibrium constant is K_{BH^+}, which is expressed as

$$K_{BH^+} = \frac{a_{H^+} \cdot a_B}{a_{BH^+}} = \frac{a_{H^+} \cdot C_B}{C_{BH^+}} \cdot \frac{f_B}{f_{BH^+}} \qquad (9)$$

in which a is the activity, C the concentration, and f the activity coefficient. From this equation, it follows that

$$\frac{C_{BH^+}}{C_B} = \frac{1}{K_{BH^+}} \cdot a_{H^+} \cdot \frac{f_B}{f_{BH^+}} \qquad (10)$$

As the first ratio represents the degree of protonation, Hammett and Deyrup[6] defined the acidity function H_0 by

$$H_0 = -\log a_{H^+} \cdot \frac{f_B}{f_{BH^+}} = -\log K_{BH^+} + \log \frac{C_B}{C_{BH^+}} \qquad (11)$$

which can be written for further discussion in the more usual form:

$$H_0 = \mathrm{p}K_{BH^+} - \log \frac{[\mathrm{BH}^+]}{[\mathrm{B}]} \qquad (12)$$

From Equation (11) it is clear that in dilute aqueous solution, as the activity coefficients tend to unity, the Hammett acidity function becomes identical with pH. On the other hand, by making the fundamental assumption that the ratio f_B/f_{BH^+} is the same for different bases in a given solution, Hammett postulated that the H_0 function was unique for a particular series of solutions of changing acidity. The first application was made for the H_2SO_4–H_2O system using a series of primary anilines as indicators. By starting with the strongest base B_1, the $pK_{B_1H^+}$ was measured in dilute aqueous solution. The pK of the next weaker base B_2 was then determined by measuring the ionization ratio of the two indicators in the same acid solution using the relation

$$pK_{B_1H^+} - pK_{B_2H^+} = \log \frac{[B_1H^+]}{[B_1]} - \log \frac{[B_2H^+]}{[B_2]} \tag{13}$$

The ionization ratio was measured by uv-visible spectroscopy. With the help of successively weaker primary aromatic amine indicators, the strongest base being *p*-nitroaniline (pK = 1.40) and the weakest trinitroaniline ($pK = -9$), Hammett explored the whole H_2O–H_2SO_4 range up to 100% sulfuric acid and the perchloric acid–water solution up to 60% of acid. Similar acidity functions such as H_-, H_+, H_{2+} were proposed related to acid-base equilibria in which the indicator is negatively, positively, or even dipositively charged. The validity of all of these functions is based on the simple assumption that the activity coefficient ratio is independent of the nature of the indicator at any given solvent composition. In this case the log $[BH^+]/[B]$ plots against H_0 should be linear with a slope of -1.00 for all neutral bases. This is not the case for groups of indicators with different structures, and especially different basic sites often show significant deviations. For this reason, it is well recognized now that the above assumption does not have a general validity. The measurement of a Hammett acidity function should be limited to those indicators for which log $[BH^+]/[B]$ plotted against H_0 gives a straight line with a negative unit slope. These indicators are called Hammett bases.

Equilibria other than proton transfer have also been used to determine acidity functions. One of these is based on the ionization of alcohols (mainly arylmethyl alcohols) in acid solution following the equilibrium

$$ROH + H^+ \rightleftharpoons R^+ + H_2O \tag{14}$$

The corresponding acidity function described as H_R is then

$$H_R = pK_{R^+} - \log \frac{[R^+]}{[ROH]} \tag{15}$$

This H_R function, also called J_0 function, has also been used to measure the acidity of the sulfuric acid-water and perchloric acid-water systems. It shows a large deviation from the H_0 scale in the highly concentrated solutions as shown in Figure 1.1. However, all these and other acidity functions are based on Hammett's

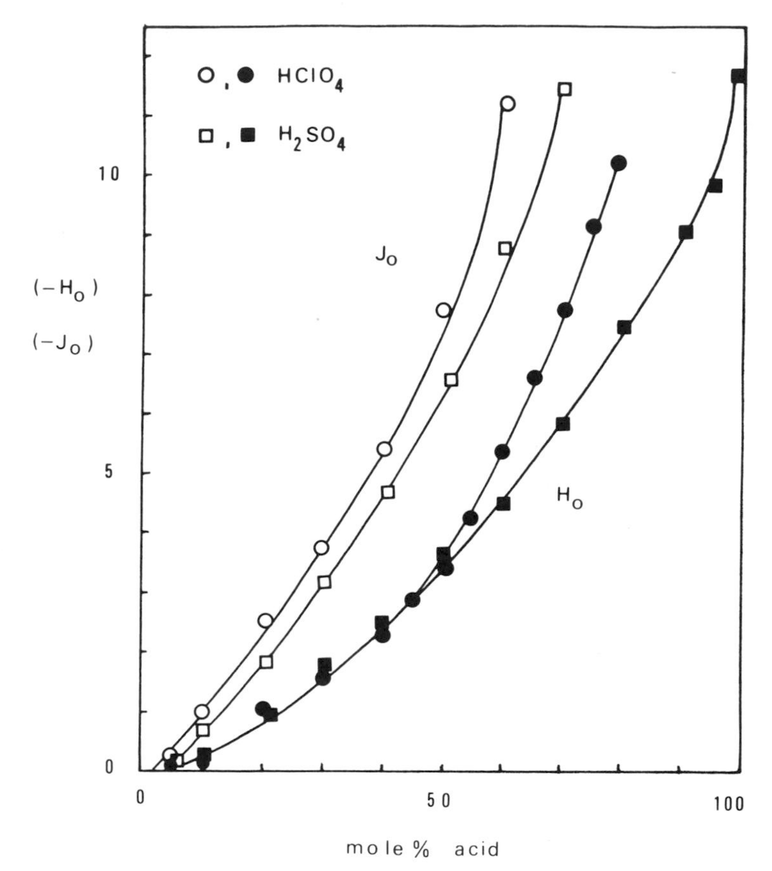

FIGURE 1.1 H_o and H_R functions for H_2SO_4—H_2O and $HClO_4$—H_2O systems. $HClO_4$: (○) Ref. 8; (●) Ref. 10. H_2SO_4: (□) Ref. 9; (■) Ref. 11.

principle and can be expressed by the equation

$$H_x = pK_A - \log \frac{A}{B} \tag{16}$$

in which B and A are the basic and the conjugate acidic form of the indicator, respectively. They become identical with the pH scale in highly dilute acid solutions. The relative and absolute validity of the different acidity functions have been the subject of much controversy and the subject has been extensively reviewed.[1,7]

Whatever may be the limitations of the concept first proposed by Hammett and Deyrup in 1932,[6a] until now no other comparable alternative has appeared to assess quantitatively the acidity of concentrated and nonaqueous strongly acidic solutions.[6b] The experimental methods that have been used to determine acidity functions are reviewed in Section 1.4.

1.2 DEFINITION OF SUPERACIDS

It was in the title of a paper published in 1927 by Conant and Hall[12] in the *American Chemical Society Journal* that the name "superacid" appeared for the first time in the chemical literature. In an electrochemical study of the hydrogen ion activity in a nonaqueous acid solution, these authors noticed that sulfuric acid and perchloric acid in glacial acetic acid were able to form salts with a variety of weak bases such as ketones and other carbonyl compounds which did not form salts with the aqueous solutions of the same acids. They ascribed this high acidity to the ionization of these acids in glacial acetic acid, increasing the concentration of $CH_3COOH_2^+$, a species less solvated than H_3O^+ in the aqueous acids. They proposed to call these solutions "superacid solutions." Their proposal was, however, not further followed up or used in the literature until the 1960s when Olah's studies of obtaining stable solutions of electron-deficient ions, particularly carbocations, focused interest on very high-acidity nonaqueous systems.[13] Subsequently, Gillespie proposed an arbitrary but since widely accepted definition of superacids,[14] defining them as any acid system that is stronger than 100% sulfuric acid, i.e., $H_0 \leqslant -12$. Fluorosulfuric acid and trifluoromethanesulfonic acid are examples of Brønsted acids that exceed the acidity of sulfuric acid with H_0 values of about -15.1 and -14.1, respectively.

To reach acidities beyond this limit, one has to start with an already strong acid ($H_0 \approx -10$) and add to it a stronger acid which increases the ionization. This can be achieved either by dissolving a strong Brønsted acid, HB, capable of ionizing in the medium

$$HA + HB \rightleftharpoons H_2A^+ + B^- \qquad (17)$$

or by adding a strong Lewis acid which will shift the autoprotonation equilibrium by forming a more delocalized counterion of the strong acid

$$2HA + L \rightleftharpoons H_2A^+ + LA^- \qquad (18)$$

In both cases, a remarkable acidity jump is noticed around the H_0 value of pure HA as shown in Figure 1.2 for HSO_3F.

It is this large acidity jump, generally more than 5 H_0 units, that raises a strong acid solution into the superacid region. Therefore, it becomes clear that the proposed reference of $H_0 = -12$ for the lower limit of superacidity is only arbitrary. It could as well be $H_0 = -11$ with HF as the solvent or $H_0 = -15.1$ with HSO_3F.

Gillespie's definition of superacids relates to Brønsted acid systems. As Lewis acids also cover a wide range of acidities extending beyond the strength of conventionally used systems, we suggested the use of anhydrous aluminum chloride as the arbitrary reference and we categorized Lewis superacids as those stronger than aluminum chloride[13b] (see, however, subsequent discussion on the difficulties to measure the strength of a Lewis acid).

It should be also noted that in biological chemistry, following a suggestion by Westheimer,[15a] it is customary to call catalysis by metal ions bound to enzyme

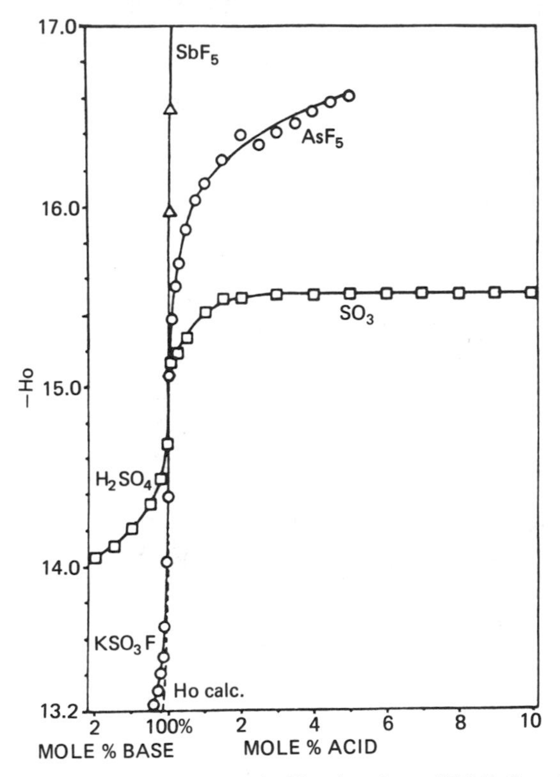

FIGURE 1.2 The acidity jump near the H_o value of neat HSO_3F (from Ref. 14b).

systems as "superacid catalysis." As the role of a metal ion is analogous to a proton, this arbitrary suggestion reflects enhanced activity and is in line with previously discussed Brønsted and Lewis superacids.

1.2.1 Range of Acidities

Despite the fact that superacids are already stronger than 100% sulfuric acid, there may be as much or more difference in acidity between different superacid systems than between neat sulfuric acid and its 0.1 *M* solution in water. The highest acidity level measured until now on the H_0 acidity scale is of the order of −27 for 90% SbF_5 in HSO_3F.[15b] Based on rate measurements of superacidic reactions, it seems that the fluoroantimonic acid system, HF:SbF_5 is even stronger. Infrared studies have shown that the H_2F^+ concentration was the highest at 80 mol% SbF_5, but until now no suitable indicator or method has been found to investigate accurately the acidity of the system beyond 2 mol% SbF_5.[16] One may reasonably expect of find acidity levels in the −30 range for the 1:1 HF:SbF_5 system. The unsolvated ("naked") proton of course is unobtainable in solution chemistry, but by comparing gas phase data with superacid solution protonation studies one could extrapolate its acidity into the -50 to -60 H_0 range. Figure 1.3 shows the comparative acidity range for the various superacids that have been investigated until now. Despite the fact that the reference scale used is the H_0 acidity scale, we must keep in mind that in most of the cases the indicators were not Hammett bases and for this reason one should consider the acidity values only in a relative sense. The various systems indicated in this figure are discussed in the following chapter.

A quantitative determination of the strength of Lewis acids to establish similar scales (H_0) as discussed in the case of protic (Brønsted type) superacids would be most useful. However, to establish such a scale is extremely difficult. Whereas the Brønsted acid-base interaction invariably involves a proton transfer reaction that allows meaningful comparison, in the Lewis acid-base interaction there is no such common denominator. Hence the term "strength" has no well-defined meaning.

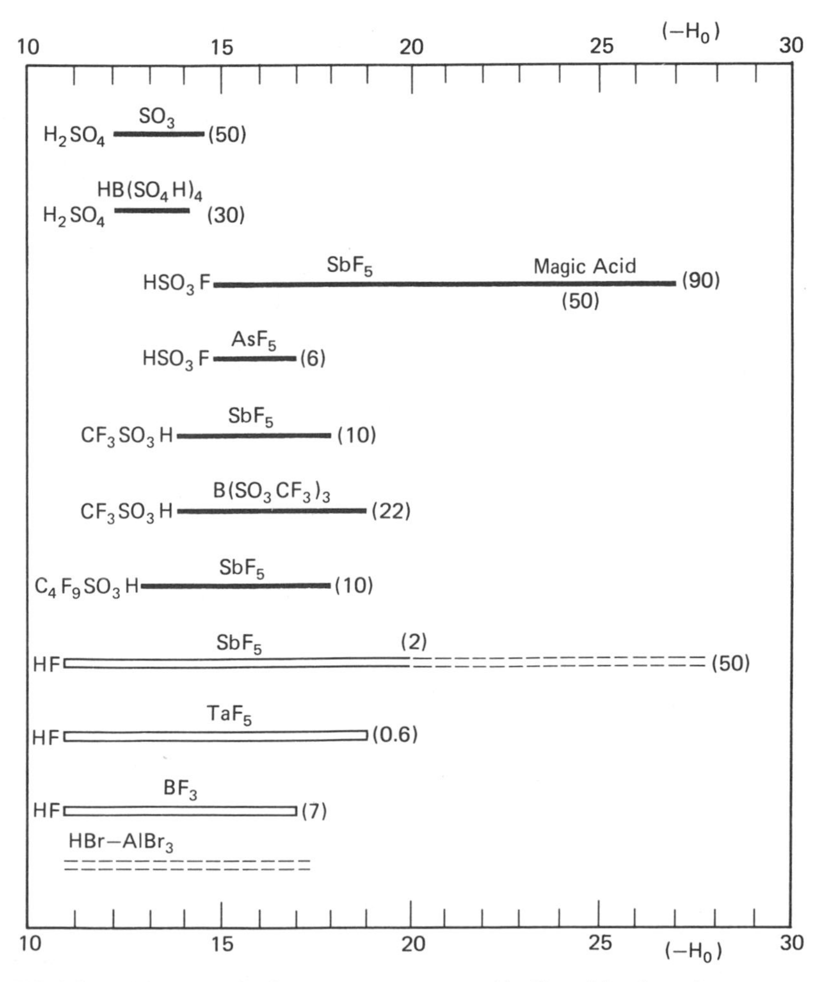

FIGURE 1.3 Acidity ranges for the most common superacids. The solid and open bars are measured using indicators; the broken bar is estimated by kinetic measurements; in ( ) mol % Lewis acid.

However, it is important to keep in mind that superacidity encompasses both Brønsted and Lewis acid systems. The qualitative picture of Lewis acid strengths will be discussed in Section 1.4.6.

The acidity range of solid superacids is not yet well known as accurate methods for their measurements have not been developed. In most cases, kinetic data of selected catalyzed reactions have to be used for comparative purposes. It has also been shown that indicators of $pK_{BH^+} = -16$ are fully ionized on some solid superacids.[17] It is still not clear to what extent the equilibrium concept used in solution acidity measurements can be extrapolated to heterogeneous solid systems, and consequently data based on such measurements may not be indicative of true acidity. Calorimetric studies on heats of protonation of weak bases on solid acids are recently gaining interest.[18] The obvious difficulty is that a small number of acid sites can bring about catalytic reactions on solid surfaces (similar to enzyme systems) and there seems to be no direct way to correlate heat measurements with selected weak bases or their absorption on the surface to catalytic activity.

1.3 TYPES OF SUPERACIDS

Superacids, similar to conventional acid systems, include both Brønsted and Lewis acids. Protic (Brønsted type) superacids include strong parent acids whose acidities can be further enhanced by various combinations with Lewis acids (conjugate acids). The following are the most frequently used superacids.

1.3.1 Brønsted and Conjugate Brønsted-Lewis Superacids

1. Perchloric acid and halosulfuric and perfluoroalkane sulfonic acids such as $HClO_4$, $ClSO_3H$, HSO_3F, CF_3SO_3H, R_FSO_3H, etc.
2. Oxygenated Brønsted acids such as H_2SO_4, HSO_3F, CF_3SO_3H, and R_FSO_3H in combination with Lewis acids such as SO_3, SbF_5, AsF_5, TaF_5, and NbF_5.
3. Hydrogen fluoride (HF) in combination with fluorinated Lewis acids as SbF_5, TaF_5, NbF_5, and BF_3.
4. Friedel-Crafts acids such as $HBr—AlBr_3$, $HCl—AlCl_3$, etc.

1.3.2 Lewis Superacids

SbF_5, NbF_5, AsF_5, TaF_5, etc.

1.3.3 Solid Superacids

Solid superacids can be divided into two groups, depending on the origin of the acid sites. The acidity may be a property of the solid as part of its chemical structure (possessing Lewis or Brønsted sites, the acidity of the latter can be further enhanced by complexing with Lewis acids). Solid superacids can also be obtained by deposition on or intercalation into an otherwise inert or low-acidity support.

1. Solid acids, the acidity of which can be enhanced:
   a. Acid-treated metal-oxides, mixed oxides, zeolites
   b. Perfluorinated resins and sulfonated cation exchange resins (complexed with Lewis acids)
2. Supported acids, intercalated superacids:
   a. Lewis acid or Brønsted acid intercalated graphite or fluorographite
   b. Supported Lewis or Brønsted superacids

As with previous classifications, these are also arbitrary and are suggested for practical discussion of an otherwise rather complex field. The individual superacid systems are discussed in Chapter 2.

1.4 EXPERIMENTAL TECHNIQUES FOR ACIDITY MEASUREMENTS (PROTIC ACIDS)

From Equation (11), it is apparent that the main experimental difficulty in determining acidities is the estimation of the ratio between the free base and the acid forms of a series of indicators, their so-called ionization ratios.

1.4.1 Spectrophotometric Method

In the early work of Hammett and co-workers,[6a] the measurement of the ionization ratio was based on the color change of the indicator. The solutions containing the indicator were compared at 25°C in a colorimeter with a standard reference. This reference was water when the indicator was colorless in its acid form and 96% sulfuric acid (or 70% perchloric acid) when the indicator was colorless in the basic form.

For example, when the indicator was colored in water the authors define a stoichiometric color intensity relative to water $I_w = C_w/C_a$, where C_a and C_w are the stoichiometric concentrations of indicator in solution A and in water. On the other hand, the specific color intensity of the colored form relative to water is defined as $S_w = [B]_w/[B]_a$. Where $[B]_w$ is the concentration of the colored base in water and $[B]_a$ is concentration in solution A. As the indicator exists only in its basic form in water $B_w = C_w$ and in the solution A, $[C]_a = [B]_a + [BH^+]_a$. The ionization ratio is given by the equation

$$\frac{[BH^+]}{[B]} = \frac{S_w - I_w}{I_w} \tag{19}$$

Despite half a century of technical and scientific progress, the original Hammett method has not become obsolete. The colorimeter has been replaced by modern spectrophotometers which can be operated at selected wavelengths extending the spectra beyond visible into the ultraviolet region of the electromagnetic spectrum. The experimental variable, which is wavelength dependent, is the optical density D. D is related to the concentration by the Beer-Lambert law

$$D_{i,\lambda} = \epsilon_{i,\lambda} C_i \cdot l \tag{20}$$

C_i is the concentration of the absorbing species, l is the length of the cell, and ϵ_i is the molar absorptivity (or extinction coefficient). If at a given wavelength λ, ϵ_{BH^+}, ϵ_B, and ϵ_λ are the extinction coefficients, respectively, of acid form of the indicator, its basic form, and of the unknown solution, the ionization ratio is given by

$$\frac{[BH^+]}{[B]} = \frac{\epsilon_B - \epsilon_\lambda}{\epsilon_\lambda - \epsilon_{BH^+}} \tag{21}$$

For a greater precision in this determination, λ should be chosen so as to have the maximum difference between ϵ_{BH^+} and ϵ_B. For this reason, the areas between the absorption line and the base line of both acidic and basic forms of the indicator should be measured and compared.

Whereas the precision of the method is generally excellent, a number of drawbacks may appear with some indicators. First, Equation (21) is true only with the assumption that ϵ is solvent independent (it is clear that ϵ_B and ϵ_{BH^+} cannot be measured separately in the same solution). The medium effect on the absorption spectrum (mainly the wavelength shift of λ_{max}) can be easily taken into account in the measurements. However, large changes in the absorption spectrum during the increase in ionization are difficult to correct. Another difficulty that might appear is the structural change of the indicator, during or after protonation. The change in temperature, however, has been shown in the H_2SO_4–H_2O system to have little effect on the H_0 value,[19] but the pK_{BH^+} and the ionization ratios are more sensitive.

The pK_{BH^+} value is easy to determine when the ionization ratio can be measured in dilute aqueous solution.

$$pK_{BH^+} = \text{limit}_{(HA \to O)} \log \frac{[BH^+]}{[B]} - \log [H_3O^+] \tag{22}$$

It is to be noted that when the acid solution is very dilute the presence of the indicator modifies the acidity: $[H_3O^+] = [HA] - [B]$, and thus the concentration of the indicator has to be taken into account.

As is apparent from Equation (11), an indicator is only useful over an acidity range where its ionization ratio can be measured experimentally with sufficient precision. For spectrophotometric method, this means approximately 2 log units per indicator. Accordingly, the direct determination of the pK_{BH^+} value of an indicator in concentrated solution is not possible. It is actually achieved by the method developed by Hammett in his early work using a series of indicators of overlapping range. Taking into account the overlapping of each indicator with the preceding and the following one, each of which is useful for 2 log units, it appears that several indicators are necessary (approximately as many indicators as the number of desired log units). This is illustrated in Figure 1.4.

Paul and Long[20] tabulated pK_{BH^+} values for indicators which were used to establish Hammett acidity functions for aqueous acids between the years 1932 and 1957. The data were summarized as a set of "best values" of pK_{BH^+} for the bases. Since then, subsequent work seems to suggest that some of these values are incorrect. This is particularly so for some of the weaker bases whose quoted pK_{BH^+} were based on a stepwise extrapolation of results of some indicators that have since been proven to be unsatisfactory based on the strict definition of H_0. These data, as well as those for weaker bases which have been studied since, covering the whole acidity range from dilute acid to the superacid media are collected in Table 1.1.

Up to a H_0 value of -10 all indicators are primary amines and are therefore suitable for the measurement of the Hammett H_0 function. For stronger acids, new indicators such as nitro compounds have to be used. Although the acidity function scale based upon nitro compounds as indicators may not be a satisfactory extension of the aniline indicator scale, Gillespie[14] has shown that the most basic nitro compound indicator *p*-nitrotoluene overlaps in a satisfactory manner with the weakest indicator in the aniline series, 2,4,6-trinitroaniline. Thus, the acidity measurements using the nitro compounds may be considered to give the best semiquantitative picture of the acidity of the various superacid systems.

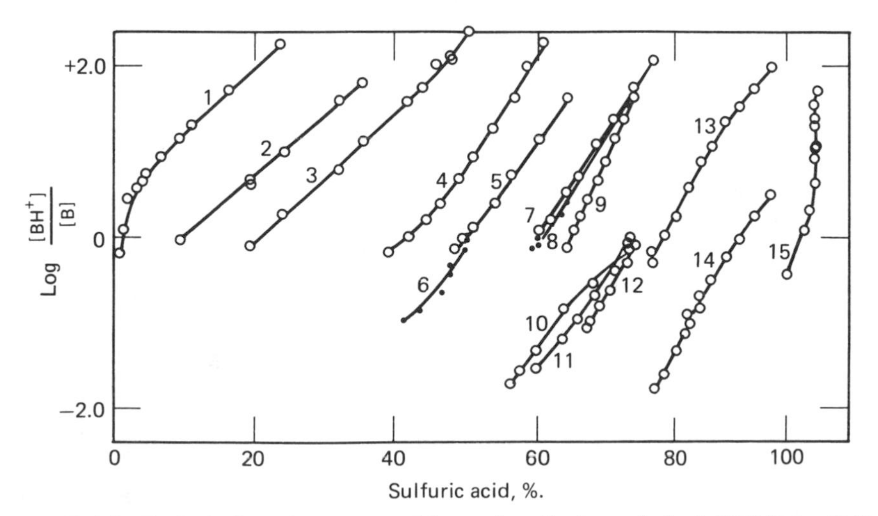

FIGURE 1.4 The ionization ratio as measured for a series of indicators in the 0–100% H_2O—H_2SO_4 system (Ref. 6a).

TABLE 1.1 Selected pK_{BH^+} Values for Extended Hammett Bases

Base	pK_{BH^+}	Reference
3-Nitroaniline	2.50	20
2-4-Dichloroaniline	2.02	20
4-Nitroaniline	0.99	20
2-Nitroaniline	−0.29	20
4-Chloro-2-nitroaniline	−1.03	20
5-Chloro-2-nitroaniline	−1.54	11
2,5-Dichloro-4-nitroaniline	−1.82	21
2-Chloro-6-nitroaniline	−2.46	21
2,6-Dichloro-4-nitroaniline	−3.24	21
2,4-Dichloro-6-nitroaniline	−3.29	20
2,6-Dinitro-4-methylaniline	−4.28	20
2,4-Dinitroaniline	−4.48	20
2,6-Dinitroaniline	−5.48	21
4-Chloro-2,6-dinitroaniline	−6.17	21
6-Bromo-2,4-dinitroaniline	−6.71	20
3-Methyl-2,4,6-trinitroaniline	−8.37	21
3-Bromo-2,4,6-trinitroaniline	−9.62	21
3-Chloro-2,4,6-trinitroaniline	−9.71	21
2,4,6-Trinitroaniline	−10.10	14
p-Nitrotoluene	−11.35	14
m-Nitrotoluene	−11.99	14
Nitrobenzene	−12.14	14
p-Nitrofluorobenzene	−12.44	14
p-Nitrochlorobenzene	−12.70	14
m-Nitrochlorobenzene	−13.16	14
2,4-Dinitrotoluene	−13.76	14
2,4-Dinitrofluorobenzene	−14.52	14
2,4,6-Trinitrotoluene	−15.60	14
1,3,5-Trinitrobenzene	−16.04	14
(2,4-Dinitrofluorobenzene)H^+	−17.57	14
(2,4,6-Trinitrotoluene)H^+	−18.36	22

1.4.2 Nuclear Magnetic Resonance Methods

NMR spectroscopy, which was developed in the late 1950s as the most powerful tool for structural analysis of organic compounds, has proven useful for acidity determinations. The measurement of the ionization ratio has been achieved by a variety of methods demonstrating the versatility of this technique. If we consider the general acid-base equilibrium Equation (23) obtained when the indicator B is dissolved in the strong acid HA, then k_p and k_d, respectively, are the rates of protonation and deprotonation,

$$HA + B \underset{k_d}{\overset{k_p}{\rightleftharpoons}} BH^+ + A^- \tag{23}$$

the thermodynamic equilibrium constant K is related to those rates by,

$$K = \frac{[BH^+]\,[A^-]}{[HA]\,[B]} = \frac{k_p}{k_d} \tag{24}$$

In nmr spectroscopy, when a species (for example here $[BH^+]$) is participating in an equilibrium, its spectrum is very much dependent on its mean lifetime (τ).[23,24] The inverse of the mean lifetime is a first-order rate constant, called the rate of exchange ($k = 1/\tau$) which can be obtained from the line-shape analysis of the nmr bands if $1\ s^{-1} \leqslant k \leqslant 10^3\ s^{-1}$. Three cases can thus be envisaged:

1. "Slow exchange" conditions: $k \leqslant 10^{-2}\ s^{-1}$. The species can be observed as if no exchange were taking place.
2. .Measurable exchange conditions: $1\ s^{-1} < k < 10^3\ s^{-1}$. The rate of exchange can be calculated from the line-shape analysis of the nmr bands of the exchanging species.
3. "Fast exchange" conditions: $k > 10^4\ s^{-1}$. The observed nmr bands appear as the weighted average of the species participating in the equilibrium.

Depending on these conditions, various nmr methods have been proposed and used to calculate the ionization ratio of weak bases in a superacid medium.

1.4.2.1 Chemical Shift Measurements

Under "slow-exchange" conditions, the ionization ratio cannot be measured. In fact one of the major advantages of the superacidic media is the ease with which weak bases can be fully protonated and directly observed by nmr. As it is known that the protonation rates are practically diffusion controlled ($\approx 10^9\ 1\ mol^{-1} \cdot s^{-1}$), under these conditions ($k \leqslant 10^{-2}\ s^{-1}$) the indicator is "totally" in the acidic form described by the nmr spectrum and no variable is available to measure the ionization ratio.

Under "fast-exchange" conditions, however, the nmr spectrum presents a weighed average of the bands of the exchanging species and with the sensitivity limits (~5–95%) the ionization ratio can be measured taking the chemical shift as a variable. The calculation is simply based on the observed chemical shift of the average line (δ_{obs}) provided that the chemical shift of the base indicator (δ_B) and of its acid form (δ_{BH^+}) are known. This is generally obtained by increasing or decreasing the acidity of the medium

$$\delta_{obs} = \frac{\delta_{BH^+}[BH^+] + \delta_B[B]}{[BH^+] + [B]} \tag{25}$$

By plotting the chemical shift variation against the acidity, one observes a typical acid-base titration curve (Fig. 1.5) and the pK_{BH^+} of the indicator can be determined this way. This nmr method which was first proposed by Grunwald et al.[25] has been applied by Levy and co-workers[26] using various ketones and α-halo-ketones for the determination of ketone basicity and evaluation of medium acidity.

Compared with spectrophotometry, the nmr method has a number of advantages: (1) The procedure is very rapid, and it can be used by observing the variation of chemical shift of diverse nuclei such as ^{1}H, ^{13}C, ^{19}F, and ^{17}O; (2) it is insensitive to colored impurities and slight decomposition of the indicator; (3) In principle it can be used over the whole range of known acidity. The medium effect which may be important in ^{1}H-nmr becomes negligible in the case of ^{13}C-nmr spectroscopy. The method can be used with a wide variety of weak bases having a lone-pair containing hetero-atoms as well as simple aromatic hydrocarbons.

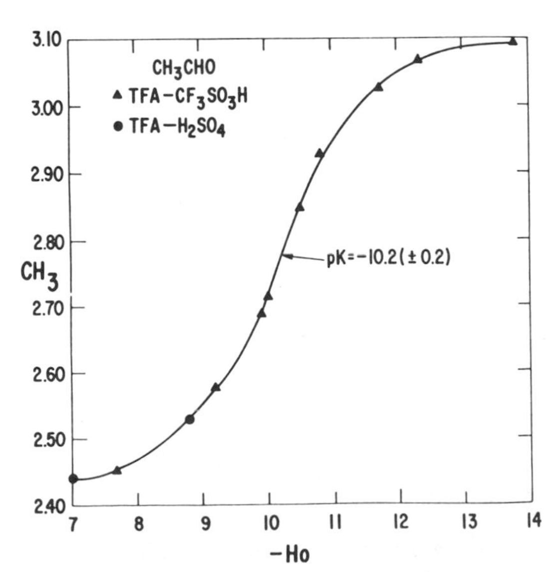

FIGURE 1.5 Acidity dependent ^{1}H chemical shift variation: protonation curve for acetaldehyde (Ref. 26).

1.4.2.2 *Exchange Rate Measurements Based on Line-Shape Analysis (DNMR: Dynamic Nuclear Magnetic Resonance)*

Under the measurable exchange rate conditions two possibilities have been considered:

1. The change in line shape can be directly related to the proton exchange.
2. The change in line shape is due to a separate exchange process related to the proton exchange.

Both methods have been exploited in the recent literature to determine ionization ratios.

Direct Exchange Rates. As mentioned earlier the line shape is directly related to the acid-base equilibrium. In Equation (23), the rate of exchange of BH^+ (k_{BH^+}) for example is related to the rate constants as shown:
The reaction rate,

$$\frac{d[BH^+]}{dt} = k_d[BH^+] \cdot [A^-] \tag{26}$$

and the rate of exchange,

$$\frac{d(BH^+]}{dt} = k_{BH^+} \cdot [BH^+] \tag{27}$$

thus from (26) and (27)

$$k_{BH^+} = k_d[A^-] \tag{28}$$

In the same way for [B], the rate of reaction is

$$\frac{d[B]}{dt} = k_p[B] \cdot [HA] \tag{29}$$

and the exchange-rate constant

$$k_B = \frac{1}{[B]}\frac{d[B]}{dt} = k_p[HA] \tag{30}$$

From Equations (24) and (30), the ionization ratio can be calculated as

$$\frac{[BH^+]}{[B]} = \frac{k_p}{k_{BH^+}}[HA] \tag{31}$$

When the indicator B behaves like a Hammett base, the ratio can be used to determine the H_0 acidity function

$$H_0 = pK_{BH^+} - \log \frac{kp \cdot [HA]}{k_{BH^+}} \quad (32)$$

With the assumption that k_p is a constant over the range of measured acidity and k_{BH^+} of a series of overlapping bases remains measurable (each base covering approximately 3 log units for a given concentration), Gold and co-workers explored the acidity of the HSO_3F-SbF_5 system containing up to 90 mol% SbF_5.[15b]

Indirect Exchange Rates. In this case, the line shape is indirectly related to the acid-base equilibrium. Besides measuring intermolecular processes like the proton exchange rates, DNMR often has been used to measure intramolecular processes like conformational changes that occur on the same time scale. When the activation energy of such a process is very different in the acidic and basic forms for an indicator, DNMR can be used to measure the ionization ratio as shown by the following example.

Aromatic carbonyl compounds have an activation energy barrier for rotation around the phenyl-carbonyl bond, the value of which is substantially increased upon protonation.[27] For example, the nmr spectrum at room temperature of the monoprotonated *para*-anisaldehyde appears as if the rotation was blocked (as in the mesomeric form **B**).

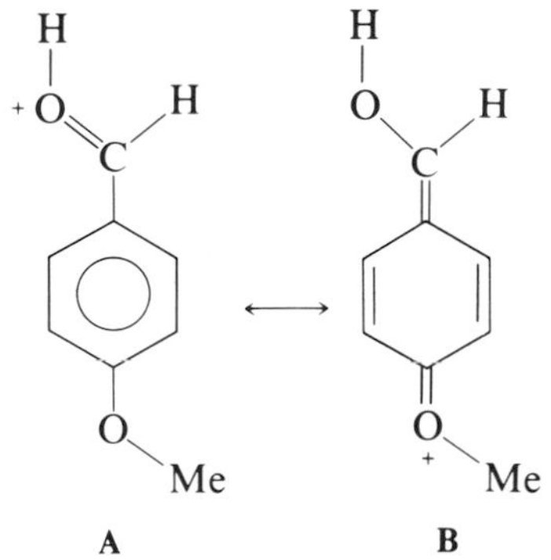

A ⟷ B

The activation energy ΔG_{BH^+} can be calculated by the line shape analysis of the temperature-dependent nmr bands. In much stronger acids, the *para*-anisaldehyde is diprotonated and, as the methoxy group is now protonated and electron withdrawing, the activation energy barrier is much lower: $\Delta G_{BH_2{}^{2+}} = 54$ kJ mol^{-1}.

When the acid composition is between these two extremes (i.e., between monoprotonated and diprotonated), the nmr spectrum will display an average picture corresponding to the following four site exchange, scheme A.[28]

Scheme A

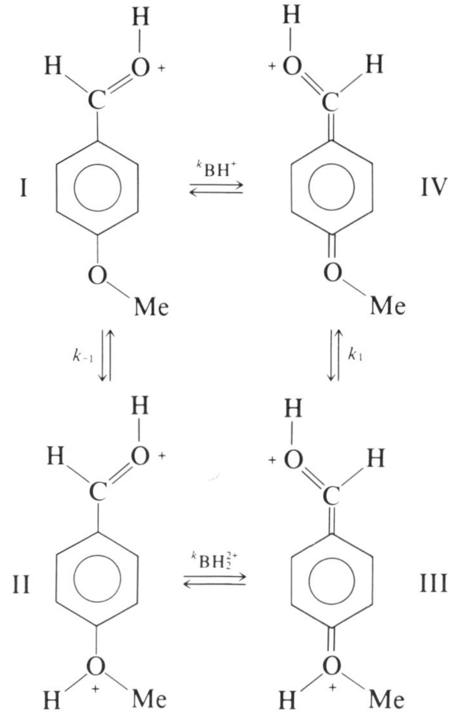

The line shape analysis of the temperature-dependent nmr spectrum yields the overall exchange rate k_{obs}, which can be related to the ionization ratios BH_2^{2+}/BH^+ in the following way.

Considering scheme **A,** the exchange process between I and IV may occur in two ways: (1) I–IV the direct exchange by internal rotation; (2) I–IV the indirect exchange (protonation, internal rotation, deprotonation). The rate of protonation and deprotonation (k_1 and k_{-1}) are very fast on the nmr time scale, as is shown by the lack of coupling even in the strongest acids, and the consequence of this is that the observed exchange rate k_{obs} can be considered as being the result of a direct competition between pathways 1 and 2:

$$k_{obs} = k_{BH^+}[BH^+] + k_{BH_2^{2+}}[BH_2^{2+}], \qquad (33)$$

with:

$$[BH^+] + [BH_2^{2+}] = 1 \qquad (34)$$

Since k_{BH^+} and $k_{BH_2^{2+}}$ can be calculated at any temperature from the values of ΔG_{BH^+} and $\Delta G_{BH_2^{2+}}$ measured above, the ionization ratio can be expressed as:

$$I = \frac{[BH_2^{2+}]}{[BH^+]} = \frac{k_{BH^+} - k_{obs}}{k_{obs} - k_{BH_2^{2+}}}. \tag{35}$$

This means that the ionization ratio can be calculated as long as $k_{BH^+} < k_{obs} < k_{BH_2^{2+}}$.

Due to the large difference in activation energy between the two processes, k_{obs} is extremely sensitive to the concentration of BH_2^{2+}.

Ionization ratios of the order of 10^{-4} can be measured easily. By the classical uv technique and the previously described nmr chemical shift method, log I can be estimated only over approximately 2 units ($1 < I < 15$). However, considering that the precision of ΔG_{obs} estimated by complete line shape analysis is around 2 kJ · mol^{-1}, the determination of I by the exchange rate method covers approximately 4 log I units ($10^{-5} < 1 < 10^{-1}$) with the same indicator.

Figure 1.6 depicts the dependence of ΔG_{obs} on the SbF_5 content in HSO_3F and shows that the method is very sensitive when I is very small. However, when the ionization ratio reaches the value of 0.5, almost all the exchange occurs through the low-energy pathway and for this reason the observed rate is too close to the upper limit to be useful for the estimation of I.

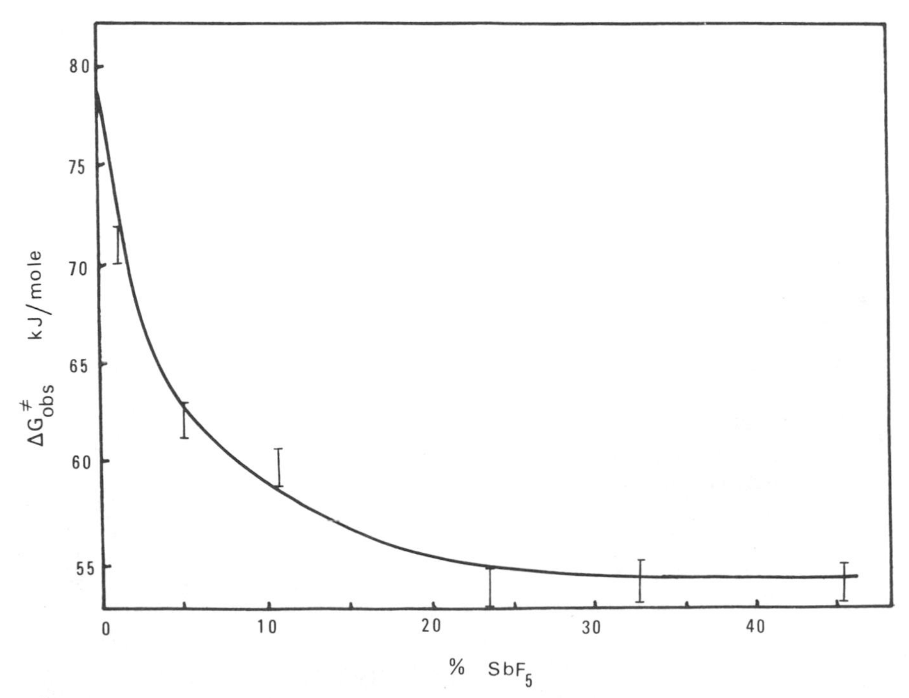

FIGURE 1.6 Acidity-dependent barrier to rotation: Dependence of $\Delta G^{\ddagger}_{obs}$ on the SbF_5 content in HSO_3F (Ref. 28).

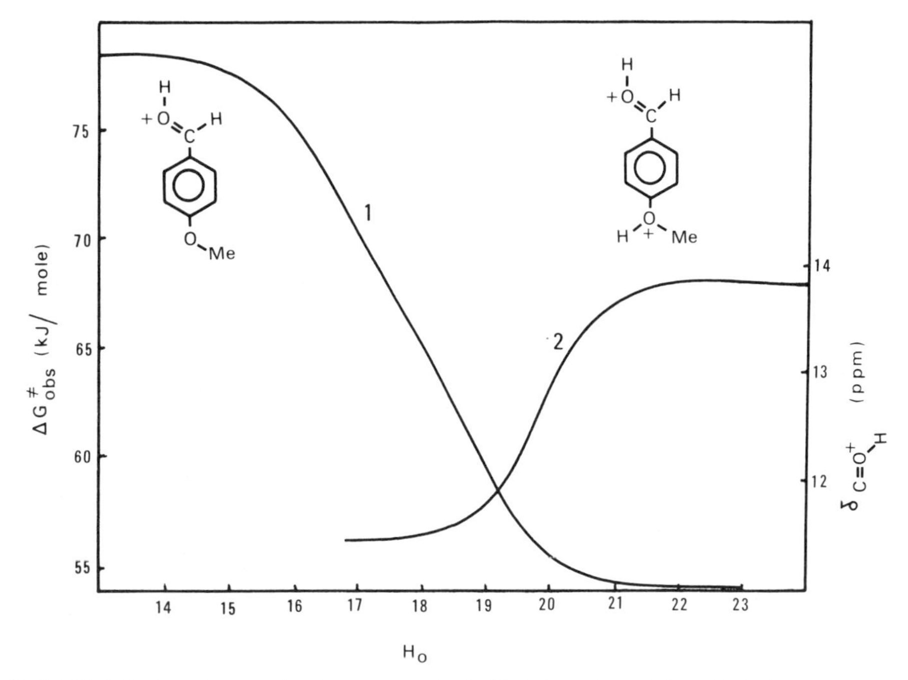

FIGURE 1.7 Complementarity of two nmr methods: Variation of the acidity with the ionization ratio of the indicator (Ref. 28). Curve 1: line-shape analysis; curve 2: chemical shift measurement.

By combining this method with the already described chemical shift method which is sensitive in the 0.05–20 range for the ionization ratio, the acidity could be measured over more than 5 H_0 units with the same indicator. Figure 1.7 shows the complementarity of both methods.

Despite the evident advantages of the nmr methods, two points have to be considered while expressing the results of the acidity measurements. First, the concentration of the indicator cannot be neglected as in the uv method and one has to take it into account in the acidity calculations, especially when the BH_2^{2+} is in low concentration. Second, aldehydes and ketones that have been generally employed for the nmr methods are not true Hammett bases and the acidity that is expressed should be taken only in a relative sense.

1.4.3 Electrochemical Methods

Electrochemistry provides a number of techniques for acidity measurements. The hydrogen electrode is the most reliable method at least in nonreducible solvents. It has been shown, however, that its reliability is limited to relatively weak acid solutions. A more general method has been proposed by Strehlow and co-workers in the early 1960s.[29] They suggested a method to measure the potential variation of a pH-dependent system with respect to a reference system whose potential was solvent independent. The measurement was made with the following cell using

Pt./H_2/H_2O-H_2SO_4, ferrocene-ferricinium (1:1)/Pt., containing sulfuric acid solution up to 100%, and Strehlow defined an acidity function R_0(H)

$$R_0(\mathrm{H}) = \frac{F}{2.303\,RT} E_x - E_1 \tag{36}$$

in which E_x and E_1 are the electromotive forces of the cell at proton activities x and unity, respectively. Like all the other acidity functions, R_0(H) equals pH in dilute aqueous solution. In strong acids, this function should be a logarithmic measure of the proton activity as long as the normal potential of the redox system ferrocene-ferricinium is constant. This was, however, not the case in very strong acid solutions because ferrocene underwent protonation. Other electrochemical pH indicators have been proposed, such as quinone-hydroquinone or semiquinone-hydroquinone, the basicity of which can be modified by substitution on the aromatic ring. These electrochemical indicators have been used with success by Tremillon, Devynck, and co-workers for acidity measurements in anhydrous HF and HF containing superacids.[30]

In principle the R_0(H) function is of limited interest for kinetic applications because the indicators are chemically very different from the organic substrates generally used. On the other hand, as the measurements are based on pH determination, the length of the acidity scale is limited by the pK value of the solvents. However, very interesting electrochemical acidity studies have been performed in HF by Devynck and co-workers, as, for example, the acidity measurement in anhydrous HF solvent and the determination of the relative strength of various Lewis acids in the same solvent. By studying the variation of the potential of alkane redox couples as a function of acidity, they provided a rational explanation of hydrocarbon behavior in superacid media.[30]

1.4.4 Chemical Kinetics

The idea that the acidity function may be useful in determining the rates of acid-catalyzed reactions was the main reason for development of the method first proposed by Hammett and Deyrup.[6a] The parallelism between reaction rate and H_0 was noticed by Hammett in the early phase of his studies.[31] Especially when the protonation of the substrate parallels the protonation of the Hammett bases, the observed rate constant can be plotted vs $-H_0$ with unit slope. The validity of this principle for a large number of acid-catalyzed reactions and its limitation due to deviations in the protonation behavior has been reviewed extensively.[32] This method has also been applied to obtain a qualitative classification of the relative acidity of various superacid solution.

Brouwer and co-workers[33] used this approach in the early 1970s. In the nmr study of the interconversion rates of alkyl tetrahydrofuryl ions, **1** and **2,** proceeding via dicarbenium ion intermediates they measured the overall rate of rearrangement in various superacid combinations of HF, HSO_3F, and SbF_5.

1 ⇌(H⁺) ⇌(H⁺) ⇌

OH OH H+ OH ⇌ ⇌ ⇌(H⁺)

2

$$\left[1 \underset{k_2}{\overset{k_1}{\rightleftharpoons}} 2\right]$$

For this system the observed overall rate k_1 was the following:

$$
\begin{aligned}
k \times 10^4 (s^{-1}) &= 20 \text{ in } 9:1 \text{ HF:SbF}_5 \quad (5°)\\
&\quad 29 \text{ in } 1:1 \text{ HSO}_3\text{F:SbF}_5 \quad (37°)\\
&\quad 3 \text{ in } 5:1 \text{ HSO}_3\text{F:SbF}_5 \quad (100°)\\
&\quad \text{no reaction in pure HSO}_3\text{F}
\end{aligned}
$$

By making the assumption that the rates were only proportional to the concentration of the dication and taking into account the temperature dependence of the rate, they could estimate the relative acidity of these systems. By repeating these experiments with closely related reactions and varying the acid composition they were able to estimate the relative acidity of 1:1 $HF:SbF_5$, 9:1 $HF:SbF_5$, 1:1 $HSO_3F:SbF_5$, and 5:1 $HSO_3F:SbF_5$ as $500:1:10^{-1}:10^{-5}$. These estimations have since been shown to be in agreement with the results obtained by other techniques. Until now, however, no H_0 value has been measured for the 1:1 $HF:SbF_5$ medium.

Another approach has been used by Kramer[34] who proposed to rank the strong acids according to a selectivity parameter $I/E = k_{iso}/k_e$, i.e., the ratio of the rate of isomerization of hydrocarbons to the rate of proton exchange that may take place simultaneously via carbenium ion deprotonation. I/E was used to rank various Lewis acids such as SbF_5, TaF_5, NbF_5, as well as $AlCl_3$ and $AlBr_3$ as 0.5 M and 2 M solution in strong Brønsted acids such as HSO_3F, CF_3SO_3H, HF, HBr, and HCl. For homogeneous dilute solutions of several Lewis acids in the same solvent, the ranking is proportional to the acidity measured on the H_0 scale. However, the scope of the I/E factor appears rather limited as it is solvent and substrate dependent.

It is interesting to note that the HBr-$AlBr_3$ system has been claimed as the strongest superacid system by Farcasiu[35] based on benzene protonation studies. Other investigations, however, do not support this claim. The Friedel-Crafts acid systems $HCl-AlCl_3$ and $HBr-AlBr_3$ are widely used superacids of great practical significance, but their acidity is lower than those of the fluoroacids discussed.

1.4.5 Heats of Protonation of Weak Bases

Arnett has shown[36] that several problems still exist in the currently available methods dealing with the behavior weak bases in solution. For example pK_A values of wide variety of carbonyl compounds given in the literature vary over an unacceptably wide range. The variations are not only due to the activity coefficient problems, but also to practical difficulties such as the effect of media on position of the uv absorption peaks.[37] The previously discussed nmr method seems to alleviate these problems. An alternative method proposed by Arnett and co-workers promises to be useful.[38] They have measured the heats of protonation of number of weak bases in HSO_3F medium for which pK_A values are known from other methods. They found a good correlation of these heats of protonation with recorded pK_A values. The heat of protonation method has the advantage over the acidity function procedure that all measurements are made in the same solvent. These studies have been recently extended to other solvent systems such as Magic Acid® ($HSO_3F:SbF_5$).[39]

1.4.6 Estimating the Strength of Lewis Acids

A quantitative method to determine the strength of Lewis acids and to establish similar scales as discussed in the case of Brønsted acids would be very useful. However, establishing such a scale is extremely difficult. Whereas the Brønsted acid-base interaction always involves proton transfer which allows a meaningful quantitative comparison, no such common relationship exists in the Lewis-base interaction. The result is that the definition of strength has no real meaning with Lewis acids.

The "strength" or "coordinating power" of different Lewis acids can vary widely against different Lewis bases. Thus, for example, in the case of boron trihalides, boron trifluoride coordinates best with fluorides, but not with chlorides, bromides, or iodides. In coordination with Lewis bases such as amines and phosphines, BF_3 shows preference to the former[40–41] (as determined by equilibrium constant measurements). The same set of bases behave differently with the Ag^+ ion. The Ag^+ ion complexes phosphines much stronger than amines.[40] In the case of halides (F^-, Cl^-, Br^-, and I^-), fluoride is the most effective base in protic acid solution. However, the order is reversed in the case of Ag^+; iodide forms the most stable complex and fluoride the least stable.[40]

Despite the apparent difficulties, a number of qualitative relationships were developed to categorize Lewis acids. Schwarzenbach[42] and Chatt[40] classified Lewis acids into two types—class *a* and class *b*. Class *a* Lewis acids form their most stable complexes with donors of first row of the periodic table—N, O, and F. Class *b* acids on the other hand complex best with donors of the second or subsequent row—Cl, Br, I, P, S, etc.[43]

Gutmann[44] introduced a series of donor numbers (DN) and acceptor numbers

(AN) for various solvents in an attempt to quantify complexing tendencies of Lewis acids. For similar reasons Drago[45] assigned parameter *E*, measuring covalent bonding potential, to each series of Lewis acids and bases.

Pearson[46] proposed a qualitative scheme in which a Lewis acid and base is characterized by two parameters, one of which is referred to as strength and the other is called softness. Thus, the equilibrium constant for a simple acid-base reaction would be a function of four parameters, two for each partners. Subsequently, Pearson introduced the *hard* and *soft* acids and bases (HSAB) principle to rationalize behavior and reactivity in a qualitative way. Hard acids correspond roughly in their behavior to Schwarzenbach and Chatt's class *a* acids. They are characterized by small acceptor atoms that have outer electrons not easily excited and that bear considerable positive charge. Soft acids, which correspond to class *b* acids, have acceptor atoms of lower positive charge, large size, and with easily excited outer electrons. Hard and soft bases are defined accordingly. Pearson's HSAB principle states that hard acids prefer to bind to hard bases and soft acids prefer to bind soft bases. The principle has proved useful in rationalizing and classifying a large number of chemical reactions involving acid-base interactions in a qualitative manner,[47] but it gives no basis for quantitative treatment.

There are many attempts made in the literature[48] to rate qualitatively the activity of Lewis acid catalysts in Friedel-Crafts-type reactions. However, such ratings largely depend on the nature of the reaction for which the Lewis acid catalyst is employed.

The classification of Lewis super acids as those stronger than anhydrous aluminum trichloride is only arbitrary. Just as in the case of Gillespie's[14] classification of Brønsted superacids, it is important to recognize that acids exist stronger than conventional Lewis acid halides with increasingly unique properties. Again the obvious difficulty is that no quantitative way is available to rate Lewis acid strengths, and reported sequences were established only against widely varying bases. Still in applications such as ionizing alkyl or cycloalkyl halides to their corresponding carbocations, in heterocation systems, catalytic activity, etc., Lewis acid halides such as SbF_5, AsF_5, TaF_5, NbF_5, etc., clearly show exceptional ability far exceeding those of $AlCl_3$, BF_3, and other conventional Lewis acid halides. Moreover, these super Lewis acid halides also show remarkable coordinating ability to Brønsted acids such as HF, HSO_3F, CF_3SO_3H, etc., resulting in vastly enhanced acidity of the resulting conjugate acids.

The determination of the strength of the Lewis acids MF_n has been carried out in various solvents using the conventional methods. Numerous techniques have been applied: conductivity measurements,[49a–f] cryoscopy,[50a–d] aromatic hydrocarbon extraction,[35,51] solubility measurements,[52] kinetic parameters determinations,[34,53] electroanalytical techniques (hydrogen electrode),[54] quinones systems as pH indicators,[55] or other electrochemical systems,[56] ir,[57] and acidity function (H_0) determinations with uv-visible spectroscopy[6,58] or with nmr spectroscopy.[15,16,27,28,59] Gas-phase measurements are also available.[60]

The results are somewhat contradictory as shown in Table 1.2, but the order of relative Lewis acid strength is now well established. The acidity scale in anhydrous

TABLE 1.2 Relative Strength of MF_n-Type Lewis Acids

Relative Strength	Solvent	Method	Reference
$BF_3 > TaF_5 > NbF_5 > TiF_4 > PF_5 > SbF_5 > WF_6 \gg SiF_4 \sim CrF_3$	HF	Xylene extraction by *n*-heptane	51a
$SbF_5 > AsF_5, BF_3 > BiF_5 > TaF_5 > NbF_5 \gg SbF_3 \sim AlF_3 \sim CrF_3$	HF	Solubility and salt formation	52b
$SbF_5, PF_5 > BF_3$	HF	Solubility	52a
$SbF_5 > TaF_5 \sim NbF_5$	HF	Conductivity and H_0 determinations	49b
$SbF_5 > AsF_5 \gg PF_5$	HF	Conductivity and Cryoscopy	50a
$TaF_5 > SbF_5 > BF_3 > TiF_4 > HfF_4$	HF	Reactions/rates parameter selectivity	34
$OsF_5 > ReF_5 > TaF_5 > MoF_5 > NbF_5 \gg MoOF_4$	HF	Conductivity and Raman spectra	50c
$SbF_5 > TaF_5 > BF_3 > SO_3$	HF	Potentiometry (quinones)	55c
$AsF_5 > TaF_5 > BF_3 > NbF_5 > PF_5$	HF	Potentiometry (hydrogen-electrodes)	54a
$SbF_5 > BF_3$	HF	Infrared spectra	57a
$SbF_5 > AsF_5 > SO_3$	HSO_3F	H_0 determination	58f
$SbF_5 > BiF_5, AsF_5 \sim TiF_4 > NbF_5 \sim PF_5$	HSO_3F	Conductivity	49c
$SbF_5 > AsF_5 > BF_3 > PF_5$	HSO_3F	Infrared and Raman Spectra	57b
$AsF_5 > BF_3 > PF_5 > SF_4, SF_5$	CH_2Cl_2	^{19}F-nmr	59b
$BF_3 > TaF_5$	Toluene	^{19}F-nmr	59c
$SbF_5 > AsF_5 > TaF_5 > NbF_5 > BF_3$	Toluene	Conductivity and cryoscopy of SeF_4-MF_5	50b
$SbF_5 > AsF_5 > BF_3 > SnF_4 > TiF_4$	Toluene	Conductivity and cryoscopy of SeF_4-MF_5	50a
$SbF_5 > AsF_5 > BF_3 > PF_5$	Gas phase		60a
$AsF_5 > PF_5 > BF_3 > SiF_4 > AsF_3$	Gas phase	Reaction rates: $SF_6^- + MF_n \rightarrow MF_{n+1}^- + SF_5\cdot$	60b
$BF_3 > SiF_4 > PF_5 > BF_3$	Gas phase	Affinity measurements for F^-	60c

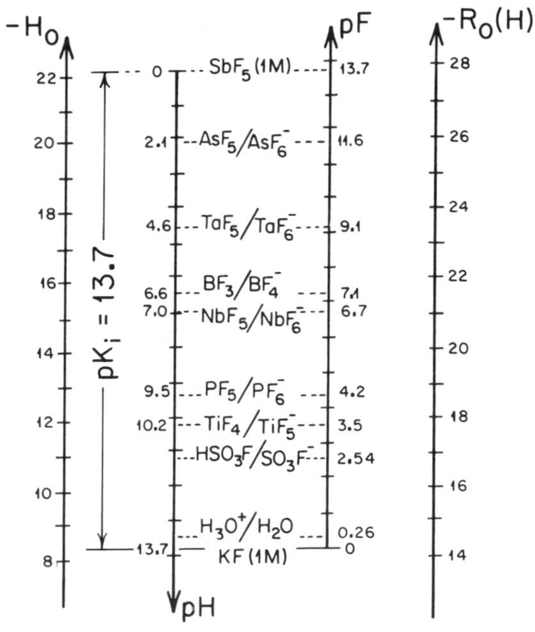

FIGURE 1.8 Relative strengths of some strong Lewis acids as measured in HF on the pH scale by electrochemical titration (Ref. 30).

hydrogen fluoride has been the subject of electrochemical investigations by Tremillon and co-workers[30] and is presented in Figure 1.8. The figure also indicates the acidity constants of various Lewis acids allowed to buffer the medium to a pH value equal to

$$\text{pH} = \text{p}K_{\text{A}} - \log \frac{{}^{a}\text{MF}_{n}}{{}^{a}\text{MF}_{n+1}} \quad (\text{p}K_{\text{A}} = -\log K_{\text{A}}) \tag{37}$$

or in dilute solution $\text{pH} = \text{p}K_{\text{A}} - \log[\text{MF}_n]/[\text{MF}^-{}_{n+1}]$.

In hydrogen fluoride, the Lewis acid strength is in the following decreasing order: $SbF_5 > AsF_5 > TaF_5 > BF_3 > NbF_5$.

1.4.7 EXPERIMENTAL TECHNIQUES APPLIED TO SOLID ACIDS

Since solid acid catalysts are used extensively in chemical industry, particularly in the petroleum field, a reliable method for measuring the acidity of solids would be extremely useful. The main difficulty to start with is that the activity coefficients for solid species are unknown and thus no thermodynamic acidity function can be properly defined. On the other hand, because the solid by definition is heterogeneous, acidic and basic sites can coexist with variable strength. The surface area available for colorimetric determinations may have widely different acidic properties from the bulk material; this is especially true for well-structured solids like zeolites. Rys and Steinegger[61] have recently developed a method for measuring the acidity function of solid bound acids. They set up a model relating the sorption of some indicators on the solid with the sorption of the same indicators in acid solution. The acidic solid material used was a sulfonated polystyrene exchange resin, Amberlyst 15, in the H^+ form, and the acidity was calculated to correspond to that of 37% H_2SO_4 solution in water ($H_0 = -2.15$). This value was confirmed by measuring the rate of hydrolysis of ethylacetate with solid acid which compared very well with the rate found in the sulfuric acid solution. Hence, this particular method seems to have great potential for use with other ion-exchange resins where the acidic sites are generally well defined. However, reported acidities for most solid acids, such as acid-treated metal oxides, supported acid systems, etc., are based on less reliable methods. The complete description of the acidic properties of a solid requires the determination of the acid strengths as well as the number of acid sites. The methods that have been used to answer these questions are basically the same as those used for the liquid acids. Three methods are generally quoted: (1) rate measurement to relate the catalytic activity to the acidity, (2) spectrophotometric method estimating the acidity from the color change of adequate indicators, (3) titration by a strong enough base for the measurement of the amount of acid.

Rate measurement is the direct way to measure the activity of solid catalyst. A close relationship between activity and acidity has been well established for liquid as well as solid acids. The method, however, is more difficult for a solid catalyst which generally possess a variety of differing activity sites available on the surface. Nevertheless, fairly good correlations have been claimed for example between acidity measurements by other techniques and the isomerization rate of 1-butene.[62]

The spectrophotometric methods for solids use the same indicators that are used for liquid acids. As the ionization ratio is experimentally very difficult to estimate, the measurement is based on the color change of the indicator. The determination is generally made by placing the powdered catalyst into a test tube and adding a nonpolar solvent containing the indicator. If the color is that of the acid form of the indicator, then the value of the H_0 function of the surface is equal to or lower than the pK_{BH^+} value of the indicator. If the color is that of the basic form, a weaker indicator base must be chosen. This method allows reliable determination of the strongest acidic sites but must be complemented by measuring the number of acidic sites present. These nonaqueous titration methods using Hammett indicators in inert solvents such as benzene have been questioned recently.[63] Generally, the methods assume that inert solvents such as benzene do not interact with the acid sites of the solid. However, it has been reported recently that benzene, a commonly used solvent in the titration methods, interacts with the acidic hydroxyl sites of zeolites and mordenites. This interaction may be a source of error in titration methods, which for example showed low acidities for some zeolites.[64–67]

The amount of acid can also be measured by titration with a strong base either in the liquid phase with the powdered solid catalyst suspended in benzene or in the gas phase by weighing the amount of base remaining chemically absorbed on the catalyst after evacuation.

These methods have been recently reviewed by Tanabe.[62] The experimental techniques vary somewhat but all the results obtained should be interpreted with caution because of the complexity of the solid acid catalysts. The presence of various sites of different activity on the same solid acid, the change in activity with temperature, and the difficulty to know the precise structure of the catalyst are some of the major handicaps in the determination of the strength of solid superacids.

REFERENCES

1a. C. H. Rochester, *Acidity Functions,* Academic Press, New York, 1970.

1b. F. Coussemant, M. Hellin, B. Torck, *Les fonctions d'acidité et leurs utilisations en Catalyse Acido-Basique,* Gordon and Breach, New York, 1969.

2. S. Arrhenius, *Z. Phys. Chem. 1,* 631 (1887).

3. J. N. Brønsted, *Recl. Trav. Chim. Pays-Bas. 42,* 719 (1923).

4a. G. N. Lewis, *Valency and Structure of Atoms and Molecules,* American Chemical Society Monographs, the Chemical Catalog Co., New York, 1923.

4b. W. B. Jensen, *The Lewis Acid-Base Concepts,* John Wiley & Sons, New York, 1980.

5. S. P. L. Sørensen, *Biochem. Z. 24,* p. 131, 201 (1909).

6a. L. P. Hammett, A. J. Deyrup, *J. Am. Chem. Soc. 54,* 2721 (1932).

6b. For discussion on other acidity functions see, T. H. Lowry, K. S. Richardson, *Mechanism and Theory in Organic Chemistry,* Harper and Row, New York, 1981, pp. 237–247.

7. D. S. Noyce, M. J. Jorgensen, *J. Am. Chem. Soc. 84,* 4312 (1962).

8. N. C. Deno, H. E. Berkheimer, W. L. Evans, H. J. Peterson, *J. Am. Chem. Soc. 81,* 2344 (1959).

9. N. C. Deno, J. J. Jaruzelski, A. Schriesheim, *J. Am. Chem. Soc. 77,* 3044 (1955).

10. K. Yates, H. Wai, *J. Am. Chem. Soc. 86,* 5408 (1964).

11. R. S. Ryabova, I. M. Medvetskaya, M. I. Vinnik, *Russ. J. Phys. Chem. 40,* 182 (1966).

12. N. F. Hall, J. B. Conant, *J. Am. Chem. Soc. 49,* 3047 (1927).

13a. For a comprehensive early review see, G. A. Olah, *Angew. Chem. Int. Ed. Engl., 12,* 173 (1973) and references cited therein.

13b. For reviews on superacids, see G. A. Olah, G. K. S. Prakash, J. Sommer, *Science 206,* 13 (1979); J. Sommer, G. A. Olah, *La Recherche 10,* 624 (1979).

14a. R. J. Gillespie, T. E. Peel, *Adv. Phys. Org. Chem. 9,* 1 (1972).

14b. R. J. Gillespie, T. E. Peel, *J. Am. Chem. Soc. 95,* 5173 (1973).

15a. F. H. Westheimer, *Spec. Publ. Chem. Soc.,* **8,** 1 (1957).

15b. V. Gold, K. Laali, K. P. Morris, L. Z. Zdunek, *Chem. Comm.* 769 (1981).

16. J. Sommer, S. Schwartz, P. Rimmelin, P. Canivet, *J. Am. Chem. Soc. 100,* 2576 (1978).

17. M. Hino, K. Arata, *Chem. Comm.* 851 (1980).

18. E. M. Arnett, unpublished results.

19. C. D. Johnson, A. R. Katritzky, S. A. Shapiro, *J. Am. Chem. Soc. 91,* 6654 (1969).

20. M. A. Paul, F. A. Long, *Chem. Rev. 57,* 1 (1957).

21. M. J. Jorgensen, D. R. Hartter, *J. Am. Chem. Soc. 85,* 878 (1963).

22. D. Brunel, A. Germain, A. Commeyras, *Nouv. J. Chim.* *2,* 275 (1978).
23. H. Kessler, *Angew. Chem.* *82,* 237 (1970).
24. H. Gunther, *NMR Spectroscopy,* Wiley, New York, 1980.
25. E. Grunwald, A. Loewenstein, S. Meiboom, *J. Chem. Phys.* *27,* 630 (1957).
26. G. C. Levy, J. D. Cargioli, W. Racela, *J. Am. Chem. Soc.* *92,* 6238 (1970).
27. J. Sommer, R. Jost, T. Drakenberg, *J. Chem. Soc., Perk. II,* 363 (1980).
28. J. Sommer, P. Canivet, S. Schwartz, P. Rimmelin, *Nouv. J. Chim.* *5,* 45 (1981).
29. H. Strehlow, H. Wendt, *Z. Phys. Chem.* (Frankfurt am Main) *30,* 141 (1960).
30. P. L. Fabre, J. Devynck, B. Tremillon, *Chem. Rev.* *82,* 591 (1982) and references therein.
31. L. P. Hammett, *Chem. Rev.* *16,* 67 (1935).
32. M. Liler, in *Reaction Mechanisms in Sulfuric Acid,* Organic Chemistry Series, Vol. 23, Academic Press, New York (1971).
33. D. M. Brouwer, J. A. Van Doorn, *Recl. Trav. Chim. Pays-Bas.* *91,* 895 (1972).
34. G. M. Kramer, *J. Org. Chem.* *40,* 298, 302 (1975).
35. D. Farcasiu, S. L. Fisk, M. T. Melchior, K. D. Rose, *J. Org. Chem.* *47,* 453 (1982).
36. E. M. Arnett, *Prog. Phys. Org. Chem.* *1,* 223 (1963).
37. J. T. Edwards, S. C. Wong, *J. Am. Chem. Soc.* *99,* 4229 (1977).
38a. E. M. Arnett, R. P. Quirk, J. J. Burke, *J. Am. Chem. Soc.* *92,* 1260 (1970).
38b. E. M. Arnett, R. P. Quirk, J. W. Larsen, *Ibid.*, *92,* 3977 (1970).
38c. E. M. Arnett, J. F. Wolf, *Ibid.*, *97,* 3262 (1975).
38d. E. M. Arnett, G. Scorrano, *Adv. Phys. Org. Chem.* *13,* 83 (1976).
39. E. M. Arnett, T. C. Hoefelich, *J. Am. Chem. Soc.* *105,* 2889 (1983) and references cited therein.
40. S. Ahrland, J. Chatt, N. R. Davies, *Q. Rev. Chem. Soc.* *12,* 265 (1958).
41. W. A. G. Graham, F. G. A. Stone, *J. Inorg. Nucl. Chem.* *3,* 164 (1956).
42a. G. Schwarzenbach, *Experientia,* Suppl. *5,* 162 (1956).
42b. G. Schwarzenbach, *Adv. Inorg. Chem. Radiochem.* *3,* 257 (1961).
42c. G. Schwarzenbach, M. Schellenberg, *Helv. Chim. Acta,* *48,* 28 (1965).
43. J. O. Edwards, R. G. Pearson, *J. Am. Chem. Soc.* *84,* 16 (1962).
44a. V. Gutmann, *Coord. Chem. Rev.* *18,* 225 (1976).
44b. A. J. Parker, U. Mayer, R. Schmid, V. Gutmann, *J. Org. Chem.* *43,* 1843 (1978).
44c. W. B. Jensen, *Chem. Rev.* *78,* 1 (1978).
45a. R. S. Drago, B. B. Wayland, *J. Am. Chem. Soc.* *87,* 3571 (1965).
45b. A. P. Marks, R. S. Drago, *Ibid.*, *97,* 3324 (1975).
46a. R. G. Pearson, *J. Am. Chem. Soc.* *85,* 3533 (1963).
46b. R. G. Pearson, J. Songstad, *J. Am. Chem. Soc.* *89,* 1827 (1967).
46c. R. G. Pearson, *Survey of Progress in Chemistry,* *5,* 1 (1969).
46d. R. G. Pearson, Ed., *Hard and Soft Acids and Bases,* Dowden, Hutchinson, and Ross, Stroudsberg, Pa., 1973.
47. T. L. Ho, *Chem. Rev.* *75,* 1 (1975).
48. G. A. Olah, *Friedel-Crafts Chemistry,* John Wiley & Sons, New York, 1973, chapter V.
49a. M. Kilpatrick, F. E. Luborsky, *J. Am. Chem. Soc.* *76,* 5865 (1954).
49b. H. H. Hyman, L. A. Quarterman, M. Kilpatrick, J. J. Katz, *J. Phys. Chem.* *65,* 123 (1961).
49c. R. J. Gillespie, K. Kouchi, *Inorg. Chem.* *8,* 63 (1969).
49d. R. J. Gillespie, K. C. Moss, *J. Chem. Soc.*, *A* 1170 (1966).
49e. M. Azeem, M. Brownstein, R. J. Gillespie, *Can. J. Chem.* *47,* 4159 (1969).

49f. M. Kilpatrick, T. J. Lewis, *J. Am. Chem. Soc. 78,* 5186 (1966).

50a. P. A. Dean, R. J. Gillespie, R. Hulme, D. A. Humphreys, *J. Chem. Soc., A* 341 (1971).

50b. R. J. Gillespie, A. Whitla, *Can. J. Chem. 48,* 657 (1970).

50c. R. T. Paine, L. A. Quarterman, *J. Inorg. Nucl. Chem., H. H. Hyman Mem. Suppl.* 85 (1976).

50d. A. A. Woolfe, *J. Chem. Soc.* 1053 (1959).

51. D. A. McCaulay, A. P. Lien, *J. Am. Chem. Soc. 73,* 2013 (1951).

52a. A. F. Clifford, S. Kongpricha, *J. Inorg. Nucl. Chem. 18,* 270 (1961).

52b. A. F. Clifford, H. C. Beachell, W. M. Jack, *Ibid., 5,* 57 (1957).

52c. A. F. Clifford, E. Samora, *Trans. Farad. Soc. 57,* 1963 (1961).

53a. G. A. Olah, J. F. Shen, R. H. Schlosberg, *J. Am. Chem. Soc. 92,* 3831 (1970), *95,* 4957 (1973).

53b. D. M. Brouwer, J. A. Van Doorn, *Recl. Trav. Chim. Pays-Bas, 89,* 553 (1970).

54a. R. Gut, K. Gautshi, *J. Inorg. Nucl. Chem. H. H. Hyman Mem. Supp.* 95 (1976).

54b. J. Devynck, A. Ben Hadid, P.-L. Fabre, *J. Inorg. Nucl. Chem. 41,* 1159 (1979).

54c. M. Herlem, A. Thiebault, *J. Electroanal. Chem. 84,* 99 (1977).

54d. J. J. Kaurova, L. M. Grubina, and T. O. A. Adzhemyan, *Elektrokhimiya 3,* 1222 (1967).

55a. A. Ben Hadid, Thesis Doct. Etat, University of Paris VI, 1980.

55b. J. Devynck, A. Ben Hadid, P.-L. Fabre, B. Tremillon, *Anal. Chim. Acta, 100,* 343 (1978).

55c. J. P. Masson, J. Devynck, B. Tremillon, *J. Electroanal. Chem., 64,* 175, 193 (1975).

55d. A. Ben Hadid, P. Rimmelin, J. Sommer, J. Devynck, *J. Chem. Soc., Perkin Trans. 2,* 269 (1982).

56a. J. Devynck, P.-L. Fabre, B. Tremillon, A. Ben Hadid, *J. Electroanal. Chem. 91,* 93 (1978).

56b. R. Gut and J. Rueede, *J. Coord. Chem. 8,* 47 (1978).

57a. M. Couzi, J. C. Cornut, P. V. Huong, *J. Chem. Soc. 56,* 42 (1972).

57b. F. O. Sladsky, P. A. Bolliner, N. Bartlett, *J. Chem. Soc., A* 2179 (1969).

58a. L. P. Hammett, A. J. Deyrup, *J. Am. Chem. Soc. 55,* 1900 (1933).

58b. R. S. Ryabova, I. M. Medvetskaya, M. I. Vinnik, *Zh. Fiz. Khim. 40,* 339 (1966).

58c. M. I. Vinnik, *Russ. Chem. Rev.* 802 (1966).

58d. H. H. Hyman, M. Kilpatrick, J. J. Katz, *J. Am. Chem. Soc. 79,* 3668 (1957).

58e. H. H. Hyman, R. A. Garber, *Ibid. 81,* 1847 (1959).

58f. R. J. Gillespie, T. E. Peel, *Ibid. 95,* 5173 (1973).

59a. J. Sommer, P. Rimmelin, T. Drakenberg, *J. Am. Chem. Soc. 98,* 2671 (1976).

59b. S. Brownstein, *Can. J. Chem. 47,* 605 (1969).

59c. S. Brownstein, *J. Inorg. Nucl. Chem. 35,* 3567 (1973).

60a. W. B. Fox, C. A. Wamser, R. Eibeck, D. K. Huggins, J. S. MacKensie, R. Juurik, *Inorg. Chem. 8,* 1247 (1969).

60b. J. C. Haartz, D. H. Daniel, *J. Am. Chem. Soc. 95,* 8562 (1973).

60c. T. C. Rhyne, J. G. Dillard, *Inorg. Chem. 10,* 730 (1971).

61. P. Rys, W. J. Steinegger, *J. Am. Chem. Soc. 101,* 4801 (1979).

62. K. Tanabe, in *Catalysis: Science and Technology,* J. R. Anderson, M. Boudart, Eds., Springer Verlag, New York (1981). Also see K. Tanabe, *Solid Acids and Bases, Their Catalytic Properties,* Tokyo Kodansha, Academic Press, New York, 1970.

63. A. K. Ghosh, G. Curthoys, *Chem. Comm.* 1271 (1983).

64. D. Barthomeuf, *J. Phys. Chem. 83,* 766 (1979).

65. W. F. Kladnig, *J. Phys. Chem. 83,* 765 (1979).

66. A. K. Ghosh, G. Curthoys, *J. Chem. Soc. Faraday Trans. I, 29,* 147 (1983).

67. R. Bezman, *J. Catal. 68,* 242 (1981).

SUPERACID SYSTEMS

The past two decades have seen a rapid development of superacids, particularly in understanding their nature as well as their chemistry. As discussed in Chapter 1, superacids encompass both Brønsted and Lewis types and their conjugate combinations. In this chapter, we will review the physical and chemical properties of the most significant superacids.

2.1 BRØNSTED SUPERACIDS

Using Gillespie's arbitrary definition, Brønsted superacids are those whose acidity exceeds that of 100% sulfuric acid ($H_0 = -12$). The physical properties of the most commonly used Brønsted superacids are summarized in Table 2.1.

2.1.1 Perchloric Acid

Historically, it was Conant's study[1] of the protonating ability of weak bases (such as aldehydes and ketones) by perchloric acid that first called attention to the "superacid" behavior of certain acid systems.

Commercially, perchloric acid is manufactured either by reaction of alkali perchlorates with hydrochloric acid[2] or direct electrolytic oxidation of 0.5 *N* hydrochloric acid.[3a,b] Another commercially attractive method is by the direct electrolysis of chlorine gas (Cl_2) dissolved in cold dilute perchloric acid.[3c] Perchloric acid is commercially available in a concentration of 70% (by weight) in water, although

TABLE 2.1 Physical Properties of Brønsted Superacids

	$HClO_4$	$ClSO_3H$	HSO_3F	CF_3SO_3H
Melting point (°C)	−112	−81	−89	−34
Boiling point (°C)	110 (Explosive)	151–152 (Decomposing)	162.7	162
Density (25°) g/cm^3	1.767[a]	1.753	1.726	1.698
Viscosity (25°) cp	—	3.0[b]	1.56	2.87
Dielectric constant	—	60 ± 10	120	
Specific conductance (20°)$ohm^{-1} \cdot cm^{-1}$	—	$0.2–0.3 \times 10^{-3}$	1.1×10^{-4}	2.0×10^{-4}
$-H_0$ (neat)	≈13.0	13.8	15.1	14.1

[a]At 20°C.
[b]At 15°C.

90% perchloric acid also had limited availability (due to its explosive hazard, it is no longer provided in this strength); 70–72% $HClO_4$, an azeotrope of 28.4% H_2O, 71.6% $HClO_4$, boiling at 203°C is safe for usual applications, however, because it is a strong oxidizing agent, it must be handled with care. Anhydrous acid (100% $HClO_4$) is prepared by vacuum distillation of the concentrated acid solution with a dehydrating agent such as $Mg(ClO_4)_2$. It is stable only at low temperatures for a few days, decomposing to give $HClO_4 \cdot H_2O$ (84.6% acid) and ClO_2.

Perchloric acid is extremely hygroscopic and a very powerful oxidizer. Contact of organic materials with anhydrous or concentrated perchloric acid can lead to violent explosions. For this reason, the application of perchloric acid has serious limitations. The acid strength, although not reported, can be estimated to be around $H_0 = -13$ for the anhydrous acid.

Formation of various perchlorate salts such as $NO_2^+ClO_4^-$, $CH_3CO^+ClO_4^-$, $R^+ClO_4^-$, etc., via the ionization of their appropriate neutral precursors in perchloric acid can also lead to serious explosions. The probable reason is not necessarily the thermal instability of the ionic perchlorates but instead their equilibria with highly unstable and explosive covalent perchlorates.

$$NO_2^+ClO_4^- \rightleftharpoons NO_2OClO_3$$

$$CH_3CO^+ClO_4^- \rightleftharpoons CH_3COOClO_3$$

$$R^+ClO_4^- \rightleftharpoons ROClO_3$$

Actually no specific advantage exists in using perchlorate salts when comparable safe conjugate fluoride salts such as BF_4^-, SbF_6^-, $Sb_2F_{11}^-$, etc., are available. The use of perchlorate salts always necessitates extreme care and precautions.

The chemistry of perchloric acid along with its applications has been well reviewed.[4] The main use of perchloric acid has been the use of its salts (such as $NH_4^+ClO_4^-$) as powerful oxidants in pyrotechniques and rocket fuels.

2.1.2 Chlorosulfuric Acid

Chlorosulfuric acid, the monochloride of sulfuric acid, is a strong acid containing a relatively weak sulfur-chlorine bond. It is prepared by the direct combination of sulfur trioxide and dry hydrogen chloride gas.[5] The reaction is very exothermic and reversible, making it difficult to obtain chlorosulfuric acid free of SO_3 and HCl. Upon distillation, even in good vacuum, some dissociation is inevitable. The acid is a powerful sulfating and sulfonating agent as well as a strong dehydrating agent and a specialized chlorinating agent. Because of the above properties, chlorosulfuric acid is rarely used for its protonating superacid properties.

Gillespie and co-workers[6] have measured systematically the acid strength of the H_2SO_4-$ClSO_3H$ system using aromatic nitro compounds as indicators. They found an H_0 value of -13.8 for 100% $ClSO_3H$.

2.1.3 Fluorosulfuric Acid

Fluorosulfuric acid, HSO_3F, is a mobile colorless liquid that fumes in moist air and has a sharp odor. It may be regarded as a mixed anhydride of sulfuric and hydrofluoric acid. It has been known since 1892[7] and is prepared commercially from SO_3 and HF in a stream of HSO_3F. It is readily purified by distillation, although the last traces of SO_3 are difficult to remove. When water is excluded, it may be handled and stored in glass containers, but for safety reasons the container should always be cooled before opening because gas pressure may have developed from hydrolysis:

$$HSO_3F + H_2O \rightleftharpoons H_2SO_4 + HF$$

Fluorosulfuric acid generally also contains hydrogen fluoride as an impurity, but according to Gillespie the hydrogen fluoride can be removed by repeated distillation under anhydrous conditions. The equilibrium $HSO_3F \leftrightarrows SO_3 + HF$ always produces traces of SO_3 and HF in stored HSO_3F samples. When kept in glass for a long time, SiF_4 and H_2SiF_6 are also formed (secondary reactions due to HF).

Fluorosulfuric acid is employed as a catalyst and chemical reagent in various chemical processes including alkylation, acylation, polymerization, sulfonation, isomerization, and production of organic fluorosulfates. It is insoluble in carbon disulfide, carbon tetrachloride, chloroform, and tetrachloroethane, but is soluble in nitrobenzene, diethyl ether, acetic acid, and ethylacetate and it dissolves most organic compounds that are potential proton acceptors. The ir, Raman, and nmr spectra of fluorosulfuric acid have been reported.[8a,b] The ^{19}F chemical shift is 44.9 ppm downfield from CCl_3F.[8c] The acid can be dehydrated to give $S_2O_5F_2$.[9] Electrolysis of fluorosulfuric acid gives $S_2O_6F_2$ or $SO_2F_2 + F_2O$, depending on conditions employed.

HSO_3F has a wide liquid range (mp $= -89.0°C$, bp $= +162.7°C$) making it advantageous as a superacid solvent for the protonation of a wide variety of weak bases.

Fluorosulfuric acid ionizes in H_2SO_4:

$$HSO_3F + H_2SO_4 \rightleftharpoons H_3SO_4^+ + SO_3F^-$$

The H_0 acidity function has been measured by Gillespie[6b] for the H_2SO_4-HSO_3F system using fluoro- and nitroaromatic bases. He found a strong increase in acidity in the vicinity of 100% HSO_3F which can be attributed to the self-ionization of the acid:

$$2HSO_3F \rightleftharpoons H_2SO_3F^+ + SO_3F^-$$

The neat acid is ascribed an H_0 value of -15.1. This ranks it as the strongest known simple Brønsted acid. Trifluoromethanesulfonic acid (HSO_3CF_3), in which fluorine is replaced by a CF_3 group, has a slightly lower acid strength (-14.1). HSO_3F, the most widely used superacidic solvent system, has a low viscosity and good thermal stability and wide liquid range (250° from mp to bp). The acidity of HSO_3F can be increased further by the addition of Lewis acid fluorides (*vide infra*).

2.1.4 Perfluoroalkanesulfonic Acids

Perfluoroalkanesulfonic acids were first reported in 1954, and, since then, have been prepared by electrochemical fluorination (ECF) of the corresponding alkanesulfonyl halides and subsequent hydrolysis.[10,12]

$$RSO_2F \xrightarrow[ECF]{HF} R_FSO_2F \xrightarrow[aq.KOH]{} R_FSO_3K \xrightarrow{H_2SO_4} R_FSO_3H$$

The boiling points, density, and H_0 values of these acids are compared in Table 2.2.

2.1.4.1 Trifluoromethanesulfonic Acid

Trifluoromethanesulfonic acid (CF_3SO_3H, triflic acid), the first member in the perfluoroalkanesulfonic acid series has been studied extensively, and excellent reviews describing its physical and chemical properties have been published.[12] Besides its preparation by electrochemical fluorination of methanesulfonyl halides,[13a] triflic acid may also be prepared from trifluoromethanesulfenyl chloride.[13b]

$$CF_3SSCF_3 \xrightarrow{Cl_2} CF_3SCl \xrightarrow[H_2O]{Cl_2} CF_3SO_2Cl \xrightarrow[aq.KOH]{} CF_3SO_3H$$

CF_3SO_3H is a stable, hygroscopic liquid that fumes in moist air and readily forms the stable monohydrate (hydronium triflate) which is a solid at room temperature, mp 34°C, bp 96°C/1 mmHg. The ^{19}F-nmr chemical shift for the neat acid is 78.5 ppm upfield from CCl_3F.[14] Conductivity measurements[15] in glacial acetic acid have shown triflic acid to be one of the strongest simple protic acids known, similar to HSO_3F and $HClO_4$. The acidity of the neat acid as measured by uv spectroscopy

TABLE 2.2 Characteristics of Perfluoroalkanesulfonic Acids

Compound	bp°C (760 mm Hg)	Density (25°C)	H_0 (22°C)
CF_3SO_3H	161	1.70	−14.1[a]
$C_2F_5SO_3H$	170	1.75	−14.0[a]
$C_4F_9SO_3H$	198	1.82	−13.2[a]
$C_5F_{11}SO_3H$	212		
$C_6F_{13}SO_3H$	222	[b]	−12.3[a,c]
$C_8F_{17}SO_3H$	249	[b]	
CF_3–F–SO_3H	241		
C_2F_5–F–SO_3H	257		

[a]Values from Ref. 11.
[b]Solid at 25°C.
[c]Recalculated from the value measured at 35°C.

with a Hammett indicator shows indeed an H_0 value of −14.1.[11] It is miscible with water in all proportions and soluble in many polar organic compounds such as dimethylformamide, dimethylsulfoxide, and acetonitrile. It is generally a very good solvent for organic compounds that are capable of acting as proton acceptors in the medium. The exceptional leaving group properties of the triflate anion, $CF_3SO_3^-$, makes triflate esters excellent alkylating agents. The acid and its conjugate base do not provide a source of fluoride ion even in the presence of strong nucleophiles. Furthermore, as it lacks the sulfonating properties of oleums and HSO_3F, it has gained a wide range of application as a catalyst in Friedel-Crafts alkylation, polymerization, and organometallic chemistry.

2.1.4.2 Higher Perfluoroalkanesulfonic Acids

Higher homologous perfluoroalkanesulfonic acids are hygroscopic oily liquids or waxy solids. They are prepared by the distillation of their salts from H_2SO_4, giving stable hydrates that are difficult to dehydrate. The acids show the same polar solvent solubilities of trifluoromethanesulfonic acid but are quite insoluble in benzene, heptane, carbon tetrachloride, and perfluorinated liquids. Many of the perfluoroalkanesulfonic acids have been prepared by the electrochemical fluorination reaction[13a] (or conversion of the corresponding perfluoroalkane iodides to their sulfonyl halides). α, ω-Perfluoroalkanedisulfonic acids have been prepared by aqueous alkaline permanganate oxidation of the compounds, $R_fSO_2(CF_2CF_2)_nSO_2F$.[16] $C_8F_{17}SO_3H$ and higher perfluoroalkanesulfonic acids are surface-active agents and form the basis for a number of commercial fluorochemical surfactants.[17]

Extreme care should be taken while handling the perfluoroalkanesulfonic acids. Studies suggest that extreme irritation and permanent eye damage could occur following eye contact, even if the eyes are flushed immediately with water. Acute inhalation toxicity studies (in albino rats) indicate that high vapor or mist concentrations can cause significant respiratory irritation. All contacts of the acids and their esters with the skin should be avoided. Usual procedures for treatment of strong acid burns should be applied.

Contact of the perfluoroalkanesulfonic acids with cork, rubber, cellulose, and plasticized materials will cause discoloration and deterioration of these materials. Samples are best stored in glass ampules or glass bottles with Kel-F or Teflon plastic screw-cap linings.

Acidity measurements on perfluoroalkanesulfonic acids have been reported by Commeyras and co-workers[11] (Table 2.2). All the acids show a strong uv absorption band around 283 nm. Up to C_4 chain length, the acids are liquids at room temperature, and the H_0 measurements using Hammett bases were carried out at 22°C. Perfluorohexanesulfonic acid $C_6F_{13}SO_3H$ melts at 33°C and the H_0 value has been corrected for the temperature difference. Perfluorooctanesulfonic acid is a waxy solid melting at 90°C and its acidity has not been measured. Surprisingly, there is relatively little decrease in acidity with increase in the perfluoroalkyl chain length, and the first CF_2 group adjacent to the sulfonic acid moiety is most responsible for the acid strength. Thus, even higher solid homologous members are capable of acting as superacidic catalysts. As will be seen in Section 2.3.6, the acidity of these acids can be further enhanced by complexation with Lewis acids such as SbF_5, TaF_5, etc.

2.2 LEWIS SUPERACIDS

In Chapter 1, we arbitrarily defined Lewis superacids as those that are stronger than anhydrous aluminum chloride, the most commonly used Friedel-Crafts catalyst. The physical properties of some of the Lewis superacids are given in Table 2.3.

2.2.1 Antimony Pentafluoride

Antimony pentafluoride (SbF_5) is a colorless, very viscous liquid at room temperature. It is hygroscopic and fumes in moist air. Its viscosity at 20°C is 460 cP which is close to that of glycerol. The pure liquid can be handled and distilled in

TABLE 2.3 Physical Properties of Some Lewis Superacids

	SbF_5	AsF_5	TaF_5	NbF_5
mp °C	7.0	−79.8	97	72–73
bp °C	142.7	−52.8	229	236
Specific gravity (15°C) (g/cc)	3.145	2.33[a]	3.9	2.7

[a]At the bp.

glass if moisture is excluded. Commercial antimony pentafluoride is shipped in steel cylinders or in perfluoroethylene bottles for laboratory quantities.

The polymeric structure of the liquid SbF_5 has been established by ^{19}F-nmr spectroscopy[18] and is shown to have the following frameworks:

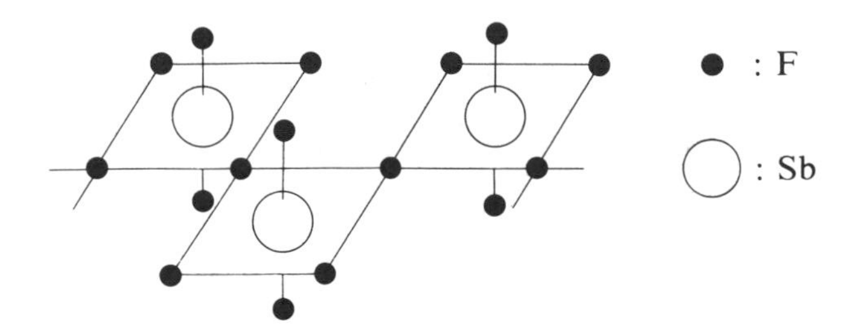

A *cis*-fluorine bridged structure is found in which each antimony atom is surrounded by six fluorine atoms in an octahedral arrangement.

Antimony pentafluoride is a powerful oxidizing and a moderate fluorinating agent. It readily forms stable intercalation compounds with graphite (*vide infra*), and it spontaneously inflames phosphorus and sodium. It reacts with water to form $SbF_5 \cdot 2H_2O$, an unusually stable solid hydrate (probably a hydronium salt, $H_3O^+\bar{S}bF_5OH$) that reacts violently with excess water to form a clear solution. Slow hydrolysis can be achieved in the presence of dilute NaOH and forms $Sb(OH)_6^-$. Sulfur dioxide and nitrogen dioxide form 1:1 adducts, $SbF_5:SO_2$ and $SbF_5:NO_2$[19,20] as do practically all nonbonded electron pair donor compounds. The exceptional ability of SbF_5 to complex and subsequently ionize nonbonded electron pair donors (such as halides, alcohols, ethers, sulfides, amines, etc.) to carbocations, recognized first by Olah (Chapter 3), has made it one of the most widely used Lewis halides in the study of ionic intermediates and catalytic reactions.

Vapor density measurements suggest a molecular association corresponding to $(SbF_5)_3$ at 150°C and $(SbF_5)_2$ at 250°C. On cooling, SbF_5 gives a nonionic solid composed of trigonal bipyramidal molecules. Antimony pentafluoride is prepared by the direct fluorination of antimony metal or antimony trifluoride (SbF_3). It can also be prepared by the reaction of $SbCl_5$ with anhydrous HF, but the exchange of the fifth chloride is difficult and the product is generally SbF_4Cl.[21]

As shown by conductometric, cryoscopic, and related acidity measurements, it appears that antimony pentafluoride is by far the strongest Lewis acid known. Thus, it is preferentially used in preparing stable ions and conjugate superacids (*vide infra*). Antimony pentafluoride is also a strong oxidizing agent, allowing, for example, preparation of arene dications. At the same time, its easy reducibility to antimony trifluoride represents a limitation in many applications, although it can be easily refluorinated.

2.2.2 Arsenic Pentafluoride

Arsenic pentafluoride (AsF_5) is a colorless gas at room temperature, condensing to a yellow liquid at -53°C. Vapor density measurements indicate some degree of association, but it is a monomeric covalent compound with a high degree of coordinating ability. It is prepared by reacting fluorine with arsenic metal or arsenic trifluoride. As a strong Lewis acid fluoride, it is used in the preparation of ionic complexes and in conjunction with Brønsted acids forms conjugate superacids. It also forms with graphite stable intercalation compounds that show electrical conductivity comparable to that of silver.[22] Great care should be exercised in handling any arsenic compound because of potential high toxicity.

2.2.3 Tantalum and Niobium Pentafluoride

The close similarity of the atomic and ionic radii of niobium and tantalum are reflected by the similar properties of tantalum and niobium pentafluorides. They are thermally stable white solids that may be prepared either by the direct fluorination of the corresponding metals or by reacting the metal pentachlorides with HF. Surprisingly, even reacting metals with HF gives the corresponding pentafluorides. They both are strong Lewis acids complexing a wide variety of donors such as ethers, sulfides, amines, halides, etc. They both coordinate with fluoride ion to form anions of the type $(MF_6)^-$. TaF_5 is a somewhat stronger acid than NbF_5 as shown by acidity measurements in HF. The solubility of TaF_5 and NbF_5 in HF and HSO_3F is much more limited than that of SbF_5 or other Lewis acid fluorides, restricting their use to some extent. At the same time, their high redox potentials and more limited volatility make them catalysts of choice in certain hydrocarbon conversions, particularly in combination with solid catalysts. Their chemistry has been extensively reviewed.[23,24]

2.3 CONJUGATE BRØNSTED-LEWIS SUPERACIDS

2.3.1 Oleums: Polysulfuric Acids

SO_3 containing sulfuric acid (oleum) has been long considered as the strongest mineral acid and one of the earliest superacid systems to be recognized. The concentration of SO_3 in sulfuric acid can be determined by weight or by electrical conductivity measurement.[25a] The density and the viscosity increase with increasing SO_3 content is shown in Table 2.4 for dilute oleums. The vapor pressure of oleum rises rapidly with the increase in concentration of SO_3 and increase in temperature[25b] as shown in Figure 2.1.

Lewis and Bigeleisen,[26] who first determined the Hammett acidity function values for oleums, used a method to derive them from the vapor pressure measurements. Brandt et al.,[27] however, subsequently by the use of nitro compound indicators showed that the H_0 values are not directly related to the vapor pressure. The most accurate H_0 values for oleums so far have been published by Gillespie and coworkers[6] (Table 2.5).

The increase in acidity on addition of SO_3 to sulfuric acid is substantial, and an H_0 value of -14.5 is reached with 50 mol% SO_3. The main component up to this

TABLE 2.4 Density and Viscosity of Dilute Oleum Solutions at 25°C (Ref. 25a)

$m_{H_2S_2O_7}$	d_4^{25} (g · cm^{-3})	$\eta(10^{-7}\ N\ s \cdot cm^{-2})$
0.0190	1.8270	24.54
0.0470	1.8280	24.54
0.2250	1.8330	24.57
0.3550	1.8360	24.66
0.5360	1.8407	24.74
0.6920	1.8439	24.78
0.8350	1.8480	24.82

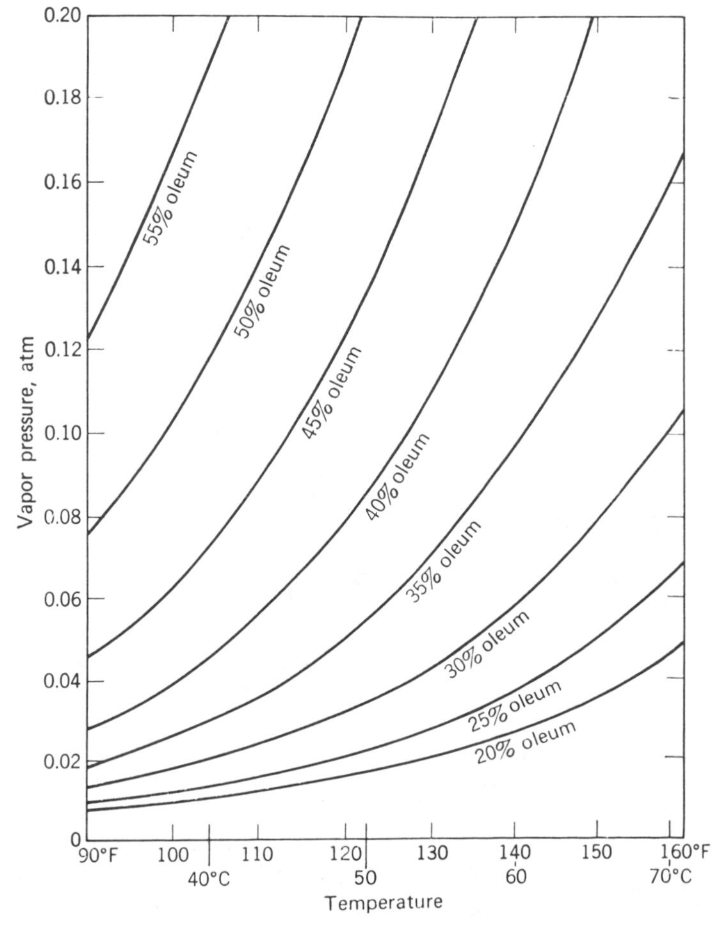

FIGURE 2.1 Vapor pressure of oleum (Ref. 25b).

TABLE 2.5 H_0 **Values for the** H_2SO_4**-**SO_3 **System**

Mol% SO_3	H_0	Mol% SO_3	H_0	Mol% SO_3	H_0
1.00	−12.24	25.00	−13.58	55.00	−14.50
2.00	−12.42	30.00	−13.76	60.00	−14.74
5.00	−12.73	35.00	−13.94	65.00	−14.84
10.00	−13.03	40.00	−14.11	70.00	−14.92
15.00	−13.23	45.00	−14.28	75.00	−14.90
20.00	−13.41	50.00	−14.44		

SO_3 concentration is pyrosulfuric (or disulfuric) acid $H_2S_2O_7$. Upon heating or in the presence of water, it decomposes and behaves like a mixture of sulfuric acid and sulfur trioxide. In sulfuric acid, it ionizes as a stronger acid:

$$H_2S_2O_7 + H_2SO_4 \rightleftharpoons H_3SO_4^+ + HS_2O_7^- \quad (K = 1.4 \times 10^{-2})$$

At higher SO_3 concentration, a series of higher polysulfuric acids such as $H_2S_3O_{10}$, $H_2S_4O_{13}$, etc., are formed and a corresponding increase in acidity occurs. However, as can be seen from Table 2.5, the acidity increase is very small after reaching 50 mol% of SO_3 and no data are available beyond 75%.

Despite its high acidity, oleum has found little application as a superacid catalyst, mainly because of its strong oxidizing power. Also, its high melting point and viscosity have considerably hampered its use for spectroscopic study of ionic intermediates and in synthesis, except as an oxidizing or sulfonating agent.

2.3.2 Tetra(Hydrogen Sulfato)Boric Acid–Sulfuric Acid

$HB(HSO_4)_4$ prepared by treating boric acid $[B(OH)_3]$ with sulfuric acid ionizes in sulfuric acid as shown by acidity measurements.[6]

$$HB(HSO_4)_4 + H_2SO_4 \rightleftharpoons H_3SO_4^+ + B(HSO_4)_4^-$$

The increase in acidity is, however, limited to $H_0 = -13.6$ as a result of insoluble complexes that precipitate when the concentration of the boric acid approaches 30 mol%. Figure 2.2 shows the composition-related acidity increase for the system in comparison with oleum.

2.3.3 Fluorosulfuric Acid–Antimony Pentafluoride ("Magic Acid")

Of all the superacids, "Magic Acid," a mixture of fluorosulfuric acid and antimony pentafluoride, is probably the most thoroughly investigated concerning measurements of acidity and also the most widely used medium for the spectroscopic

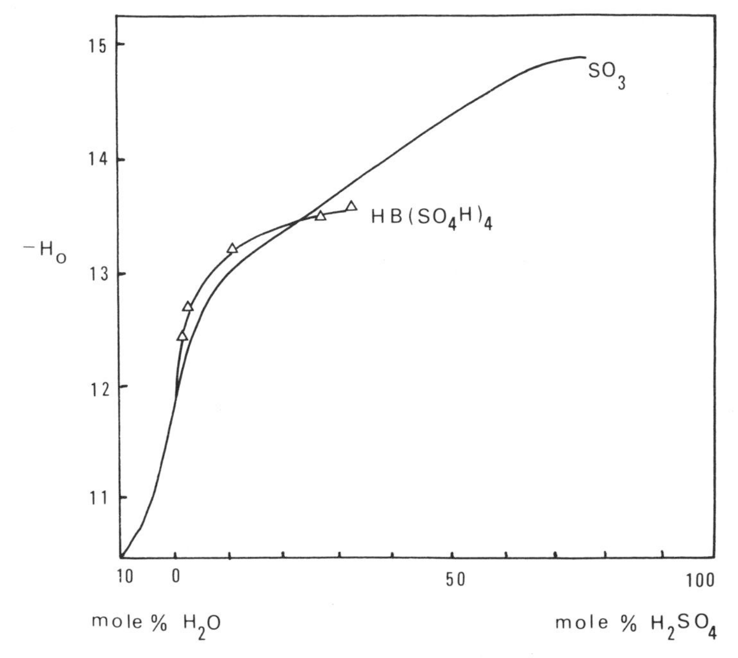

FIGURE 2.2 H_0 acidity function values for H_2SO_4-$HB(SO_4H)_4$ and H_2SO_4-SO_3 from ref. 6.

observation of stable carbocations (see Chapter 3). The fluorosulfuric acid–antimony pentafluoride system was developed in the early 1960s by Olah for the study of stable carbocations and was studied by Gillespie for electron-deficient inorganic cations. The name Magic Acid originated in Olah's laboratory at Case Western Reserve University in the winter of 1966. The $HSO_3F:SbF_5$ mixture was extensively used in his group to generate stable carbocations. J. Lukas, a German postdoctoral fellow, put a small piece of Christmas candle left over from a lab party into the acid system and found that it dissolved readily. He then ran a ^{1}H-nmr spectrum of the solution. To everybody's amazement, he obtained a sharp spectrum of the *t*-butyl cation. The long-chain paraffin, of which the candle is made, had obviously undergone extensive cleavage and isomerization to the more stable tertiary ion. It impressed Lukas and others in the laboratory so much that they started to nickname the acid system Magic Acid. The name stuck and soon others started to use it too. It is now a registered trade name and has found its way into the chemical literature.

The acidity of the Magic Acid system as a function of SbF_5 content has been measured successively by Gillespie,[6] Sommer,[28] Gold,[29] and their co-workers. The increase in acidity is very sharp at low SbF_5 concentration ($\approx$10%) and continues up to the estimated value of $H_0 = -26.5$ for the 90% SbF_5 content as shown in Figure 2.3.

The initial ionization of $HSO_3F:SbF_5$ is as shown:

$$2HSO_3F + SbF_5 \rightleftharpoons H_2SO_3F^+ + SbF_5(SO_3F)^-$$

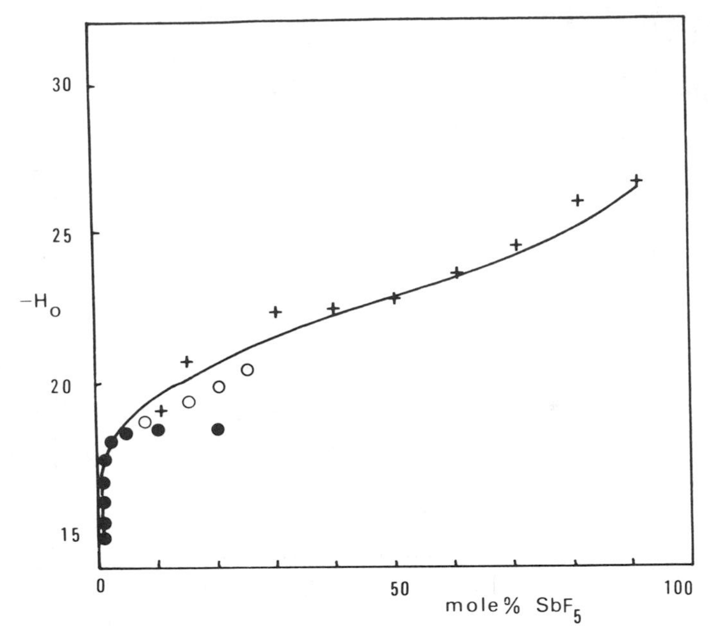

FIGURE 2.3 Dependence of H_0 acidity function values for the HSO_3F-SbF_5 system upon SbF_5 addition (up to 90 mol%) from refs. 6: (●), 28: (○) and 29: (+).

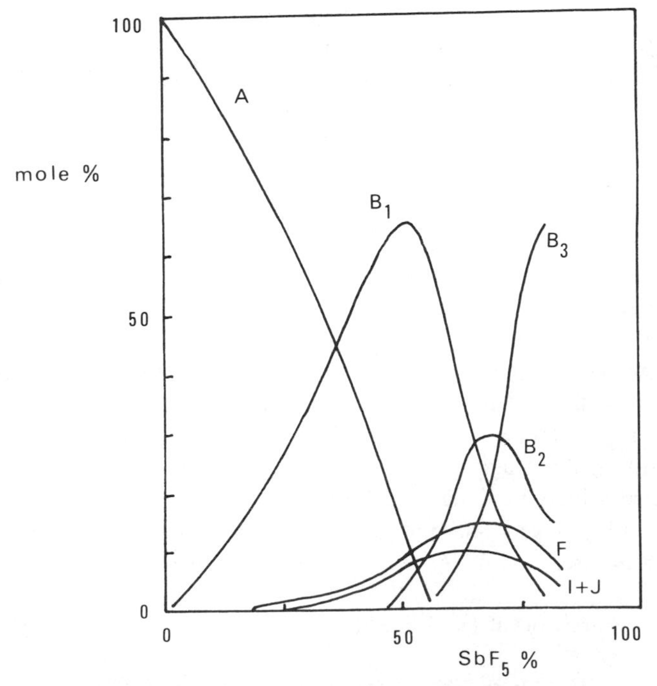

FIGURE 2.4 Variation of the composition of HSO_3F-SbF_5 depending on the SbF_5 content from ref. 8c.

At higher concentrations of SbF_5, larger polyantimony fluorosulfate ions are formed.

$$SbF_5 + SbF_5(SO_3F)^- \rightleftharpoons Sb_2F_{10}(SO_3F)^-$$

Due to these equilibria, which have been discussed by several authors,[30,31] the composition of the $HSO_3F{:}SbF_5$ system is very complex and depends on the SbF_5 content. A thorough ^{19}F-nmr study by Commeyras and co-workers[8c] has shown such dependence and is presented in Figure 2.4 for the major components that are listed in Table 2.6.

This shows that the 1:1 Magic Acid has the following counterion composition: 70% of $SbF_5SO_3F^-$, 5% of $Sb_2F_{10}SO_3F^-$, 12% of the uncomplexed SO_3F^- ion, and 7% of SbF_6^- as well as 7% of $SbF_4(SO_3F)_2$. The formation of the latter two ions in equal amounts is explained by the ligand redistribution reaction

$$2FSO_3SbF_5^- \rightleftharpoons (FSO_3)_2SbF_4^- + SbF_6^-$$

Free SbF_5 is not observed as it complexes the Brønsted acid and leads to its initial ionization:

$$HSO_3F + SbF_5 \rightleftharpoons HFSO_3SbF_5$$

$$HSO_3F + HFSO_3SbF_5 \rightleftharpoons H_2SO_3F^+ + FSO_3SbF_5^-$$

TABLE 2.6 Main Components of the Magic Acid System

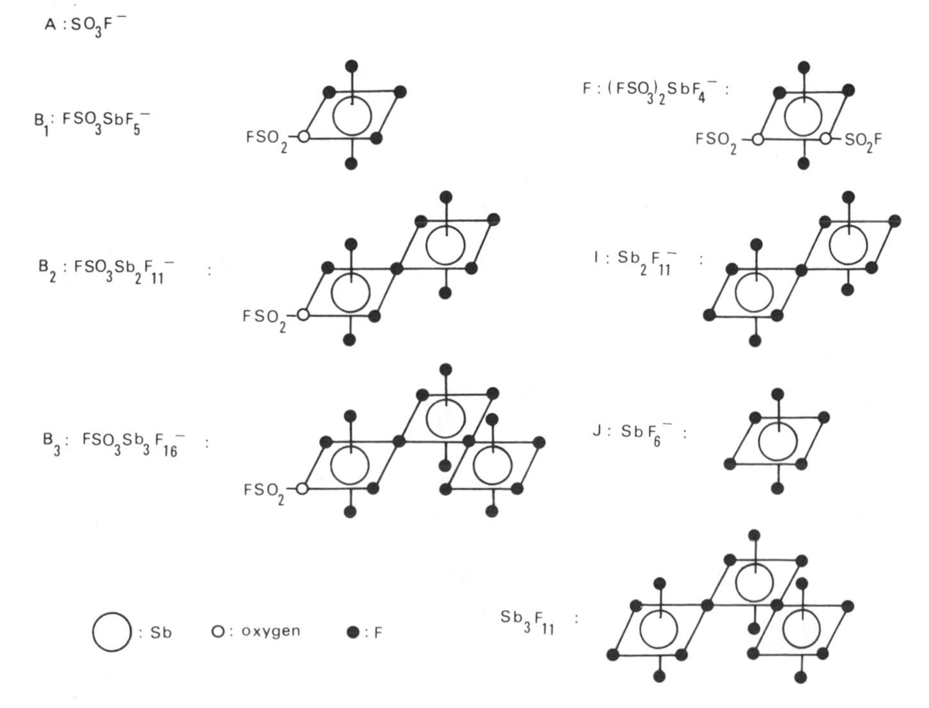

As shown in Figure 2.4, the B_1 and B_2 species reach their maximum at their stoichiometric values of 0.5 and 0.66, respectively. The autoprotolysis of the corresponding acids $HFSO_3SbF_5$ and $HFSO_3Sb_2F_{11}$ are responsible for the increased acidity (as observed in Figure 2.3). The major reason for the wide application of this superacid system compared with others (besides its very high acidity) is probably the large temperature range in which it can be used. In the liquid state, nmr spectra have been recorded from temperatures as low as −160°C[32] (acid diluted with SO_2F_2 and SO_2ClF) and up to +80°C (neat acid in sealed nmr glass tube).[28] Glass is attacked by the acid very slowly when moisture is excluded. The Magic Acid system can also be an oxidizing agent that results in reduction to antimony trifluoride and sulfur dioxide. On occasion this represents a limitation.

2.3.4 Fluorosulfuric Acid–Sulfur Trioxide

Freezing point and conductivity measurements[33] show that SO_3 behaves as a nonelectrolyte in HSO_3F. Acidity measurements show a small increase in acidity that is attributed to the formation of polysulfuric acids HS_2O_6F and HS_3O_9F up to $HS_7O_{21}F$.[34] Evidence for the existence of these acids has been obtained by ^{19}F-nmr measurements in SO_2ClF solutions at −100°C.[30] The acidity of these solutions reaches a maximum of −15.5 on the H_0 scale for 4 mol% SO_3 and does not increase any further.

2.3.5 Fluorosulfuric Acid–Arsenic Pentafluoride

AsF_5 ionizes in HSO_3F.[35] The $AsF_5SO_3F^-$ anion has the octahedral structure as shown for the antimony analog (B_1; Table 2.6). Values for the H_0 acidity function up to 4 mol% AsF_5 show a larger increase as compared with SO_3 but smaller when compared with SbF_5[6] (Figure 2.5).

2.3.6 $HSO_3F:HF:SbF_5$

When Magic Acid is prepared from fluorosulfuric acid not carefully distilled (which always contains HF), upon addition of SbF_5 the ternary superacid system HSO_3F: $HF:SbF_5$ is formed. Because HF is a weaker Brønsted acid, it ionizes fluorosulfuric acid, which, upon addition of SbF_5, results in a high-acidity superacid system at low SbF_5 concentrations. ^{19}F-nmr studies on the system have indicated the presence of SbF_6^- and $Sb_2F_{11}^-$ anions, although these can result from the disproportionation of $SbF_5(FSO_3)^-$ and $Sb_2F_{10}(FSO_3)^-$ anions.

The ternary $HSO_3F:HF:SbF_5$ system was recognized as a highly efficient superacid catalyst in its own right by McCauley.[36]

2.3.7 $HSO_3F:SbF_5:SO_3$

When sulfur trioxide is added to a solution of SbF_5 in HSO_3F, there is a marked increase in conductivity that continues until approximately 3 mol of SO_3 have been

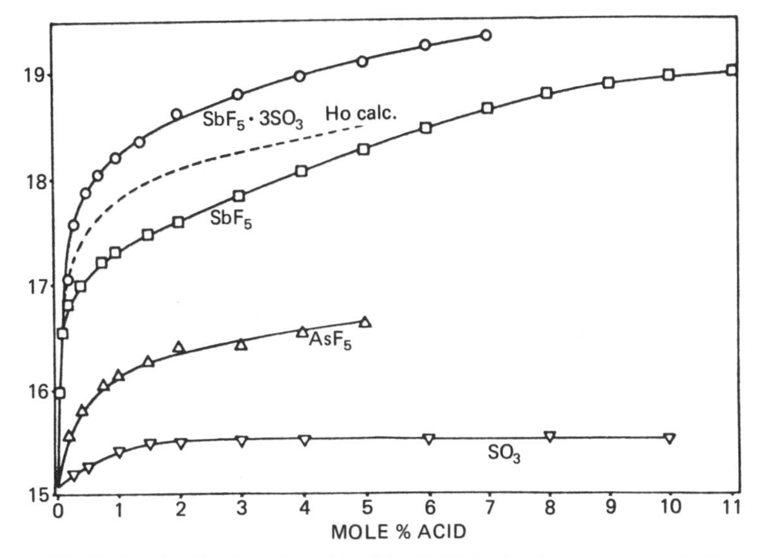

FIGURE 2.5 H_0 values for acids of the HSO_3F solvent system from ref. 6b.

added per mol of SbF_5.[37] This increase in conductivity has been attributed to an increase in $H_2SO_3F^+$ concentration arising from the formation of a much stronger acid than Magic Acid. Acidity measurements have confirmed the increase in acidity with $SO_3{:}SbF_5$ in the HSO_3F system (Fig. 2.5). This has been attributed to the presence of a series of acids of the type $H[SbF_4(SO_3F)_2]$ (see Table 2.6, anion F), $H[SbF_3(SO_3F)_3]$, $H[SbF_2(SO_3F)_4]$ of increasing acidity. Thus, of all the fluorosulfuric acid-based superacid systems, sulfur trioxide-containing acid mixtures are, however, difficult to handle and cause extensive oxidative side reactions when contacted with organic compounds.

2.3.8 Perfluoroalkanesulfonic Acid-Based Systems

2.3.8.1 $C_nF_{2n+1}SO_3H{:}SbF_5$

$CF_3SO_3H{:}SbF_5$ ($n = 1$) was introduced by Olah as an effective superacid catalyst for isomerizations and alkylations. The composition and acidity of systems where $n = 1,2,4$ have been thoroughly studied by Commeyras and co-workers.[8c] The ^{19}F-nmr spectra are very similar for all of these systems and closely resemble those obtained with fluorosulfuric acid, as described in the preceding section. The change in composition of the triflic acid–antimony pentafluoride system depending on the SbF_5 content has been studied. For the 1:1 composition, the main counteranion is $[CF_3SO_3SbF_5]^-$ and for the 1:2 composition $[CF_3SO_3(Sb_2F_{11})]^-$ is predominant, with the following structures proposed for the anions. With increasing SbF_5 concentration, the anionic species grow larger and anions containing up to 5 SbF_5 units have been found. Under no circumstances could free SbF_5 be detected.

It has not been possible to measure the acidity of the $CF_3SO_3H{:}SbF_5$ system by spectrophotometry because of its very strong absorption in the uv-visible spectrum.

In comparison with the strength of the related perfluoroalkanesulfonic acids, its acidity should be very close or slightly higher (by 1 H_0 unit) than the acidity of the perfluoroethane– and perfluorobutanesulfonic acid–SbF_5 mixtures, which was meas-

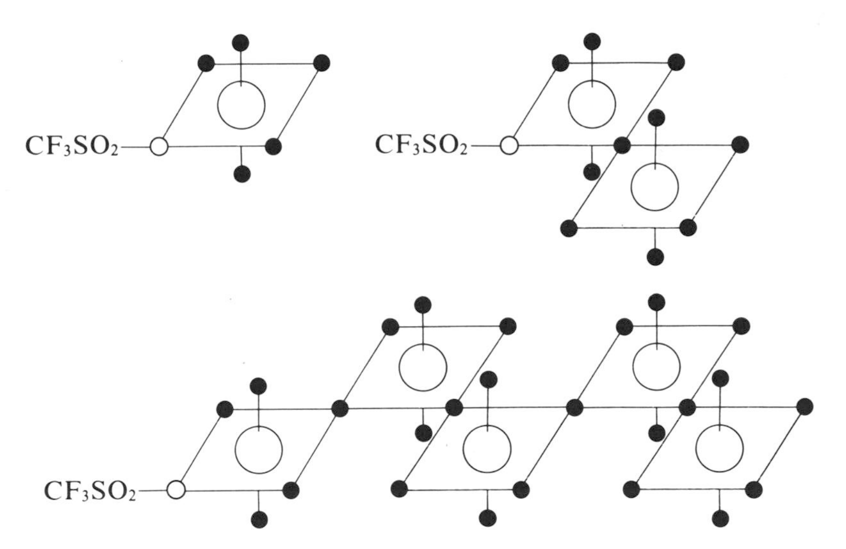

ured by Commeyras and co-workers[8c] (Fig. 2.6). As these authors were using the same spectroscopic technique as previously used by Gillespie, the measurements were limited to $H_0 \geqslant -18.5$ due to lack of suitable weak indicator bases. However, on the basis of ^{19}F-nmr structural studies, a moderate increase in acidity is expected beyond the 50 mol% SbF_5 concentration corresponding to the following autoprotolysis equilibrium:

$$2\ HR_FSO_3SbF_5 \rightleftharpoons R_FSO_3^- + H_2R_FSO_3SbF_5^+$$

2.3.8.2 CF_3SO_3H: $B(SO_3CF_3)_3$

The acidity of triflic acid can also be substantially increased by addition of a boron triflate $B(OSO_2CF_3)_3$ as indicated by Engelbrecht and Tschager[38] (in Fig. 2.6). The increase in acidity is explained by the ionization equilibrium:

$$B(OSO_2CF_3)_3 + 2HSO_3CF_3 \rightleftharpoons H_2SO_3CF_3^+ + \bar{B}(SO_3CF_3)_4$$

The measurements were again limited for the lack of suitable indicator base and even 1,3,5-trinitrobenzene, the weakest base used, was fully protonated ($H_0 \approx -18.5$) in the 22 mol% solution of boron triflate.

2.3.9 Hydrogen Fluoride–Antimony Pentafluoride (Fluoroantimonic Acid)

The HF:SbF_5 (fluoroantimonic acid) system is considered the strongest liquid superacid and also the one that has the widest acidity range. Due to the excellent

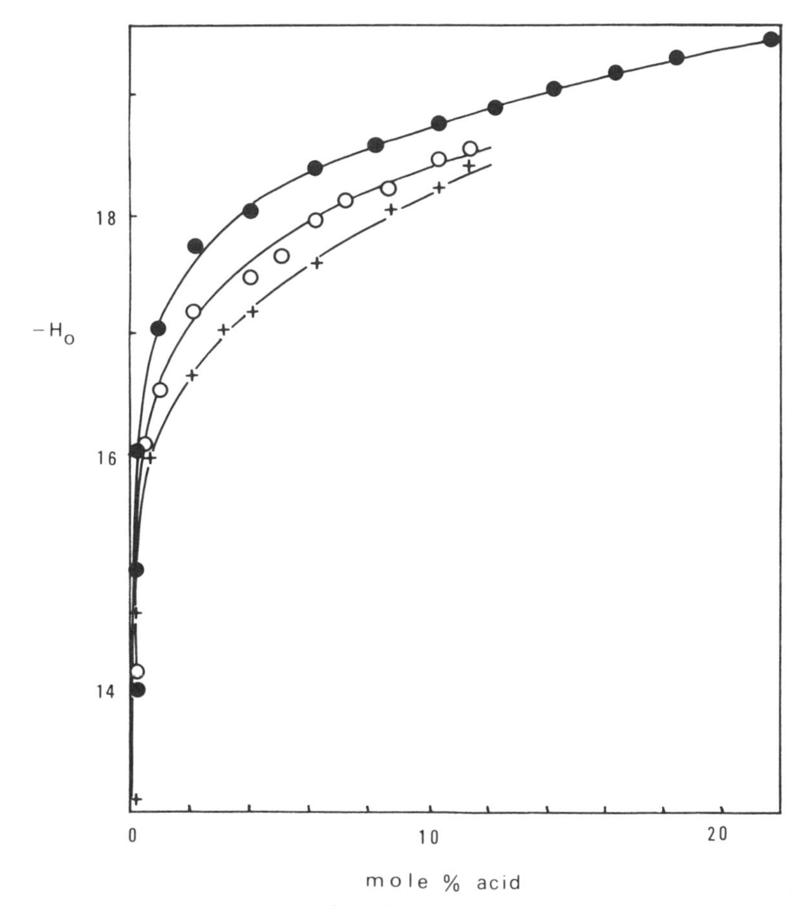

FIGURE 2.6 H_0 acidity function values for various perfluoroalkanesulfonic acid based systems. (○): $C_2F_5SO_3H$-SbF_5, (+): $C_4F_9SO_3H$-SbF_5, (●): $CF_3SO_3H(CF_3SO_3)_3B$, (○), (+): ref. 8c; (●): ref. 38.

solvent properties of hydrogen fluoride, HF:SbF_5 is used advantageously for a variety of catalytic and synthetic applications (see Chapter 5). Anhydrous hydrogen fluoride is an excellent solvent for organic compounds with a wide liquid range. The relatively weak acidity of HF ($H_0 \sim -11$) is compensated by a drastic increase in acidity on addition of a strong Lewis acid. Electrochemical pH determination[39] and acidity measurements using uv spectroscopy[40] and nmr spectroscopy[28] all show the dramatic increase in acidity (approximately 10 H_0 units, when 1 mol% SbF_5 is added to anhydrous HF; Fig. 2.7). Although HSO_3F is a stronger acid than HF (the H_0 values are -15 and -11, respectively), to achieve the same acidity of $H_0 \approx -20.5$ one has to add 25 times more SbF_5 in HSO_3F than in HF. A 1 *M* solution of SbF_5 in HF is about 10^4 times stronger than HSO_3F containing the same amount of SbF_5.

For more concentrated solutions, only kinetic data are available mainly from the work of Brouwer and co-workers[41,42] who estimated the relative acidity ratio of 1:1

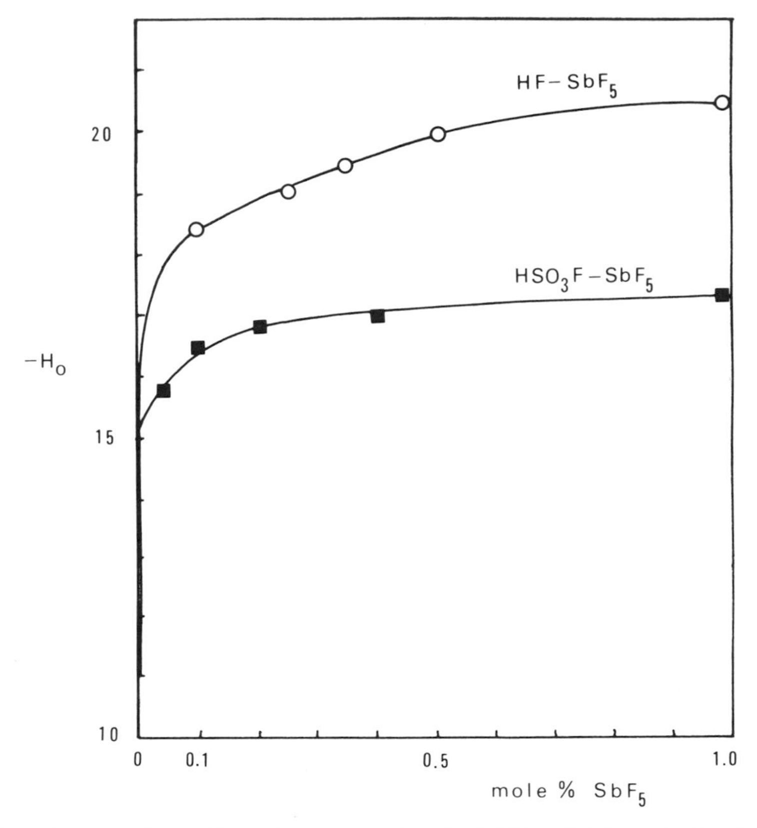

FIGURE 2.7 Comparison of H_0 acidity function values for HF-SbF_5: (○) from ref. 28 and HSO_3F-SbF_5 (■) from ref. 6.

HF:SbF_5 and 5:1 HSO_3F:SbF_5 to be $5 \times 10^8:1$. This means a H_0 value in excess of -30 on the Hammett scale for the 1:1 composition. The acidity may increase still further for higher SbF_5 concentrations. It has been shown by ir measurements that 80% SbF_5 solution has the maximum concentration of H_2F^+.[43] In any case, even for the composition range of 0–50% SbF_5, this is the largest range of acidity (-30 to -11) known. The same ir study has also shown that the predominant cationic species (i.e., solvated proton) in the 0–40 mol% of SbF_5 is the $H_3F_2^+$ ions. The H_2F^+ ion is only observed in highly concentrated solutions (40–100 mol% of SbF_5), contrary to the widespread belief that it is the only proton-solvated species in the HF:SbF_5 solutions.

Ionization of SbF_5 in dilute HF solutions (0–20% SbF_5) thus is

$$SbF_5 + 3HF \rightleftharpoons SbF_6^- + H_3F_2^+$$

The structure of the hexafluoroantimonate and of its higher homologous anions $Sb_2F_{11}^-$ and $Sb_3F_{16}^-$, which are formed when the SbF_5 content is increased, have been determined by ^{19}F-nmr studies as already indicated in Table 2.6 (Section

2.3.3).[44] The predominant anionic species in the 20–40% solution is $Sb_2F_{11}^-$ associated with $H_3F_2^+$.

2.3.10 Hydrogen Fluoride–Tantalum Pentafluoride

HF:TaF_5 is a catalyst for various hydrocarbon conversions of practical importance. In contrast to antimony pentafluoride, tantalum pentafluoride is stable in a reducing environment. The HF:TaF_5 superacid system has attracted attention mainly through the studies concerning alkane-alkylation and aromatic protonation.[45] Generally, heterogeneous mixtures such as the 10:1 and 30:1 HF:TaF_5 have been used because of the low solubility of TaF_5 in HF (0.9% at 19°C and 0.6% at 0°C). For this reason, acidity measurements have been limited to very dilute solutions, and an H_0 value of -18.85 has been found for the 0.6% solution.[40] Both electrochemical studies[46] and aromatic protonation studies[47] indicate that the HF:TaF_5 system is a weaker superacid than HF:SbF_5.

2.3.11 Hydrogen Fluoride–Boron Trifluoride (Tetrafluoroboric Acid)

Boron trifluoride ionizes in anhydrous HF as follows:

$$BF_3 + 2HF \rightleftharpoons BF_4^- + H_2F^+$$

The stoichiometric compound exists only in excess of HF or in the presence of suitable proton acceptors. The HF:BF_3 (fluoroboric acid) catalyzed reactions cover many of the Friedel-Crafts-type reactions.[48] One of the main advantages of this system is the high stability of HF and BF_3. Both are gases at room temperature and are easily recovered from the reaction mixtures.

The large number of patents in this field demonstrates the industrial interest in this superacid system (such as isomerization of xylenes and carbonylation of toluene). The HF:BF_3 system in the presence of hydrogen has also been found to be an effective catalyst for ionic hydrodepolymerization of coal to liquid hydrocarbons.[49] The basicity of many aromatic hydrocarbons has been measured in the HF:BF_3 system by nmr[50] and vapor pressure determinations.[51] Acidity measurements of the system have been limited to electrochemical determinations, and a 7 mol% BF_3 solution was found to have an acidity of H_0 $-$ 16.6.[52] This indicates that BF_3 is a much weaker Lewis acid as compared with either SbF_5 or TaF_5. Nevertheless, the HF:BF_3 system is strong enough to protonate many weak bases and is an efficient and widely used catalyst.[53]

2.3.12 Conjugate Friedel-Crafts Acids (HBr:$AlBr_3$, HCl:$AlCl_3$ etc.)

The most widely used Friedel-Crafts catalyst systems are HCl:$AlCl_3$ and HBr:$AlBr_3$. These systems are indeed superacids by the present definition. However, experiments directed toward preparation from aluminum halides and hydrogen halides of the composition $HAlX_4$ were unsuccessful in providing evidence that such conjugate acids are not formed in the absence of proton acceptor bases.

The hydrogen chloride–aluminum chloride system has been investigated by Brown[54] under a variety of conditions, including temperatures as low as −120°C by vapor pressure measurements. No evidence was found for a combination of the two components.

The catalytic activity of hydrogen bromide itself has been recognized for alkylation, acylation, and polymerization reactions. The experimental difficulty associated with its narrow liquid range is probably the reason why there is no acidity measurements available for neat HBr. H_0 measurements on a 1 *M* solution in anhydrous sulfolane[55] have shown the following sequence of decreasing acidity: $HBF_4 > HClO_4 > HSO_3F > HBr > H_2SO_4 > HCl$. This sequence is in accord with conductometric measurements performed in glacial acetic acid by Gramstad.[15] Recently Gold and co-workers[56] have shown that HBr in CF_2Br_2 is capable of protonating a variety of ketones at sufficiently low temperatures ($T \sim -93°C$) at which the conjugate acids can be observed by nmr under slow-exchange conditions. This indicates that the acidity of HBr is between −10 and −13 on the H_0 scale. The acidity can be increased further by addition of a Lewis acid such as SbF_5[56b] or $AlBr_3$.[57] The relative acidity of the Lewis acids in HBr has been found to be in the order $AlBr_3 > GaBr_3 > TaF_5 > BBr_3 > BF_3$, a sequence deduced from selectivity parameter measurements in 2 *M* solutions[57] assuming an ideal behavior of the individual components. An advantage of HBr as a Brønsted acid is its nonoxidizing nature, but at the same time the bromide ion is a strong nucleophile and neat HBr is a very poor solvent for most of the organic substrates.

Aluminum bromide is sparingly soluble in HBr (1.77 g/mol at −80°C) but its solubility increases substantially in the presence of aromatic or aliphatic hydrocarbons, in which case two phases separate—the upper layer is pure HBr and the lower layer consists of a sludge, for which the following composition has been found: (R^+, $Al_2Br_7^-$):$AlBr_3$:HBr in the ratio 1:0.8:0.7.[47] The commonly encountered sludges or "red oil" in Friedel-Crafts hydrocarbon conversion processes can thus be considered as a solution of carbocations in a superacid HBr:$AlBr_3$ system. The system is generally very complex, containing organic oligomers and alkylates. Similar results were obtained for the HCl:$AlCl_3$ system. In the ethylation of benzene, the sludge was shown by nmr spectroscopy to contain the heptaethylbenzenium ion. Protonation studies based on ^{13}C chemical-shift measurements have shown that the HBr:$AlBr_3$ system is capable of protonating benzene at 0°C. Comparing these results with those obtained using HF:TaF_5, Farcasiu and co-workers[47] claimed that HBr-$AlBr_3$ was an acid of comparable strength to HF:SbF_5, a ranking also proposed by Kramer based on his selectivity parameter.[57] It is, however, highly unlikely that HBr:$AlBr_3$ has comparable strength with HF:SbF_5 as it is incapable of protonating many weak bases or ionize precursors to long-lived carbocations (such as alkyl cations). An extensive discussion of Friedel-Crafts superacids is available and will not be repeated here.[48]

2.4 SOLID SUPERACIDS

Considering the exceptional activity of liquid superacids and their wide application in hydrocarbon chemistry, it is not surprising that work was also extended to solid superacids. The search for solid superacids has become an active area since the early 1970s, as reflected primarily by the extensive patent literature.

Solid acid catalysts such as mixed oxides (chalcides) have been used extensively for many years in the petroleum industry and organic synthesis. Their main advantage compared with liquid acid catalysts is the ease of separation from the reaction mixture, which allows continuous operation, as well as regeneration and reutilization of the catalyst. Furthermore, the heterogeneous solid catalysts can lead to high selectivity or specific activity. Due to the heterogeneity of solid superacids, accurate acidity measurements are difficult to carry out and to interpret. Up until now, the most useful way to estimate the acidity of a solid catalyst is to test its catalytic activity in well-known acid-catalyzed reactions.

Solid acidic oxide catalysts generally do not show intrinsic acidity comparable with liquid superacids and therefore generally high temperatures are required to achieve catalytic activity. To obtain solid superacid catalyst, two approaches can be utilized: either enhancing the intrinsic acidity of a solid acid by treatment with a suitable co-acid, or physically or chemically binding a liquid superacid to an otherwise inert surface. In Sections 2.4.1 and 2.4.2, wc will review the nature and properties of the most significant solid superacid systems known at present.

2.4.1 Acidity Enhancement of Solid Acids

The acidic sites of solid acids may be either of the Brønsted (proton donor, often OH group) or Lewis type (electron acceptor). Both types have been identified by ir studies of solid surfaces absorbed with pyridine. Various solids displaying acidic properties, whose acidities can be enhanced to the superacidity range, are listed in Table 2.7.

The natural clay minerals are hydrous aluminum silicates with iron or magnesium replacing aluminum wholly or in part, and with alkalies or alkaline earth metals present as essential constituents in some others. Their acidic properties and natural abundance have favored their use as catalysts for cracking of heavy petroleum fractions. With the exception of zeolites[58] and some specially treated mixed oxides for which superacid properties have been claimed, the acidity as measured by the color changes of absorbed Hammett bases is generally far below the superacidity range. They are inactive for alkane isomerization and cracking below 100°C and need coacids to reach superacidity.

2.4.1.1 Brønsted Acid-Treated Metal Oxides: TiO_2:H_2SO_4, ZrO_2:H_2SO_4

By exposing freshly prepared $Ti(OH)_4$ to 1 *N* sulfuric acid followed by calcination

TABLE 2.7 Solid Acids

1. Natural clay minerals: kaolinite, bentonite, attapulgite, montmorillonite, clarit, fuller's earth, zeolites, and synthetic clays or zeolites
2. Metal oxides and sulfides: ZnO, CdO, Al_2O_3, CeO_2, ThO_2, TiO_2, ZrO_2, SnO_2, PbO, As_2O_3, Bi_2O_3, Sb_2O_5, V_2O_5, Cr_2O_3, MoO_3, WO_3, CdS, ZnS
3. Metal salts: $MgSO_4$, $CaSO_4$, $SrSO_4$, $BaSO_4$, $CuSO_4$, $ZnSO_4$, $CdSO_4$, $Al_2(SO_4)_3$, $FeSO_4$, $Fe_2(SO_4)_3$, $CoSO_4$, $NiSO_4$, $Cr_2(SO_4)_3$, $KHSO_4$, $(NH_4)_2SO_4$, $Zn(NO_3)_2$, $Ca(NO_3)_2$, K_2SO_4, $Bi(NO_3)_3$, $Fe(NO_3)_3$, $CaCO_3$, BPO_4, $AlPO_4$, $CrPO_4$, $FePO_4$, $Cu_3(PO_4)_2$, $Zn_3(PO_4)_2$, $Mg_3(PO_4)_2$, $Ti_3(PO_4)_4$, $Zr_3(PO_4)_4$, $Ni_3(PO_4)_2$, AgCl, CuCl, $CaCl_2$, $AlCl_3$, $TiCl_3$, $SnCl_2$, CaF_2, BaF_2, $AgClO_4$, $Mg(ClO_4)_2$
4. Mixed oxides: $SiO_2:Al_2O_3$, $SiO_2:TiO_2$, $SiO_2:SnO_2$, $SiO_2:ZrO_2$, $SiO_2:BeO$, $SiO_2:MgO$, $SiO_2:CaO$, $SiO_2:SrO$, $SiO_2:ZnO$, $SiO_2:Ga_2O_3$, $SiO_2:Y_2O_3$, $SiO_2:La_2O_3$, $SiO_2:MoO_3$, $SiO_2:WO_3$, $SiO_2:V_2O_5$, $SiO_2:ThO_2$, $Al_2O_3:MgO$, $Al_2O_3:ZnO$, $Al_2O_3:CdO$, $Al_2O_3:B_2O_3$, $Al_2O_3:ThO_2$, $Al_2O_3:TiO_2$, $Al_2O_3:ZrO_2$, $Al_2O_3:V_2O_5$, $Al_2O_3:MoO_3$, $Al_2O_3:WO_3$, $Al_2O_3:Cr_2O_3$, $Al_2O_3:Mn_2O_3$, $Al_2O_3:Fe_2O_3$, $Al_2O_3:Co_3O_4$, $Al_2O_3:NiO$, $TiO_2:CuO$, $TiO_2:MgO$, $TiO_2:ZnO$, $TiO_2:CdO$, $TiO_2:ZrO_2$, $TiO_2:SnO_2$, $TiO_2:Bi_2O_3$, $TiO_2:Sb_2O_5$, $TiO_2:V_2O_5$, $TiO_2:Cr_2O_3$, $TiO_2:MoO_3$, $TiO_2:WO_3$, $TiO_2:Mn_2O_3$, $TiO_2:Fe_2O_3$, $TiO_2:Co_3O_4$, $TiO_2:NiO$, $ZrO_2:CdO$, $ZnO:MgO$, $ZnO:Fe_2O_3$, $MoO_3:CoO:Al_2O_3$, $MoO_3:NiO:Al_2O_3$, $TiO_2:SiO_2:MgO$, $MoO_3:Al_2O_3:MgO$
5. Cation exchange resins, polymeric perfluorinated resinsulfonic acids

in air at 500°C for 3 hr, Hino and Arata[59a] have obtained a solid catalyst active for isomerization of butane in low yields at room temperature. The acid strength as measured by the color change of a Hammett indicator (2,4-dinitrobenzene) was found to be as high as $H_0 \geqslant -14.5$.

Subsequently, the same authors[59b] described the preparation of a solid superacid catalyst with acid strength of $H_0 = -16$ with a sulfuric acid-treated zirconium oxide. They exposed $Zr(OH)_4$ to 1 *N* sulfuric acid and calcinated it in air at approximately 600°C. The obtained catalyst was able to isomerize (and crack) butane at room temperature. The acidity was examined by the color change method using Hammett indicators added to a powdered sample placed in sulfuryl chloride. The existence of both Brønsted and Lewis sites was shown by the ir spectra of absorbed pyridine.

The X-ray photoelectron and ir spectra showed that the catalyst possessed bidentate sulfate ion coordinated to the metal. The specific surface areas were much larger than those of the zirconium oxides, which had not undergone the sulfate treatment. The interesting feature of these catalysts is the high temperature at which they are prepared which means that they maintain their acidity at temperatures as high as 500°C and should thus be easy to regenerate and reuse.

Considering the high activity obtained after treatment of these oxides with H_2SO_4, it is rather surprising that a large variety of oxides and mixed oxides show almost no increase in activity after treatment with much stronger HSO_3F.[60] $SiO_2:Al_2O_3$ treated with Magic Acid was, however, found to be moderately active at room temperature in cracking butane, but its activity was much less than the Lewis acid-treated oxides (*vide infra*).

2.4.1.2 Lewis Acid-Treated Metal Oxides

In 1976 Tanabe and Hattori[61] reported the preparation of solid superacids such as $SbF_5:TiO_2:SiO_2$, $SbF_5:TiO_2$ and $SbF_5:SiO_2:Al_2O_3$ whose acid strengths were in the range of -16 on the H_0 scale. In a subsequent work, the same authors reported a thorough study of the preparation and activity measurement of a large number of oxides and mixed oxides, treated with a variety of superacids. The SbF_5-treated oxides such as TiO_2, SiO_2, and mixed oxides like $TiO_2:ZrO_2$, $TiO_2:SiO_2$, $SiO_2:Al_2O_3$ were found to be most effective in the isomerization and cracking reactions of butane[61] and pentanes.[62] The catalysts were prepared by exposing the powdered metal oxides to SbF_5 vapor at room temperature followed by degassing. The adsorption-desorption cycle was repeated a number of times and finally the catalyst was subjected to high vacuum to remove last traces of free SbF_5 at a given temperature. The amount of SbF_5 remaining on the catalyst, as measured by the weight increase of the catalyst, for typical preparations are given in Table 2.8.

The highest activity for butane cracking at room temperature was obtained with the TiO_2- and SiO_2-containing systems. For pentane and 2-methylpentane isomerizations at 0°C, the most active catalyst was $SbF_5:TiO_2:ZrO_2$ with a selectivity close to 100% for skeletal isomerization.

TABLE 2.8 Amount of SbF_5 Retained in Catalyst (wt%)

Catalyst[a]	SbF_5 Ads. Temp/°C	Evacuation Temp.			
		0°C	30°C	50°C	100°C
$SbF_5:SiO_2:Al_2O_3$	0	49.7	37.6	27.7	
	10	53.2	40.9	34.5	30.7
	30		34.1	32.2	
	100				17.2
$SbF_5:TiO_2:ZrO_2$	0	49.9	30.0	32.8	
	10		33.5	27.2	
	30			29.8	
	100				30.9
$SbF_5:SiO_2$	0	33.1	18.8	12.8	
	10			18.2	
	30			17.2	
$SbF_5:Al_2O_3$	0	37.1			
$SbF_5:TiO_2$	0	28.1			
$SbF_5:ZrO_2$	0	24.8			

[a]Adsorption-desorption cycle in SbF_5 treatment, repeated four times.

The high activity of these mixed oxides was ascribed to oxygen coordination of SbF_5 enhancing the acidity of both Brønsted and Lewis acid sites.

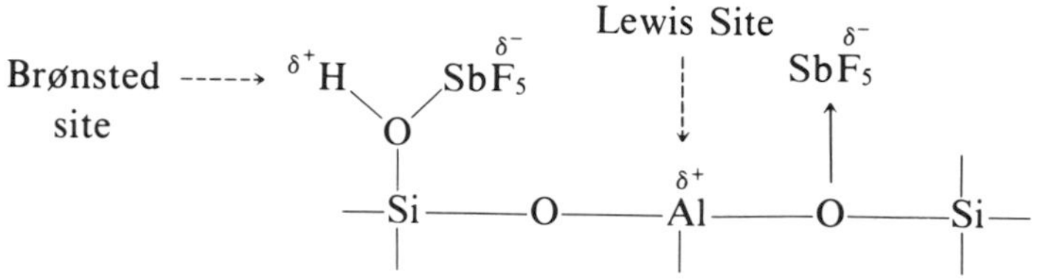

Brønsted site -----→ $^{\delta^+}H$ $\overset{\delta^-}{SbF_5}$ Lewis Site $\overset{\delta^-}{SbF_5}$

$$-Si-O-\overset{\delta^+}{Al}-O-Si-$$

The ir study of $SbF_5:Al_2O_3$ after addition of pyridine to the catalyst showed an absorption at 1460 cm^{-1}, which was assigned to pyridine coordinated with the Lewis acid site, and another at 1540 cm^{-1}, which was attributed to the pyridinium ion resulting from the protonation of pyridine by the Brønsted acid sites. When the catalyst was heated to 300°C, the ir band of the pyridinium ion disappeared, whereas the absorption band for the Lewis acid sites was still present. The fact that the catalyst was still active suggests that the Lewis acid sites are the active sites for catalysis.

2.4.1.3 Aluminum Halide-Treated Metal Salts

Mixtures of aluminum chloride and metal chloride are known to be active for the isomerization of paraffins at room temperature.[63] Ono and co-workers have shown that the mixtures of aluminum halides with metal sulfates[64–67] are much more selective for the same reactions at room temperature.

$AlCl_3$–Metal Sulfates. Room temperature isomerization of pentane has been carried out with a series of mixtures containing aluminum chloride with sulfates of metals such as Ti, Fe(III), Ni, Cu, Mn, Fe(II), Al, Zr, Co, Mg, Ca, Ba, Pd, with conversions in excess of 10% after 3 hr.[64] The most effective catalyst was an equimolar mixture of $AlCl_3$ and $Ti_2(SO_4)_3$ with a conversion of 46% and a selectivity to isopentane of 84%. The catalysts were prepared by kneading a mixture of aluminum chloride and a dehydrated metal salt in a porcelain mortar in a dry nitrogen atmosphere.

The $AlCl_3:CuSO_4$ mixture has been more thoroughly investigated[67] in 5–23°C temperature range. The acidity was measured as approximately $H_0 = -14$ and the activity was found to be proportional to the amount of $CuSO_4$ and also to the specific surface area of the $CuSO_4$ used. It was claimed that the addition of water had no effect on the catalytic activity which seems to indicate that the active species are essentially different from those in the $AlCl_3:H_2O$ system.

$AlCl_3$–Metal Chlorides. The catalytic activity of metal chloride–aluminum chloride mixtures for pentane isomerization has been studied by Ono.[66] The highest conversion was found when $AlCl_3$ was combined with $TiCl_3$ (31% after 3 hr at room temperature), $MnCl_2$ (12%), and $CuCl_2$ (11%) but no such catalyst reached the activity of metal sulfate–aluminum halide mixtures.

2.4.1.4 Aluminum Halide-Treated Oxides and Mixed Oxides

The activity and stability of aluminum chloride-treated alumina and silica-alumina as alkane isomerization catalysts has been investigated by Oelderik and Platteuw.[68] The degree of hydration of the carrier and the effects of hydrogen chloride and hydrogen pressure on the carrier have been studied. The authors[68] concluded that $AlCl_3$ reacts with the hydrated carrier to give a surface bound compound that is converted to an acidic site by HCl absorption. However, prolonged treatment of the carrier with excess $AlCl_3$ results in HCl gas evolution, until no acidic site is left on the catalyst. The activity of the catalyst system for hexane isomerization, however, declines exponentially with time.

2.4.1.5 Sulfonic Acid Resins Treated with Lewis Acids

In most of the acidic ion exchangers, the active sites are the sulfonic acid groups which are attached to a solid backbone such as sulfonated coal, sulfonated phenolformaldehyde resins, or sulfonated styrene-divinylbenzene cross-linked polymers. The latter sulfonated resins are the only type acidic enough to gain importance as a catalyst for acid-catalyzed electrophilic reactions. The most widely used are the Dowex 50 (Dow Chemicals) and Amberlite IR-120 (Rohm and Haas) type resins which are made by sulfonation of cross-linked polystyrene divinylbenzene beads. Dowex 50 is comparable in acid strength to HCl[69] and therefore is considered as a solid acid of moderate strength. The acidity of these resins increases significantly by treating them with Lewis acid halides.

$AlCl_3$-Complexed Polystyrene Sulfonic Acid. Gates and co-workers[70] have prepared a superacid catalyst from $AlCl_3$ and beads of macroporous, sulfonated polystyrene divinylbenzene. The catalyst was prepared by exposing the macroporous beads of the polymeric Brønsted acid at 110°C to a stream of nitrogen containing sublimed $AlCl_3$. HCl was liberated and the resulting polymer had a S:Al:Cl ratio of 2:1:2. Electron microprobe X-ray analysis has shown that Al and Cl were uniformly dispersed throughout each polymer bead. By analogy with the known structure of liquid superacids the following structure was suggested.

$$2\left[\begin{array}{c} | \\ CH_2 \\ | \\ CH{-}C_6H_4{-}\overset{O}{\underset{O}{\overset{\|}{\underset{\|}{S}}}}{-}OH \\ | \\ CH_2 \\ | \end{array}\right] + AlCl_3 \longrightarrow$$

$$\left[\begin{array}{c} | \\ CH_2 \\ | \\ CH{-}C_6H_4{-}\overset{O}{\underset{O}{\overset{\|}{\underset{\|}{S}}}}{-}O{-}\overset{Cl}{\underset{Cl}{\overset{|}{\underset{|}{Al}}}}{-}O{-}\overset{O}{\underset{O}{\overset{\|}{\underset{\|}{S}}}}{-}C_6H_4{-}CH \\ | \\ CH_2 \\ | \end{array}\right]^{-} H^{+} + HCl$$

The catalyst was found to be capable of isomerizing and cracking *n*-hexane at 85°C in a flow reactor with an overall initial conversion of 80%.

2.4.2 Perfluorinated Polymer Sulfonic Acids

The considerable success of Magic Acid and related superacids in solution chemistry and interest to extend the scope and utility of acid catalyzed reactions, particularly hydrocarbon transformations, logically led to the attempts to adopt this chemistry to solid systems allowing heterogeneous catalytic processes.

On the basis of the realization that HSO_3F and CF_3SO_3H are of comparable acid strength and their acidities can be substantially enhanced by complexing with Lewis acids fluorides, such as antimony pentafluoride, Olah and his co-workers developed the chemistry of solid superacidic catalysts related to perfluorinated resinsulfonic acids. In these systems, the sulfonic acid group is attached to a —CF_2— or —CF— group in a perfluorinated backbone. This structure provides high acidity sites and at the same time the perfluorinated polymer itself is highly inert and resists acid cleavage.

A convenient solid perfluorinated sulfonic acid can be made readily from Du Pont's commercially available Nafion brand ion membrane resins (a copolymer of perfluorinated epoxide and vinylsulfonic acid). Nafion-H-catalyzed reactions gained substantial interest and are reviewed in Chapter 5. Similarly, (C_8 to C_{18}) perfluorinated sulfonic acids, such as perfluorodecanesulfonic acid (PDSA) are very strong acids and are useful catalysts for many organic syntheses.[71]

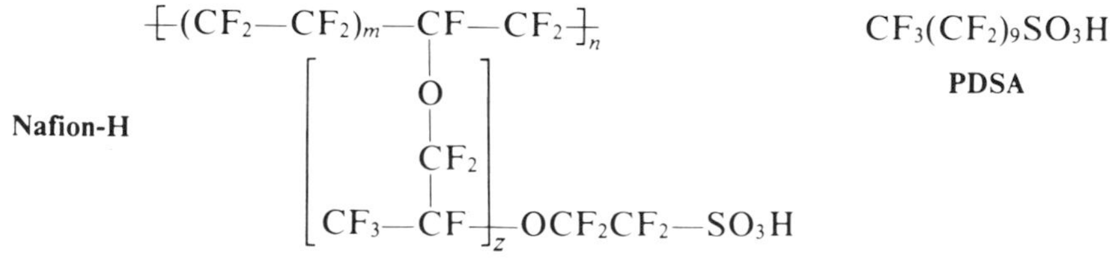

$$\left[(CF_2{-}CF_2)_m{-}CF{-}CF_2\right]_n \quad (\text{CF bears } {-}[O{-}CF_2{-}CF(CF_3)]_z{-}OCF_2CF_2{-}SO_3H)$$

Nafion-H

$$CF_3(CF_2)_9SO_3H$$

PDSA

The acidity of these Brønsted acids, which is around $H_0 \sim -10$ to -12, can be increased further by complexation with Lewis acid fluorides, such as SbF_5, TaF_5, and NbF_5.[72] They have bccn found to be effective catalysts for *n*-hexane, *n*-heptane isomerization, alkylation of benzene, and transalkylation of alkylbenzenes.

2.4.3 Immobilized Superacids (Bound to Inert Supports)

Ways have been found to immobilize and/or to bind superacidic catalysts to an otherwise inert solid support. Several such types are described in this section.

2.4.3.1 Graphite-Intercalated Superacids

The last decade has witnessed a renewal of interest and a virtual explosion in research in the preparation, structure, and properties of graphite intercalation compounds. A number of recent reviews[73] have been devoted to this subject showing that intercalation compounds can be useful materials as catalysts, reagents, electronic conductors, battery components, and lubricating agents.

Graphite possesses a layered structure that is highly anisotropic. It consists of sheets of sp^2 carbon atoms in hexagonal arrays with a C—C bond distance of 1.42 Å consistent with a one-third double bond and two-thirds single bond character (Fig. 2.8). The distance between the layers is 3.35 Å and is in accord with the fact that the graphite sheets are held together only by weak Van der Waals forces. The atomic layers are not directly superimposable but alternate in the pattern A-B-A-B- as shown in Figure 2.8 for the most stable form, called hexagonal graphite. Two special types of synthetic graphite are considered of great interest—highly oriented pyrolytic graphite (HOPG) and graphite fibers. HOPG is a highly crystalline material in which all the carbon layers are nearly parallel in the whole sample, which may be of sizeable dimension. The graphite fibers are another form of synthetic graphite that show unusual mechanical properties in the composite material (used extensively in aviation industry).

The intercalation process has been a much disputed topic, but there is now sufficient experimental proof that the intercalation step is generally a redox process in which ions and neutral species enter the interlamellar space without disrupting the layered topology of graphite. X-ray studies show that the compounds at equilibrium favor the existence of a sequence of filled and empty interlamellar voids. This forms the basis for the concept of staging. In the first stage or stage-1 compound, all the interlamellar regions are filled by the intercalant molecules (Fig. 2.9). In the second- and third-stage compounds, each second or third interlamellar region, respectively, is filled by the intercalant and so on.

AsF_5 and SbF_5 Graphite Intercalates. Graphite reacts with AsF_5 (at 3 atm, 25°C) and SbF_5 (at 100°C) to form the first-stage compounds $C_8:AsF_5$ and $C_{6.5}:SbF_5$,

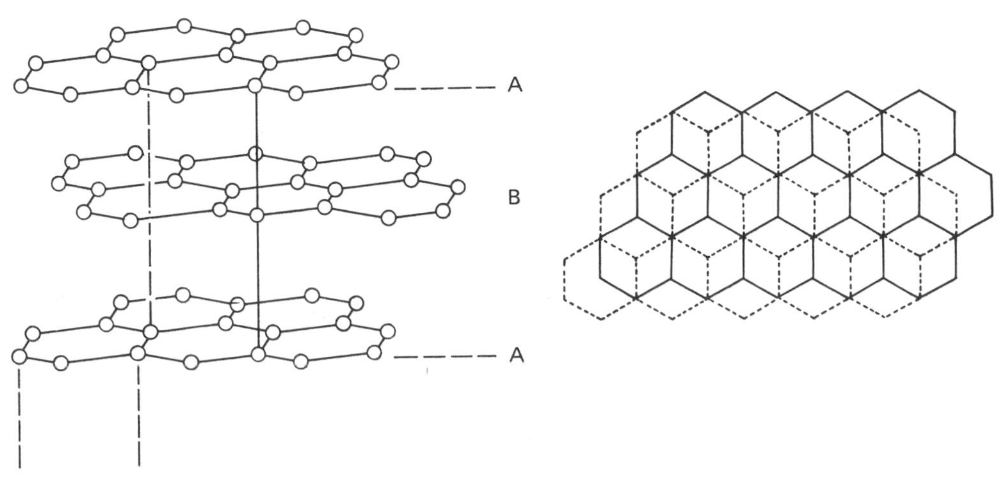

FIGURE 2.8 Graphite structure.

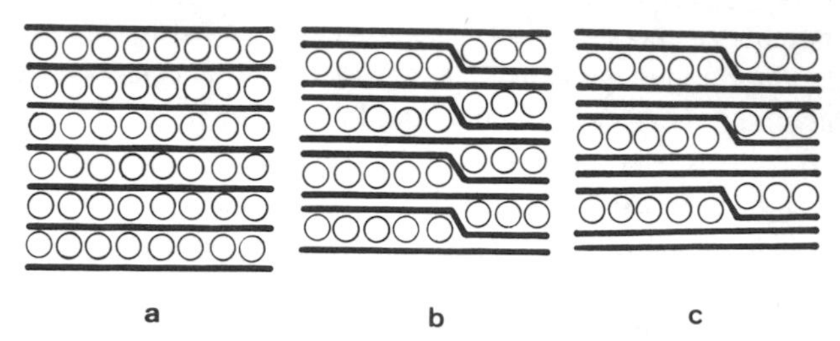

FIGURE 2.9 Staging in graphite intercalates: (*a*) first-stage; (*b*) second stage; (*c*) third stage.

respectively. At higher temperatures, higher-stage compounds can be prepared. For the formation of the AsF_5 compound, the following equation has been proposed.[74]

$$8C + AsF_5 \longrightarrow (C_8{}^+)_{0.33}(AsF_6{}^-)_{0.33}(AsF_5)_{0.5}(AsF_3)_{0.17}(C_8)_{0.67}$$

During the insertion process, the metal is partially reduced from M_V to the M_{III} oxidation state and at the same time the neutral species are inserted. The same assumptions hold good for the SbF_5 inserted graphite for which ^{19}F-nmr[75] and ^{121}Sb Mossbauer[76] and X-ray studies[77] have been conducted. Similar to the reactions of SbF_5 in liquid phase, an additional equilibrium is believed to take place in SbF_5-graphite intercalates, that is of significance in catalytic applications.

$$C_n{}^+MF_6{}^- + MF_5 \longrightarrow C_n{}^+(M_2F_{11})^-$$

The catalytic properties of SbF_5–graphite have been investigated.[78,79] The chemistry is basically the same as that of SbF_5 itself with two major differences: (1) as a solid it can be more easily separated from the reaction mixture, (2) the activity is that expected from a very dilute solution of the superacid. The activity decreases, however, quite rapidly and the catalyst becomes deactivated after 10–20 turnovers. Three possible reasons for such a deactivation have been given by Heinemann and Gaaf:[79]

1. Leaching out of SbF_5
2. Reduction of Sb_V to Sb_{III}
3. Poisoning of the acidic sites by carbonaceous products

Consequently, it is considered that the chemistry is not due to the graphite intercalate itself but to the SbF_5 present at the exposed surface areas. The reaction takes place at these reactive sites in the first step of the formation of the intermediate carbenium ion and SbF_5 is reduced to SbF_3. The carbenium ion itself may alkylate or undergo cleavage, and higher molecular polymeric material with increasing C:H ratio builds up on the catalytic site deactivating the catalyst slowly.[79] X-ray diffraction studies and elemental analysis show that large quantities of SbF_5 exist in the deactivated catalyst[80] and the catalyst is not easily exfoliated (the lamellar structure is conserved and large regions of the first stage compound can still be detected).

Miscellaneous Superacid Intercalates. The intercalation of $AlCl_3$ and $AlBr_3$ in graphite in the presence of chlorine or bromine has been long known.[81] The intercalates have been found to be milder catalysts than the pure Lewis acids.[82] However, it was shown that the Lewis acids are readily leached out from the intercalates during Friedel-Crafts-type reactions, and it is not clear whether the intercalates themselves or more probably the surface-exposed Lewis acids are the de facto catalysts. Other superacids such as NbF_5, TaF_5, $HF:SbF_5$, $AlBr_3:Br_2$ have also been used as graphite intercalates for various superacid-catalyzed reactions such as reduction of alkyl halides,[83] alkylation of aromatics,[84] and isomerization and cracking of C_5, C_6 alkanes.[85]

2.4.3.2 SbF_5-fluorinated graphite: SbF_5-fluorinated Al_2O_3

It has been shown that improved catalytic activity and stability is obtained with SbF_5-treated fluorinated graphites instead of graphite.[79] Fluorinated graphites are not uniform in structure and contain many imperfections and unfluorinated areas. It has also been demonstrated that SbF_5 is not intercalated into higher fluorinated graphite with a F:C ratio of 1.1; similar catalytic activity can be obtained by reacting SbF_5 with fluorinated γ-alumina.[79] In less fluorinated samples it seems that SbF_5 is intercalated only in the nonfluorinated regions. The catalytic activity, however, is due to SbF_5 bound to the surface-exposed areas. The isomerization of pentane and hexane has been studied as a model reaction to assess the potential of SbF_5:graphite and SbF_5:fluorographite. A kinetic measurement of the deactivation process showed that nonfluorinated graphite–SbF_5 deactivated much more rapidly than fluorographite–SbF_5. It was suggested that in SbF_5–fluorographite, because of strong interaction with the carbon-bound fluorine and SbF_5, the catalyst is strongly immobilized. Better catalytic activity could be obtained by using fluorographite containing some graphite (optimum $CF_{0.8}$), and it has been assumed that the initially intercalated SbF_5 leaches out of the layer during the isomerization and binds to the surface of fluorinated graphite, thereby creating a new active site. The stability of the catalyst has been found comparable to SbF_5 bound to fluorinated γ-alumina. The characteristics of the hydroisomerization of pentane over SbF_5 on fluorinated graphite or fluorinated alumina, such as disproportionation reactions, influence of hydrogen pressure, stability, and deactivation mechanism are similar to those found with the homogeneous $HF:SbF_5$ liquid acid system.[79]

REFERENCES

1. N. F. Hall, J. B. Conant, *J. Am. Chem. Soc. 49*, 3047 (1927).
2. U. S. Pat. 2,392,861 (Jan. 15, 1946), J.C. Pernert (to Oldbury Electrochemical Co.).
3a. H. M. Goodwin, E. C. Walker, *Chem. Metall. Eng. 25*, 1093 (1921).

3b. H. M. Goodwin, E. C. Walker, *Trans. Electrochem. Soc. 40,* 157 (1921).

3c. W. Muller, P. Jonck, *Chem. Eng. Tech. 35,* 78 (1963).

4. G. N. Dorofeenko, S. V. Krivun, V. I. Dulenko, Yu, A. Zhadanov, *Russ. Chem. Rev. 34,* 88 (1965).

5. A. W. Williamson, *Proc. Royal. Soc.* (*London*) *7,* 11 (1854).

6a. R. J. Gillespie, T. E. Peel, E. A. Robinson, *J. Am. Chem. Soc. 93,* 5083 (1971).

6b. R. J. Gillespie, T. E. Peel, *Ibid., 95,* 5173 (1973).

7. T. E. Thorpe, W. Kirman, *J. Chem. Soc.* 921 (1892).

8a. R. J. Gillespie, E. A. Robinson, *Can. J. Chem. 40,* 675 (1967).

8b. R. Savoie, P. A. Giguere, *Can. J. Chem. 42,* 277 (1964).

8c. D. Brunel, A. Germain, A. Commeyras, *Nouv. J. Chim. 2,* 275 (1978).

9. R. C. Paul, K. K. Paul, K. C. Malhotra, *J. Inorg. Nucl. Chem. 31,* 2614 (1969).

10. T. Gramstad, R. N. Haszeldine, *J. Chem. Soc.* 173 (1956).

11. J. Grondin, R. Sagnes, A. Commeyras, *Bull. Soc. Chim. Fr.* 1779 (1976).

12a. R. D. Howells, J. D. McCown, *Chem. Rev. 77,* 69, 1977.

12b. P. J. Stang, M. R. White, *Aldrichimica Acta 16,* 15 (1983).

13a. U. S. Patent 2,732,398 (Jan. 24, 1956), T. J. Brice, P. W. Troff (to 3M Co.).

13b. R. N. Haszeldine, J. M. Kidd, *J. Chem. Soc.* 2901 (1955).

14. C. H. Dungon, J. R. van Wazer, *Compilation of reported ^{19}F nmr Chemical Shifts 1951–1967,* Wiley-Interscience, New York, 1970.

15. T. Gramstad, *Tidsskr. Kemi. Bergres. Metall. 19,* 62 (1959); *Chem. Abstr. 54,* 14103 (1960).

16. U. S. Patent 3,346,606 (Oct. 10, 1967), R. B. Ward (to E. DuPont).

17. R. A. Guenthner, M. L. Victor, *Ind. Eng. Chem. Prod. Res. Div. 1,* 165 (1962).

18. G. J. Hoffman, B. F. Holder, W. L. Jolly, *J. Phys. Chem. 62,* 364 (1958).

19. E. E. Ainsley, R. D. Peacock, P. L. Robinson, *Chem. Ind.* 1117 (1951).

20. R. D. Peacock, I. L. Wilson, *J. Chem. Soc. A,* 2030 (1969).

21. O. Ruff, W. Plato, *Ber. 37,* 673 (1904).

22. E. R. Palardeau, G. M. T. Foley, C. Zeller, F. L. Vogel, *Chem. Commun.* 389 (1977).

23. F. Fairbrother, *The Chemistry of Niobium and Tantalum,* Elsevier, London, 1967.

24. D. Brown, "Chemistry of Niobium and Tantalum," in *Comprehensive Inorganic Chemistry,* J. C. Bailer et al. Eds., Vol. 3, Pergamon Press, Compendium Publishers, Elmsford, New York, 1973, p. 565.

25a. R. J. Gillespie, S. Wasif, *J. Chem. Soc.* 215 (1953).

25b. F. D. Miles, H. N. Bloch, and D. Smith, *Trans. Farad. Soc. 36,* 350 (1940).

26. G. N. Lewis, J. Bigeleisen, *J. Am. Chem. Soc. 65,* 1144 (1943).

27. J. C. D. Brandt, W. Horning, J. D. Thornley, *J. Chem. Soc.* 1374 (1952).

28. J. Sommer, P. Canivet, S. Schwartz, P. Rimmelin, *Nouv. J. Chim. 5,* 45 (1981).

29. V. Gold, K. Laali, K. P. Morris, L. Z. Zduneck, *Chem. Commun.* 769 (1981).

30. P. A. W. Dean, R. J. Gillespie, *J. Am. Chem. Soc. 92,* 2362 (1970).

31. G. A. Olah, A. Commeyras, *J. Am. Chem. Soc. 91,* 2929 (1969).

32. G. A. Olah, G. K. S. Prakash, M. Arvanaghi, F. A. L. Anet, *J. Am. Chem. Soc. 104,* 7105 (1982).

33. R. J. Gillespie, E. A. Robinson, *Inorg. Chem. 8,* 63 (1969).

34. R. J. Gillespie, E. A. Robinson, *Can. J. Chem. 40,* 675 (1962).

35. R. J. Gillespie, T. E. Peel, *Adv. Phys. Org. Chem. 9,* 1 (1972).

36. D. A. McCauley, U. S. Pat. 4144,282 (1979).

37. R. C. Thompson, J. Barr, R. J. Gillespie, J. B. Milne, R. A. Rothenburg, *Inorg. Chem. 4,* 1641 (1965).
38. A. Engelbrecht, E. Tschäger, *Z. Anorg. Allg. Chem. 433,* 19 (1977).
39. J. Devynck, P. L. Fabre, B. Tremillon, A. Ben Hadid, *J. Electroanal. Chem. 91,* 93 (1978).
40. R. J. Gillespie, unpublished results.
41. D. M. Brouwer, J. A. Van Doorn, *Recl. Trav. Chim. Pays-Bas. 89,* 553 (1970).
42. D. M. Brouwer, J. A. Van Doorn, *Recl. Trav. Chim. Pays-Bas. 91,* 895 (1972).
43. B. Bonnet, G. Mascherpa, *Inorg. Chem. 19,* 785 (1980).
44. J. Bacon, P. A. W. Dean, R. J. Gillespie, *Can. J. Chem. 48,* 3413 (1970).
45. D. Farcasiu, *Acc. Chem. Res. 15,* 46 (1982) and references therein.
46a. R. Gut, K. Gautshi, *J. Inorg. Nucl. Chem.*, H. H. Hyman Mem. Supp., 95 (1976).
46b. J. Devynck, A. Ben Hadid, P.-L. Fabre, *J. Inorg. Nucl. Chem. 41,* 1159 (1979).
46c. M. Herlem, A. Thiebault, *J. Electroanal. Chem. 84,* 99 (1977).
46d. J. J. Kaurova, L. M. Grubina, T. O. A. Adzhemyan, *Elektrokhimiya, 3,* 1222 (1967).
47. D. Farcasiu, S. L. Fisk, M. T. Melchior, K. D. Rose, *J. Org. Chem. 47,* 453 (1982).
48. For a review see G. A. Olah, *Friedel-Crafts and Related Reactions,* John Wiley & Sons, New York, 1963, Vol. 1, and G. A. Olah, *Friedel and Crafts Chemistry,* John Wiley & Sons, New York, 1973.
49. G. A. Olah, M. R. Bruce, E. H. Edelson, A. Husain, *Fuel,* 63(8), 1130 (1984).
50. C. MacLean, E. L. Mackor, *Disc. Farad. Soc. 34,* 165 (1962).
51. E. L. Mackor, A. Hoftra, J. H. Van der Waals, *Trans. Farad. Soc. 54,* 186 (1958).
52. D. A. McCauley, private communication.
53. P. M. Brouwer, E. L. MacKor, C. MacLean, *Recl. Trav. Chim. Pays-Bas. 84,* 1564 (1965).
54. H. C. Brown, H. W. Pearsall, *J. Am. Chem. Soc. 73,* 4681 (1951).
55. R. W. Alder, G. R. Chalkley, M. C. Whiting, *Chem. Commun.* 405 (1966).
56a. J. Emsley, V. Gold, M. J. B. Jais, L. Z. Zdunek, *J. Chem. Soc. Perk. II* 881 (1982).
56b. J. Emsley, V. Gold, M. J. B. Jais, *Chem. Commun.* 961 (1979).
57. G.M. Kramer, *J. Org. Chem. 40,* 298, 302 (1975).
58. C. Mirodatos, D. Barthomeuf, *Chem. Commun.* 39 (1981).
59a. M. Hino, K. Arata, *Chem. Commun.* 1148 (1979).
59b. M. Hino, K. Arata, *Ibid.*, 851 (1980).
60. H. Hattori, O. Takahashi, M. Takagi, K. Tanabe, *J. Catal. 68,* 132 (1981).
61. K. Tanabe, H. Hattori, *Chem. Letters,* 625 (1976).
62. O. Takahashi, T. Yamauchi, T. Sakuhara, H. Hattori, K. Tanabe, *Bull. Chem. Soc. Japan, 53,* 1807 (1980).
63. L. Schmerling, J. A. Vesely, U. S. Patents 3,846,503 and 3,846,504 (1972).
64. Y. Ono, T. Tanabe, N. Kitajima, *Chem. Letters,* 625 (1978).
65. Y. Ono, S. Sakuma, T. Tanabe, N. Kitajima, *Chem. Letters,* 1061 (1978).
66. Y. Ono, T. Tanabe, N. Kitajima, *J. Catal. 56,* 47 (1979).
67. Y. Ono, K. Yamaguchi, N. Kitajima, *J. Catal. 64,* 13 (1980).
68. J. M. Oelderik, J. C. Platteuw, in *Proceedings of the 3rd International Congress on Catalysis,* W. H. M. Sachtler, G. C. A. Schmit, P. Zwietering, Eds., Vol. 1, North Holland, Amsterdam, 736 (1965).
69. Ref. 48b, p. 356.
70a. V. L. Magnotta, B. C. Gates, G. C. A. Schuit, *Chem. Commun.* 342 (1972).
70b. V. L. Magnotta, B. C. Gates, *J. Catal. 46,* 266 (1977).

70c. G. A. Fuentes, B. C. Gates, *Ibid., 76,* 440 (1982).

71. For a review see G. A. Olah, *Acc. Chem. Res. 13,* 330 (1980).

72. G. A. Olah, U. S. Patent application 1983.

73a. N. Bartlett, B. W. McQuillan, *Intercalation Chemistry,* in Material Science and Technology Series, M. S. Whittingham, A. J. Jacobson, Eds., Academic Press, New York, 1982.

73b. W. C. Forsman, T. Dziemianowicz, K. Leong, D. Carl, *Synth. Met. 5,* 77 (1983).

74. L. B. Ebert, *J. Mol. Catal. 15,* 275 (1982).

75a. L. Facchini, J. Bouar, H. Sfihi, A. P. Legrand, G. Furdin, J. Melin, R. Vangelisti, *Synth. Met. 5,* 11 (1982).

75b. L. B. Ebert, R. A. Huggins, J. I. Brauman, *Chem. Commun.* 924 (1974).

76a. J. Ballard, T. Birchall, *J. C. S. Dalton Trans.* 1859 (1976).

76b. J. M. Friedt, L. Soderholm, Pont-a-Moussan, R. Vangelisti, *Synth. Met.* Proceedings of G. I. C. Conference, France, May 20, 1983.

77. P. Touzain, *Synth. Met. 1,* 3 (1979).

78a. F. Le Normand, F. Fajula, F. Gault, J. Sommer, *Nouv. J. Chim. 6,* 411 (1982) and references therein.

78b. K. Laali, J. Sommer, *Ibid., 5,* 469 (1981).

78c. K. Laali, M. Muller, J. Sommer, *Chem. Commun.* 1088 (1980).

79. J. J. L. Heinerman, J. Gaaf, *J. Mol. Catal. 11,* 215 (1981).

80. J. Sommer, R. Vangelisti, unpublished results.

81a. W. Rudorff, R. Zeller, *Z. Anorg. Chem. 279,* 182 (1955).

81b. T. Sasa, Y. Takahashi, T. Mukaibo, *Bull. Chem. Soc. Jpn. 45,* 2250 (1972).

82. J. M. Lalancette, M. J. Fournier-Breault, R. Thiffault, *Can. J. Chem. 52,* 589 (1979).

83. G. A. Olah, J. Kaspi, *J. Org. Chem. 42,* 3046 (1977).

84. G. A. Olah, J. Kaspi, J. Bukala, *J. Org. Chem. 42,* 4187 (1977).

85. N. Yoneda, T. Fukuhara, T. Abe, A. Suzuki, *Chem. Lett.* 1485 (1981).

CARBOCATIONS IN SUPERACIDS

3.1 INTRODUCTION

3.1.1 Development of the Carbocation Concept: Early Kinetic and Stereochemical Studies

At the turn of the century, the pioneering work of Norris, Kehrmann, Baeyer, and others on triarylcarbenium salts generated wide interest in these species.[1] They were, however, long considered only of interest to explain the color of these dyes. One of the most daring and fruitful ideas born in organic chemistry was the suggestion that in the course of reactions carbocations might be intermediates that start from nonionic reactants and lead to nonionic covalent products. It was H. Meerwein[2] who in 1922, while studying the kinetics of the rearrangement of camphene hydrochloride to isobornyl chloride reported the important observation that the reaction rate increased in a general way with the dielectric constant of the solvent. Furthermore, he found that metallic chlorides such as $SbCl_5$, $SnCl_4$, $FeCl_3$, $AlCl_3$, and $SbCl_3$ (but not BCl_3 or $SiCl_4$), as well as dry HCl, which promote the ionization of triphenylmethyl chloride by formation of ionized complexes, considerably accelerated the rearrangement of camphene hydrochloride. Meerwein concluded that the conversion of camphene hydrochloride to isobornyl chloride actually does not proceed by way of migration of the chlorine atom but by a rearrangement of a cationic intermediate. Thus, the modern concept of carbocation intermediates was born.

Ingold, Hughes, and their collaborators in England, starting in the late 1920s, carried out detailed kinetic investigations on what later became known as nucleophilic substitution at saturated carbon and polar elimination reactions.[3] The well-known work relating to S_N1 and later E1 reactions established the carbocation concept in these reactions. Whitmore,[4] in a series of papers which began in 1932, generalized Meerwein's rearrangement theory to many organic chemical reactions, although he cautiously avoided writing positive signs on any intermediates.

Kinetic and stereochemical evidence helped to establish carbocation intermediates in organic reactions. These species, however, were generally too short lived and could not be directly observed by physical means.

3.1.2. Observation of Stable, Long-Lived Carbocations

The transient nature of carbocations arises from their extreme reactivity with nucleophiles. The use of low nucleophilicity counter ions, particularly tetrafluoroborates ($\bar{B}F_4$), enabled Meerwein in the 1940s to prepare a series of oxonium and carboxonium ion salts, such as $R_3O^+BF_4^-$ and $HC(OR)_2{}^+BF_4^-$, respectively.[5a] These Meerwein salts are effective alkylating agents, and they transfer alkyl cations in S_N2-type reactions. However, simple alkyl cation salts ($R^+BF_4^-$) were not obtained in Meerwein's studies. The first acyl tetrafluoroborate, i.e., acetylium tetrafluoroborate, was obtained by Seel[5b] in 1943 by reacting acetyl fluoride with boron trifluoride at low temperature.

$$CH_3COF + BF_3 \longrightarrow CH_3C\overset{+}{O}BF_4^-$$

In the early 1950s, Olah and co-workers[6] started a study of the intermediates of Friedel-Crafts reactions and inter alia carried out a systematic investigation of acyl fluoride–boron trifluoride complexes. They were able to observe a series of donor-acceptor complexes as well as stable acyl cations. Subsequently, the investigations were also extended to other Lewis acid halides. In the course of these studies, they increasingly became interested in alkyl halide–Lewis acid halide complexes. The study of alkyl fluoride–boron trifluoride complexes by electrical conductivity measurements indicated the formation of ionic complexes in the case of tertiary butyl and isopropyl fluoride at low temperature, whereas methyl and ethyl fluoride formed molecular coordination complexes. Subsequently Olah and co-workers initiated a systematic study of more suitable acid and low-nucleophilicity solvent systems. This resulted in the discovery of antimony pentafluoride as a suitable very strong Lewis acid and related superacid systems such as Magic Acid[7] [$HSO_3F:SbF_5$], and low-nucleophilicity solvent systems such as SO_2, SO_2ClF, and SO_2F_2, which finally allowed obtaining alkyl cations as stable, long-lived species.[8] Subsequently, by the use of a variety of superacids, a wide range of practically all conceivable carbocations became readily available for structural and chemical studies.[9,10]

3.1.3 General Concept of Carbocations

Electrophilic reactions since Meerwein's pioneering studies are generally considered to proceed through cationic (i.e., carbocationic) intermediates. The general concept[11] of carbocations encompasses all cations of carbon-containing compounds, which sometimes were differentiated into two limiting classes (1) trivalent (''classical'') carbenium ions and (2) five or higher coordinate (''nonclassical'') carbonium ions. Whereas the differentiation of limiting trivalent carbenium and pentacoordinate carbonium ions serves a useful purpose to establish the significant differences between these ions, it is also clear that in most specific systems there exists a continuum of charge delocalization. In fact, in all carbocations (even in the parent CH_3^+), there is a continuum of the degree of charge delocalization, and thus to think in limiting terms is rather meaningless. Participation by neighboring groups cannot only be by *n*- and π-donors, as most generally recognized, but also by σ-ligands. There is in principle no difference between these. σ-Participation in properly oriented systems is not only possible, but it is unavoidable. The only question is its degree, and not whether it exists.[12]

It is well known that trivalent carbenium ions play an important role in electrophilic reactions of π- and *n*-donor systems. Similarly, pentacoordinate carbonium ions are the key to electrophilic reactions of σ-donor systems (single bonds). The ability of single bonds to act as σ-donor lies in their ability to form carbonium ions via delocalized two-electron, three-center bond formation. Consequently, there seems to be in principle no difference between the electrophilic reactions of π- and σ-bonds except that the former react more easily even with weak electrophiles, whereas the latter necessitate more severe conditions.

On the basis of the study of carbocations by direct observation of long-lived species, it became increasingly apparent that the carbocation concept is much wider than previously realized and necessitated a general definition.[11] Therefore, such a definition was offered based on the realization that two distinct, limiting classes of carbocations exist.

1. Trivalent (''classical'') carbenium ions contain an sp^2-hybridized electron-deficient carbon center that tends to be planar in the absence of constraining skeletal rigidity or steric interference. (It should be noted that *sp*-hybridized, linear acyl cations and vinyl cations also show substantial electron deficiency on carbon.) The carbenium carbon contains six valence electrons, and thus is highly electron deficient. The structure of trivalent carbocations can always be adequately described by using two-electron, two-center bonds (Lewis valence bond structures).
2. Penta- (or higher) coordinate (''nonclassical'') carbonium ions, which contain five (or higher) coordinate carbon atoms, cannot be described by two-electron single bonds alone, but also necessitates the use of two-electron, three (or multi-) center bond(s). The carbocation center is always surrounded by eight electrons, although two (or more) of them are involved in multicenter bonds,

and thus the ions overall are electron deficient (due to electron sharing of two binding electrons between three (or more) centers).

Lewis's concept that a chemical bond consists of a pair of electrons shared between two atoms became the foundation of structural chemistry, and chemists still tend to name compounds as anomalous when their structures cannot be depicted in terms of such bonds alone. Carbocations with too few electrons to allow a pair for each "bond" came to be referred to as "nonclassical," a label still used even though it is now recognized that, like many other substances, they adopt the delocalized structures appropriate for the number of electrons they contain.

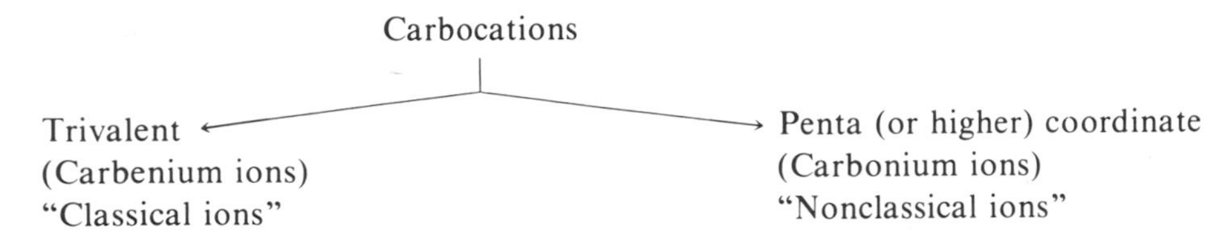

Expansion of the carbon octet via 3*d*-orbital participation does not seem possible; there can be only eight valence electrons in the outer shell of carbon.[13,14] Thus, the covalency of carbon cannot exceed four. Penta- (or higher) coordination implies a species with five (or more) ligands within reasonable bonding distance from the central atom.[15] The transition states long ago suggested for S_N2 reactions represent such cases, but involving 10 electrons around the carbon center. Charge–charge repulsions in the S_N2 transition state forces the entering and leaving substituents as far apart as possible leading to a trigonal bypyramid with a long 4*e*-3*c* bond allowing little possibility of a stable intermediate. In contrast, S_E2 substitution reactions involve 2*e*-3*c* interactions but have been in the past mainly restricted to organometallic compounds, e.g., organomercurials.[16]

The direct observation of stable penta- (or higher) coordinate species with eight electrons around the carbon center in solution was not reported until recent studies of long-lived "nonclassical" ions in superacid solvent systems.

Neighboring group interactions with the vacant *p* orbital of the carbenium ion center can contribute to ion stabilization via charge delocalization. Such phenomena can involve atoms with unshared electron pairs (*n*-donors), C—H and C—C hyperconjugation, bent σ-bonds (as in cyclopropylcarbenium ions) and π-electron systems (direct conjugative or allylic stabilization). Thus, trivalent carbenium ions can show varying degrees of delocalization without becoming pentacoordinate carbonium ions. The limiting classes defined do not exclude varying degrees of delocalization, but in fact imply a spectrum of carbocation structures.

In contrast to the rather well-defined trivalent ("classical") carbenium ions, "nonclassical ions"[17] have been more loosely defined. In recent years, a lively controversy has centered on the classical–nonclassical carbonium ion problem.[18] The extensive use of "dotted lines" in writing carbonium ion structures has been (rightly) criticized by Brown,[18d] who carried, however, the criticism to question the existence of any σ-delocalized (nonclassical) ion. For these ions, if they exist, he stated ". . . a new bonding concept not yet established in carbon structures is required."

Clear, unequivocal experimental evidence has by now been obtained for nonclassical ions such as the norbornyl cation.[19,20] The bonding concept required to define "nonclassical ions" is simply to consider them as penta- (or higher) coordinated carbonium ions involving at least one two electron three (or multi-) center bond, of which CH_5^+ (the methonium ion–carbonium ion) is the parent, as CH_3^+ (methenium ion, methyl cation, carbenium ion) is the parent for trivalent carbenium ions. An example of a hexacoordinate carbonium ion is the pyramidal dication of Hogeveen.[21]

Concerning the carbocation concept, it is regrettable that for a long time in the Anglo-Saxon literature the trivalent planar ions of the CH_3^+ type were named as carbonium ions. If the name is considered analogous to other -onium ions (ammonium, sulfonium, phosphonium ions, etc.), then it should relate to the higher valency or coordination state carbocation. The higher bonding state carbocations, however, clearly are not the trivalent but the penta- (or higher) coordinate cations. The German and French literatures indeed frequently use the "carbenium ion" naming for the trivalent cations. If one considers these latter ions as protonated carbenes, the naming is indeed correct.[22a] It should be pointed out, however, that the "carbenium ion" naming depicts only trivalent ions and thus should not be a general name for all carbocations. IUPAC's Organic Chemistry Division has reviewed the nomenclature of physical organic chemistry[22b,c] and recommends the use of the "carbocation" for naming all positive ions of carbons. "Carbenium" or "carbonium ion" naming, similar to the "carbinol" naming of alcohols, is discouraged.

3.2 METHODS OF GENERATING CARBOCATIONS IN SUPERACIDS

Common superacid systems that are generally employed in the preparation of carbocations are the Brønsted acids such as HSO_3F and HF and Lewis acids such as SbF_5. Also Lewis and Brønsted acid combinations such as $HSO_3F:SbF_5$ (Magic Acid), $HSO_3F:SbF_5$ (4:1), $HF:SbF_5$ (fluoroantimonic acid) have been used. Other superacid systems such as $HF:BF_3$, AsF_5, CF_3SO_3H, $HF:PF_5$, $HCl:AlCl_3$, $HBr:AlBr_3$, H_2SO_4, and $HClO_4$ have been successfully adopted. Sometimes a metathetic reaction involving a halide precursor and $AgBF_4$ or $AgSbF_6$ has also been successful. The most convenient nonnucleophilic solvent systems used in the preparation of carbocations are SO_2, SO_2ClF, and SO_2F_2. To be able to study ionic solutions at very low temperatures (ca. $-160°C$) by nmr spectroscopy, Freons like $CHCl_2F$ may be used as a cosolvent to decrease the viscosity of the solution.

The success of a carbocation preparation in superacid generally depends on the technique employed. For most of the stable systems, Olah's method[23] of mixing precooled progenitor dissolved in appropriate solvent system along with the superacid using a simple vortex stirrer to mix the components is generally convenient.

However, care should be taken to avoid moisture and local heating. Low temperatures between −78°C (using acetone–dry ice) or −120° (using liquid N_2–ethanol slush) are most commonly employed which suppress side reactions like dimerization and oligomerizations.

Methods for generation of very reactive carbocations have been developed most notably by Ahlberg et al.[24] and Saunders et al.[25] The former group describes an ion-generation apparatus which consists of a Schlenk tube-like vessel attached to an nmr tube in which the carbocation is prepared at low temperatures. This methodology has been further improved.[26]

The method of Saunders,[25] however, is more sophisticated. It uses deposition of the starting reagents from the gas phase (using high vacuum on a surface cooled to liquid nitrogen temperature) to produce stable solutions of carbocations. The details of the method have been published in detail.[27] Yannoni and Myhre[28] have successfully utilized the above method to generate carbocations in a SbF_5 matrix at very low temperatures for their solid-state ^{13}C-nmr spectroscopic work.

Kelly and Brown[29] describe a method for the preparation of concentrated solutions of carbocations (~1 *M*) in $SbF_5:SO_2ClF$. This method, which employs a syringe technique, allows quantitative conversion of precursors soluble in SO_2ClF at −78°C into the corresponding carbocations.

3.3 METHODS AND TECHNIQUES IN THE STUDY OF CARBOCATIONS

3.3.1 Nuclear Magnetic Resonance Spectra in Solution

One of the most powerful tools in the study of carbocations is nuclear magnetic resonance (nmr) spectroscopy. This method yields direct information, through chemical shifts, coupling constants, and the temperature dependence of band shapes about the structure and dynamics of cations.

Although the initial developments in the observation of stable carbocations in solution relied heavily upon ^1H-nmr spectroscopy, ^{13}C-nmr spectroscopy has proven to be the single most useful technique. ^{13}C-nmr permits the direct observation of the cationic center, and the chemical shifts and coupling constants can be correlated to the cation geometry and hybridization. At the beginning of the ^{13}C-nmr studies in the early 1960s, Olah and co-workers used the INDOR technique to obtain spectra of carbocations at natural ^{13}C abundance. In the past 20 years or so, the development of pulse Fourier transform nmr spectrometers has made available ^1H and ^{13}C-nmr spectra of very dilute solution of cations. The signal-to-noise ratio (S:N) has been further improved by the introduction of high-field superconducting magnets. With such high sensitivity, high-resolution spectrometers degenerate rearrangement reaction rate constants of the order $10^7\ s^{-1}$, which corresponds to a free energy of activation of 3.3 kcal · mol^{-1} at −160°C, have been measured.

Degenerate rearrangement of carbocations, if they are fast enough, result in temperature-dependent nmr spectra. At slow exchange, the signals of the exchanging nuclei show up as separate absorptions. If the exchange rate is increased by raising the temperature, the signals first broaden and, upon a further rise in the temperature, they coalesce. Still further increase of the exchange rate results in sharpening of the broad coalesced signals (Fig. 3.1).

The formula in Figure 3.1*c* shows that the larger is the shift difference (ν_{AB}) between the exchanging signals, the larger is the rate constant needed to get coalescence at a specific temperature. Thus, since chemical shift differences in ^{13}C-nmr are usually much larger (in Hz) than in ^{1}H-nmr, ^{13}C-nmr spectroscopy permits the quantitative study of much faster processes than can be investigated by ^{1}H-nmr spectroscopy.

Very slow exchange could be detected and the rate measured by transfer of spin saturation, a tool that could be useful in the elucidation of reaction mechanisms causing exchange. One of the signals participating in very slow exchange is saturated by an extra RF field while the rest of the spectrum is observed. If exchange of the spin-saturated nuclei takes place at a rate comparable to that of the nuclear-spin relaxation (T_1) then transfer of the spin saturation by the degenerate reaction will partially saturate the other exchanging nuclei. From the degree of transferred spin saturation and measured T_1, the rate of exchange could be evaluated. This technique was devised by Forsen and Hoffman[30,31] using ^{1}H-nmr spectroscopy and it has been utilized in the study of carbocations. The method has some limitations with ^{1}H-nmr due to the small shift differences and large couplings between the protons. However, the method has recently been applied using ^{13}C-nmr spectroscopy (proton decoupled) for study of complex carbocation rearrangements.[32]

3.3.2 ^{13}C-nmr Chemical Shift Additivity

An empirical criterion based on additivity of ^{13}C chemical shifts for distinguising trivalent and higher coordinate carbocations has been developed by Schleyer, Olah, and their co-workers.[33] In this method, the sum of all ^{13}C chemical shifts of carbocations and their respective hydrocarbon precursors are compared. A trivalent carbocation has a sum of ^{13}C chemical shifts of at least 350 ppm higher than the sum for the corresponding neutral hydrocarbon. This difference can be rationalized

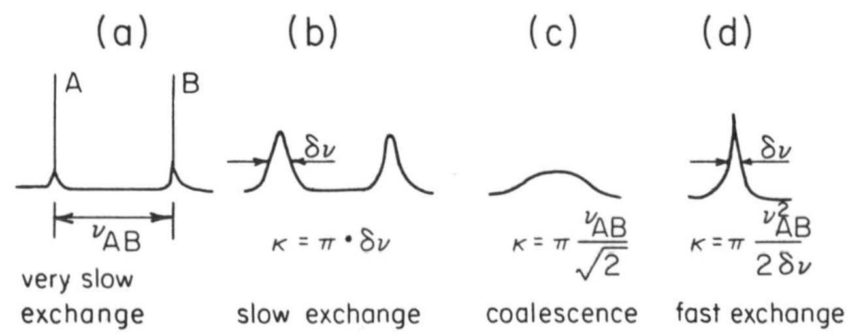

FIGURE 3.1 Results in the nmr spectrum of exchange of two equally populated sites at different rates. The rate constants (k) for the exchange at the conditions (b), (c), and (d) could be approximately estimated using the attached formulas and corrected values of $\delta\nu$.

by partly attributing it to the hybridization change to sp^2 and to the deshielding influence of an unit positive charge in the trivalent carbocation. Since higher coordinate carbocations (nonclassical ions) have penta- and hexacoordinate centers, the sum of their chemical shifts relative to their neutral hydrocarbons is much smaller, often by less than 200 ppm.

3.3.3 Isotopic Perturbation Technique

The deuterium isotopic perturbation technique developed by Saunders and co-workers is capable of providing a convenient means to differentiate the rapidly equilibrating or bridged nature of carbocations.

Saunders and Vogel in 1971 discovered[34,35] that by asymmetrically introducing deuterium into some reversible rearrangement processes of carbocations, large splittings were produced in their nmr spectra. Although the ions were interconverting extremely rapidly and thus gave averaged spectra, the isotope made the energies of the two interconverting species slightly different and thus the equilibrium constant between them was no longer 1. Each ion therefore spends a little more time on one side of the equilibrium barrier than on the other side. The weighted-average peaks of the two carbon atoms that were interchanged by the rearrangement process no longer coincided. Splitting values in the ^{13}C-nmr spectrum of over 100 ppm were observed as a result of perturbation by deuterium.[36]

However, when deuterium was introduced into systems that are recognized as static, single minimum, nonequilibrating species,[37] such as the cyclohexenyl cation, no large splittings were observed and, in contrast to the behavior of the equilibrating systems, there were no observable changes in the spectra with temperature. In fact, the isotope-induced changes in the spectra were not very different from changes that occur in any simple molecule on introducing deuterium, and were roughly 50 times smaller than the effects produced in the equilibrating systems.

These observations led to the method of observing changes in nmr spectra produced by asymmetric introduction of isotopes (isotopic perturbation) as a means of distinguishing systems involving equilibrating species passing rapidly over a low barrier from molecules with single energy minima, intermediate between the presumed equilibrating structures. Application of this method for the individual carbocations will be discussed later. This method also allows accurate determination of equilibrium isotope effects.

3.3.4 Solid-State ^{13}C-nmr

The use of magic-angle spinning and cross-polarization has enabled high-resolution ^{13}C-nmr spectra to be obtained in the solid state. Yannoni, Myhre and Fyfe have obtained solid state nmr spectra of frozen carbocation solutions (such as alkyl cations, the norbornyl cation, etc.) at very low temperatures[28,38] (even at 5°K). At such low temperatures rearrangements and most molecular motions can be frozen out. Olah's group[38c] has studied a series of stable carbocations, including the 1-adamantyl cation, acyl cations, etc. by ^{13}C CPMAS at ambient temperature. In these studies solid state effects seem to be unimportant on chemical shifts (as compared to solution data) but are significant on rates of exchange processes.

3.3.5 Tool of Increasing Electron Demand

The nature of electronic effects in cationic reactions has been probed by application of the Gassman-Fentiman tool of increasing electron demand.[39] Aryl-substituted cationic centers can be made more electron demanding, i.e., electrophilic, by introduction of electron-withdrawing substituents into the aryl ring.

When a cationic center becomes sufficiently electrophilic, it may draw on electrons from neighboring π- and σ-bonds and delocalize positive charge density. The onset of participation of π- and σ-bonds can be detected as a departure from linearity in a Hammett type plot as the electron-withdrawing ability of the aryl substituent increases. In stable ion studies, ^{13}C chemical shift is generally used as a structural probe reflecting the charge density at the cation center (in closely related homologous cations, other factors that may affect chemical shift may be assumed constant).

The tool of increasing electron demand has proved useful in detecting the onset of π- and σ-delocalizations under stable ion conditions. It is, however, an extremely coarse technique since the aryl group can still delocalize charge into its π-system, even with strongly electron-withdrawing substituents. Only in cases where neighboring σ- and π-bonds can effectively compete with the aryl ring in stabilizing the cationic center are significant deviations from linearity observed in a Hammett-type plot.

3.3.6 Core Electron Spectroscopy

X-ray photoelectron spectroscopy measures the energy distribution of the core electrons emitted from a compound on irradiation with X-rays.[40] The electron binding energy (E_b) is a function of the chemical environment of the atom. In particular, the atomic charge on carbon can be directly correlated to the carbon 1*s* electron binding energy.[41] The cationic center of a classical carbocation, e.g., *t*-butyl cation has a carbon 1*s* E_b approximately 4 eV *higher* than a neutral sp^3 carbon atom.[41] Electron deficiencies of different degrees in different carbocations give different carbon core electron 1*s* binding energies; the delocalization of charge from a cationic center lowers the carbon 1*s* binding energy.

Core electron spectroscopy for chemical analysis (ESCA) is perhaps the most definitive technique applied to the differentiation of nonclassical carbocations from equilibrating classical species. The time scale of the measured ionization process is of the order of 10^{-16} s so that definite species are characterized, regardless of (much slower) intra- and intermolecular interactions, e.g., hydride shifts, Wagner-Meerwein rearrangements, proton exchange, etc.

3.3.7 Infrared and Raman Spectra

Infrared and Raman spectra of stable carbocations have been obtained[42] and are in complete agreement with their electron-deficient structures. Infrared spectra of alkyl cations and their deuterated analogs correspond to the spectra predicted by calculations based on molecular models and force constants. Thus, these spectra can be used in the identification of stable carbocations.

Laser Raman spectroscopy, particularly with helium–neon lasers, is another powerful tool in the study of carbocations. Because Raman spectra give valuable information on symmetry, these spectra help to establish, in detail, structures of the ions and their configurations.

3.3.8 Electronic Spectra

The observation of stable alkyl cations in antimony pentafluoride solutions also opened up the possibility of investigating the electronic spectra of these solutions. It has been reported[43] that solutions of alkyl cations in $HSO_3F:SbF_5$ solution at $-60°C$ showed no absorption maxima above 210 nm. In view of this observation, it was resolved that previous claims relating to a 290-nm absorption of alcohols and olefins in sulfuric acid solutions were due to condensation products or cyclic allylic ions and not to the simply alkyl cations.[44]

3.3.9 Low-Temperature Solution Calorimetric Studies

Arnett and co-workers,[45,46] in a series of investigations, have determined heats of ionization (ΔHi) of secondary and tertiary chlorides and alcohols in $SbF_5:SO_2ClF$ and $HSO_3F:SbF_5:SO_2ClF$ solutions, respectively, at low temperatures. They have also measured heats of isomerizations of secondary to tertiary carbocations in the superacid media. These measured thermochemical data have been useful to determine the intrinsic thermodynamic stability of secondary and tertiary carbocations.

3.4 TRIVALENT CARBOCATIONS

3.4.1 Alkyl Cations

3.4.1.1 Early Unsuccessful Attempts

Until the early 1960s simple alkyl cations were considered only as transient species.[6] Their existence has been inferred from the kinetic and stereochemical studies of reactions. No reliable physical measurements, other than electron impact measurements in the gas phase (mass spectrometry), were known. The formation of gaseous organic cations under electron bombardment of alkenes, haloalkanes, and other precursors has been widely investigated in mass spectrometric studies.[47] No direct observation of carbocations in solutions was achieved prior to the early 1960s.

The observation of alkyl cations like that of *t*-butyl cation (trimethylcarbenium ion), $(CH_3)_3C^+$ **1** of the isopropyl cation (dimethylcarbenium ion), $(CH_3)_2CH^+$ **2** in solution was a long-standing challenge. The existence of alkyl cations in systems containing alkyl halides and Lewis acids has been inferred from variety of observations, such as vapor pressure depressions of CH_3Cl and C_2H_5Cl in the presence of $GaCl_3$,[48] conductivity of $AlCl_3$ in alkyl chlorides,[49] and of alkyl fluorides in BF_3,[50] as well as the effect of ethyl bromide on the dipole moment of $AlBr_3$.[51] However, in no case had well-defined, stable alkyl cation complexes been established even at very low temperatures.

Electronic spectra of alcohols and olefins in strong proton acids (H_2SO_4) were obtained by Rosenbaum and Symons.[52] They observed, for a number of simple aliphatic alcohols and olefins, absorption maxima around 290 nm and ascribed this absorption to the corresponding alkyl cations.

Finch and Symons,[53] on reinvestigation of the absorption of aliphatic alcohols and olefins in sulfuric acid solution, showed that the condensation products formed with acetic acid (used as solvent for the precursor alcohols and olefins) were responsible for the sepctra and not the simple alkyl cations. Moreover, protonated mesityl oxide was identified as the absorbing species in the system of isobutylene, acetic acid, and sulfuric acid.

Deno and his co-workers[54] have carried out an extensive study of the fate of alkyl cations in undiluted H_2SO_4 and oleum. They obtained equal amounts of a saturated hydrocarbon mixture (C_4–C_{19}) insoluble in H_2SO_4 and a mixture of cyclopentenyl cations (C_9–C_{10}) in the H_2SO_4 layer. These cations exhibit strong uv absorption around 300 nm.

Therefore, it must be concluded that earlier attempts to prove the existence of stable, well-defined alkyl cations were unsuccessful in experiments using sulfuric acid solutions and inconclusive in the interaction of alkyl halides with Lewis acid halides. Proton elimination reactions or dialkyl halonium ion formation may have affected the early conductivity studies.

3.4.1.2 Preparation from Alkyl Fluorides in Antimony Pentafluoride Solution and Spectroscopic Studies

In 1962 Olah et al.[42,55–57] first directly observed alkyl cations in solution. They obtained the *t*-butyl cation **1** (trimethylcarbenium ion) when *t*-butyl fluoride was dissolved in excess antimony pentafluoride, which served both as Lewis acid and solvent. Later, the counterion was found to be, under these conditions, primarily the dimeric Sb_2F_{11} anion[58] whereas in $SbF_5:SO_2$ or $SbF_5:SO_2ClF$ solutions, SbF_6^- and $Sb_2F_{11}^-$ are both formed.

$$(CH_3)_3CF + (SbF_5)_2 \longrightarrow \underset{\mathbf{1}}{(CH_3)_3C^+Sb_2F_{11}^-}$$

The possibility of obtaining stable alkyl fluoroantimonate salts from alkyl fluorides (and subsequently other halides) in antimony pentafluoride solution (neat or diluted with sulfur dioxide, sulfuryl chloride fluoride, or sulfuryl fluoride) or in other superacids[59] such as $HSO_3F:SbF_5$ (Magic Acid), $HF:SbF_5$ (fluoroantimonic acid), $HF:TaF_5$ (fluorotantalic acid), and the like was evaluated in detail, extending studies to all isomeric C_3–C_8 alkyl halides, as well to a number of higher homologs.[60,61]

Propyl, butyl, and pentyl fluorides with antimony pentafluoride gave the isopropyl, *t*-butyl, and *t*-amyl cations (as their fluoroantimonate salts) **2, 1,** and **3.**

$$C_3H_7F + (SbF_5)_2 \longrightarrow CH_3{-}\overset{+}{C}H{-}CH_3\ Sb_2F_{11}^-$$

2

$$C_4H_9F + (SbF_5)_2 \longrightarrow CH_3{-}\underset{\displaystyle CH_3}{\overset{+}{C}}{-}CH_3\ Sb_2F_{11}^-$$

1

$$C_5H_{11}F + (SbF_5)_2 \longrightarrow CH_3{-}\underset{\displaystyle CH_3}{\overset{+}{C}}{-}CH_2CH_3\ Sb_2F_{11}^-$$

3

The secondary butyl and amyl cations can be observed only at very low temperatures, and they rearrange readily to the more stable tertiary ions. Generally, the most stable tertiary or secondary carbocations are observed from any of the isomeric alkyl fluorides in superacidic solvent system.

The main feature of the proton nmr spectra of alkyl fluorides in antimony pentafluoride is the substantial deshielding of the protons in the carbocations as compared with the starting alkyl fluorides (Fig. 3.2 and Table 3.1).

To prove that stable alkyl cations, and not exchanging donor acceptor complexes, were obtained, Olah and co-workers also investigated as early as 1962 the ^{13}C-nmr of the potentially electropositive carbenium carbon atom in alkyl cations.[60] The $^{13}C^+$shift in the *t*-butyl cation $(CH_3)_3C^+$ **1** in $SO_2ClF:SbF_5$ solution at $-20°C$ is at $\delta\ ^{13}C$ 335.2 (all ^{13}C-mr shifts are from ^{13}C tetramethylsilane) with a long-range coupling to the methyl protons of 3.6 Hz.

The $^{13}C^+$ shift in the isopropyl cation **2** under identical conditions, is $\delta\ ^{13}C$ 320.6 with a long-range coupling to the methyl protons of 3.3 Hz. The direct ^{13}C–H coupling is 169 Hz (indicating sp^2 hybridization of the carbenium carbon atom), whereas the long-range proton–proton coupling constant is 6.0 Hz (see Fig. 3.2). Substitution of the methyl group in the *t*-butyl cation by hydrogen thus causes an upfield shift of 10.4 ppm. Although the ^{13}C-nmr shift of the carbocation center of the *t*-butyl cation is more deshielded than that of the isopropyl cation (by about 10 ppm), this can be explained by the methyl substituent effect, which may sometimes amount up to 22 ppm. The tertiary butyl cation thus is more delocalized and stable than the secondary isopropyl cation.

The $^{13}C^+$ shift in the *t*-amyl cation $[C_2H_5\overset{+}{C}(CH_3)_2]$ **3** is at $\delta\ ^{13}C$ 335.4 which is similar to that of *t*-butyl cation. The shift difference is much smaller than the 17 ppm found in the case of the related alkanes,although the shift observed is in the same direction. The ^{13}C-nmr chemical shifts and coupling constants J_{C-H} of C_3 to C_8 alkyl cations **1–13** are shown in Tables 3.2 and 3.3.[61]

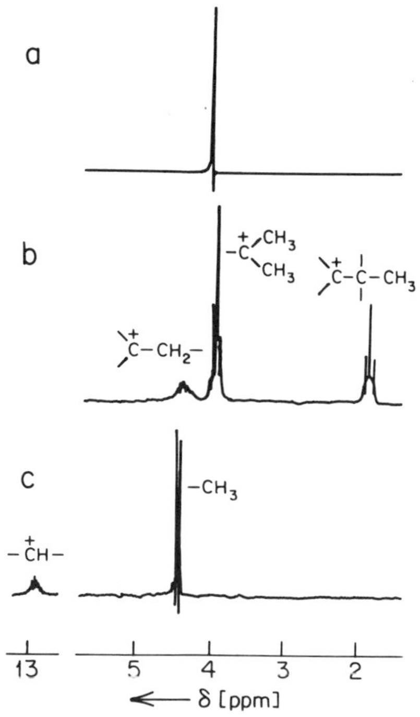

FIGURE 3.2 ^{1}H NMR spectra of: a) the *t*-butyl cation *1* [trimethylcarbenium ion, $(CH_3)_3C^+$]; b) the *t*-amyl cation *3* [dimethylethylcarbenium ion, $(CH_3)_2C^+$—C_2H_5]; c) the isopropyl cation *2* [dimethylcarbenium ion, $(CH_3)_2C^+H$]. (60 MHz, in SbF_5:SO_2 ClF solution, −60°C).

It is difficult to interpret these large deshieldings in any way other than as a direct proof that (1) the hybridization state of the carbon atom at the carbenium ion center is sp^2; and (2) at the same time, the sp^2 center carries a substantial positive charge.

Data in Tables 3.2 and 3.3 are characterized by substantial chemical shift deshieldings and coupling constants (J_{CH}) that indicate sp^2 hybridization. Also recently Myhre and Yannoni[28] have obtained ^{13}C-nmr spectrum of *t*-butyl cation **1** in the solid state, which agrees very well with the solution data.[61]

The temperature-dependent ^{1}H-nmr spectrum of isopropyl cation **2** (prepared from isopropyl chloride in SbF_5:SO_2ClF solution) demonstrated[62] rapid interchange of two types of protons. Line-shape analysis showed the reaction to be intramolecular, with an activation energy barrier of 16.4 ± 0.4 kcal · mol^{-1}. Based on these observations, Saunders and Hagen[62] suggested that the rearrangement involves *n*-propyl cation **14** as an intermediate and that the activation energy provides an estimate of the energy difference between primary and secondary carbocations. Another mechanism involving protonated cyclopropane intermediates could not be excluded with the above result.

$$\underset{\mathbf{2}}{CH_3{-}\overset{+}{C}H{-}CH_3} \rightleftharpoons \underset{\mathbf{14}}{CH_3{-}CH_2{-}CH_2^+} \rightleftharpoons \underset{\mathbf{2}}{CH_3{-}\overset{+}{C}H{-}CH_3} \quad \mathbf{(I)}$$

TABLE 3.1 Characteristic ^{1}H-nmr Parameters of Alkyl Cations in $SbF_5:SO_2ClF$ Solution at −70°C

Ion		$\delta_{^1H}$					
		CH^+	$J_{^+CH}$	$J_{^+CCH}$	α-CH_2	α-CH_3	β-CH_3
$(CH_3)_2CH^+$	**2**	13	169	3.3		4.5	
$(CH_3)_3C^+$	**1**			3.6		4.15	
$(CH_3)_2C^+CH_2CH_3$	**3**				4.5	4.1	1.94
$CH_3C^+(CH_2CH_3)_2$	**4**				4.44	4.16	1.87

Strong support for the involvement of protonated cyclopropane intermediates came from the work of Olah and White.[60] The isopropyl cation obtained from [2-^{13}C]2-chloropropane (50% ^{13}C) was studied by ^{1}H-nmr. The ^{13}C label scrambled uniformly over 1 and 2 positions at $-60°C$ within a few hours (half life $\approx$ 1 h).

The relative rates of hydrogen and carbon interchange have been measured by Saunders et al.[63] using a mixture of (1,1,1-D_3)- and (2^{13}C)-labeled isopropyl cations at $-88°C$. The changes in the relative areas of different peaks as well as ^{13}C-satellites were observed, and the time dependence of the concentrations of different labelled isomers were simulated using a computer [using mechanisms **(I)–(III)**]. A combination of mechanisms **(I)** and **(II)** or mechanisms **(I)** and **(III)** could match the measurements. The rate for **(I)** was found to be 1.5 ± 0.5 times that of **(II)** or **(III)**. Thus, proton interchange is only slightly faster than the carbon interchange. Quenching of the D_3-isopropyl ion by methylcyclopentane and nmr analysis of the D_3-propane product mixture gave preliminary results consistent with mechanisms

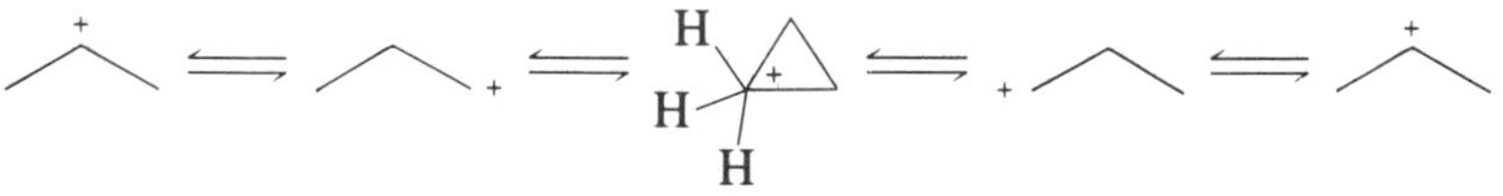

(II)

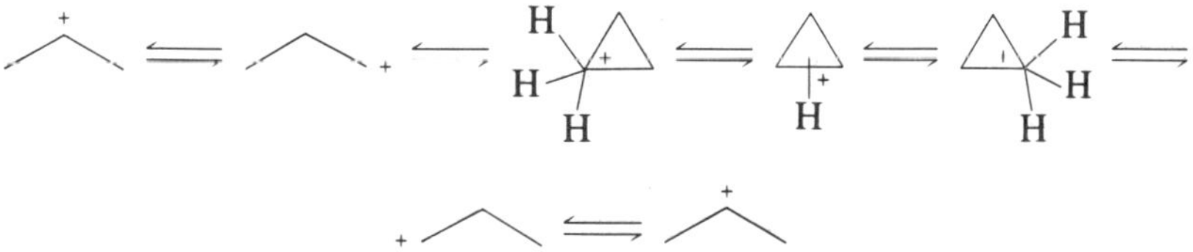

(III)

(I) and **(II)** alone. Labeling experiments indicating the intermediacy of protonated cyclopropanes have also been performed by Lee and Woodcock[64] and by Karabatsos et al.[65]

Theoretical calculations of the *ab initio* type by Pople et al.[66,67] and at the semiempirical level by Bodor and Dewar[68,69] did not give consistent results. Recent *ab initio* calculations, including electron correlation by Lischka and Kohler,[70] are inconsistent with earlier *ab initio* work. Their calculations have confirmed the stability of the isopropyl cation **2** and the instability of face-protonated cyclopropane **15.** Edge-protonated cyclopropane **16** is found to be a saddle point on the potential energy surface of lower energy than the corner-protonated species **17.**

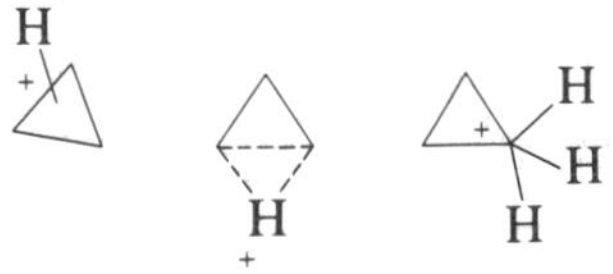

15 **16** **17**

Intermolecular secondary–secondary hydride transfer between **2** and propane in $SbF_5:SO_2ClF$ solution has been observed by Hogeveen and Gaasbeek.[71] The reaction

TABLE 3.2 ^{13}C **Chemical Shifts of the Static C_3 to C_8 Alkyl Cations**

Cation		1	2	3	4	5	6	7	8
	2	51.5 (*q*)	320.6 (*s*)						
	1	47.5 (*q*)	335.2 (*s*)						
	3	44.6 (*q*)	335.4 (*s*)	57.5 (*t*)	9.3 (*q*)				
	4	41.9 (*q*)	336.4	54.5 (*t*)	8.9 (*q*)				
	5	45.0 (*q*)	333.4	64.4 (*t*)	20.9 (*t*)	12.6 (*q*)			
	6	44.9 (*q*)	332.9 (*s*)	62.8 (*t*)	29.3 (*t*)	22.6 (*t*)	13.0 (*q*)		
	7	42.1 (*q*)	334.7 (*s*)	55.1 (*t*)	9.1 (*q*)	61.6 (*t*)	20.2 (*t*)	12.5 (*q*)	

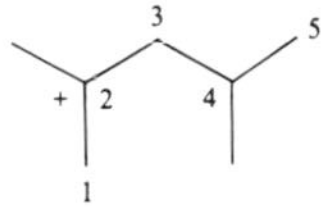	**8**	45.4 (*q*)	332.1 (*s*)	70.1 (*t*)	31.4 (*d*)	21.7 (*q*)			
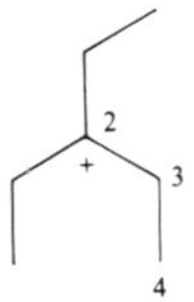	**9**		336.8 (*s*)	51.8 (*t*)	8.6 (*q*)				
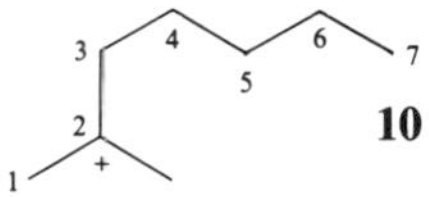	**10**	44.6 (*q*)	332.5 (*s*)	62.7 (*t*)	27.4 (*t*)	31.1 (*t*)	22.3 (*t*)	13.1 (*q*)	
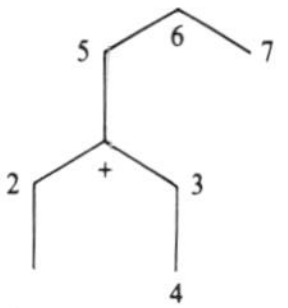	**11**		334.7 (*s*)	51.6 (*t*)	8.1 (*q*)	58.8 (*t*)	18.3 (*t*)	11.6 (*q*)	
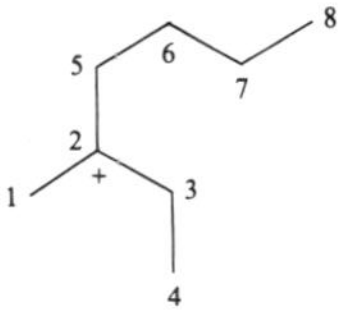	**12**	42.0 (*q*)	334.3 (*s*)	54.7 (*t*)	8.7 (*q*)	59.7 (*t*)	28.5 (*t*)	22.0 (*t*)	12.9 (*q*)
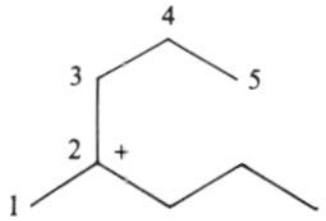	**13**	42.1 (*q*)	332.8 (*s*)	61.9 (*t*)	19.7 (*t*)	12.1 (*q*)			

TABLE 3.3 $J_{^{13}C-H}$ **Coupling Constants**[a] **of the Static C_3 to C_8 Alkyl Cations**

Cation		1	2	3	4	5	6	7	8
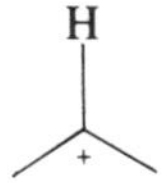	**1**	131.7[a]	171.3						
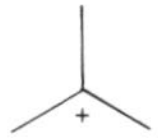	**2**	130.8							
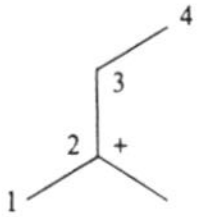	**3**	131.8		127.4	130.8				
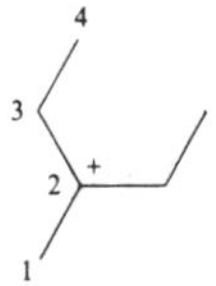	**4**	131.7		124.8	129.6				
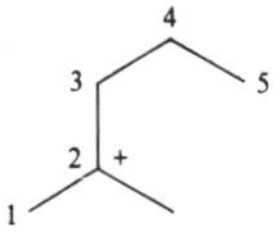	**5**	132.1		126.6	131.8	129.1			
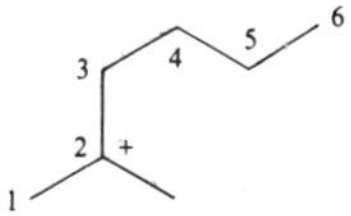	**6**	131.4		126.7	131.4	127.5	126.2		
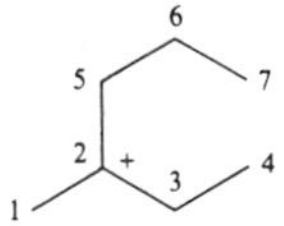	**7**	131.5		124.0	128.8	119.2	126.2	123.9	

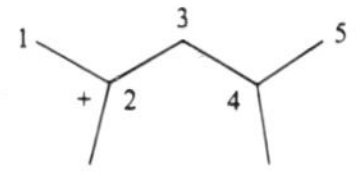 **8**	131.6		124.7	137.2	124.1			
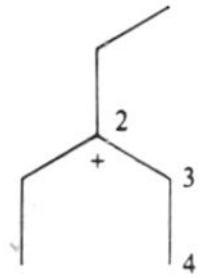 **9**			123.2	129.8				
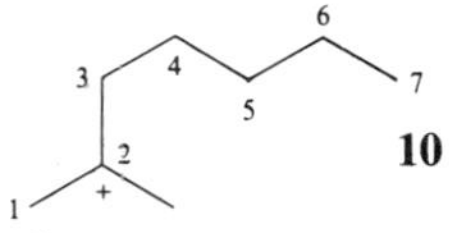 **10**	131.4		127.2	~131	~131	133.1	126.4	
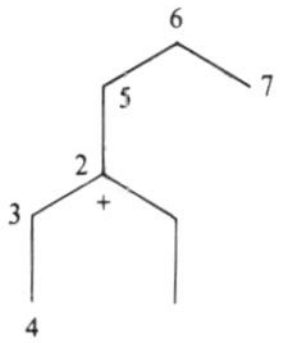 **11**			121.6	130.3	121.1	125.6	124.3	
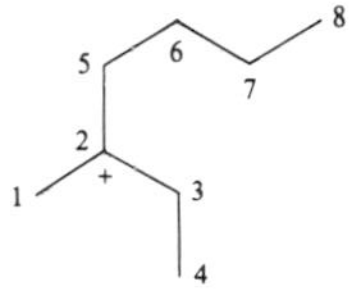 **12**	132.5		~122	129.2	122.4			127.0
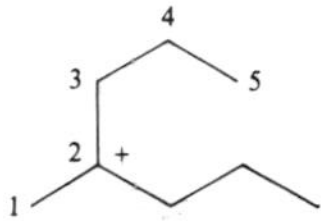 **13**	132.0		126.2	132.5	129.5			

[a]All values are measured in hertz.

was rapid on the nmr time scale, and a single peak was obtained from the two types of methyl groups down to at least $-100°C$ ($\Delta G^{\ddagger} \leqslant 6$ kcal · mol^{-1}).

Similar scramblings have been documented in the case of secondary butyl **18** (see later discussion) and *t*-amyl and 1-methyl-1-cyclopentyl cations **4** and **19.**[63,72] Recently, it has been reported that even *t*-butyl cation **1** undergoes such scrambling.[73] The *t*-butyl cation **1** (60% ^{13}C enriched at the cationic center) prepared from 2-chloro-2-methyl propane in $HSO_3F:SbF_5:SO_2ClF$ solution undergoes complete carbon scrambling in about 20 h at $+70°C$ (Fig. 3.3). The most likely mechanism one can invoke for the scrambling process is the rearrangement through primary isobutyl cation **20** via the delocalized protonated methyl-cyclopropane intermediate (or transition state) **21** to the secondary-butyl cation **18** (Scheme **IV**).

$$CH_3-{}^{13}\overset{+}{C}(CH_3)_2 \;(\mathbf{1}) \underset{\text{shift}}{\overset{1,2\text{ H}}{\rightleftharpoons}} \left[(H_3C)_2{}^{13}CH-\overset{+}{C}H_2\right] \;(\mathbf{20})$$

$$\mathbf{20} \rightleftharpoons \left[\text{protonated methylcyclopropane: } H_3C-{}^{13}CH-CH_2 \text{ bridged by } CH_3\right]^+ \;(\mathbf{21})$$

$$H_3C-{}^{13}\overset{+}{C}H-CH_2-CH_3 \;(\mathbf{18}) \rightleftharpoons \mathbf{21}$$

$$\mathbf{18} \rightleftharpoons (H_3C)_2\overset{+}{C}-{}^{13}CH_3 \;(\mathbf{1})$$

IV

These results seem to indicate a lower limit of $E_a \sim 30$ kcal · mol^{-1} [63] for the scrambling process which could correspond to the energy difference between *t*-butyl and primary isobutyl cation.

Infrared and Raman spectra of alkyl cations give valuable information of their structures (Table 3.4).[42,43] The Raman spectroscopic data provide strong evidence that *t*-butyl cation in Magic Acid solution prefers a conformation leading to overall C_{3v} point group symmetry (Table 3.4 and Fig. 3.4). Thus the $\overset{+}{C}(CH_3)_3$ ion exists in these solutions with a planar $\overset{+}{C}(C_3)$ carbon skeleton and with one hydrogen atom of each CH_3 group above the $\overset{+}{C}-C_3$ plane. The other two hydrogen atoms are arranged symmetrically below the $\overset{+}{C}-C_3$ plane to the right and left of the C_3 axis. Raman spectra observed for the *t*-amyl cation, the pentamethylethyl cation, and the tetramethylethyl cation also show similar structures. The Raman spectroscopic stud-

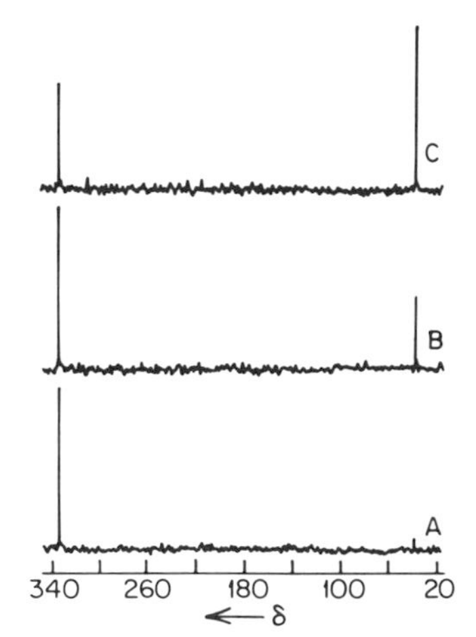

FIGURE 3.3 ^{13}C scrambling in *t*-butyl cation (A) immediately after preparation (B) after 2.5 hrs at +70°C (C) after 9.5 hrs at +70°C.

ies thus provide, in addition to ^{13}C-nmr data, direct evidence for the planar carbenium center of alkyl cations.

Evidence for planarity or near planarity of the sp^2 center of trivalent alkyl cations thus comes from the combined results of nmr (^1H and ^{13}C), ir, and Raman spectroscopy.[42,43,61]

In the electronic spectra (uv), alkyl cations in $HSO_3F:SbF_5$ at −60°C show no absorption maxima above 210 nm.[44]

X-ray photoelectron (ESCA) spectra of carbenium ions have also been obtained in frozen superacid solutions, generally in $SbF_5:SO_2$ solution or as isolated salts. Sulfur dioxide was subsequently removed by the usual freeze-thaw procedure. A thin layer of the viscous SbF_5 solution was deposited on the precooled sample holder, in a dry nitrogen atmosphere. The spectra were recorded at liquid nitrogen temperature.[74] The photoelectron spectrum of *t*-butyl cation **1** is shown in Figure 3.5. The lower traces in Figure 3.5 represent the result given by a curve resolver. The peak area ratio is close to 1:3.

The experimental carbon 1*s* binding energy difference (3.9 eV) between the carbenium ion center and the remaining three carbon atoms is in the limit of that predicted by *ab initio* calculation (4.45 eV). Comparable results were obtained for the *t*-amyl cation ($dE_{b^+_{c\text{-}c}} = 4 \pm 0.2$ eV).

3.4.1.3 Preparation from Other Precursors

Alkyl cations can be formed not only from halide precursors (the earlier investigation of generation from alkyl fluorides was later extended to alkyl chlorides, bromides, and even iodides) but also from olefins in superacids like $HF:SbF_5$.

$$RCH{=}CH_2 \xrightarrow[HF:SbF_5]{HSO_3F:SbF_5} R\overset{+}{C}H{-}CH_3$$

TABLE 3.4 Raman and IR Frequencies of the *t*-Butyl Cation and [D_9]-*t*-Butyl Cation and Their Correlation with Those of $(CH_3)_3B$ and $(CD_3)_3B$

Species	Frequency of Vibration [cm^{-1}]										
	$\nu_1, \nu_{12}, \nu_7, \nu_{19}$	ν_2, ν_{13}	ν_{21}	ν_{14}	ν_{15}	ν_{17}	ν_5	ν_{16}	ν_6	ν_9	ν_{10}	ν_{18}
$(CH_3)_3C^{\oplus}$ (**2**)	2947	2850		1450		1295			667		347	306
$(CH_3)_3B$	2975	2875	1060	1440	1300	1150	906	866	675	973 (486?)	336[a]	320
$(CD_3)C^{\oplus}$	2187	2090		1075		980			720		347	300
$(CD)_3B$	2230	2185		1033	1018	1205			620	870	(289)[b]	(276)[b]

[a] IR frequency.
[b] Calculated.

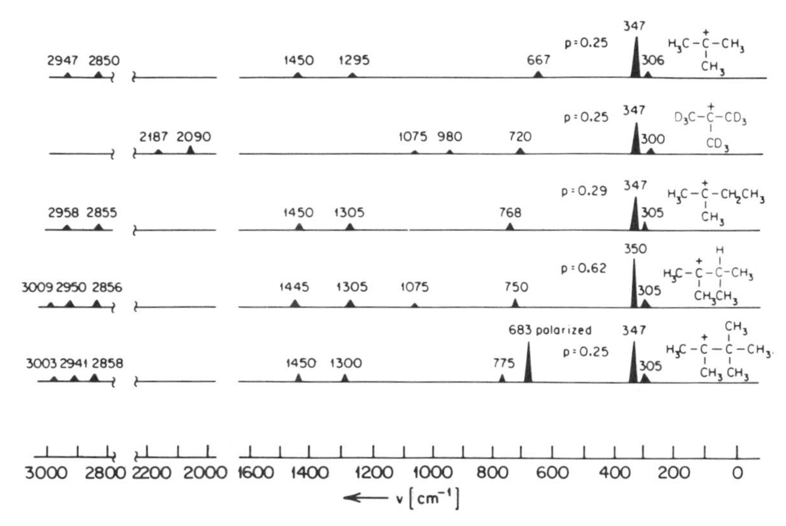

FIGURE 3.4 Schematic representation of Raman spectra of alkyl cations

Tertiary and reactive secondary alcohols in superacids like $HSO_3F:SbF_5$ (Magic Acid, HSO_3F, and SbF_5—$SO_2:SO_2ClF$) also ionize to the corresponding carbocations.[75] The generation of alkyl cations from alcohols indicates the great advantages of increasing acidity and of using acid systems with low freezing points. Deno showed that the use of sulfuric acid and oleum results in formation of cyclized allylic ions from simple aliphatic alcohols.[54] With the use of extremely strong acid, $HSO_3F:SbF_5$, tertiary and many secondary alcohols can be ionized to the corre-

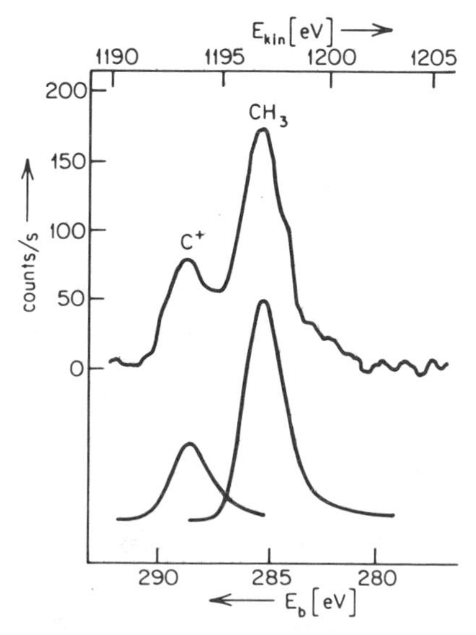

FIGURE 3.5 Carbon 1s photoelectron spectrum of *t*-butyl cation *1*

sponding alkyl cations.

$$(CH_3)_3COH \xrightarrow{HSO_3F:SbF_5} (CH_3)_3C^+$$

1

Primary and less reactive secondary alcohols are protonated in $HSO_3F:SbF_5$ solution at low temperatures (−60°C) and show very slow exchange rates.[76]

$$CH_3CH_2CH_2OH \longrightarrow CH_3CH_2CH_2\overset{+}{O}H_2$$

$$CH_3—\underset{OH}{\underset{|}{C}H}—CH_3 \longrightarrow CH_3—\underset{\underset{+}{OH_2}}{\underset{|}{C}H}—CH_3$$

Temperature-dependence studies of the nmr spectra of protonated alcohols allow the kinetics of dehydration to be followed.[77]

$$CH_3CH_2CH_2CH_2OH \xrightarrow[-60°C]{HSO_3F:SbF_5:SO_2} CH_3CH_2CH_2CH_2\overset{+}{O}H_2$$

$$\xrightarrow[-H_3O^+]{\Delta, H^+} [CH_3CH_2CH_2CH_2^+] \longrightarrow \text{Rearrangement} \longrightarrow (CH_3)_3\overset{+}{C}$$

1

Antimony pentafluoride itself (neat or in SO_2 or SO_2ClF solution) also ionizes alcohols to form alkyl carbocations.

$$R—OH + SbF_5 \longrightarrow R^+SbF_5OH^-$$

To overcome difficulties and achieve ionization of primary (and less reactive secondary) alcohols at low temperatures, it was found, in some cases, that it is advantageous to convert them to the corresponding haloformates or halosulfites with carbonyl halides or thionyl halides. These in turn ionize readily in $SbF_5:SO_2$ solution and lose CO_2 or SO_2.[78]

$$ROH \xrightarrow{COX_2} ROCOX \xrightarrow{SbF_5:SO_2} R^+SbF_5X^- + CO_2$$

$$ROH \xrightarrow{SOX_2} ROSOX \xrightarrow{SbF_5:SO_2} R^+SbF_5X^-$$

$$X = Cl, F$$

Aliphatic ethers are protonated in strong acids and, at low temperatures, the exchange rates of the acidic proton are slow enough to permit direct observation by nmr spectroscopy.[79] Temperature dependent nmr spectral studies allow again to follow the kinetics of ether cleavage to form alkyl cations.

$$CH_3OCH_2CH_2CH_2CH_3 \xrightarrow[-60^\circ C]{HSO_3F:SbF_5} CH_3\overset{+}{O}(H)CH_2CH_2CH_2CH_3$$

$$\xrightarrow{H^+} CH_3\overset{+}{O}H_2 + [CH_3CH_2CH_2CH_2^+] \xrightarrow{\text{rearrangement}} \mathbf{1}$$

Protonation and ionization of mercaptans (thiols) and sulfides have been similarly studied.[80]

Superacids such as $HSO_3F:SbF_5$ also act as very effective hydrogen-abstracting agents, allowing the generation of carbocations from saturated hydrocarbons.[81]

$$CH_3-CH(CH_3)-CH_3 \xrightarrow[SO_2ClF]{HSO_3F:SbF_5} \mathbf{1}$$

$$(CH_3)_3CCOF + SbF_5 \longrightarrow \underset{\mathbf{22}}{(CH_3)CC\overset{+}{O}SbF_6^-} \xrightarrow{\Delta} \mathbf{1} + CO$$

Alkyl cations can also be generated by decarbonylation of tertiary acylium ions, like the pivaloyl cation **22.**[57] This reaction corresponds to the reverse of the Koch-Haaf acid synthesis, which is known to involve carbocation intermediates. Indeed the reaction of the *t*-butyl cation with carbon monoxide gives the pivaloyl cation.[57,82]

Thiols and thioesters (sulfides) can also be used, similar to their oxygen analogs, as precursors for alkyl cations.[83] Ionization with SbF_5-type superacids generally necessitates somewhat more forcing conditions (higher temperatures). Alkyl thiohaloformates also form alkyl cations via fragmentative ionization.[84]

$$RSH \xrightarrow[HF:SbF_5 \text{ or } HSO_3F:SbF_5]{SbF_5 \text{ or}} R^+SbF_5SH^- \quad (SbF_6^-) \quad (SbF_5FSO_3^-)$$

$$RSCOCl\ (F) \xrightarrow{SbF_5:SO_2ClF} R^+SbF_5Cl\ (F) + COS$$

Amines also can be used as precursors for the generation of alkyl cations. The classic method of deaminative formation of carbocations involves some type of diazotization reaction producing an equimolar amount of water.

$$RNH_2 \xrightarrow[-H_2O]{HNO_2} [RN_2^+] \longrightarrow R^+ + N_2$$

Newer methods overcome this difficulty. The corresponding sulfinylamine or isocyanate is first prepared and then reacted with stable nitrosonium salts to give the corresponding carbocation.[85]

$$RNH_2 \xrightarrow{SOCl_2} RNSO \xrightarrow{NO^+SbF_6^-} R^+SbF_6^- + N_2 + SO_2$$

$$RNH_2 \xrightarrow{COCl_2} RNCO \xrightarrow{NO^+SbF_6^-} R^+SbF_6^- + N_2 + CO_2$$

3.4.1.4 Observation in Different Superacids

Whereas antimony pentafluoride-containing superacids (such as $HF:SbF_5$, $HSO_3F:SbF_5$, $CF_3SO_3H:SbF_5$, etc.) are the preferred solvents for obtaining alkyl cations, other nonoxidizing superacids such as $HF:BF_3$, $HF:TaF_5$, etc., can also be on occasions used successfully. The stability of carbocations in these solvents is generally somewhat lower.

3.4.2 Cycloalkyl Cations

Tertiary cycloalkyl cations, such as the 1-methyl-1-cyclopentyl cation **19,** show high stability in strong acid solutions. This ion can be obtained from a variety of precursors (Fig. 3.6).[86] It is noteworthy to mention that not only cyclopentyl- but also the cyclohexyl-type precursors give the 1-methylcyclopentyl cation **19.** This indicates that cyclopentyl cation has higher stability, which causes isomerization of the secondary cyclohexyl cation to the tertiary methylcyclopentyl ion.

The methylcyclopentyl cation **19** undergoes both carbon and hydrogen scrambling.[63,72] An activation energy of 15.4 ± 0.5 kcal·mol^{-1} was measured for a process that interchanges α and β hydrogens in **19.** Above 110°C, coalescence to a single peak was observed in the ^{1}H-nmr spectrum of **19.** An activation energy

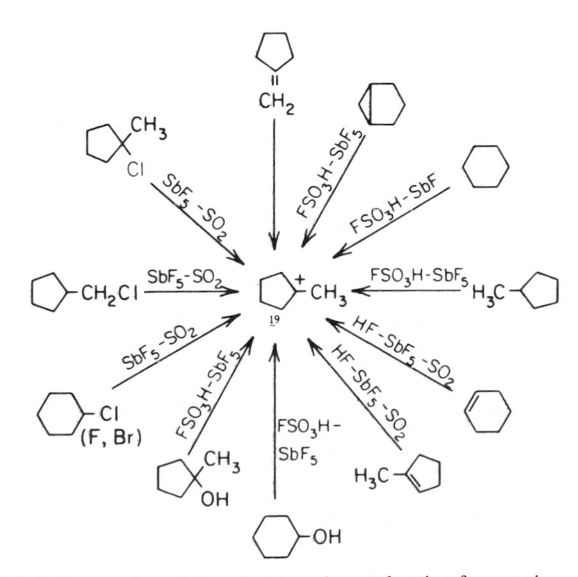

FIGURE 3.6 Preparation of 1-methyl-1-cyclopentyl cation from various precursors

barrier of 18.2 ± 0.1 kcal·mol^{-1} was found for the methyl and ring hydrogen interchange. The protonated cyclopropane intermediate **24** required for the process might undergo an additional reversible ring opening to secondary cyclohexyl cation **26.** To investigate this possibility, Saunders and Rosenfeld[87] measured deuterium and carbon scrambling rates simultaneously using a mixture containing ^{13}C- and deuterium-labeled methyl groups in methylcyclopentyl cation **19.** The two observed rates were related in exactly the manner predicted if the protonated bicyclohexane **25** opens to **26** more rapidly than it returns to **24** via a corner-to-corner hydrogen shift.

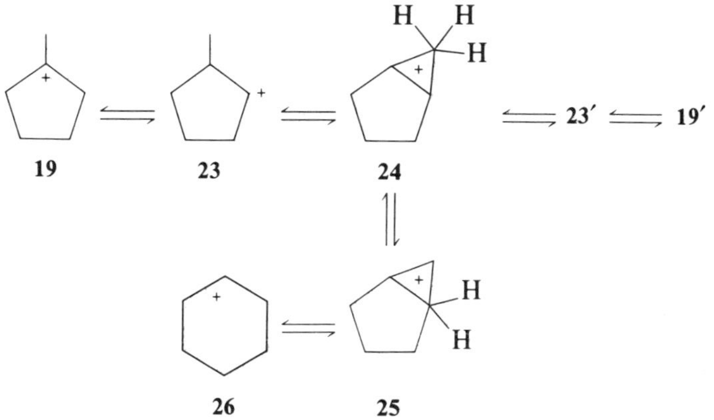

H H
H
⇌ 23′ ⇌ 19′
19 23 24
H
H
26 25

The cyclopentyl cation **27** shows in its proton nmr spectrum in $SbF_5:SO_2ClF$ solution, even at $-150°C$, only a single absorption line at δ 4.75.[86,88] This observation indicates a completely degenerate ion with a very low barrier to the secondary-secondary hydrogen shift (see later discussion). Although stable secondary cyclohexyl cation **26** is unknown under long-lived stable ion conditions, a static secondary cyclohexyl cation **28** with an α-spiro-cyclopropyl group has been prepared and studied by various routes.[89] However, this ion derives its stability by substantially delocalizing its positive charge into the neighboring cyclopropyl group (i.e., it is also a cyclopropylcarbinyl cation).

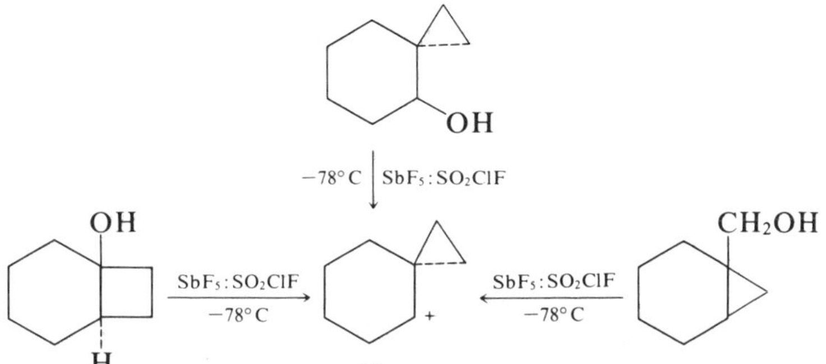

OH
$-78°C$ $SbF_5:SO_2ClF$
OH
$SbF_5:SO_2ClF$
$-78°C$
$SbF_5:SO_2ClF$
$-78°C$
CH_2OH
H
28

Sorensen and co-workers[90] have prepared tertiary cycloalkyl cations of different ring sizes, $n = 4$ (small ring), $n = 5$–7 (common rings), $n = 8$–11 (medium rings), and $n = 12$–20 (large rings). These ions were in general found to undergo ring expansion or contraction reactions, often in multiple or repetitive steps as shown in the following sequence.

$$
\begin{aligned}
4^+\text{-Pr} &\rightarrow 5^+\text{-Et} + 6^+\text{-Me} \\
5^+\text{-Et} &\rightarrow 6^+\text{-Me} \\
7^+\text{-Pr} &\rightarrow 6^+\text{-But} \\
8^+\text{-Et} &\rightarrow 6^+\text{-But} \\
9^+\text{-Me} &\rightarrow 8^+\text{-Et} + 7^+\text{-Pr} + 6^+\text{-But} \\
10^+\text{-Me} &\rightarrow 6^+\text{-Pen} \\
10^+\text{-Et} &\leftrightarrows 11^+\text{-Me} \\
11^+\text{-Me} &\leftrightarrows 10^+\text{-Et} \leftrightarrows 6^+\text{-Hex} \\
10^+\text{-Pr} &\rightarrow 12^+\text{-Me} + 6^+\text{-Hept} \\
11^+\text{-Et} &\rightarrow 12^+\text{-Me} + 10^+\text{-Pr} \\
12^+\text{-Pr} &\rightarrow 13^+\text{-Et} \\
13^+\text{-Et} &\rightarrow 14^+\text{-Me} \\
14^+\text{-Et} &\rightarrow 15^+\text{-Me} \\
12^+\text{-R} &\rightarrow \text{ring expansions}
\end{aligned}
$$

Some of these cycloalkyl cations ($n = 8$–11), at very low temperature show μ-hydrido bridging (see later discussion). A series of aryl substituted cycloalkyl cations **29** have been studied in connection with the application of the tool of increasing electron demand.[91]

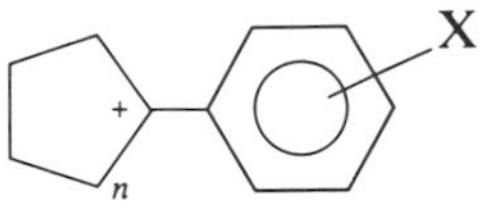

29

$n = 0,1,2,3$

X = 4-OCH_3,4-CH_3,H, 4-Br,4-Cl, 4-F, 3-CF_3,4-CF_3, 3,5(CF_3)$_2$ etc.,

The direct observation of the cyclopropyl cation **30** has evaded all attempts, owing to its facile ring opening to the energetically more favorable allyl cation **31**.[92]

30 → **31**

Many cyclopropyl cation precursors indeed readily rearrange to allyl cations under stable ion conditions.[93] However, a distinct cyclopropyl cation **32** showing a significant 2π aromatic nature has been prepared by the ionization of 11-methyl-11-bromotricyclo[$4.4.1.0^{1,6}$] undecane in SO_2ClF:SbF_5 at $-120°C$.[94]

CH_3 Br

$SbF_5:SO_2ClF$
$-120°C$

CH_3 +

32

The direct observation of ion **32** is of particular interest in that it clearly does not involve a significantly opened cyclopropane ring, which could lead to the formation of an allylic cation. Thus, it must be considered as a bent cyclopropyl cation.[95,96] It is, however, clear that the C_1–C_6 σ-bond must to some degree interact with the empty *p* orbital at C_{11} and that "homoconjugation" between them becomes the important factor in stabilizing such a "bent" cyclopropyl species.

The parent secondary cyclobutyl cation **33** undergoes three-fold degenerate rearrangement via σ-bond delocalization involving nonclassical bicyclobutonium ion-like system (see discussion on nonclassical ions).[97,98] Similar behavior is also observed for the 1-methylcyclobutyl cation **34.**[99–101] The 1-phenyl-cyclobutyl cation **35** on the other hand is a trivalent tertiary carbocation.[97,101]

3.4.3 Bridgehead Cations

Bredt's rule in its original form[102] excluded the possibility of carbocation formation at bridgehead positions of cycloalkanes. Indeed, bridgehead halides, such as apocamphyl chloride, proved extremely unreactive under hydrolysis conditions.[103] However, 1-bromoadamantane very readily gives the bridgehead carboxylic acid under the usual conditions of Koch-Haaf acid synthesis.[104] 1-Fluoroadamantane is ionized in SbF_5 to give the stable bridgehead adamantyl cation **36.**[105,106]

The proton nmr spectrum of the 1-adamantyl cation **36** in SbF_5 solution at 25°C consists of resonances at δ 5.40, δ 4.52, and δ 2.67 with peak areas of 3:6:6 (Fig. 3.7*a*). The ^{13}C-nmr spectrum (Fig. 3.7*b*) shows the γ-carbons more deshielded than the β-*s*, indicating strong C—C bond hyperconjugation with the empty *p* orbital. The bridgehead 1-adamantyl cation **36** can also be prepared from 2-adamantyl as well as trimethylenenorbornyl precursors.[106] A recent solid-state ^{13}C-nmr spectral study of 1-adamantyl hexafluoroantimonate salt using magic-angle spinning and cross-polarizations techniques indicates similarities between solid-state spectra and previously obtained solution spectra.[106] Several methyl-substituted bridgehead adamantyl cations **37** have been prepared and characterized. The bridgehead homoad-

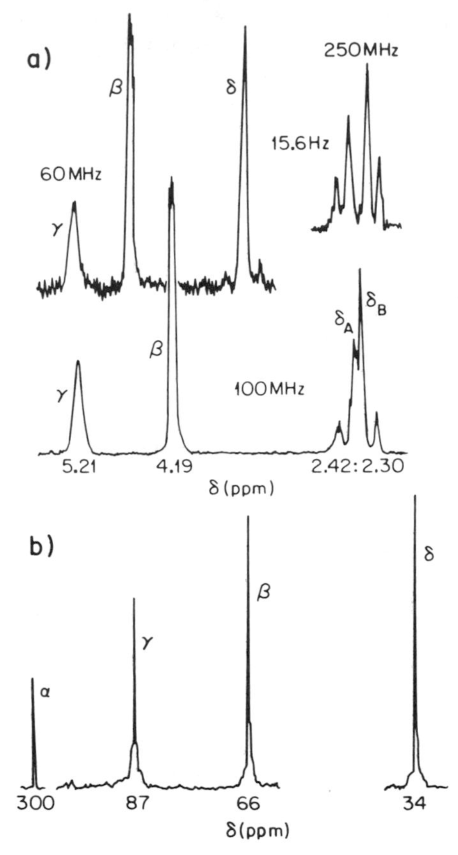

FIGURE 3.7: a) ^{1}H NMR spectrum of the 1-adamantyl cation at 60 MHz, 100 MHz, and 250 MHz; b) Fourier-transform ^{13}C-NMR of the 1-adamantyl cation (in HSO_3F-SbF_5)

amantyl cation **38** has been obtained[107] from both adamantylcarbinyl and homoadamantyl precursors.

H

X

$SbF_5:SO_2ClF$

or

$HSO_3F:SbF_5:SO_2ClF$

−78° C or −120° C

36

− 78° C or −120° C

$SbF_5:SO_2$

X

H

X = OH, Cl, Br, F

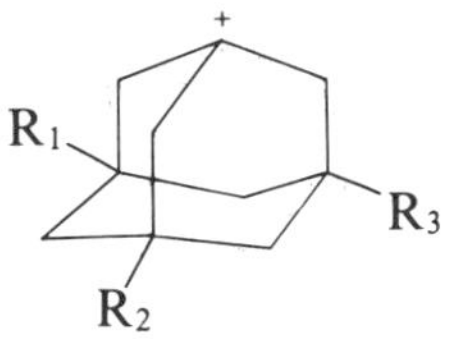

$R_1 = CH_3, R_2,R_3 = H$
$R_1 = R_2 = CH_3, R_3 = H$
$R_1 = R_2 = R_3 = CH_3$

37

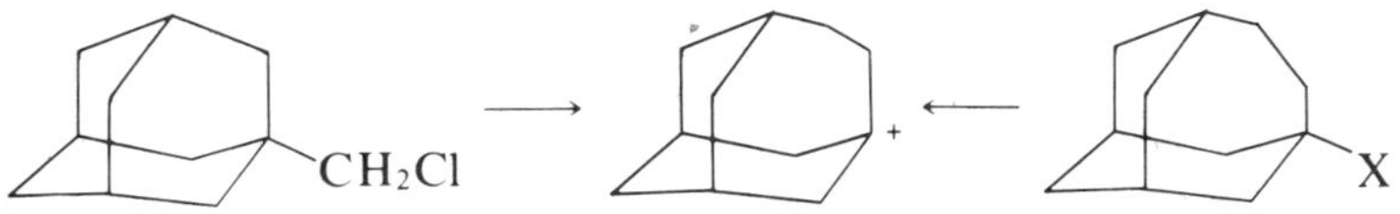

CH_2Cl → **38** ← X

$X = OH, Cl, Br$

Bridgehead bicyclo[4.4.0]decyl, bicyclo[4.3.0]nonyl, and bicyclo[3.3.0]octyl cations **39, 40,** and **41** are found to be rapid equilibrating ions.[108] The isomeric bridgehead congressane (diamantane) cations **42** and **43** have been prepared and observed.[106] The 4-diamantyl cation **42** rapidly rearranges to the 1-diamantyl cation **43** at $-60°C$, possibly through intermolecular hydride shifts. Bridgehead bicyclo[3.3.3]undecyl cation **44** has also been observed by ^1H- and ^{13}C-nmr spectroscopy.[109]

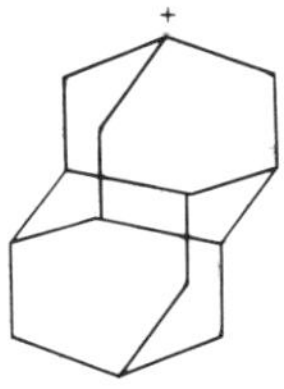

42 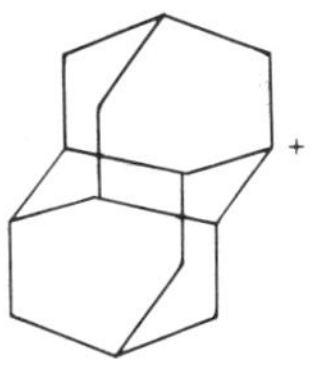 **43**

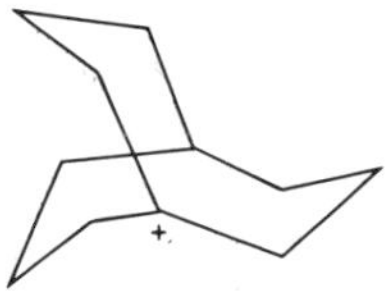

44

Bridgehead bicyclo[2.2.1]heptyl cation (1-norbornyl cation) has not been directly observed; 1-chloronorbornane yields the stable 2-norbornyl cation in $SbF_5:SO_2$ solution.[110] Thus, ionization to the bridgehead carbocation must be followed by a fast shift of hydrogen from C-1 to C-2 (either intramolecular or intermolecular), the driving force for which is obviously the tendency to relieve strain in the carbocation.

3.4.4 Cyclopropylmethyl Cations

Solvolysis studies of Roberts[111] and Hart[112] showed both the unusual stability of cyclopropylmethyl cations and the ease with which such ions rearrange. Cyclopropyl

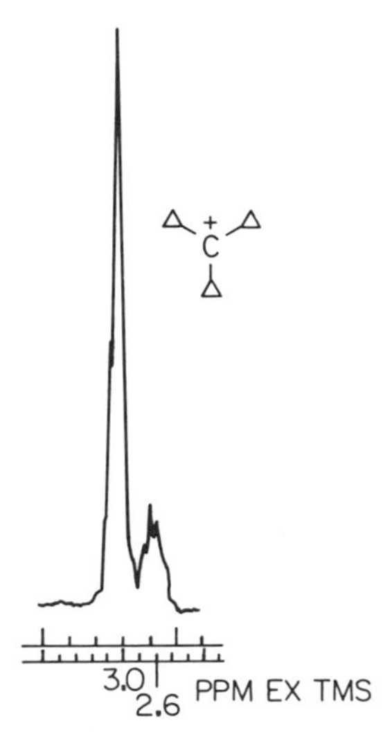

FIGURE 3.8 ^{1}H-NMR spectrum (300 MHz) of the tricyclopropylmethyl cation ion in HSO_3F-SbF_5-SO_2ClF at −60°C

groups have a strong stabilizing effect on neighboring carbocation center by delocalizing charge through bent σ-bonds. The direct observation[113] of a variety of cyclopropylmethyl cations in cyclic, acylic, and polycyclic systems by nmr spectroscopy provides one of the clearest examples of delocalization of positive charge into a saturated system.

The first cyclopropylmethyl cation directly observed was the tricyclopropylmethyl cation **45** by Deno.[113] Its ^{1}H-nmr spectrum in H_2SO_4 consists of a single sharp line at δ 2.26. In the 300 MHz ^{1}H-nmr spectrum in SO_2ClF:SbF_5 solution, however, the methine and methylene protons are well resolved[114] (Fig. 3.8). Since then, a wide variety of cyclopropylmethyl cations have been prepared and studied by ^{13}C- and ^{1}H-nmr spectroscopy.[114–116] These studies have led to the conclusion that cyclopropylmethyl cations adopt bisected geometry and are static in nature with varying degrees of charge delocalization into the cyclopropane ring. Most interesting of these ions is the dimethylcyclopropylmethyl cation **46** (Fig. 3.9). The methyl groups are nonequivalent and show a ^{1}H shift difference of 0.54 ppm. The energy difference between bisected and eclipsed structures is estimated to be 13.7 kcal/

CH_3 CH_3

CH_3 CH_3

Eclipsed **46** Bisected

mol[117] (by temperature-dependent nmr studies) and is quite close to 12.3 kcal/mol energy obtained by molecular orbital calculations at the minimal basis set STO-3G.[99b]

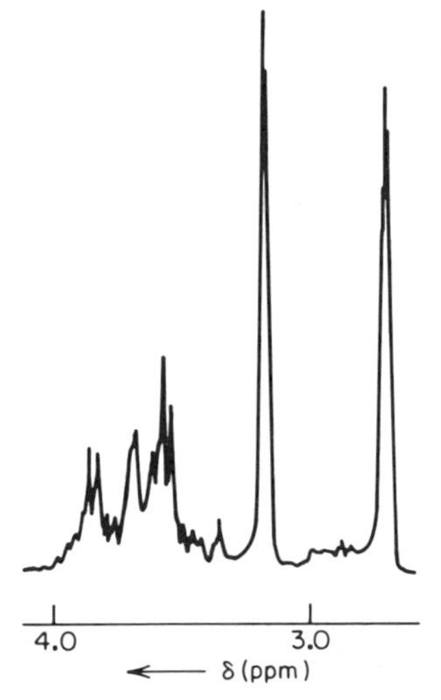

FIGURE 3.9 100-MHz ^{1}H-NMR spectrum of the dimethylcyclopropylmethyl cation *46*.

Previously discussed spiro[2.5]oct-4-yl cation **28**[89] is also a cyclopropylmethyl cation with substantial positive-charge delocalization into the spiro cyclopropyl group. Other representative cyclopropylmethyl cations that have been prepared in the superacid media and characterized are the following.

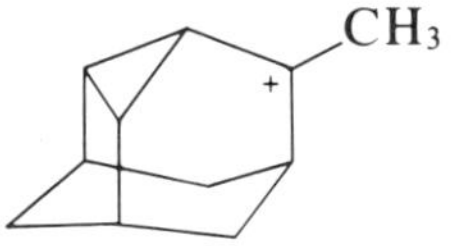

47[116] 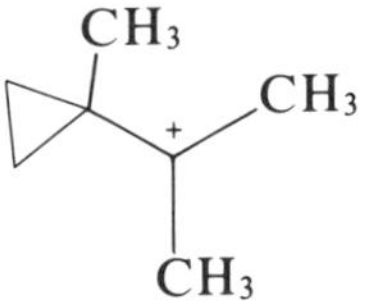 **48**[99b] 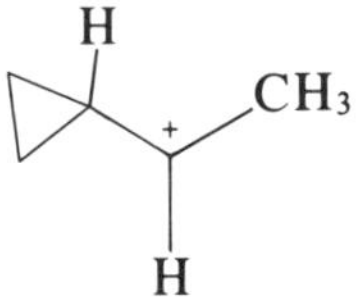 **49**[99]

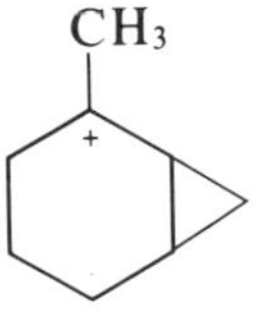

50[118]

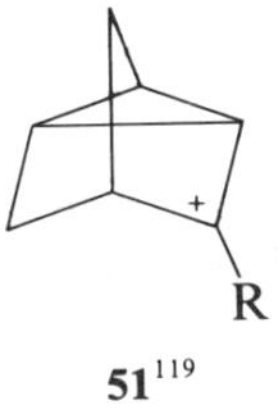

$R = CH_3, H, C_6H_5$

51[119]

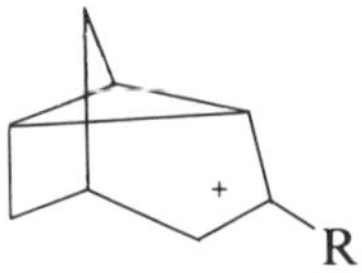

$R = CH_3, C_6H_5$

52[120]

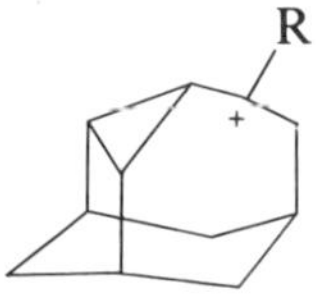

$R = H, CH_3, C_6H_5$

53[121]

A wide range of studies[114,122–124] have indicated that a cyclopropyl group is equal or more effective than a phenyl group in stabilizing an adjacent carbocation center.

In contrast to "classical" tertiary and secondary cyclopropylmethyl cations (showing substantial charge delocalization into the cyclopropane ring but maintaining its identity), primary cyclopropylmethyl cations show completely σ-delocalized non-classical carbonium ion character (see subsequent discussion). Also, some of the secondary cyclopropylmethyl cations undergo rapid degenerate equilibrium (see later discussion).

3.4.5 Alkenyl Cations

Many alkenyl cations have now been directly observed particularly by Deno and Richey,[54,125] Sorensen,[126] Olah,[127–131] and Carpenter.[132] Deno has reviewed the chemistry of these ions.[133] Allylic cations particularly show great stability with generally insignificant 1,3-overlap, except in the case of cyclobutenyl cations[134] (*vide infra*). Representative observed alkenyl cations are the following:

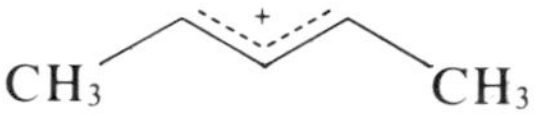
CH3
CH3

54[133]

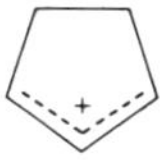

55[128]

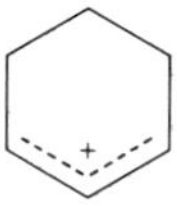

56[128]

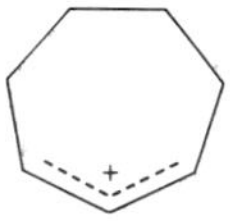

57[128]

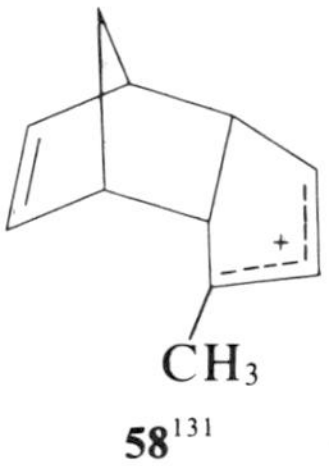
CH3

58[131]

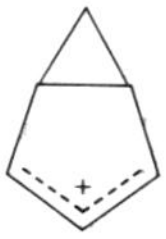

59[128c]

The formation of allyl cations from halocyclopropanes via ring opening of the unstable cyclopropyl cations also has been investigated.[93]

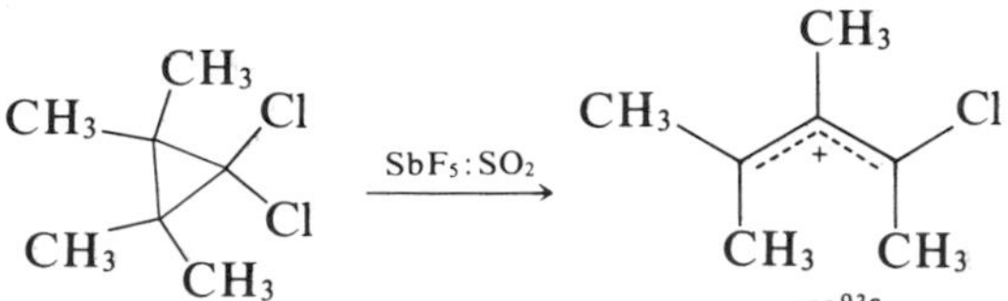
CH3
CH3
Cl
Cl
CH3
CH3
SbF5 : SO2
CH3
CH3
Cl
CH3
CH3

60[93a]

Protonation of allenes also leads to allyl cations, allowing one to obtain ions that are otherwise difficult from allylic precursors.[93a,135]

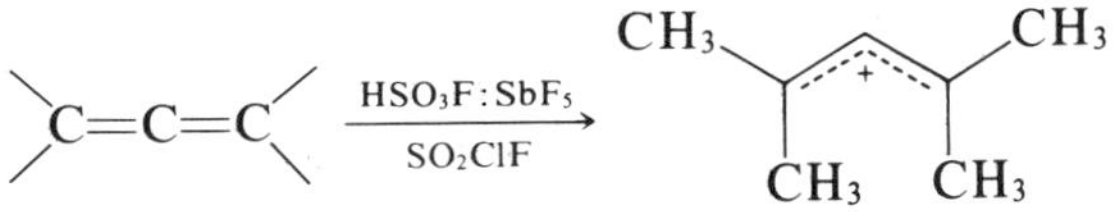

$>C=C=C<$ $\xrightarrow[SO_2ClF]{HSO_3F:SbF_5}$ CH_3 CH_3 CH_3 CH_3

61[93a]

3.4.6 Alkadienyl and Polyenylic Cations

Deno, Richey, and their co-workers[54] have observed a substantial number of alkadienyl cations. Sorensen[136] has observed divinyl and trivinyl cations **62** and **63.**

62[136] 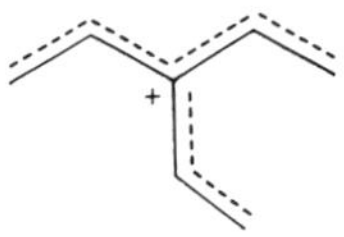 **63**[136]

Alkadienyl cations show great tendency to cyclize and these reactions have been followed by nmr.[137] Recently, several novel fulvenes have been protonated to their corresponding dienyl cations[138] (**65, 66,** etc.).

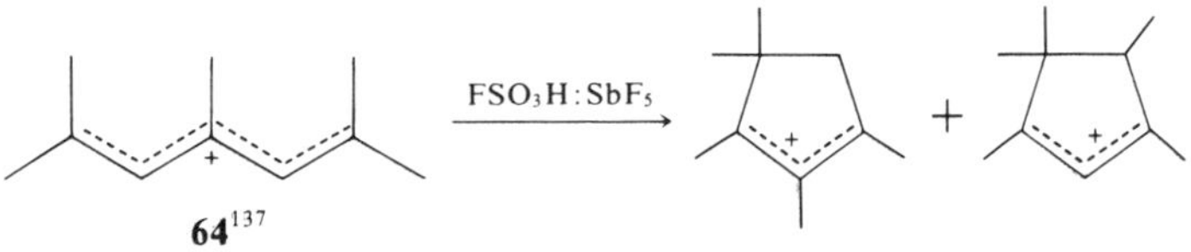

64[137] $\xrightarrow{FSO_3H:SbF_5}$ +

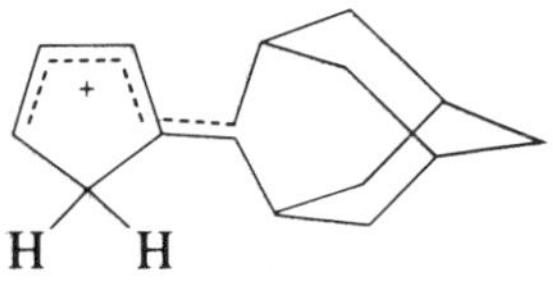

H H **65**[138] H H **66**[138]

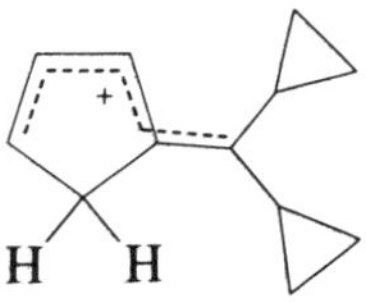

More recently, Sorensen and co-workers[139] have studied the stereochemistry of ring closure of arylallyl cations to bicyclic trienyl cations. Similar studies on 1-phenyl allyl cations have been carried out by Olah et al.[140]

3.4.7 Arenium Ions

Cycloalkadienyl cations, particularly cyclohexadienyl cations (benzenium ions), the intermediate of electrophilic aromatic substitution, frequently show remarkable stability. Protonated arenes can be readily obtained from aromatic hydrocarbons[141–143] in superacids and studied by ^{1}H- and ^{13}C-nmr spectroscopy.[144,145] Olah et al.[145] have

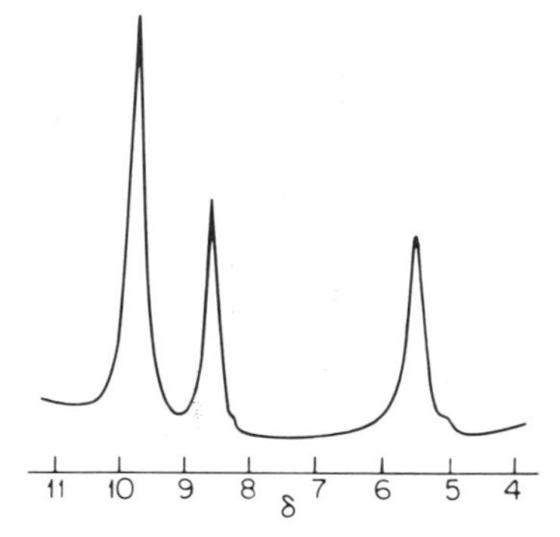

FIGURE 3.10 The 270-MHz ^{1}H-NMR spectrum of the "static" benzenium ion *67* in SbF_5-HSO_3F-SO_2ClF-SO_2F_2 solution at −140°C

even prepared and studied the parent benzenium ion ($C_6H_7^+$) **67.** Representative ^{1}H-nmr spectra of benzenium and naphthalenium ions[146] **67** and **68** are shown in Figures 3.10 and 3.11.

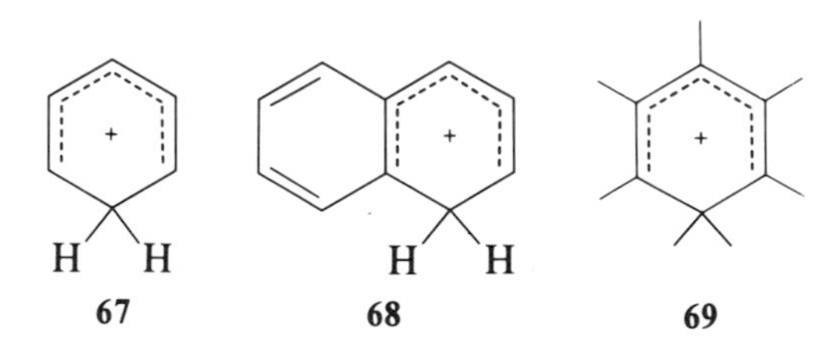

Anthracenium ions[147] as well as isomeric mono-, di-, tri-, tetra-, penta-, and hexaalkylbenzenium and halobenzenium ions have been observed.[146,147] Alkylation, nitration, halogenation, etc., of hexamethylbenzene give the related ions. Doering and Saunders have studied the dynamic nmr spectra of heptamethylbenzenium ion **69.**[148] Fyfe and co-workers[149] have obtained the solid-state ^{13}C-nmr spectrum of **69.**

Winstein and co-workers[150] have obtained the stable monocation **70** by protonating 1,6-methano[10]annulene in HSO_3F. Cram and co-workers have been successful[151] in protonating [2.2]-*para*-cyclophane. The initial *para*-protonated species **71** rear-

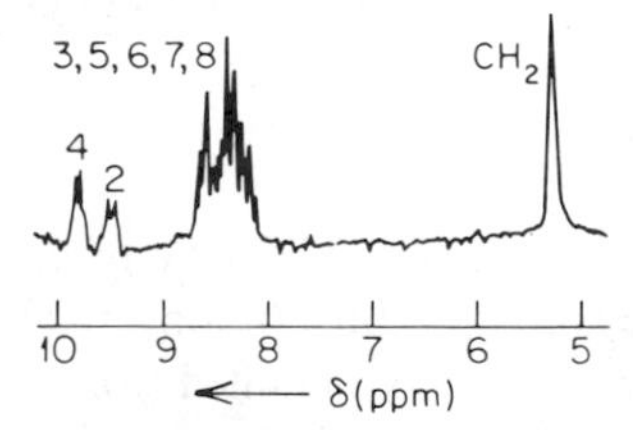

FIGURE 3.11 100-MHz ^{1}H-NMR spectrum of the naphthalenium ion *68* at −90°C

ranges to the *meta*-protonated species **72.**

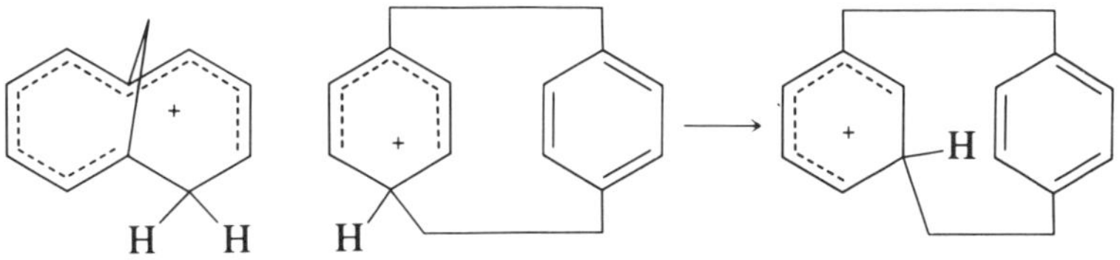

70 **71** **72**

3.4.8 Ethylenearenium Ions

The classical–nonclassical ion controversy[18] initially also included the question of the so-called ''ethylenephenonium'' ions.

Cram's original studies[152] established, based on kinetic and stereochemical evidence, the bridged ion nature of β-phenylethyl cations in solvolytic systems. Spectroscopic studies (particularly ^{1}H and ^{13}C-nmr)[153] of a series of stable long-lived ions proved the symmetrically bridged structure, and at the same time showed that these ions do not contain a pentacoordinate carbocation center (thus are not ''nonclassical ions''). They are spiro[2.5]-octadienyl cations **73** (spirocyclopropylbenzenium ions), in other words cyclopropyl-annulated cations in which the cationic center belongs to a cyclohexadienyl cation (benzenium ion).

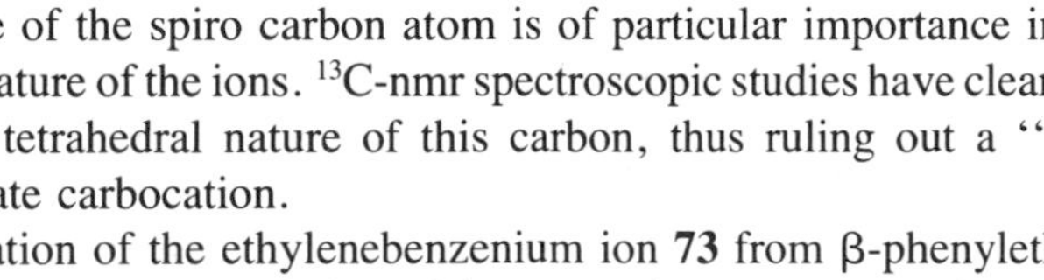

73

The nature of the spiro carbon atom is of particular importance in defining the carbocation nature of the ions. ^{13}C-nmr spectroscopic studies have clearly established the aliphatic tetrahedral nature of this carbon, thus ruling out a ''nonclassical'' pentacoordinate carbocation.

The formation of the ethylenebenzenium ion **73** from β-phenylethyl precursors can be depicted as cyclialkylation of the aromatic π-systems and not of the C_i—C_α

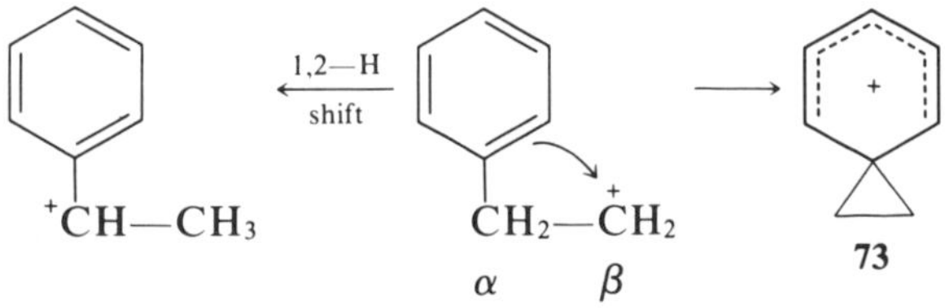

$^+CH—CH_3$ ←(1,2—H shift) $CH_2—\overset{+}{C}H_2$ (α, β) → **73**

bond which would give the tetracoordinate ethylenephenonium ion. Rearrangement of the β-phenylethyl to a α-phenylethyl (styryl ions), on the other hand, takes place through a regular 1,2-hydrogen shift. Rearrangement and equilibria of ions formed from side-chain-substituted β-phenylethylchlorides have also been explored.[153d]

The benzonorbornenyl cation **74** can be considered as the spiro cation **74.**[154] Olah and Singh[155a,b] have prepared and characterized a series of 4-substituted ethylenenaphthalenium ions **76.** Winstein et al. have reported the ^{1}H-nmr spectrum of ethyleneanthracenium ion **77** (R = H).[156] A series of 9-substituted anthracenium ions have been studied by ^{13}C-nmr spectroscopy.[155c]

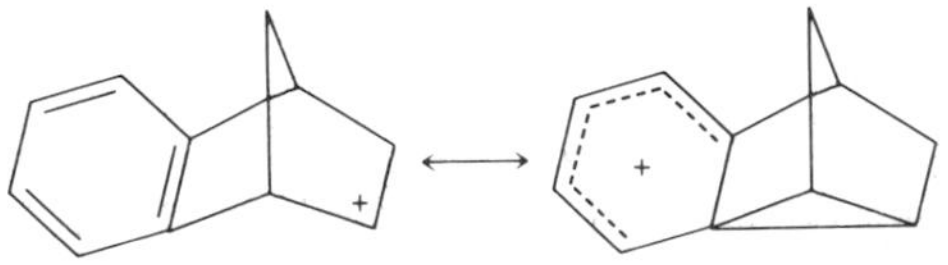

74 ⟷ **75**

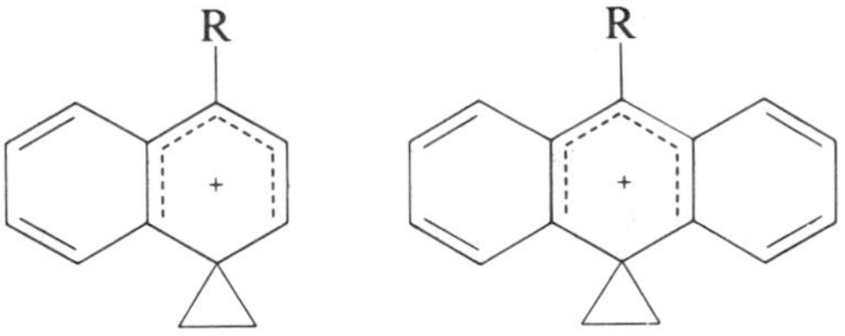

76 **77**

3.4.9 Propargyl and Allenylmethyl Cations (Mesomeric Vinyl Cations)

No *bona fide* vinyl cations have yet been experimentally observed under long-lived stable ion conditions.[157] Propargyl cations **78** however, exist in mesomeric allenyl forms **79,** which can serve as models for vinyl cations.

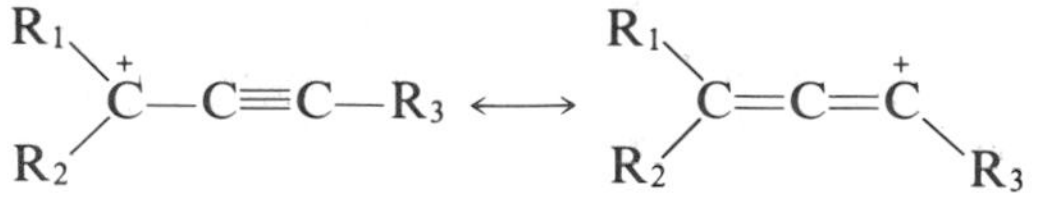

$$R_1R_2\overset{+}{C}{-}C{\equiv}C{-}R_3 \longleftrightarrow R_1R_2C{=}C{=}\overset{+}{C}{-}R_3$$

78 **79**

Extensive work has been carried out on these propargyl cations with a wide variety of substituents.[124,158]

Recently Siehl et al.[159] have prepared two allenylmethyl cations **80** and **81,** which also exhibit extensive mesomeric vinyl cation character.

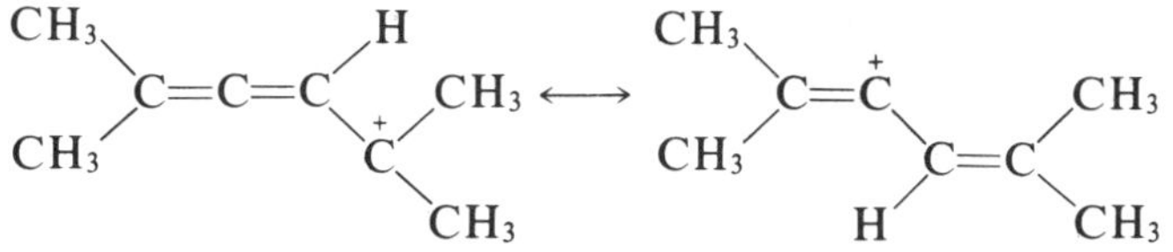

80

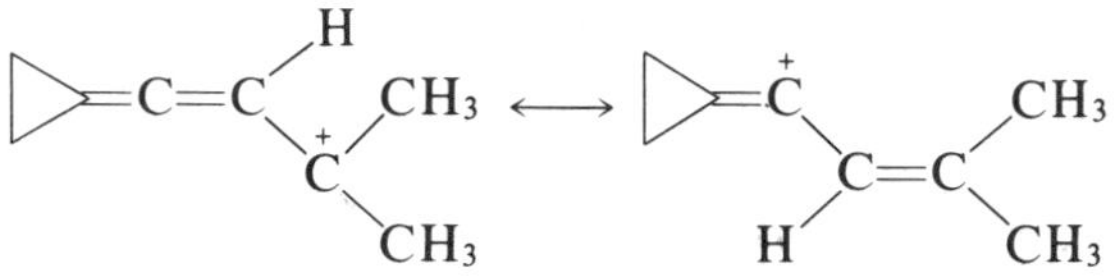

81

3.4.10 Aryl- and Alkylarylmethyl Cations

The first stable, long-lived carbocation observed was the triphenylmethyl cation **82.**[1]

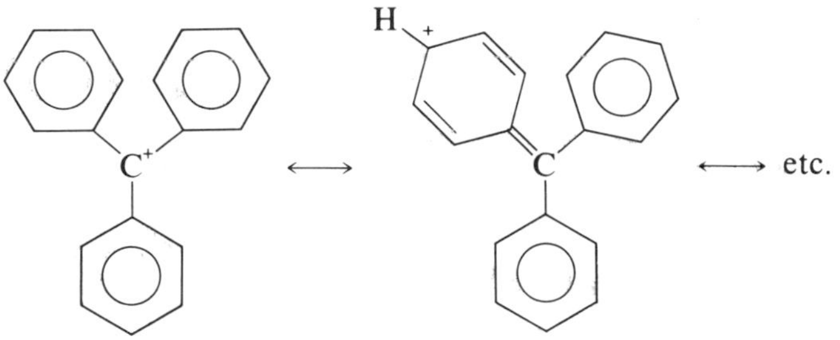

82

This ion is still the best-investigated carbocation and its propeller-shaped structure is well recognized. Strong contribution from *para*- (and *ortho*-) quinonoidal resonance forms are responsible for much of the reactivity of the ion. Diphenylmethyl cations (benzhydryl cations) are considerably less stable than their tertiary analogs. Although uv spectra in dilute sulfuric acid solutions have been obtained,[160] only recently has the benzhydryl ion **83** been observed in higher concentrations in superacid solutions [$ClSO_3H$,[161] HSO_3F,[162] and $HSO_3F:SbF_5$[163]].

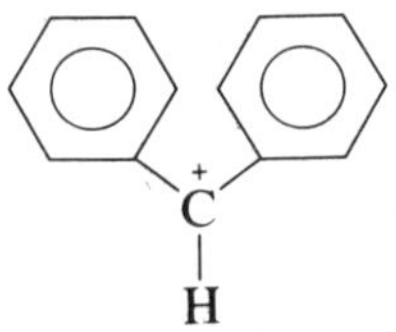

83

Mono- and dialkylarylmethyl cations can be obtained readily from the corresponding alcohols, olefins, or halides in superacid solutions, such as $HSO_3F:SbF_5$,[163] $ClSO_3H$ and HSO_3F,[161,162] and oleum.[54] Representative alkylarylmethyl cations are:

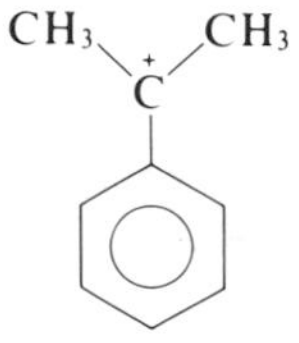

84 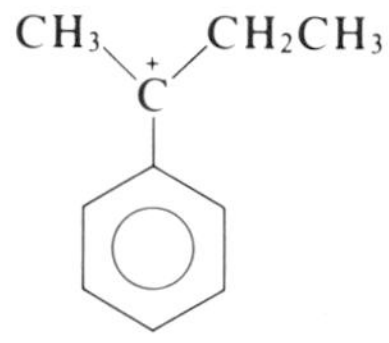 85 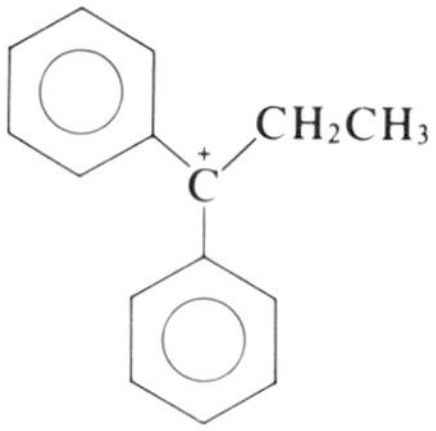 86 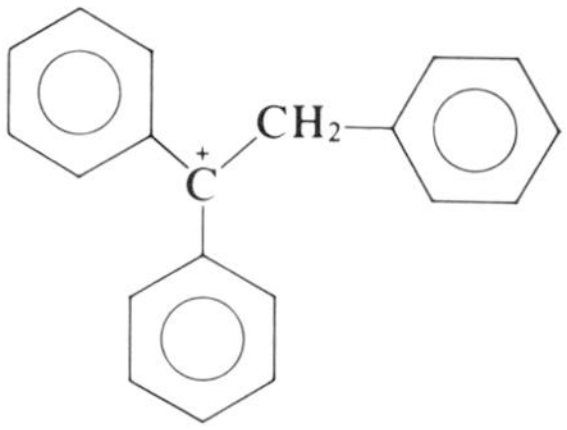 87

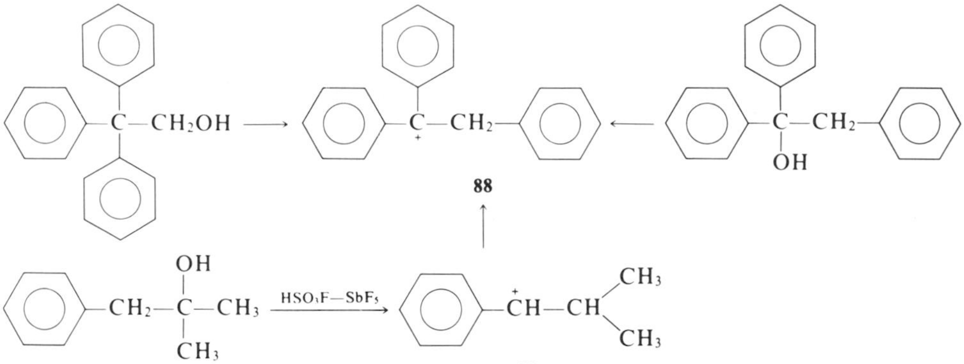

88

89

Because of the high stability of the tertiary ions, these are preferentially formed in the superacid systems from both tertiary, secondary, and even primary precursors.[164] If, however, the tertiary carbocation is not benzylic, rearrangement to a secondary, benzylic ion can be observed:[163,164] With suitable substituent groups (which also prevent transalkylations), secondary styryl cations were found as stable, long-lived ions.[165,166,153d]

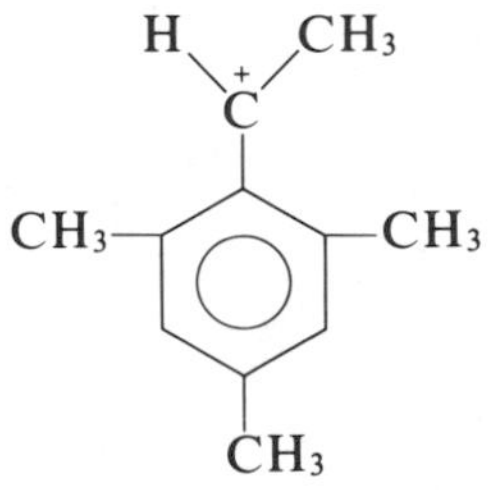

90 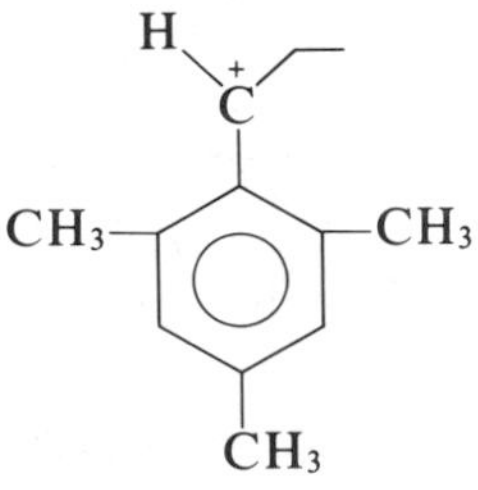 91 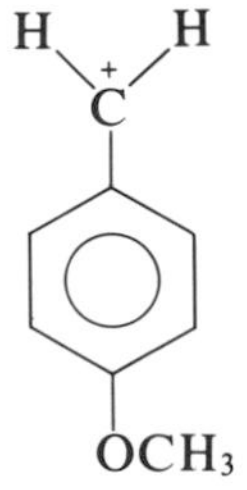 92

Although the unsubstituted benzyl cations still elusive, many substituted derivatives have been observed.[165–167]

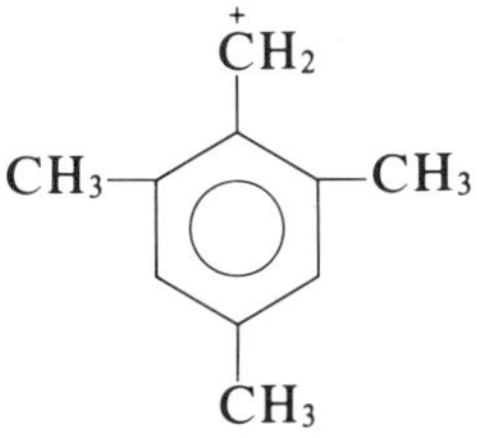

93 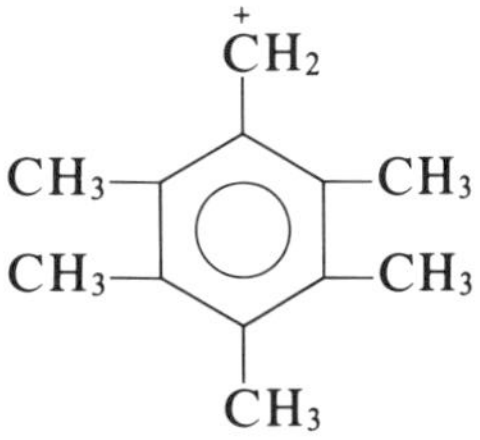 **94**

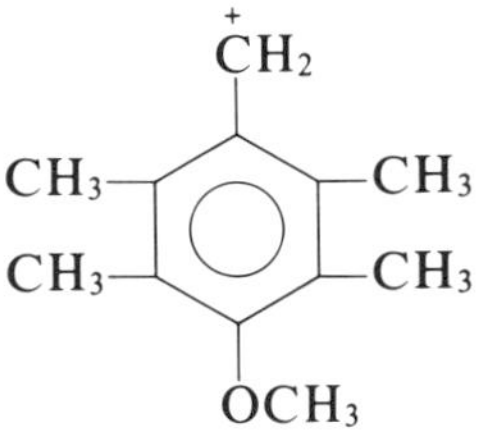

95 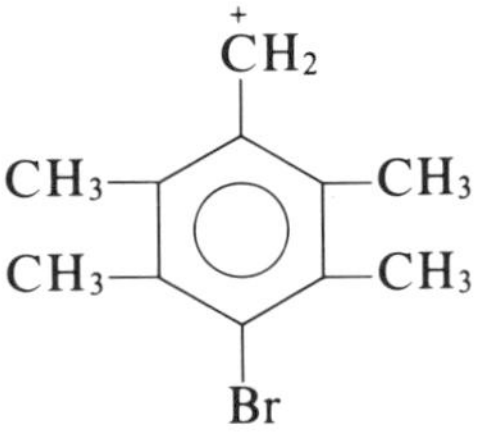 **96**

In a cation such as the (2,4-dimethyl-6-*t*-butyl)benzyl cation **97,** a high rotational barrier around sp^2-hybridized atom is observed. The methylene protons are found magnetically nonequivalent in the ^{1}H-nmr spectrum.[166]

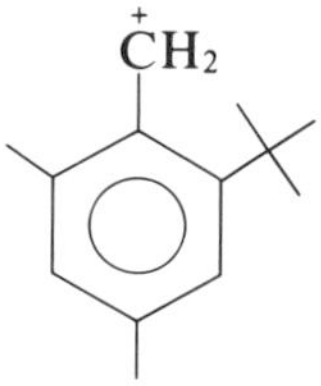

97

No rearrangement of benzyl cations in acid solution to tropylium ions has been found, although this rearrangement is observed in the gas phase (mass spectrometry).[168]

3.4.11 Carbodications

Interest in carbocations has not been confined to monopositive carbon species (carbomonocations). The study of carbodications has more recently been of substantial interest and the topic has been recently reviewed.[169]

Early reports[170] that a carbodication had been observed from pentamethyltrichloromethyl benzene turned out to be incorrect. The species obtained was the dichloropentamethylbenzyl cation **98**.[171–173]

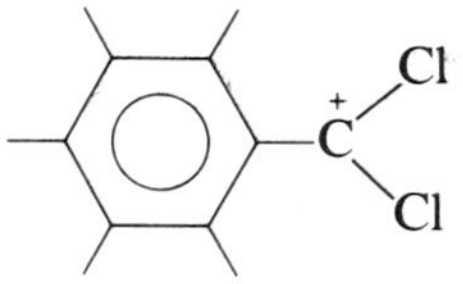

98

If two carbocation centers are separated by a phenyl ring, a variety of carbodi- and trications can be obtained.[174–176]

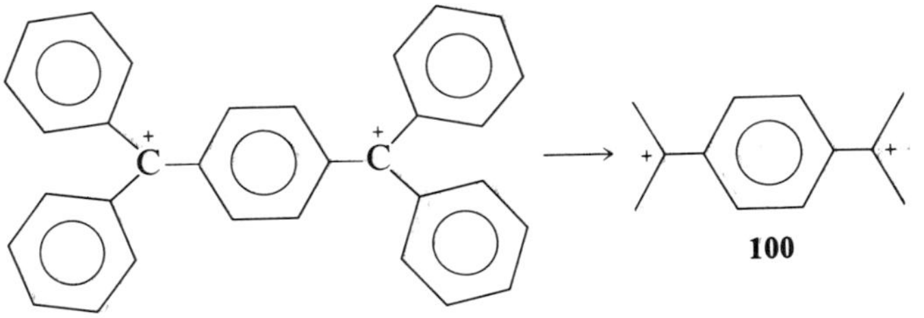

99 → **100**

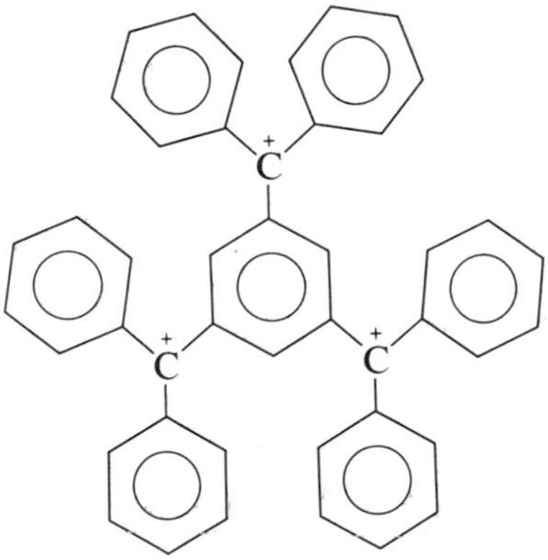

101

Separation of two carbocation centers by at least two methylene groups in open-chain carbodications renders them stable and observable.[177] Indeed, such stable carbodications **102** have been subjected to a comprehensive nmr spectroscopic study.[176]

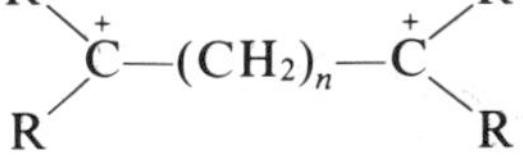

$$R_2\overset{+}{C}-(CH_2)_n-\overset{+}{C}R_2$$

102 $n \geq 2$

Carbodications have also been observed in more rigid systems such as the apical, apical congressane (diamantane) dication **103,**[106] bicyclo[2.2.2]octyl dication **104,**[178] and bicyclo[3.3.3]undecyl dication **105**[109] (manxyl dication—^{13}C-nmr spectrum in Fig. 3.12). Other worth mentioning polycyclic bridgehead dications are **106** and **107.**[106]

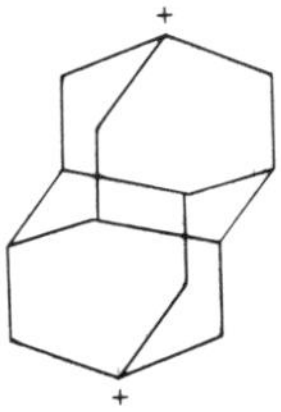

103 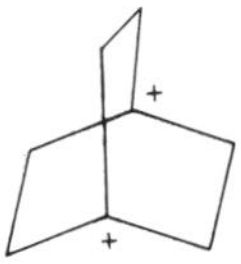 **104** 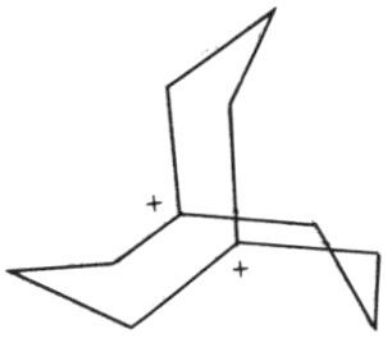 **105** 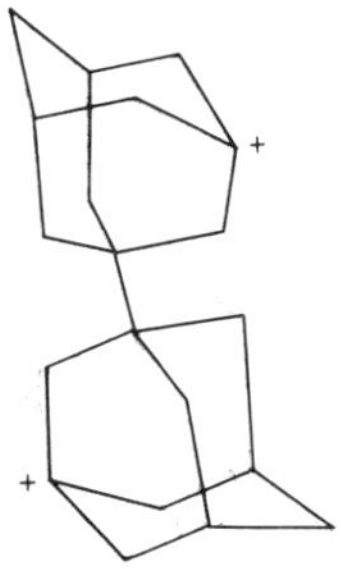 **106**

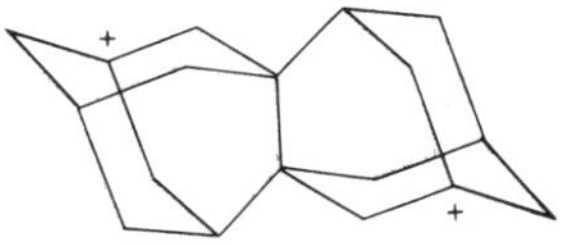

107

Attempts to observe circumambulatory rearrangement in the 2,6-*anti*-tricyclo[5.1.0.0^{3,5}]octan-2,6-diyl dication **108** have been unsuccessful.[179] The dication **108** would appear to rearrange instantaneously to the homotropylium ion **109** by proton elimination. However, substituted dications of type **108,** e.g., **110,** are quite stable; they are static and a substantial part of the charge is delocalized into the cyclopropane rings.[179]

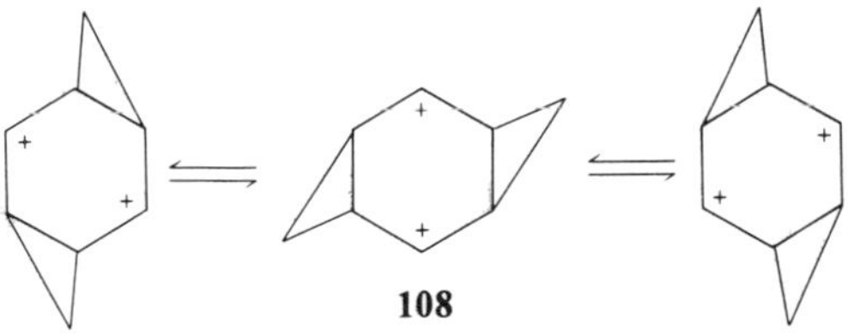

108

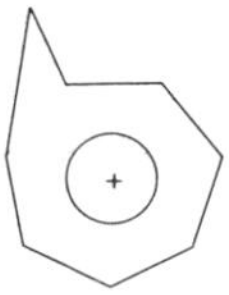

109 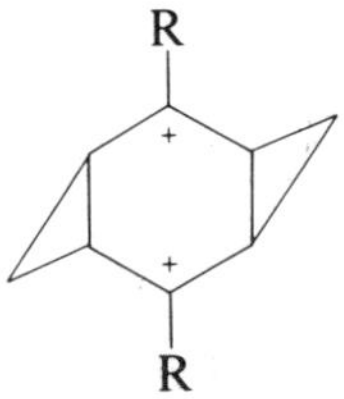 **110**

R = CH_3, C_6H_5,

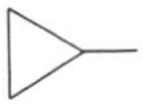

Recently several 2,6-disubstituted 2,6-adamantdiyl dications **111** have been prepared and characterized.[180] They are stable only when they contain charge delo-

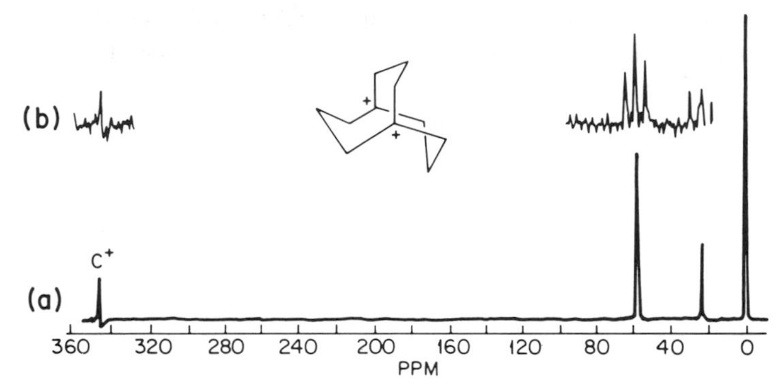

FIGURE 3.12 The 25 MHz ^{13}C-NMR spectrum of the bicyclo[3.3.3] undecyl dication *105* in SbF_5-SO_2ClF solution, a) proton decoupled, b) proton coupled.

calizing substituents such as cyclopropyl or phenyl groups.

R

R

R = C_6H_5,

111

The effect of the introduction of two electron-deficient centers into the bicyclo[2.2.1]heptyl skeleton (norbornyl framework) has been explored.[181] A ^{13}C-nmr spectroscopic study of several substituted 2,5-diaryl-2,5-norbornandiyl dications **112** reveals the regular dicarbenium ion nature of the system.[181]

X

112

X

X = H, 4-CH_3, 4-OCH_3, 4-CF_3, 3,5$(CF_3)_2$, etc.

The bicyclic bisallylic dication **113** has been obtained from a variety of precursors under strong acid conditions.[182]

113

The *endo*-3,10-dimethyltricyclo[5.2.1.$0^{2,6}$]deca-4,8-dien-3,10-diyl dication **114** has been prepared[183] in $HSO_3F:SbF_5:SO_2ClF$ at −120°C. It possesses a novel bishomoaromatic/allylic dication structure. At higher temperatures, **114** rearranges stereospecifically to the symmetrical *cis-anti-cis*-3,10-dimethyltricyclo[5.3.0.$0^{2,6}$]deca-4,8-dien-3,10-diyl dication **115.**

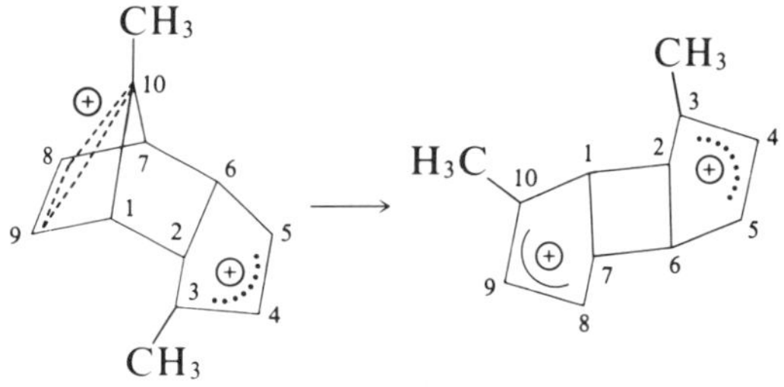

114 **115**

Lammertsma and Cerfontain[184] have obtained the cyclopropyldicarbinyl dication **116** by diprotonating the 1,6-methano [10] annulene in much stronger Magic Acid at −60°C. If the same protonation was carried out at −120°C, they were able to obtain the previously discussed monocation **70.**[150]

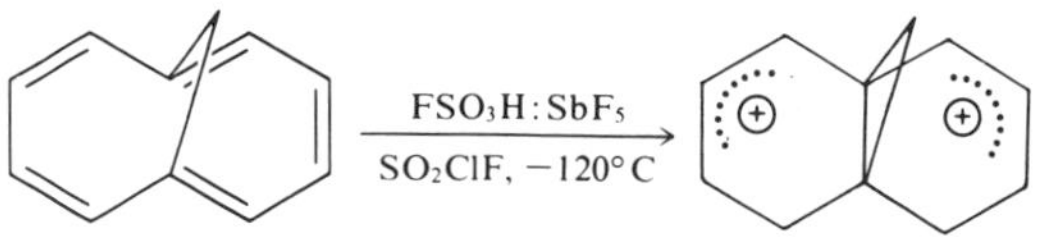

$FSO_3H:SbF_5$ / SO_2ClF, −120°C

116

Recently, Lammertsma[185] has also obtained dication **117** by the protonation of hexahydropyrene in Magic Acid solution. This is the first example of a β,β-diprotonated naphthalene derivative. Koptyug and co-workers[186] have studied α,α′-diprotonated 1,2,3,5,6,7-hexa- and octamethylnaphthalenes **118** under superacid conditions.

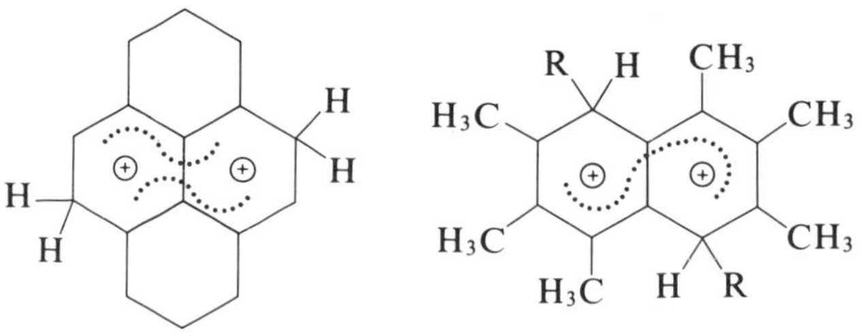

117 **118**

Many aromatic stabilized dications have been prepared and characterized by nmr spectroscopy (see subsequent discussion).

3.4.12 Aromatic Stabilized Cations and Dications

If a carbocation or a dication at the same time is also a Huckeloid $(4n + 2)\pi$ aromatic system, resonance can result in substantial stabilization. The simplest 2π aromatic system is the Breslow's cyclopropenium ion **119.**[187] The benzo analog of **119,** benzocyclopropenium ions **120,** are also known.[188] Although the 2π aromatic parent cyclobutadiene dication is still elusive, substituted analogs **121** have been prepared and well characterized.[189]

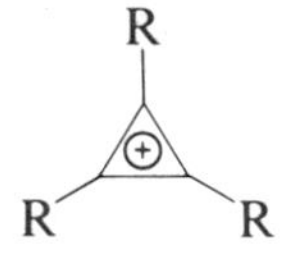

X = Cl,F

119 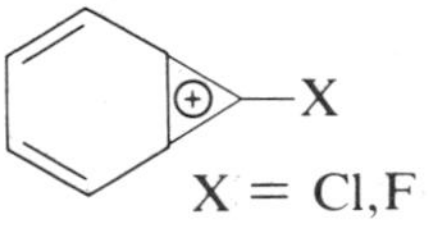 **120** 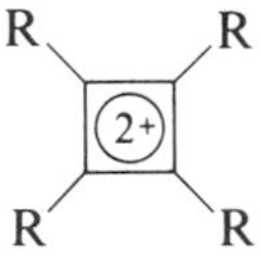 **121**

Among 6π aromatic systems, tropylium ions **122–125** cyclooctatetraene dications **126** and benzocyclobutadiene dications **127** are well studied. The parent cyclooctatetraene dication is still elusive, despite repeated attempts to prepare it under a variety of superacid conditions.[182]

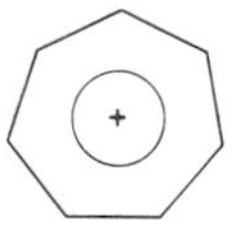

122[190] 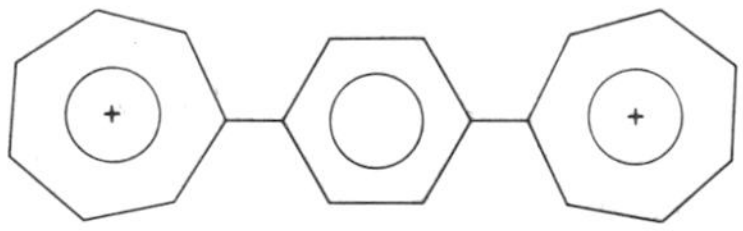 **123**[190]

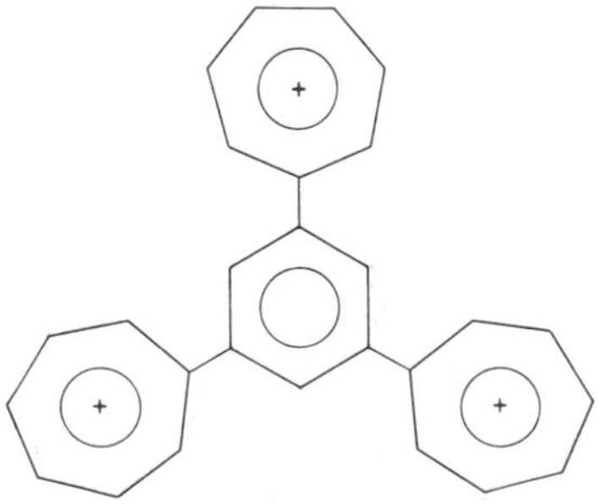

124[190] 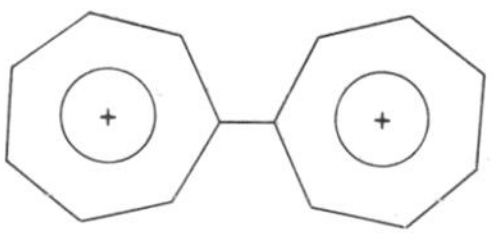 **125**[191]

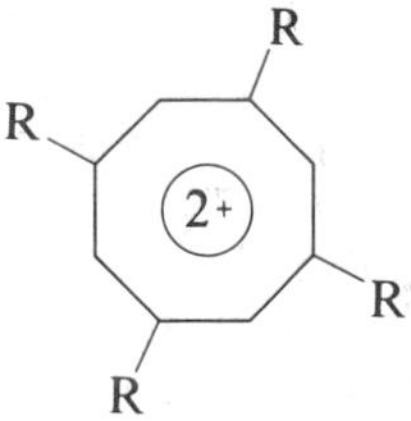

126[182] 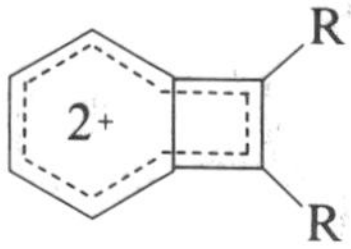 **127**[192]

The 10π dibenzocyclobutadiene dication **128**[193] has been prepared by the two-electron oxidation of biphenylenes in excess of $SbF_5:SO_2ClF$ solutions at $-10°C$.

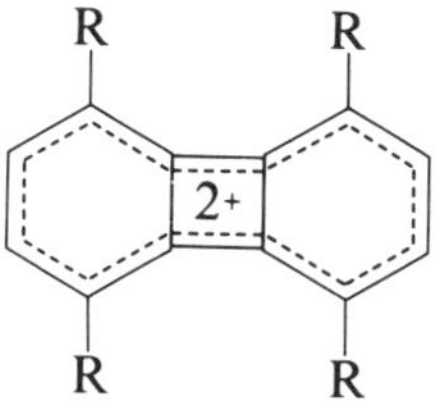

128

Since the protonation of cyclooctatetraene is known to yield the homotropylium ion (see discussion on homoaromatic systems), Schröder and co-workers reasoned that the homo [15] annulenyl cation **129** can be formed by the protonation of the [16] annulene.[194] Instead the [16] annulenediyl dication **130** was obtained along with polymeric products in $HSO_3F:SO_2:CD_2Cl_2$ media at $-80°C$. The ^1H- and ^{13}C-nmr data are consistent with the formation of the 14π dication **130,** however, attempts to trap the dication with $NaOAc:CH_3OH$ gave only polymeric products. Two possible mechanisms for the formation of dipositive ion **130** have been considered.[194] The first involves initial protonation to the homo [15] annulenyl cation **129** which further undergoes protolytic cleavage to the dication **130** (by loss of a molecule of hydrogen). The second route is associated with the stepwise oxidation of annulene by the proton or the conjugate acid of sulfur dioxide ($O{=}S{=}\overset{+}{O}H$), where a radical cation **131** is involved as an intermediate. Schröder et al.[194] seem to prefer the second mechanism since no hydrogen gas has been detected in the reaction.

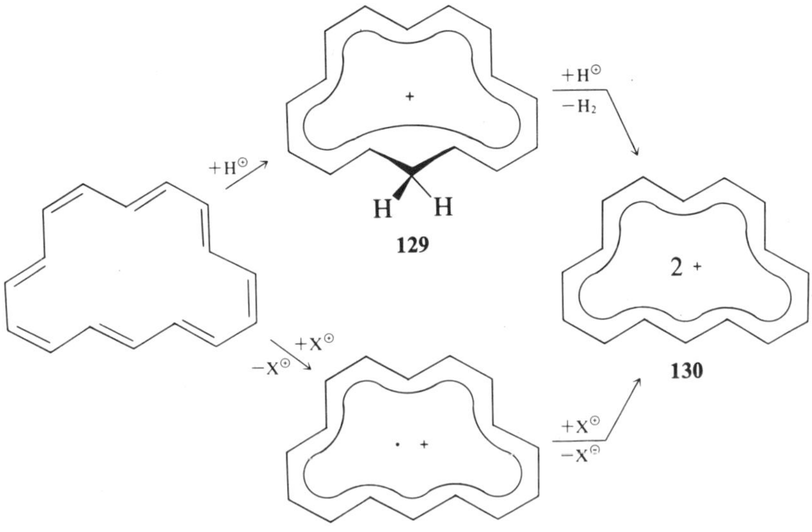

129

130

131

The aromatic nature of the presently discussed carbocations and dications have been further established by subjecting their nmr parameters to charge density-chemical shift relationship[189c,195] originally developed by Spiesecke and Schneider.[196,197]

3.4.13 Polycyclic Arene Dications

The ease of oxidation of polycyclic aromatic hydrocarbons in the gas phase[198] as well as in solution is well documented.[199–208] In strong acid solutions, monopositive radical ions and/or dipositive ions have been reported.[207,208] Similar species have been observed in anodic oxidations of aromatic compounds in nonnucleophilic solvents.[199] Simple Hückel molecular orbital theory predicts that arenes whose highest occupied molecular orbitals (HOMOs) are at higher energy levels (smaller E_{HOMO} values) should be prone to two-electron oxidation to dipositive ions. On the other hand, the arenes with low-lying HOMOs should be more difficult to ionize.[208] Certain polycyclic arenes have also been protonated to the corresponding dipositive ions.

Although benzene does not undergo two-electron oxidation reactions upon treatment with $SbF_5:SO_2ClF$, but gives instead the benzenium ion **67**[145] by protonation (due to HF impurity in the system), naphthalene has been reported to give the corresponding radical monocation upon treatment with $SbF_5:SO_2ClF$ at $-78°C$.[207,208] However, the presence of methyl substituents on the ring lowers the ionization potential to a point that stable carbodications can be formed. Thus, the tetramethyl and octamethylnaphthalene dications **132** and **133**[208] have been prepared from the corresponding arenes.

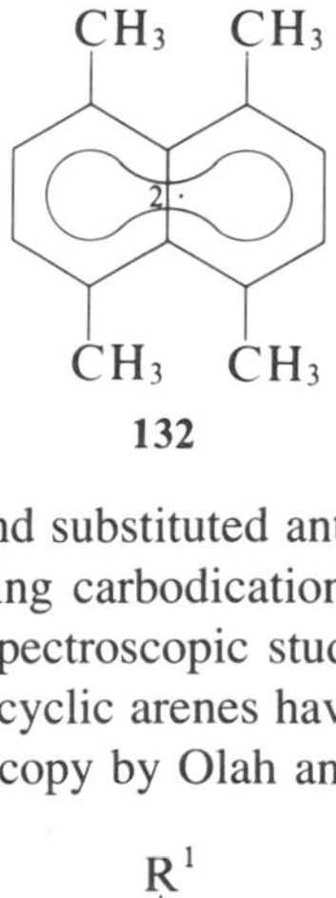

132 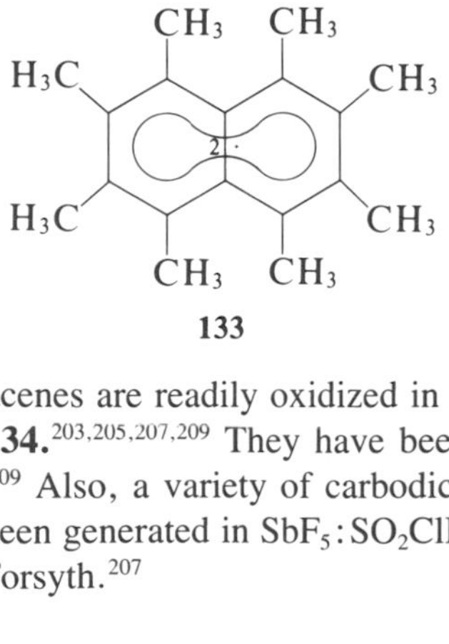 **133**

Anthracene and substituted anthracenes are readily oxidized in $SbF_5:SO_2ClF$ to their corresponding carbodications **134.**[203,205,207,209] They have been a subject of a recent ^{13}C-nmr spectroscopic study.[209] Also, a variety of carbodications of higher homologous policyclic arenes have been generated in $SbF_5:SO_2ClF$ and studied by ^{13}C-nmr spectroscopy by Olah and Forsyth.[207]

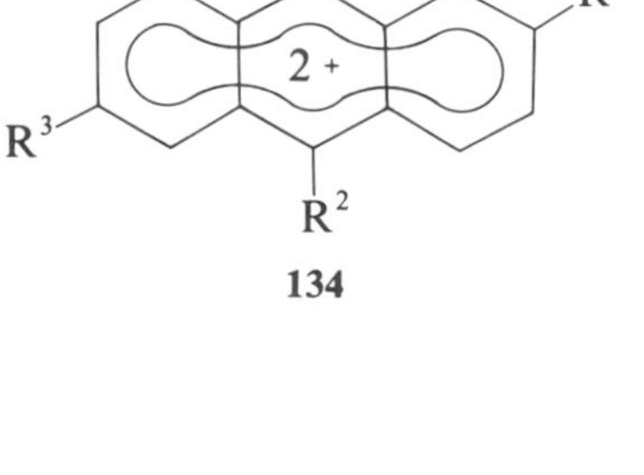

134

$R^1 = R^2 = R^3 = H$
$R^1 = R^3 = H;\ R^2 = CH_3$
$R^1 = R^2 = CH_3;\ R^3 = H$
$R^1 = R^2 = H;\ R^3 = CH_3$
$R^1 = R^3 = H;\ R^2 = Br$
$R^1 = R^3 = H;\ R^2 = Cl$

The preparation of the 8C-6π aromatic dication of **136** of the unknown pentalene **135** has proven to be unusually difficult. However, the dibenzoannulated derivatives have been prepared by Rabinowitz and co-workers.[210]

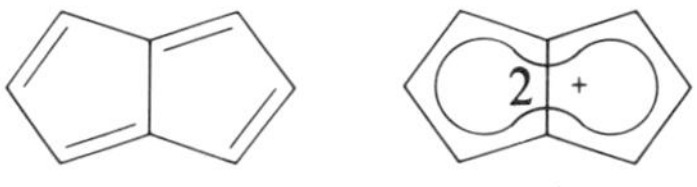

135 **136**

Upon treatment of dibenzo[*b,f*]pentalene or the 1,9-dimethyldibenzo[*b,f*] pentalene with $SbF_5:SO_2ClF$ at $-78°C$, the two-electron oxidation product dibenzo[*b,f*]pentalene dications **137** were obtained. The observed ^1H- and ^{13}C-nmr spectral deshieldings of **137** as compared with those of their progenitors clearly establish their dicationic nature.[210]

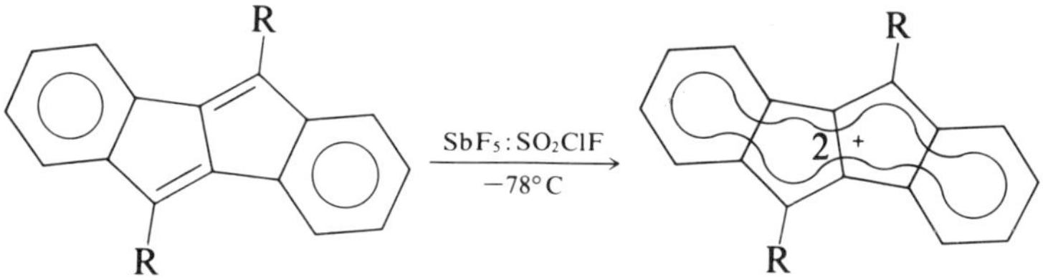

$$\xrightarrow[-78°C]{SbF_5:SO_2ClF}$$

137

3.4.14 Heteroatom-Stabilized Carbocations

In contrast to hydrocarbon cations, heteroatom-substituted carbocations are strongly stabilized by electron donation from the unshared electron pairs of the heteroatoms adjacent to the carbocation center.[8]

$$(R)_2\overset{+}{C}{-}X \longleftrightarrow (R)_2C{=}\overset{+}{X}$$

$$X = Br,\ NR_2,\ SR,\ F,\ Cl,\ OR\ \text{etc.},$$

The stabilizing effect is enhanced when two, or even three, electron-donating heteroatoms coordinate with the electron-deficient carbon atom.

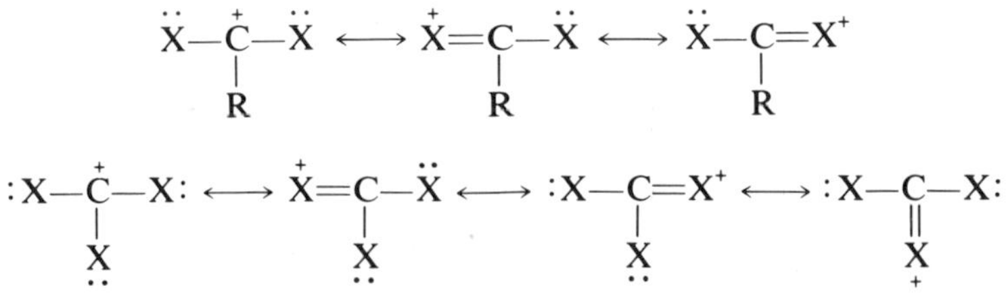

$$\ddot{X}{-}\overset{+}{C}(R){-}\ddot{X} \longleftrightarrow \overset{+}{X}{=}C(R){-}\ddot{X} \longleftrightarrow \ddot{X}{-}C(R){=}X^+$$

$$:X{-}\overset{+}{C}(\ddot{X}){-}X: \longleftrightarrow \overset{+}{X}{=}C(\ddot{X}){-}\ddot{X} \longleftrightarrow :X{-}C(\ddot{X}){=}X^+ \longleftrightarrow :X{-}C({=}\overset{+}{X}){-}X:$$

More recently, carbocations with α-heteroatom substituents such as trimethylsilyl and nitro groups that lack a stabilizing lone pair of electrons have been prepared and studied.[158c,211]

3.4.14.1 Halogen as Heteroatom

In 1965 Olah, Comisarow, and Cupas reported the first α-fluoromethyl cation.[212] Since then, a large variety of fluorine-substituted carbocations have been prepared. α-Fluorine has a particular ability to stabilize carbocations via back-coordination of its unshared electron pairs into the vacant *p* orbital of the carbocationic carbon atom. ^{19}F-nmr spectroscopy is a particularly efficient tool for the structural investigations of these ions.[213] The 2-fluoro-2-propyl and 1-phenylfluoroethyl cations **138** (nmr spectra, Figure 3.13) and **139** are representative examples of the many reported similar ions.[214]

$CH_3-\overset{+}{C}F-CH_3$

138

$\bigcirc-\overset{+}{C}F-CH_3$

139

Trifluoromethyl-[215] and perfluorophenyl-substituted carbocations have also been prepared and studied.[216,217] Because of the relatively large fluorine chemical shifts, anisotropy and ring current effect play a relatively much smaller role than they do in the case of proton shifts. Therefore, a better correlation of charge distribution with chemical shifts can be obtained. The trifluorocyclopropenium ion **140** has also

F F (+) F

140

been reported.[218] A series of chloromethyl cations were observed, including phenyldichloromethyl cations[171–173,219] and perchlorotriphenylmethyl ion **141**.[220] West has

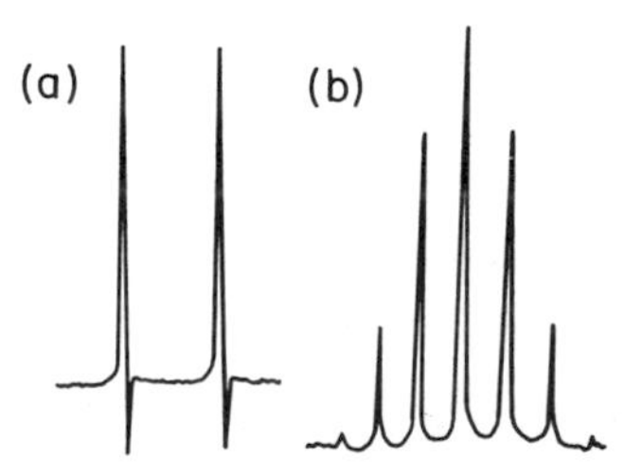

FIGURE 3.13: a) ^{1}H-NMR spectrum of the dimethylfluorocarbenium ion at 60 MHz: J_{HF} = 25.4 Hz; b) ^{19}F-NMR spectrum of the same ion at 56.4 MHz:J_{HF} = 25.4 Hz.

characterized the perchloroallyl cation **142.**[221] A series of chloro- as well as bromo-

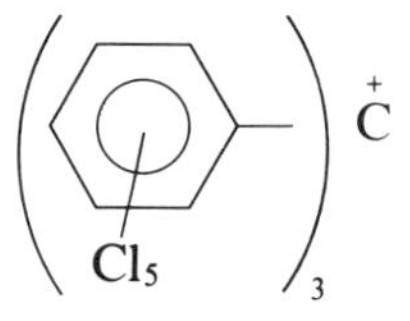

$(C_6Cl_5)_3\overset{+}{C}$

141

and iodomethyl cations have been observed and the general stabilizing effect of halogen attached to carbocation center has been demonstrated.[222] Olah, Halpern, and Mo were able to study these effects in detail using ^{13}C-nmr spectroscopy.[223]

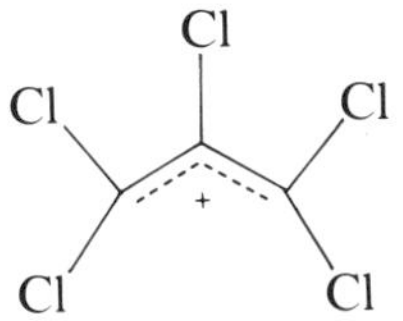

Cl, Cl, Cl, Cl, Cl (allyl cation)

142

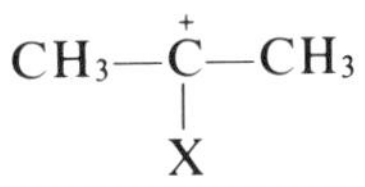

$CH_3\text{—}\overset{+}{C}(X)\text{—}CH_3$

143

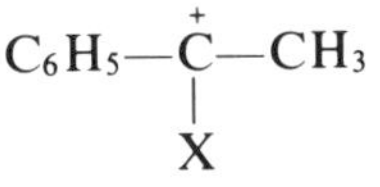

$C_6H_5\text{—}\overset{+}{C}(X)\text{—}CH_3$

144

X = I, Br, Cl, F

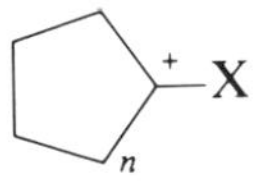

$\overset{+}{C}$—X X = Cl, F

$n = 1, 2, 3$

145

3.4.14.2 *Oxygen as Heteroatom*

The ionic structure of pyrylium salts were clearly stated by Hantzsch as early as 1922.[224] In pyrylium salts **146,** there is a contribution from carbocation structures, a fact apparent in the behavior toward strong nucleophiles leading to phenols.

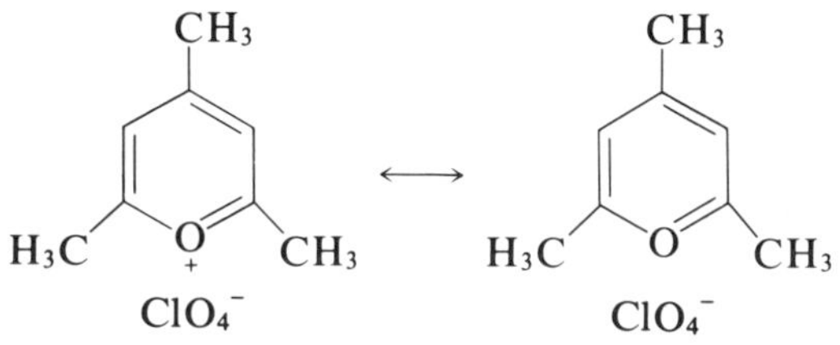

CH_3, H_3C, $\overset{+}{O}$, CH_3, ClO_4^- ⟷ CH_3, H_3C, O, CH_3, ClO_4^-

146

Alkoxy and Hydroxylated Cations. Resonance, similar to that in pyryllium salts, was shown[225] to exist between oxonium and carboxonium ion forms in alkylated ketones, esters, and lactones that were obtained via alkylation with trimethyl or triethyloxonium tetrafluoroborates. Taft and Ramsey[226] used ^{1}H-nmr spectroscopy to investigate the nature of a series of secondary and tertiary carboxonium ions.

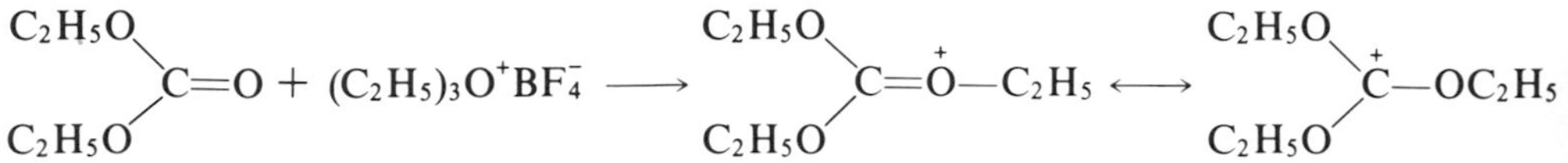

$$(C_2H_5O)_2C{=}O + (C_2H_5)_3O^+BF_4^- \longrightarrow (C_2H_5O)_2C{=}\overset{+}{O}{-}C_2H_5 \longleftrightarrow (C_2H_5O)_2\overset{+}{C}{-}OC_2H_5$$

147

$$H\overset{+}{C}(OCH_3)_2 \qquad \overset{+}{C}H(OC_2H_5)_2 \qquad CH_3{-}\overset{+}{C}(OCH_3)_2$$

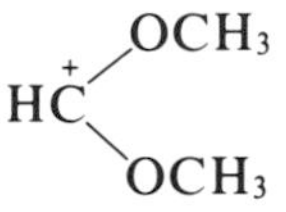

148 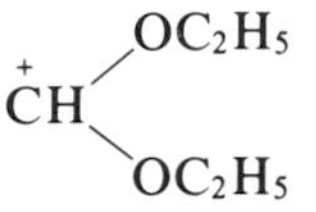 **149** 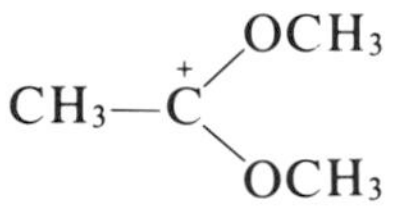 **150**

Olah and Bollinger[227] have obtained primary carboxonium ions such as methoxy and phenoxymethyl cations and their halogenated derivatives.

$$CH_3O\overset{+}{C}H_2 \longleftrightarrow CH_3\overset{+}{O}{=}CH_2 \quad \textbf{151} \qquad CH_3O\overset{+}{C}HCl \longleftrightarrow CH_3\overset{+}{O}{=}CHCl \quad \textbf{152}$$

$$ClCH_2O\overset{+}{C}H_2 \longleftrightarrow ClCH_2\overset{+}{O}{=}CH_2 \quad \textbf{153} \qquad CH_2{-}OCHF \longleftrightarrow CH_3\overset{+}{O}{=}CHF \quad \textbf{154}$$

$$C_6H_5O\overset{+}{C}H_2 \longleftrightarrow C_6H_5\overset{+}{O}{=}CH_2 \quad \textbf{155}$$

Olah et al. have also carried out a ^{1}H and ^{13}C-nmr spectroscopic investigation of a series of haloalkyl carboxonium ions.[228]

Aldehydes and ketones protonate on the carbonyl oxygen atom in superacid media at low temperatures, and the corresponding carboxonium ions can be directly observed.[229–233]

$$RCHO \xrightarrow{HSO_3F:SbF_5:SO_2} RCH{=}\overset{+}{O}H$$

$$(R)_2CO \xrightarrow{HSO_3F:SbF_5:SO_2} (R)_2C{=}\overset{+}{O}H$$

Even protonated formaldehyde has been observed. Protonated acetaldehyde shows two isomeric forms, the proton on oxygen being *syn*- or *anti*- to the methine proton:

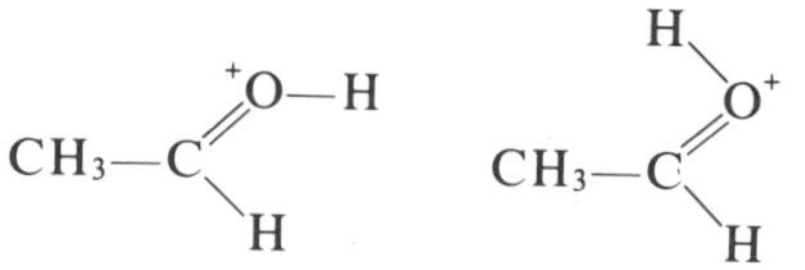

$$CH_3{-}C(H){=}\overset{+}{O}{-}H \qquad CH_3{-}C(H){=}O^+{-}H$$

80% *syn* 20% *anti*

156

The hydroxymethyl cation forms of protonated ketones and aldehydes contribute to the resonance hybrid. Based on ^{13}C-nmr studies,[233,234a] the degree of contribution of the hydroxymethyl cation forms can be quite accurately estimated. Similar studies have been carried out using ^{17}O-nmr spectroscopy.[235a] Recently, Olah and co-workers have measured one-bond ^{13}C—^{13}C coupling constants in a series of protonated benzaldehydes and acetophenones.[235b]

$$(R)_2C{=}\overset{+}{O}H \longleftrightarrow (R)_2\overset{+}{C}{-}OH$$

$$RCH{=}\overset{+}{O}H \longleftrightarrow R\overset{+}{C}H{-}OH$$

Alkylated carboxonium ions have also been prepared by direct electrophilic oxygenations of alkanes, alcohols, etc., by ozone or hydrogen peroxide in superacidic media.[236]

$$(CH_3)_3CH \xrightarrow[HSO_3F:SbF_5]{H_2O_2 \text{ or } O_3} [(CH_3)_3C\overset{+}{O}] \longrightarrow (CH_3)_2C{=}\overset{+}{O}CH_3$$

Smit et al. have isolated a series of cyclic carboxonium salts **157–159** by acylation of cycloalkenes.[237] ^{13}C-nmr spectral investigations have been extended[234b] to the study of heteroaromatic stabilized 6π, 3-dioxolium and 10π benzo-1,3-dioxolium

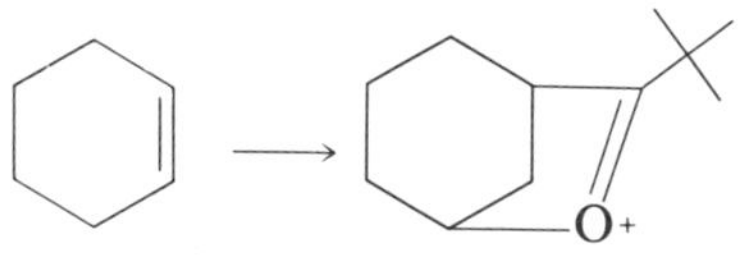

O+

157

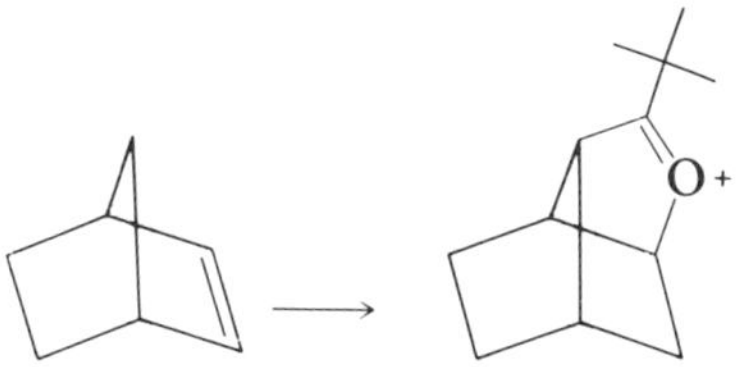

O+

158

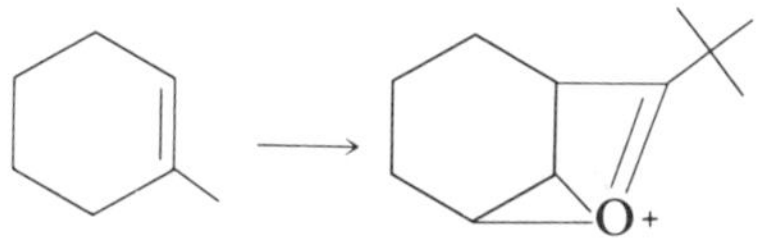

O+

159

ions **160** and **161.**

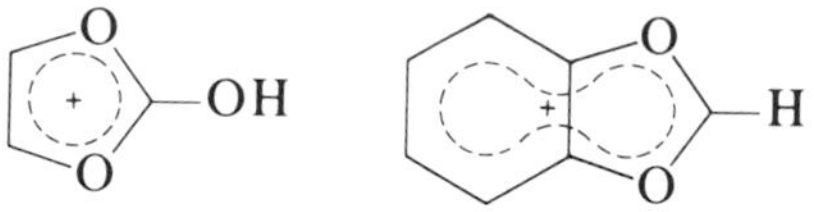

160 **161**

Carboxylic acids are protonated in superacid media, such as $HSO_3F:SbF_5:SO_2$, $HF:SbF_5$, or $HF:BF_3$.[238] The nmr spectrum of acetic acid in such media at low temperature shows two OH resonances indicating that carbonyl protonation is favored and that hindered rotation about the resultant COH bond occurs. The predominant conformer observed in the *syn, anti,* although about 5% of the *syn, syn* isomer has also been seen. These isomers can be readily identified from the mag-

$$CH_3{-}CO_2H \xrightarrow{H^+} CH_3{-}C(OH)_2^+ \; + \; CH_3{-}C(OH)_2^+$$

syn, anti 95% *syn, syn* 5%

162

nitudes of the proton vicinal coupling constants. No evidence for the *anti, anti* isomer has been found in either protonated carboxylic acids, esters, or their analogs.

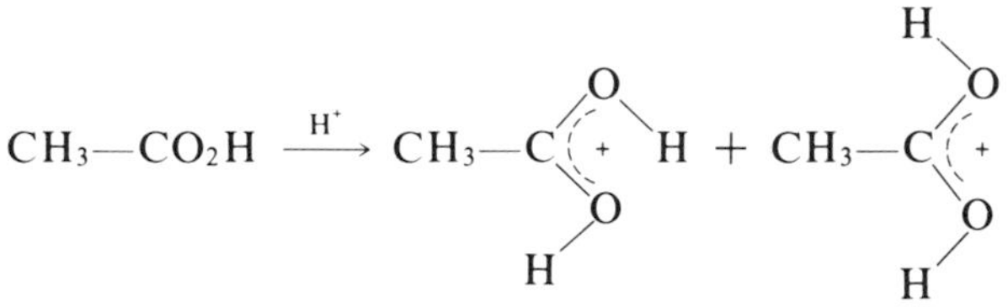

Esters behave in an analogous fashion, with carbonyl protonation being predominant. Thus, protonated methyl formate **163** is present in $HSO_3F:SbF_5:SO_2$ solution as two isomers in a ratio of 90% to 10%.[239]

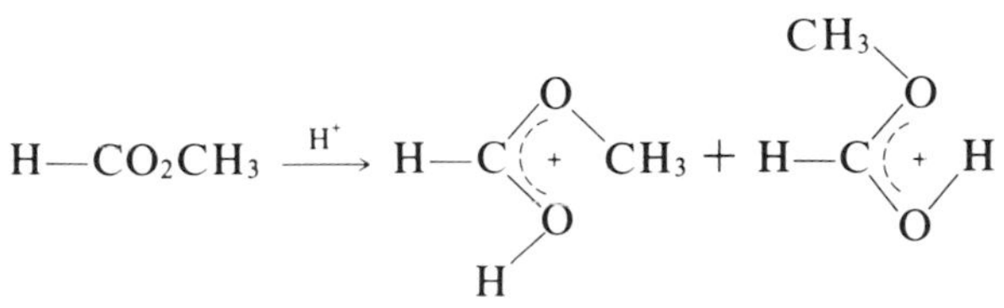

$$H{-}CO_2CH_3 \xrightarrow{H^+} H{-}C(OH)(OCH_3)^+ \; + \; H{-}C(OH)(OCH_3)^+$$

90% 10%

163

By raising the temperature of solutions of protonated carboxylic acids and esters, unimolecular cleavage reactions are observed. These reactions can be considered within the framework of the two unimolecular reaction pathways for acid-catalyzed hydrolyses of esters, either involving alkyl or alkyl-oxygen cleavage. The advantage of studies of these reactions in superacid media, as compared with solvolytic con-

ditions, is that the cleavage step can be isolated and studied in detail because the cleavage products generally do not undergo any further reaction.

For example, in the case of protonated carboxylic acid in $HSO_3F:SbF_5:SO_2$ solution, a reaction analogous to the rate-determining step in the unimolecular cleavage of esters is observed leading to the acyl cation and oxonium (hydronium)ion.

$$RCO\overset{+}{O}H_2 \xrightarrow[H^+]{-20°C} RC\overset{+}{O} + H_3\overset{+}{O}$$

Unimolecular cleavage in this case corresponds to dehydration of the acid, but, in the case of protonated esters, the cleavage pathway depends on the nature of the alkoxy group. Dialkyl carbonates have been studied in $HSO_3F:SbF_5$ solution and

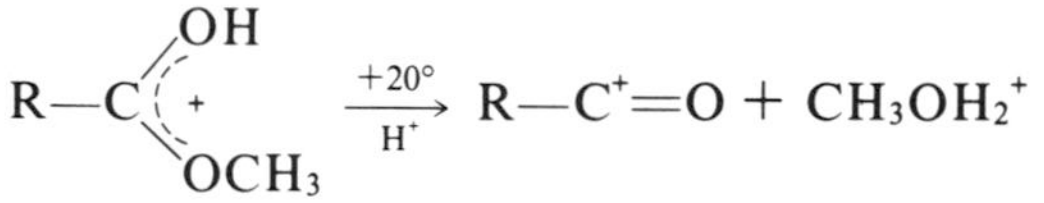

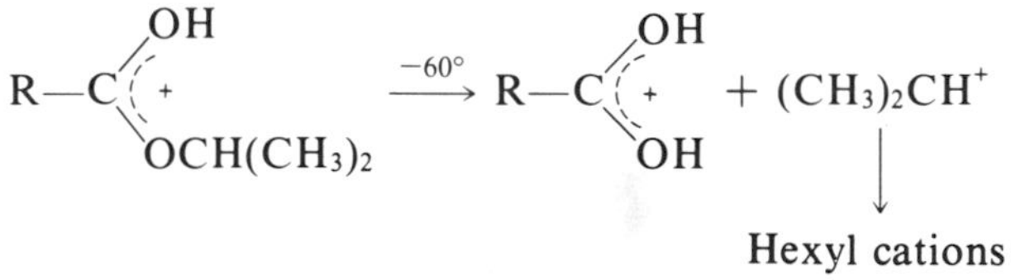

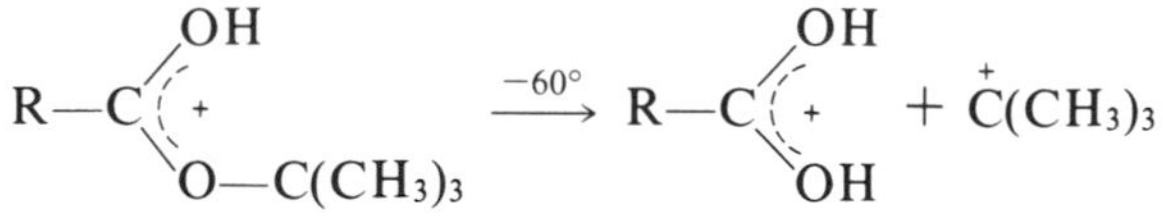

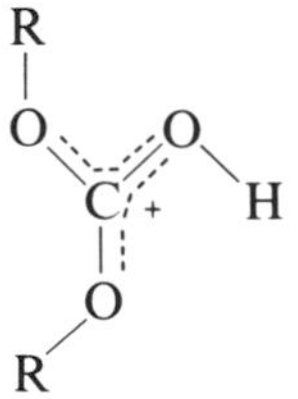

$$R-C^{+}(OH)(OCH_3) \xrightarrow[H^+]{+20°} R-C^{+}{=}O + CH_3OH_2^{+}$$

$$R-C^{+}(OH)(OCH(CH_3)_2) \xrightarrow{-60°} R-C^{+}(OH)(OH) + (CH_3)_2CH^{+} \rightarrow \text{Hexyl cations}$$

$$R-C^{+}(OH)(O-C(CH_3)_3) \xrightarrow{-60°} R-C^{+}(OH)(OH) + \overset{+}{C}(CH_3)_3$$

$$(RO)(RO)C^{+}(OH)$$

164

$$((CH_3)_3CO)_2\overset{+}{C}-OH \xrightarrow{H^+} (HO)_2\overset{+}{C}-OH + 2(CH_3)_3\overset{+}{C}$$

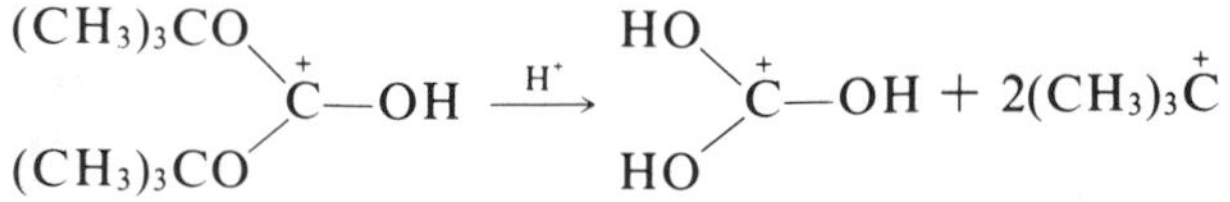

165

have been shown to be protonated on the carbonyl group giving the dialkoxyhydroxy methyl cation.[240] Di-*t*-butyl carbonate cleaves immediately at $-80°$ with alkyl-oxygen fission, giving the *t*-butyl cation **1** and protonated carbonic acid **165.** The structure of the latter has been established from the ^{13}C-nmr spectrum of the central carbon atom which shows a quartet, 4.5-Hz long-range C—H coupling constant, with the three equivalent hydroxyl protons.[240] Diisopropyl and diethyl carbonate cleave at a higher temperature, also via alkyl-oxygen cleavage, with initial formation of protonated alkyl hydrogen carbonates. The alkyl hydrogen carbonates are also formed by protonation of their metal salts. Protonated carbonic acid **165** can also

$$(RO)_2\overset{+}{C}-OH \xrightarrow[\text{heat}, -R^+]{H^+} (HO)(RO)\overset{+}{C}-OH \xrightarrow[\text{heat}, -R^+]{H^+} (HO)_2\overset{+}{C}-OH \longleftrightarrow (HO)_2C=\overset{+}{O}H$$

165

be obtained by dissolving inorganic carbonates and hydrogen carbonates in $HSO_3F:SbF_5$ at $-80°C$. It is stable in solution to about 0°C, where it decomposes to the hydronium ion and carbon dioxide.

$$CO_3^{2-} \text{ or } HCO_3^- \xrightarrow{HSO_3F-SbF_5} \underset{\textbf{165}}{H_3CO_3^+} \xrightarrow[\text{heat}]{} CO_2 + H_3O^+$$

It is worthwhile to point out the close similarity of protonated carbonic acid (trihydroxymethyl cation) with the guanidinium ion, its triaza-analog. Both are highly resonance stabilized through their onium forms. The observation of protonated carbonic acid as a stable chemical entity with substantial resonance stabilization also may have implications in understanding of some of the more fundamental biological carboxylation processes. Obviously, the in vitro observation in specific, highly acidic solvent systems cannot simply be extrapolated to different environments (biological systems). However, it is possible that on the active receptor sites of enzyme systems (for example, those of the carbonic anhydrase type) local hydrogen ion concentration may be very high, as compared with the overall "biological pH." In addition, on the receptor sites a very favorable geometric configuration may help to stabilize the active species, a factor that cannot be reproduced in model systems in vitro.

Acylium Ions (Acyl Cations). Seel[5b] observed in 1943 the first stable acyl cation. Acetyl fluoride with boron trifluoride gave a complex (decomposition point

20°C) which was characterized as the acetyl tetrafluoroborate salt

$$CH_3COF + BF_3 \longrightarrow CH_3\overset{+}{C}{=}O\ BF_4^-$$

166

The identification was based on analytical data and chemical behavior. Only in the 1950s were physical methods like infrared and nmr spectroscopy applied, making further characterizations of the complex possible. Since 1954, a series of other acyl and substituted acyl cations have been isolated and identified.[241–243] The hexafluoroantimonate and hexafluoroarsenate complexes were found to be particularly stable.[243] Deno and his co-workers investigated solutions of carboxylic acids in sulfuric acid and oleum.[244] They observed protonation at lower acid concentrations and dehydration, giving acyl cations, at higher acidities.

$$RCOOH \xrightarrow{H_2SO_4} RCOOH_2 \xrightarrow{\text{oleum}} R\overset{+}{CO} + H_3\overset{+}{O}$$

The investigation of acyl cations in subsequent work was substantially helped by nmr. Not only ^1H, but also ^2H, ^{13}C, ^{19}F, and ^{17}O resonance studies established the structure of these ions.[235a,243–246] These investigations based on ^{13}C and proton resonance, showed that acyl cations, such as the CH_3CO^+ ion, are not simple oxonium ions (acylonium complexes), but are resonance hybrids of the oxonium ion, acyl cation, and the ketene-like nonbonded mesomeric forms.

$$CH_3{-}C{\equiv}O^+ \longleftrightarrow CH_3{-}\overset{+}{C}{=}O \longleftrightarrow \underset{\overset{+}{H}}{CH_2}{=}C{=}O$$

166

The X-ray crystallographic study of the $CH_3CO^+SbF_6^-$ salt[247] substantiated this suggestion and provided convincing evidence for the linear structure of the crystalline complex.

Investigation of acyl cations has been extended to be study of cycloacylium ions,[248] diacylium ions (dications),[249] and unsaturated acylium ions.[250]

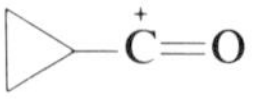

cyclopropyl–$\overset{+}{C}{=}O$ **167**; cyclobutyl–$\overset{+}{C}{=}O$ **168**; cyclopentyl–$\overset{+}{C}{=}O$ **169**; cyclohexyl–$\overset{+}{C}{=}O$ **170**; 1-adamantyl–$\overset{+}{C}{=}O$ **171**

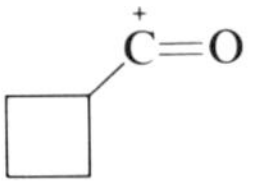

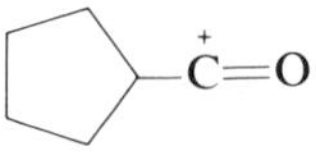

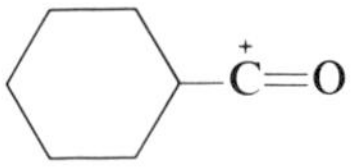

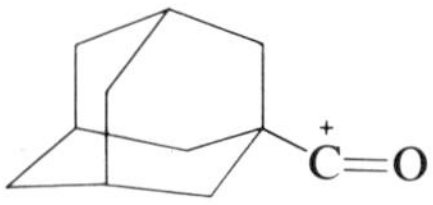

$O{=}\overset{+}{C}(CH_2)_3\overset{+}{C}{=}O$ **172**; $O{=}\overset{+}{C}(CH_2)_4\overset{+}{C}{=}O$ **173**; $O{=}\overset{+}{C}{-}C_6H_4{-}\overset{+}{C}{=}O$ **174**

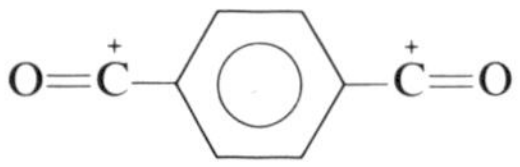

$CH_2{=}C(CH_3){-}\overset{+}{C}{=}O$ **175**; $CH_2{=}CH\overset{+}{C}{=}O$ **176**; $CH_3CH{=}C(CH_3)\overset{+}{C}{=}O$ **177**

3.4.14.3 Sulfur As Heteroatom

Thiols and sulfides are protonated on sulfur in superacid media and give mono- and dialkylsulfonium ions, respectively.[83] Thiocarboxylic acids, *S*-alkyl esters, thioesters, dithioesters, and thiocarbonates in similar media also form stable protonated ions[251] such as the following.

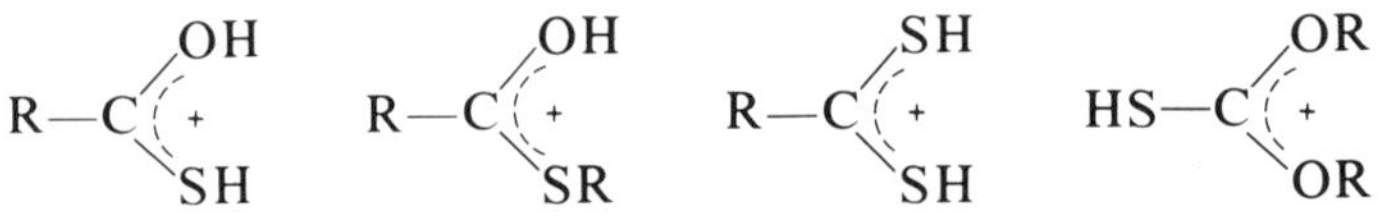

$$R-C(OH)(SH)^+ \quad \textbf{178} \qquad R-C(OH)(SR)^+ \quad \textbf{179} \qquad R-C(SH)(SH)^+ \quad \textbf{180} \qquad HS-C(OR)(OR)^+ \quad \textbf{181}$$

$$R = CH_3, C_2H_5 \qquad HS-C(SR)(SR)^+ \quad \textbf{182} \qquad HS-C(SH)(SH)^+ \quad \textbf{183}$$

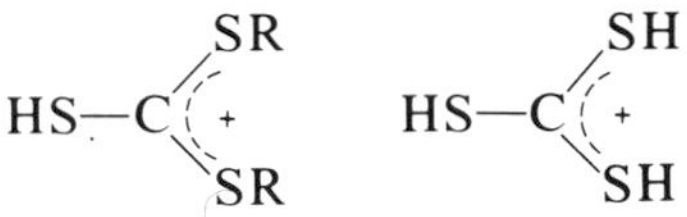

Sulfur-stabilized heteroaromatic species such as **184** and **185** are also known.[234b]

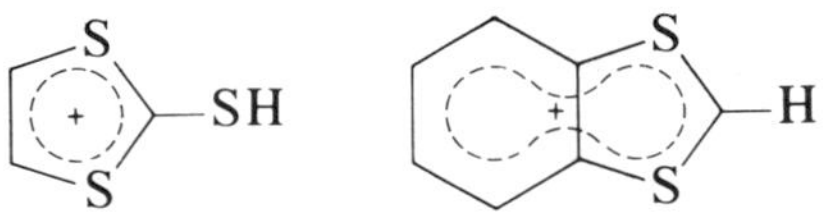

S, S, SH (+) **184** S, S, H (+) **185**

A series of thioketones **186** have also been protonated and studied by ^{13}C-nmr spectroscopy.[252] It appears that there is significant mercaptocarbenium **186b** contribution to the overall protonated thioketone structure. However, this is not the case in the case of protonated ketones.

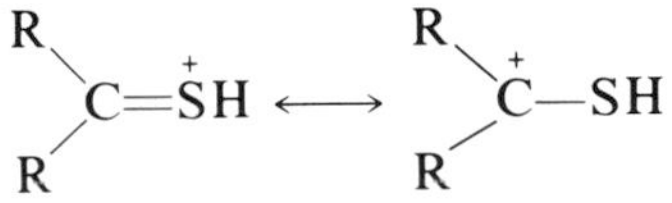

$$R_2C=\overset{+}{S}H \longleftrightarrow R_2\overset{+}{C}-SH$$

a *b*

186

Recently, a series of thiobenzoyl cations **187** have been prepared and studied[253] using a metathetic silver salt reaction.

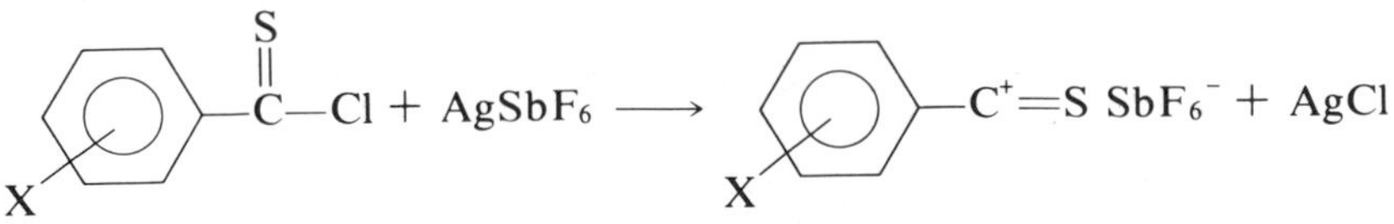

$$X-C_6H_4-C(=S)-Cl + AgSbF_6 \longrightarrow X-C_6H_4-C^+=S\ SbF_6^- + AgCl$$

187

3.4.14.4 Nitrogen as Heteroatom

Amides are protonated on the carbonyl oxygen atom in superacid media at low temperatures, as shown first by Gillespie.[254]

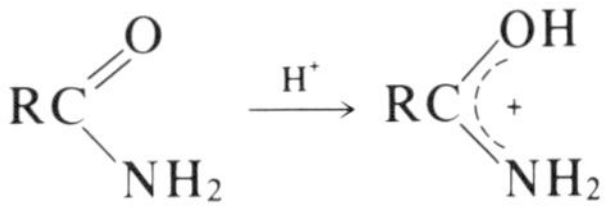

$$\mathrm{RC(=O)NH_2} \xrightarrow{H^+} \mathrm{RC(\overset{+}{\cdots}OH)(\cdots NH_2)}$$

188

It was claimed that protonation of ethyl *N,N*-diisopropylcarbamate, a hindered amide, takes place on nitrogen and not on oxygen.[255] A reinvestigation, however, established that at low temperature initial *O*-protonation takes place (kinetic control) with the *O*-protonated amide subsequently rearranging to the more stable *N*-protonated form (thermodynamic control).[256]

$$R_2C{=}\overset{+}{N}R_2 \longleftrightarrow R_2\overset{+}{C}{-}NR_2$$

189

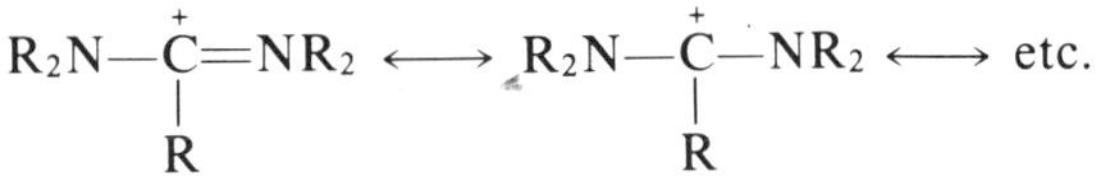

$$R_2N{-}\overset{+}{C}(R){=}NR_2 \longleftrightarrow R_2N{-}\overset{+}{C}(R){-}NR_2 \longleftrightarrow \text{etc.}$$

190

$$(R_2N)_2C{=}\overset{+}{N}R_2 \longleftrightarrow (R_2N)_2\overset{+}{C}{-}NR_2 \longleftrightarrow \text{etc.}$$

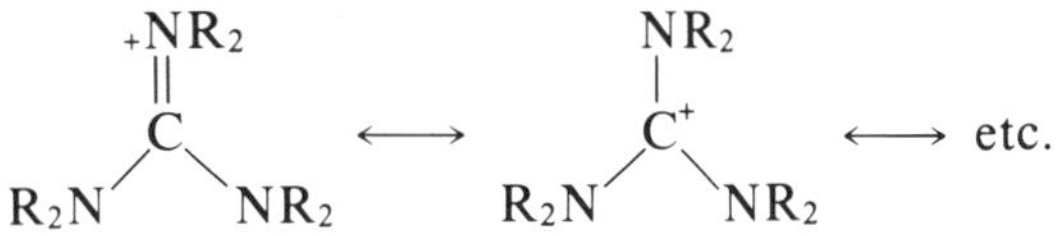

191

The possibility of observing the protonated amide linkage in strong acid media has particular relevance in the study of peptides and proteins.[257,258]

Since nitrogen is a better electron donor than oxygen, the contribution of aminomethyl cation structures in acid salts of imines, amidines, and guanidines is small.[259] In protonated nitriles, however, the contribution from the iminomethyl cation resonance form becomes important.[260]

$$R{-}C{\equiv}\overset{+}{N}H \longleftrightarrow R{-}\overset{+}{C}{=}NH$$

192

Even diazomethane has been protonated[261] in the superacid media. In Magic Acid media, both methyldiazonium ion **193** as well as *N*-protonated diazomethane **194** are formed.

$$CH_2N_2 + HSO_3F{:}SbF_5 \longrightarrow \underset{\mathbf{193}}{CH_3N_2^+} + \underset{\mathbf{194}}{CH_2{=}N\overset{+}{=}NH}$$

3.4.15 Carbocations Complexed to Metal Atoms

Organometallic cations, in which an organic ligand is coordinated to a metal atom bearing a unit positive charge, constitute a significant class of compounds.[262] Most common of these are the π-allylic, π-dienyl, π-cycloheptatrienyl, and π-cyclooctatetraenyl systems. Generally, metal atoms of low oxidation states are involved in complexing the electron-deficient ligands. The usual metals involved are transition metals such as Fe, Cr, Co, Rh, Ir, Pd, Pt, etc. Some of the representative metal-complexed carbocations are shown below. There are also several carbocations that

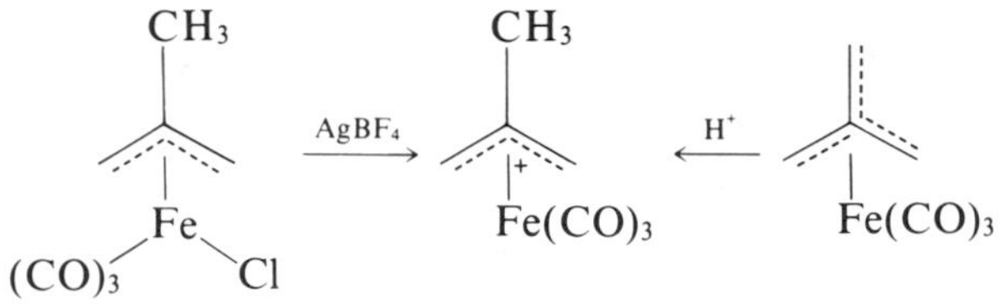

CH_3 CH_3 $AgBF_4$ H^+ Fe $(CO)_3$ Cl $Fe(CO)_3$ $Fe(CO)_3$

195

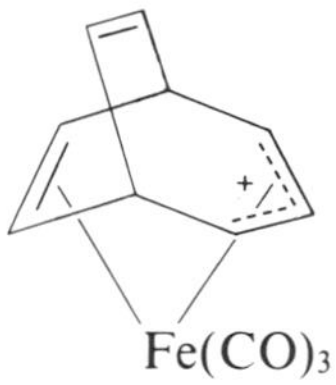

$Fe(CO)_3$ **196** $Fe(CO)_3$ **197**

$Cr(CO)_3$ **198** $Fe(CO)_3$ $Fe(CO)_3$ **199**

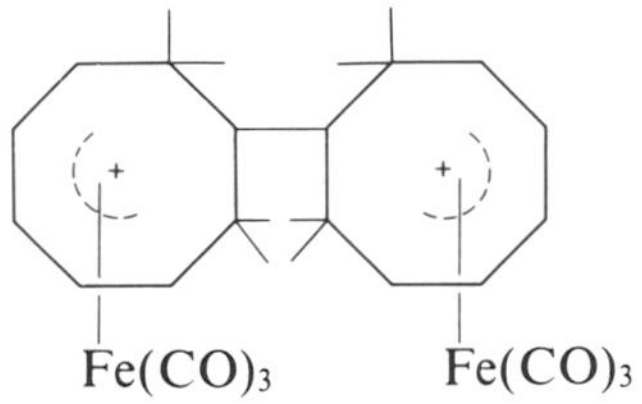

have an α-π-complexed organometallic system. The most notable of these are the α-ferrocenyl carbenium ion **201.**

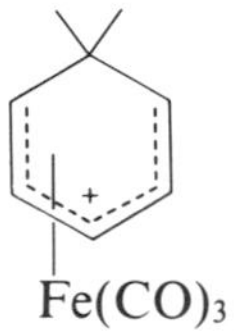

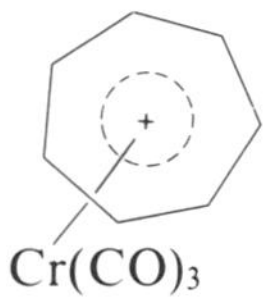

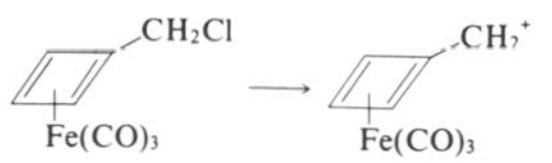

CH_2Cl $Fe(CO)_3$ → CH_2^+ $Fe(CO)_3$ **200**

Fe C H Fe **201**

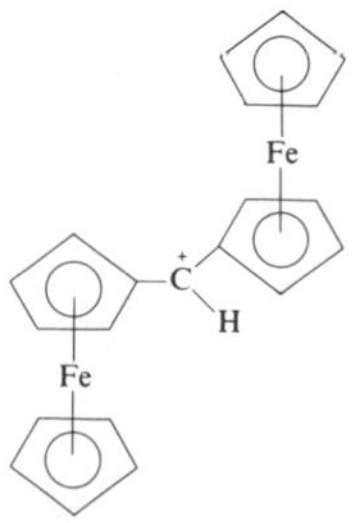

3.5 EQUILIBRATING (DEGENERATE) AND HIGHER (FIVE OR SIX) COORDINATE (NONCLASSICAL) CATIONS

3.5.1 Alkonium Ions (Protonated Alkanes $C_nH_{2n+3})^+$

As recognized in the pioneering works of Meerwein, Ingold, and Whitmore, trivalent alkyl cations $(C_nH_{2n+1})^+$ play important roles in the acid-catalyzed transformations of hydrocarbons as well as various electrophilic and Friedel-Crafts-type reactions. Trivalent alkyl cations can directly be formed only by ionization of lone-pair (non-bonded electron pair) containing precursors (*n*-bases) such as alkyl halides, alcohols, thiols, etc., or by protonation of singlet carbenes or olefins.

Protonated alkanes $(C_nH_{2n+3})^+$ also play a significant role in alkane reactions. Saturated hydrocarbons can be protonated to alkonium ions, of which the methonium ion (CH_5^+) is the parent, and formation of these pentacoordinate carbocations involves two-electron, three-center bonds. The dotted lines in the structure symbolize the bonding orbitals of the three-centered bonds; their point of junctions does not represent an additional atom.

$$CH_4 + H^+ \rightleftharpoons [CH_5]^+ \equiv \left[H_3C\cdots\langle^{H}_{H}\right]^+$$

3.5.1.1 *The Methonium Ion* (CH_5^+)

The existence of the methonium ion, CH_5^+ **202** was first indicated by mass spectrometric studies[263] of methane at relatively high source pressures, i.e., molecular-ion reaction between neutral CH_4 and a proton. Isotope exchange and collisonal associations in the reactions of CH_3^+ and its deuterated analogs with H_2, HD, and D_2 have also been studied by mass spectrometry using a variable-temperature, ion-flow method.[264a] The chemistry of methane and homologous alkanes, e.g., hydrogen-deuterium exchange and varied electrophilic substitutions in superacidic media pointed out the significance of alkonium ions in condensed state chemistry.

Direct spectroscopic observation of CH_5^+ in the condensed state is difficult, as the concentration of the ion even in superacidic media at any time is extremely low. The matrices of superacids, such as $HSO_3F:SbF_5$ or $HF:SbF_5$ saturated with methane, were studied by ESCA[264b] at $-180°C$, and the observed carbon 1*s*, binding energy differing by less than 1 eV from that of methane, is attributed to CH_5^+. Neutral methane has practically no solubility in the superacids at such low temperature of the experiment and at the applied high vacuum (10^{-9} torr) would be pumped out of the system. The relatively low 1*s* carbon binding energy in CH_5^+ is in good accord with theoretical calculations,[265] indicating that charge density is heavily on the hydrogen atoms and the five coordinate carbon carries relatively little charge.

Of the possible structures for the methonium ion (D_{3h}, C_{4v}, C_s, D_{2h}, or C_{3v} symmetry) Olah, Klopman, and Schlosberg suggested[265] preference for the C_s front-side protonated form. Preference for this form was based on consideration of the

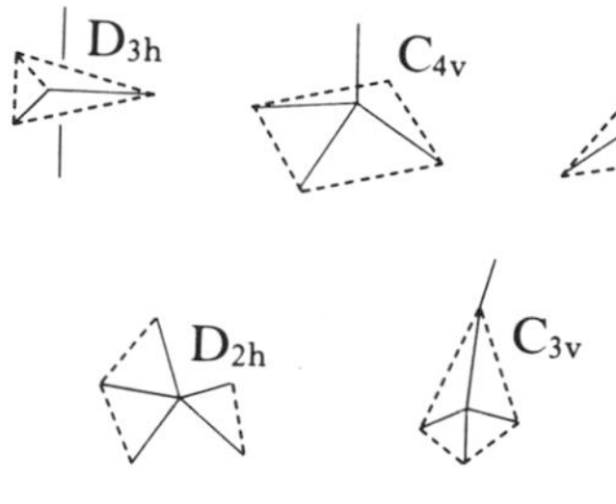

D_{3h} C_{4v} C_s

D_{2h} C_{3v}

observed chemistry of methane in superacids (hydrogen-deuterium exchange and, more significantly, polycondensation indicating ease of cleavage to CH_3^+ and H_2) and also on the basis of self-consistent field (SCF) calculations.[265] More extensive calculations, including *ab initio* methods[266] utilizing an "all geometry" parameter search, confirmed the favored structure with C_s symmetry. This structure is about 2 kcal/mol more stable than the structure with C_{4v} (or D_{2h} or C_{3v}) symmetry, which in turn is about 8 kcal/mol more stable than the structure with trigonal bypyramidal D_{3h} symmetry. Interconversion of stereoisomeric forms of CH_5^+ is obviously possible by a pseudorotation process. Muetterties[266e] suggested that stereoisomerization processes of this type in pentacoordinated compounds could be termed "polytopal rearrangements." However, it is preferable to call intramolecular carbonium ion rearrangements as "bond-to-bond rearrangements" since these are not limited to equivalent bonds in the case of higher homologs of CH_5^+ (see subsequent discussion).

3.5.1.2 The Ethonium Ion ($C_2H_7^+$)

The next higher alkonium ion, the ethonium ion (protonated ethane) $C_2H_7^+$ **203,** is analogous to its parent CH_5^+ ion.[265] Protonation of ethane can take place initially either at a C—H bond or the C—C bond, but the interconversion of the resulting ions is a facile low-energy process. This is consistent with the observed H-D exchange in labeled systems as well as the formation of methane as a by-product in the protolytic cleavage of ethane in superacids.

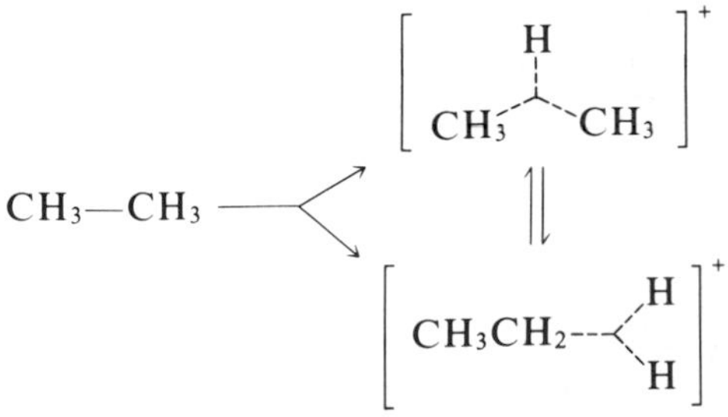

$CH_3—CH_3$ → $[CH_3\cdots H\cdots CH_3]^+$ ⇌ $[CH_3CH_2\cdots(H)(H)]^+$

203

3.5.1.3 Higher Alkonium Ions

The higher homologous alkonium ions ($C_3H_9^+$, $C_4H_{11}^+$, etc.) have been observed in the gas phase by high-pressure mass spectrometry.[263] In solution, the higher hydrocarbons show an increasing tendency to form C⋯H⋯C (three-center bonds) on protonation as evidenced by the increasing tendency to form C—C bond cleavage products.

The acid-induced H-D exchange in isobutane at conventional acidities, e.g., with deuterated sulfuric acid was studied by Otvos[267] and his associates. All nine methyl hydrogens are readily exchanged, but not the methine hydrogen. The mechanistic explanation for this observation must involve formation of the trivalent *t*-butyl cation, probably in an oxidative ionization step. The *t*-butyl cation then undergoes reversible deuteration, the process repeating itself and thus accounting for exchange of the methyl hydrogens with deuterated sulfuric acid. The *t*-butyl cation **1** reforms

$$\underset{\mathbf{1}}{(CH_3)_3C^+} \xrightarrow{-H^+} (CH_3)_2C{=}CH_2 \xrightarrow{+D^+} (CH_3)_2\overset{+}{C}{-}CH_2D, \text{ etc.,}$$

isobutane via hydride abstraction from isobutane involving the tertiary C—H bond only and thus not exchanging the methine hydrogen.

In contrast, Olah et al.[268] have shown that in deuterated superacid media, e.g., $DSO_3F:SbF_5$ or $DF:SbF_5$ at low temperature only the methine hydrogen is exchanged indicating no deprotonation-protonation equilibria. The latter reaction proceeds through a five-coordinate $C_4H_{11}^+$ ion arising by protonation (deuteration) of the tertiary C—H bond.

$$(CH_3)_3CH \xrightarrow{D^+} \left[(CH_3)_3C{\cdots}\!\!<^{H}_{D}\right]^+ \xrightarrow{-H^+} (CH_3)_3CD$$

Evidence for a C—C protonated $C_4H_{11}^+$ ion was obtained by Siskin.[269] When studying the $HF:TaF_5$-catalyzed ethylation of a large excess of ethane with ethylene in a flow system, *n*-butane was obtained as the only four-carbon product, free from isobutane. This remarkable result can only be explained by C—H bond ethylation of ethane, through the five-coordinate carbocation intermediate which subsequently by proton elimination gives *n*-butane. Use of a flow system that limits the contact

$$CH_3CH_3 + C_2H_5^+ \rightleftharpoons \left[CH_3CH_2{\cdots}\overset{H}{\vdots}{\cdots}CH_2CH_3\right]^+ \xrightarrow{-H^+} CH_3CH_2CH_2CH_3$$

of the product *n*-butane with the acid catalyst is essential, because on more prolonged contact, isomerization of *n*-butane to isobutane occurs.

Alternatively, if the reaction involved trivalent butyl cation (from ethylation of ethylene) the ion would inevitably rearrange via 1,2-hydrogen shift to *s*-butyl cation **18,** which in turn would isomerize to the *t*-butyl cation **1** and thus give isobutane.

$$CH_2{=}CH_2 + CH_3CH_2^+ \longrightarrow [CH_3CH_2CH_2CH_2^+] \xrightarrow{1,2\ H} CH_3\overset{+}{C}HCH_2CH_3 \ (\mathbf{18})$$

$$\longrightarrow (CH_3)_3C^+ \ (\mathbf{1}) \longrightarrow (CH_3)_3CH$$

3.5.2 Equilibrating and Bridged Carbocations

Some carbocations, because of their flat potential energy surfaces, show great tendency to undergo fast degenerate rearrangements, through intramolecular hydrogen or alkyl shifts leading to the corresponding identical structures.[8–10,270] The question arises whether these processes are true equilibria between the limiting trivalent carbocations (''classical ion intermediates'') separated by low-energy level transition states or whether they are hydrogen- or alkyl-bridged higher-coordinate (nonclassical) carbocations. Extensive discussion of the kinetic and stereochemical results in these systems has been made, and it is not considered to be within the scope of this chapter to recapitulate the arguments. The reader is referred to reviews[8–10,270] and the original literature.

3.5.2.1 Degenerate 1,2-Shifts in Carbocations

Many acyclic and monocyclic tertiary and secondary cations, undergoing degenerate 1,2-hydrogen (and alkyl) shift give average proton and carbon absorptions in their nmr spectra, even at low temperatures. If the exchange rate process is rapid on the nmr time scale, single sharp resonances will appear for the exchanging nuclei at frequencies that are weighted averages of the frequencies being exchanged. It has been possible to freeze exchange processes in some equilibrating ions by observing ^{13}C-nmr spectra at low temperature (ca. $-160°C$) at high magnetic field strength (to enhance signal separation). Typically, the barriers for such migrations range from 2.4 to 10 kcal · mol^{-1}. Another important technique that has been immensely useful is the low-temperature solid-state ^{13}C-nmr spectroscopy.

The *s*-butyl cation **18** has been prepared from 2-chlorobutane in SbF_5:SO_2ClF at $-100°C$ on a vacuum line by Saunders and Hagen[62,271] with very little contamination from the *t*-butyl cation. Even at $-110°C$, only two peaks from the 2,3 and 1,4 protons are observed in the ^1H-nmr spectrum of **18** (δ 6.7 and δ 3.2). This is consistent with an *s*-butyl cation averaged by very rapid 1,2-hydride shifts ($\Delta G^{\ddagger} \sim 6$ kcal · mol^{-1}). Warming the sample from $-110°C$ to $-40°C$ first causes line broadening and then coalescence of the two peaks, revealing a rearrangement process making all protons equal on the ^{1}H-nmr time scale (indicating the formation of **1**). Line-shape analysis gave an activation barrier of 7.5 ± kcal · mol^{-1} for the process. This low barrier is not compatible with a mechanism involving primary cations as suggested for the corresponding rearrangement of the isopropyl cation. It appears necessary to invoke protonated methylcyclopropanes **21** as intermediates. The barrier for the irreversible rearrangement to **1** was measured to be ca. 18 kcal · mol^{-1}, indicating that this rearrangement involves primary cationic structures as intermediates.

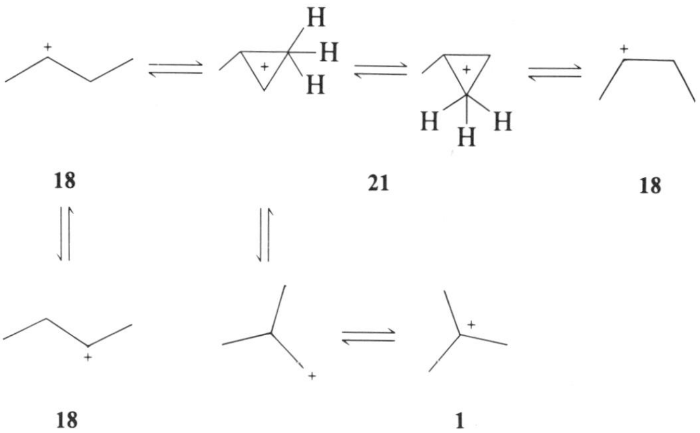

18 **21** **18**

18 **1**

Olah and White[60] obtained an early ^{13}C-nmr INDOR spectrum of **18** that showed a single peak from the two central carbon atoms in reasonable agreement with values calculated from model equilibrating ions. It was therefore concluded that **18** is a classical equilibrating ion rather than being bridged as in **204.**

204

In a comprehensive ^{13}C-nmr spectroscopic study of alkyl cations, Olah and Donovan[61] applied the constancy of ^{13}C methyl substituent effects to the study of equilibrating cations and their rearrangements. They calculated the chemical shifts of the 2-butyl cation **18** from both isopropyl cation and *t*-amyl cation using methyl group substituent effects and reached practically the same result in both cases. The observed chemical shifts deviate from the calculated ones by 9.2 and 19.8 ppm for the equilibrating methyl and carbocation carbons, respectively. Therefore, a hydrogen-bridged intermediate **204** was suggested to be involved. A static hydrogen-bridged *s*-butyl cation was excluded by the observation of two quartets in the fully

^{1}H-coupled ^{13}C-nmr spectrum. Comparison with bridged halonium ions indicates that equilibrating hydrogen-bridged ions have more shielded carbons [C(2), C(3)] than are observed experimentally for the 2-butyl cation. Therefore, it was suggested that the open-chain 2-butyl cation is of similar thermodynamic stability to the hydrogen-bridged **204** and that these intermediates in equilibrium may contribute to the observed average ^{13}C shifts. However, the percentage of different structures could not be calculated owing to lack of accurate models to estimate ^{13}C chemical shifts of hydrogen-bridged structures.

In a study of rates of degenerate 1,2-shifts in tertiary carbocations, Saunders and Kates[272] used high-field (67.9 MHz) ^{13}C-nmr line broadening in the fast-exchange limit. The *s*-butyl cation showed no broadening at $-140°C$. Assuming the hypothetical "frozen out" chemical shift difference between C(2) and C(3) to be 277 ppm, an upper limit for $\Delta G^{\ddagger}$ was calculated to be 2.4 kcal · mol^{-1}.

Application of the isotopic perturbation technique by Saunders et al.[34] to the *s*-butyl cation **18** showed it to be a mixture of equilibrating open-chain ions since a large splitting of the ^{13}C resonance [C(2), C(3)] is obtained upon deuterium substitution.[273]

The cross-polarization, magic-angle spinning method has been applied by Yannoni and Myhre[28] to **18** in the solid state at very low temperatures using ^{13}C-nmr spectroscopy. In the initial study, no convincing evidence for a frozen out *s*-butyl cation was obtained even at $-190°C$. However, recently they[275] have managed to freeze out the equilibration of *s*-butyl ion **18** at $-223°C$. It behaves like a normal secondary trivalent carbocation.

As mentioned earlier (Section 3.4.2), cyclopentyl cation **27** shows a single peak in the ^{1}H-nmr spectrum of δ 4.75 even at $-150°C$.[86] In the ^{13}C-nmr spectrum,[274] a ten-line multiplet centered around δ ^{13}C 99.0 with J_{C-H} = 28.5 Hz was observed. This is in excellent agreement with values calculated for simple alkyl cations and cyclopentane and supports the complete hydrogen equilibration by rapid 1,2-shifts.

etc.,

27

More recently, Yannoni and Myhre[275] have succeeded in freezing out the degenerate hydride shift in **27** in the solid state at $-203°C$. The obtained ^{13}C chemical shifts at δ 320.0, 71.0, and 28.0 indicate the regular trivalent nature of the ion and are in good agreement with the estimated shifts in solution based on the average shift data.

The 2,2,3-trimethyl 2-butyl cation **205** (triplyl cation) consists of a single proton signal at δ 2.90 for all the methyl groups.[81] This indicates that all five methyl groups undergo rapid interchange through 1,2-methyl shifts. The chemical shift of the singlet is similar to that of 2,3-dimethyl-2-butyl cation **206,** another equilibrating ion that undergoes rapid 1,2-hydride shifts.[81]

205

206

The ^{13}C-nmr spectroscopic data of the average cationic center in **205** and **206** were found to be at δ ^{13}C 205 and 197 ($J_{C-H_{av}} \approx 65$ Hz), respectively, indicating their regular trivalent carbenium nature. From studies of methyl substituent effects, Olah and Donovan[61] reached the same conclusions, which are also supported by laser Raman and ESCA studies.[74,276] Saunders and Vogel[35] have introduced deuterium into a methyl group of **205** and thereby perturbed the statistical distribution of the otherwise degenerate methyl groups and split the singlet into a doublet. The CD_3 group prefers to be attached to the tertiary carbon **207.**

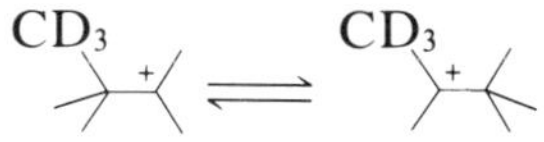

CD_3 CD_3

207 **208**

Saunders and Kates, have been successful[272] in measuring the rates of degenerate 1,2-hydride and 1,2-methide shifts of simple tertiary alkyl cations employing high-field (67.9 MHz) ^{13}C-nmr spectroscopy. From line broadening in the fast-exchange limit, the free energies of activation ($\Delta G^{\ddagger}$) were determined to be 3.5 ± 0.1 kcal · mol^{-1} at −136°C for **205** and 3.1 ± 0.1 kcal · mol^{-1} at −138°C for **206.** The rapid equilibrium process in cations **205** and **206** have been frozen out in the solid state at −165°C and −160°, respectively, by Yannoni and Myhre.[275]

Many more cyclic and polycyclic equilibrating carbocations have been reported.[277–280] Some representative examples are the following:

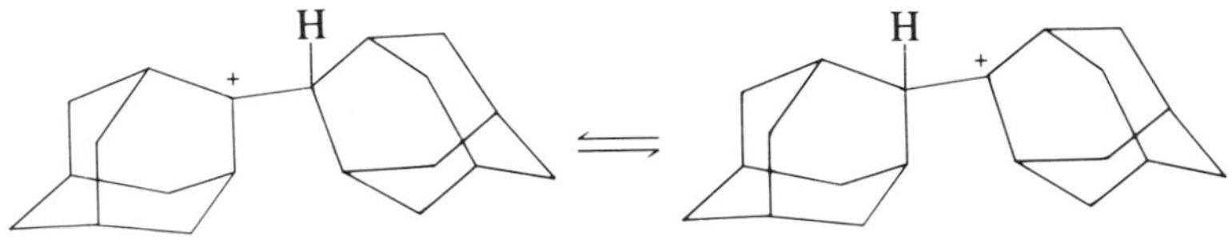

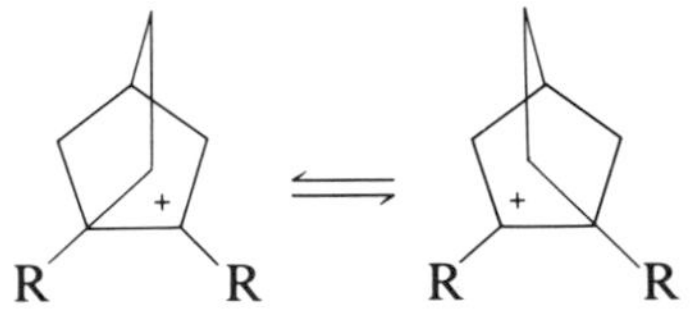

R = CH_3, C_6H_5 etc.,

209[277]

210[278]

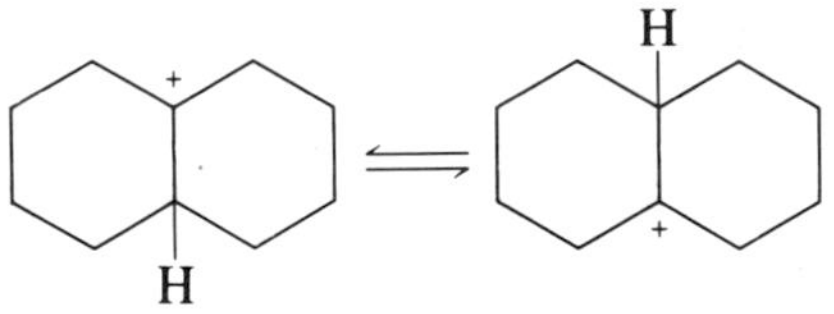

211[279]

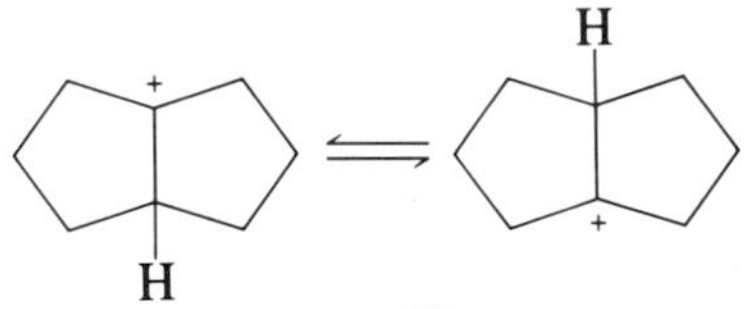

212[279]

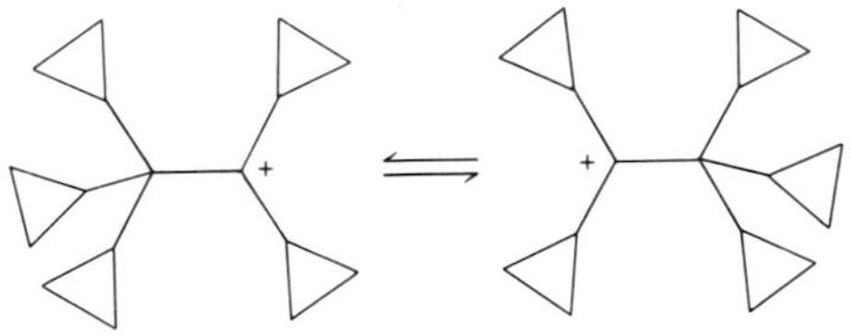

213[280]

3.5.2.2 2-Norbornyl Cation

The 2-norbornyl cation holds a unique position in the history of organic chemistry because of the important role it has played in the bonding theory of carbon compounds. Since Winstein's early solvolytic work,[281] the norbornyl cation was at the heart of the so-called nonclassical ion problem, and no other system has been studied so much by various chemical and physical methods and by so many investigators. The controversy[18,20,282–286] about this ion is well known and the question has been whether it has a symmetrically bridged nonclassical structure **214** with a pentacoordinate carbon atom or if it is a rapidly equilibrating pair of classical trivalent ions **215.**

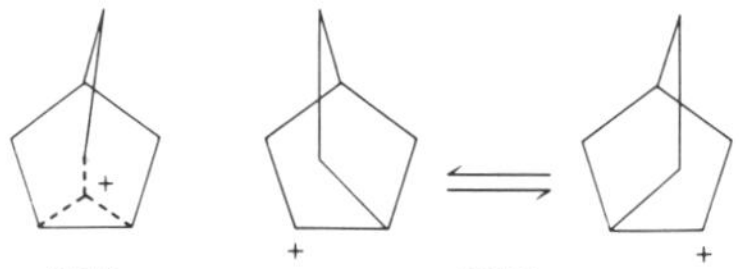

214 **215**

This controversy has been instrumental in the development of important structural methods in physical organic chemistry with respect to critical evaluation of results as well as to concepts behind the methods.

The methods that were developed in the early 1960s to generate and observe stable carbocations in low-nucleophilicity solutions[9] were successfully applied to direct observation of the norbornyl cation. Preparation of the ion by the "σ route" from 2-norbornyl halides, by the "π route" from β-Δ^3-cyclopentenylethyl halides, and by the protonation of nortricyclene ("bent σ-route") all led to the same 2-norbornyl cation.

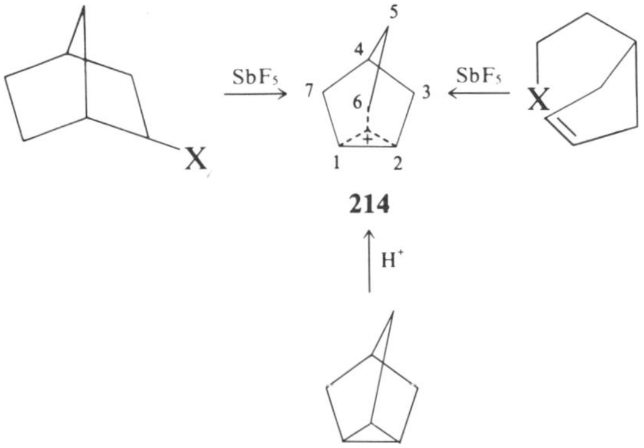

SbF_5 SbF_5 X X

214

H^+

The method of choice for the preparation of the norbornyl cation (giving the best-resolved nmr spectra, free of dinorbornylhalonium ion equilibration) is from *exo*-2-fluoronorbornane in $SbF_5:SO_2$ (or SO_2ClF) solution.

In a joint effort, Olah, Saunders and Schleyer[287] first investigated the 60-MHz ^{1}H-nmr spectrum of the 2-norbornyl cation in the early 1960s. Subsequently, Olah et al.[19] carried out detailed 100-MHz ^{1}H- and 25-MHz ^{13}C-nmr spectroscopic studies in the early 1970s at successively lower temperatures. From the detailed ^{1}H-nmr investigation at various temperatures (RT to -154°C), the barrier for the 2,3-hydrogen shift, as well as the 6,1,2-hydrogen shift was determined by line-shape analysis and found to be 10.8 ± 0.6 kcal · mol^{-1} and 5.9 ± 0.2 kcal · mol^{-1}, respectively (Scheme **V**).[19]

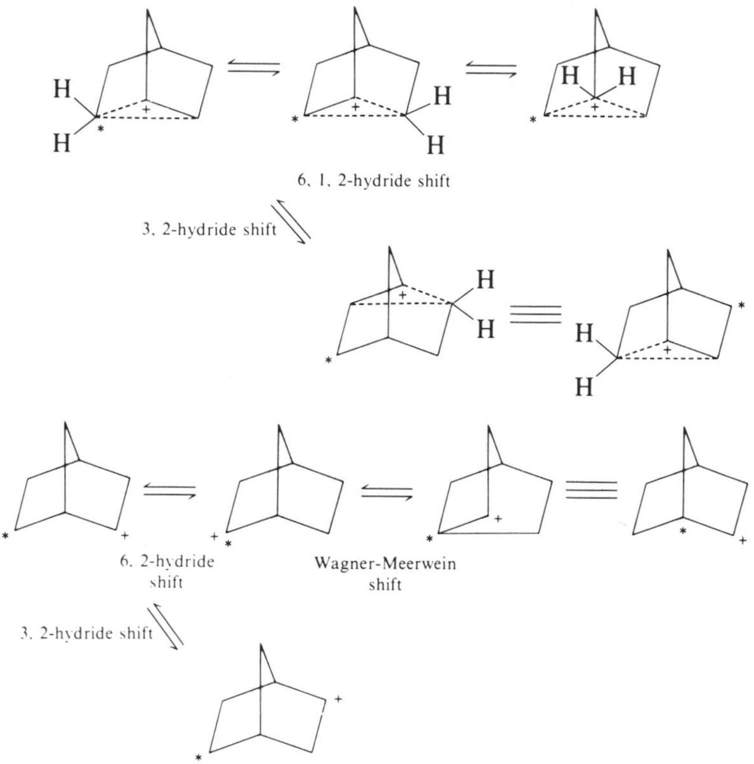

6, 1, 2-hydride shift

3, 2-hydride shift

6, 2-hydride shift

Wagner-Meerwein shift

3, 2-hydride shift

DEGENERATE SHIFTS IN 2-NORBONYL CATION (SCHEME V)

The 60-MHz ^{1}H-nmr spectrum of the 2-norbornyl cation at room temperature shows a single peak at δ 3.10 for all protons indicating fast 2,3-hydrogen, 6,1,2-hydrogen, and Wagner-Meerwein shifts.[287] Cooling the solution of the 2-norbornyl cation in the SbF_5:SO_2ClF:SO_2F_2 solvent system down to −100°C at 395 MHz (Fig. 3.14) results in three peaks at δ 4.92 (4 protons), 2.82 (1 proton), and 1.93 (6 protons) indicating that the 2,3-hydrogen shift is fully frozen, whereas the 6,1,2-hydrogen and Wagner-Meerwein shifts are still fast on the nmr time scale.[288] Cooling the solution down further to −158°C results in significant changes in the spectrum. The peak at δ 4.92 splits into two peaks at δ 6.75 and 3.17 with a ratio of 2:2. The high-field peak broadens and splits into two peaks at δ 2.13 and 1.37 and in the ratio of 4:2. The peak at δ 2.82 remains unchanged. The line width (~60 Hz) observed at 395 MHz was found to be rather small[288] as compared with the one obtained at 100 MHz (~30 Hz).[19] This has some important implications. If the line width were due to any slow exchange process occurring at this temperature, the line width should have broadened 15.6 times at 395 MHz over the one observed at 100 MHz. The observation of comparably narrow line widths at 395 MHz indicates that either the 6,1,2-hydrogen shift and the Wagner-Meerwein shift (σ-bond shift) are completely frozen and the 2-norbornyl cation has the symmetrically

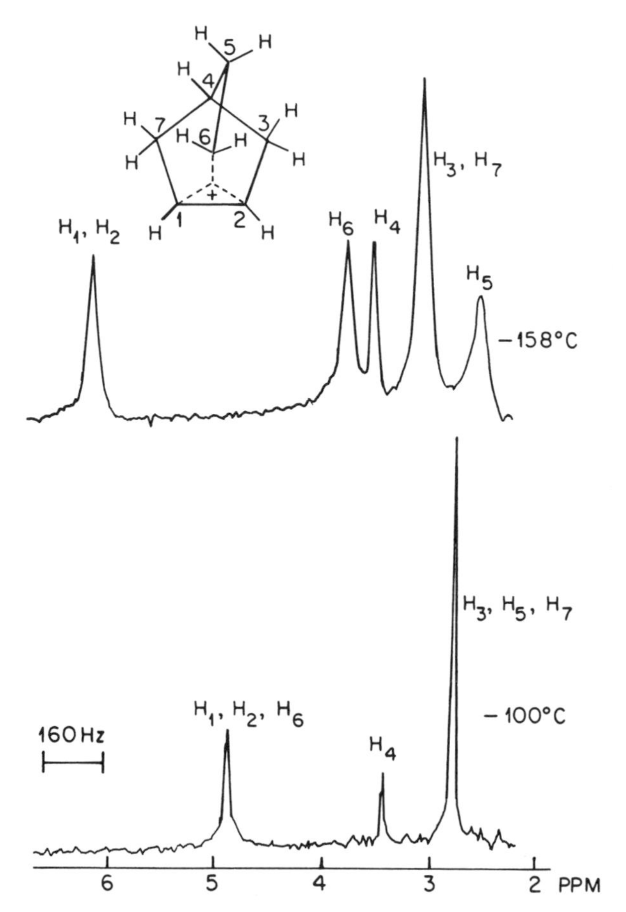

FIGURE 3.14 395 MHz ^{1}H NMR Spectra of 2-Norbornyl cation in $SbF_5/SO_2ClF/SO_2F_2$ solution.

bridged structure **214** or the 6,1,2-hydrogen shift is frozen and the so-called Wagner-Meerwein shift (if any) is still fast on the nmr time scale through a very shallow activation energy barrier (less than 3 kcal/mol). The second possibility raises the question as to the nature of the ion undergoing equilibration through an extremely low-activation energy barrier. It has been pointed out[18g] that if such a process occurs it must be exclusively between unsymmetrically bridged ions **216** equilibrating through the intermediacy of the symmetrically bridged species **214.**

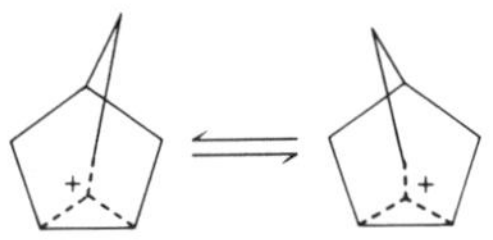

216

The unsymmetrically bridged ions **216** would be indistinguishable from the symmetrically bridged system **214** in solution nmr experiments. (However, see subsequent discussion of solid-state low-temperature ^{13}C-nmr as well as ESCA studies). It is important to recognize that equilibrating open classical cations **215** cannot explain the nmr data and thus cannot be involved as populated species.

The 50-MHz ^{13}C-nmr spectrum of the 2-norbornyl cation has also been obtained in the mixed $SbF_5:SO_2ClF:SO_2F_2$ solvent system at $-159°C$.[288] To obtain a well-resolved ^{13}C-nmr spectrum, the cation was generated from 15% ^{13}C-enriched *exo*-2-chloronorbornane (the label corresponds to one carbon per molecule randomly distributed over the C_1, C_2, and C_6 centers). The ionization of the ^{13}C-enriched *exo*-2-chloronorbornane in $SbF_5:SO_2ClF:SO_2F_2$ solution at $-78°C$ results in the 2-norbornyl cation wherein the ^{13}C label is distributed evenly over all the seven carbons as a result of slow 2,3-hydrogen and fast 6,1,2-hydrogen and Wagner-Meerwein shifts.[288]

At $-80°C$, the 50-MHz ^{13}C-nmr spectrum of the cation (Fig. 3.15) shows three absorptions at δ 91.7 (quintet, $J_{C—H}$ = 55.1 Hz), 37.7 (doublet, $J_{C—H}$ = 153.1 Hz) and 30.8 (triplet, $J_{C—H}$ = 139.1 Hz), indicating that the 2,3-hydrogen shift is frozen, but the 6,1,2-hydrogen and the Wagner-Meerwein shift is still fast on the nmr time scale. Cooling the solution down results in broadening and slow merger into the base line of the peaks at δ 91.7 and 30.8, but the peak at δ 37.7 remains relatively sharp. At $-159°C$, the peaks at δ 91.7 and 30.8 separate into two sets of two peaks at δ 124.5 (doublet, $J_{C—H}$ = 187.7 Hz), 21.2 (triplet, $J_{C—H}$ = 147.1

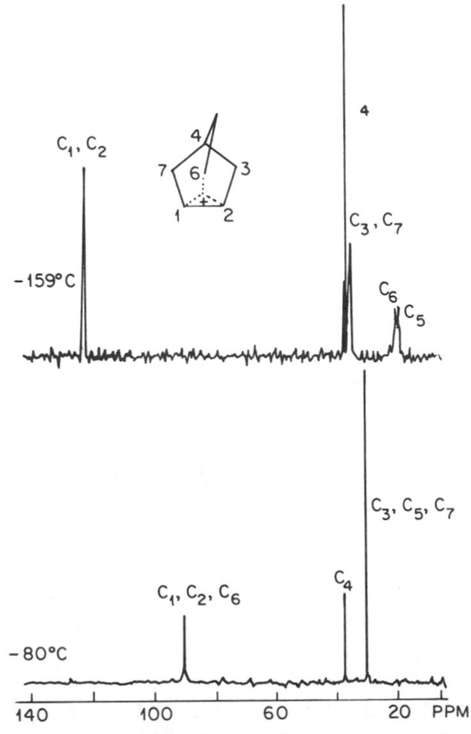

FIGURE 3.15 50 MHz ^{13}C NMR Spectra of 2-Norbornyl cation in SbF_5 SO_2ClF/SO_2F_2 solution.

Hz), and δ 36.3 (triplet, J_{C-H} = 131.2 Hz), 20.4 (triplet, J_{C-H} = 153.2 Hz), respectively. The observed ^{13}C-nmr spectral data at −159°C complement well the 395-MHz ^{1}H-nmr data at −158°C. The observation of the C_1 and C_2 carbons at δ 124.5 and the C_6 carbon at δ 21.2 clearly supports the bridged structure for the ion. Five (or higher) coordinate carbons generally show shielded (upfield) ^{13}C-nmr shifts.[33]

Applying the additivity of the chemical shift analysis[33] to the 2-norbornyl cation also supports the bridged nature of the ion. A chemical shift difference of 168.0 ppm is observed between the ion and its parent hydrocarbon, i.e., norbornane, whereas ordinary trivalent carbocations such as cyclopentyl cation reveal a chemical shift difference of ~360 ppm.[33]

Yannoni and Myhre[289] have obtained magic-angle spinning cross-polarization ^{13}C-nmr spectra of the ^{13}C-enriched 2-norbornyl cation in SbF_5 solid matrix down to −196°C. Their solid-state chemical shifts and measured barriers for 6,1,2-hydrogen shift of 6.1 kcal · mol^{-1} correlate well with the discussed solution data. More recently they even have obtained ^{13}C-nmr spectra in the solid state at −268°C (5°K),[290] a remarkable achievement indeed. A fortuitous combination of large isotropic chemical shifts and small chemical shift anisotropies permitted them to obtain reasonable resolution of the positively charged carbon resonance without the need for magic-angle spinning. Comparison of their previous MAS spectra[289] down to −196°C shows that the nonspinning spectra reflect slowing of 6,2,1-hydride shift. Since no changes were observed in the positively charged carbon resonance (at ~δ 125) between −173°C and −268°C (Fig. 3.16), the authors[290] concluded that if the hypothetical 1,2-Wagner-Meerwein shift is still occurring, then it should be rapid, and an upper limit for the barrier for such a process (involving structures **215**) can be estimated to be no greater than 0.2 kcal/mol^{-1}. This can be taken as

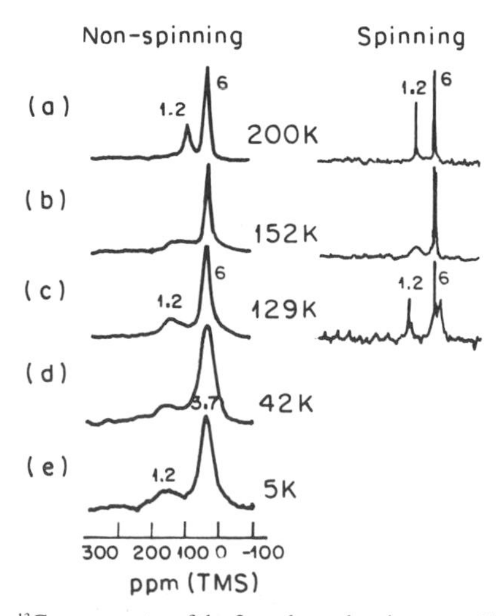

FIGURE 3.16 Solid state ^{13}C nmr spectra of the 2-norbornyl cation according to Yannoni and Myhre[290].

the most definitive spectroscopic evidence besides the ESCA studies for the symmetrical σ-bridged structure of the 2-norbornyl cation (*vide infra*).

As discussed earlier, the method of observing changes in nmr spectra produced by asymmetric introduction of isotopes (isotopic perturbation) as a means of distinguishing systems involving equilibrating species passing rapidly over a low barrier from molecules with single-energy minima, intermediate between the presumed equilibrating structures has been developed by Saunders and co-workers.[34] Applying this method[291] to the 2-norbornyl cation clearly supports its bridged nature. In the ^{13}C-nmr spectrum of the 2-norbornyl cation, even at low temperatures, besides Wagner-Meerwein rearrangement, the 6,1,2-hydride shift has a barrier of only 5.9 kcal/mol^{-1} and results in a certain amount of line broadening of the lowest field signal observed. Even in the ion with no deuterium, the downfield signal at ~δ 124.5 (C-2 and C-6 cyclopropane-like carbons) is found to be 2 ppm wide. Nevertheless, no additional isotopic splitting or broadening was observed with either 2-monodeutero or 3,3-dideutero cations, and therefore the isotopic splitting can be no more than 2 ppm. This is true even if a slow 6,2-hydride shift converts part of the latter ions to a symmetrical 5,5-dideutero system that lacks an equilibrium isotope effect. This result, when compared with the significantly larger splitting observed[36] for deuterated dimethylcyclopentyl and dimethylnorbornyl cations (known to be equilibrating ions) is in accordance with the nonclassical nature of 2-norbornyl cation. A similar conclusion was recently reached[292] based on high-temperature deuterium isotopic perturbation effect in 2-norbornyl cation.

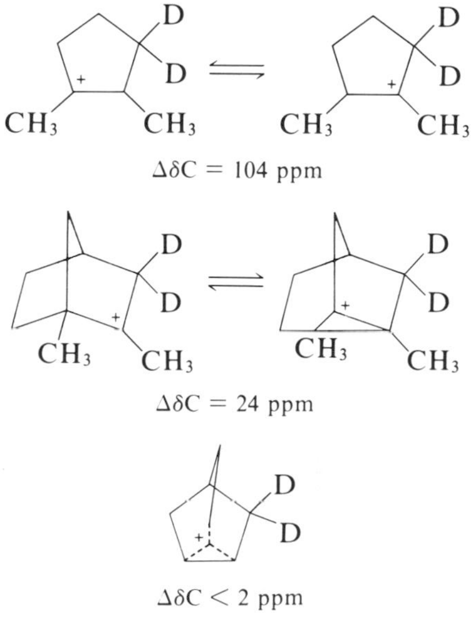

$\Delta\delta C = 104$ ppm

$\Delta\delta C = 24$ ppm

$\Delta\delta C < 2$ ppm

Farnum and Olah's groups, respectively, have extended the so called Gassman-Fentiman tool of increasing electron demand coupled with ^{1}H- and ^{13}C-nmr spectroscopy as the structural probe under stable ion conditions to show the onset of

π, $\pi\sigma$, and σ-delocalization in a variety of systems.[91,292–300] The ^{13}C-nmr chemical shifts of the cationic carbon of a series of regular trivalent 1-aryl-1-cyclopentyl, 1-aryl-1-cyclohexyl, 2-aryl-2-adamantyl, 6-aryl-6-bicyclo[3.2.1]octyl, 7-aryl-7-norbornyl cations (so called classical cations) correlate linearly with the observed cationic chemical shifts of substituted cumyl cations over a range of substituents[298–300] [generally from the most electron releasing p-OCH_3 to the most electron withdrawing 3,5$(CF_3)_2$ groups].

However, 2-aryl-norbornyl cations **217** show deviations from linearity in such chemical shift plots with electron-withdrawing substituents (Fig. 3.17) indicative of onset of nonclassical σ-delocalization fully supporting the nonclassical nature of the parent secondary cation. These conclusions were criticized by Brown.[300] In a recent paper,[39] Olah and Farnum have shown major flaws in such criticisms.

Ar

217

As mentioned earlier, since in electron spectroscopy the time scale of the ionization processes is on the order of 10^{-16} s, definite ionic species are characterized, regardless of their possible intra- and intermolecular rearrangements (e.g., Wagner-Meerwein rearrangements, hydride shifts, etc.), even at rates equalling or exceeding those of vibrational transitions. Thus, electron spectroscopy can give an unequivocal answer to the long-debated question of the "classical" or "nonclassical" nature of the norbornyl cation, regardless of the rate of any possible equilibration processes.

Olah, Mateescu, and Riemenschneider[19b] succeeded in observing the ESCA spectrum of the norbornyl cation and compared it with those of the 2-methyl-2-norbornyl

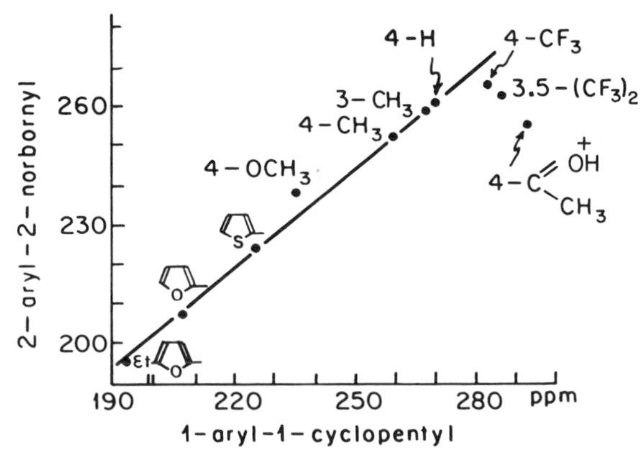

FIGURE 3.17 Plot of the ^{13}C nmr chemical shifts of the cationic center of 2-aryl-2-norbornyl cations vs. those of model 1-aryl-1-cyclopentyl cations.

cation and other trivalent carbenium ions, such as the cyclopentyl and methylcyclopentyl cations. The 1*s* electron spectrum of the norbornyl cation shows no high-binding energy carbenium center, and a maximum separation of less than 1.5 eV is observed between the two "cyclopropyl"-type carbons, to which bridging takes place from the other carbon atoms (including the penta-coordinated bridging carbon). In contrast, the 2-methylnorbornyl cation shows a high-binding energy carbenium center, deshielded with the $\delta\ E_b$ of 3.7 eV from the other carbon atoms. Typical ESCA shift differences are summarized in Table 3.5.

Recently, Grunthaner has reexamined the ESCA spectrum of the norbornyl cation on a higher-resolution X-ray photoelectron spectrometer using highly efficient vacuum techniques.[301] The spectrum closely matches the previously published spectra. Furthermore, the reported ESCA spectral results are consistent with the theoretical studies of Allen and Goetz[302] on the classical and nonclassical norbornyl cation at the STO-3G and STO-4.31G levels. Using the parameters obtained by Allen and Goetz, Clark et al.[303] were able to carry out a detailed interpretation of the experimental ESCA data for the core-hole state spectra at SCF STO-4.31G level and calculated equivalent cores at STO-3G level. Agreement between experimentally obtained spectra and those calculated for the nonclassical cation are good and dramatically different from those for the classical cation.

If the classical structure were correct, the 2-norbornyl cation would be a usual

TABLE 3.5 Binding Energy Differences of Carbocation Centers from Neighboring Carbon Atoms $dE_b{}^{+}{}_{C—C}$

Ion	$dE_b{}^{+}{}_{C—C}$	Approx Rel C^+:C Intensity
$(CH_3)_3{}^+C$	3.9 ± 0.2	1:3
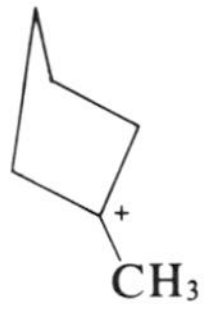 + CH_3	4.2 ± 0.2	1:5
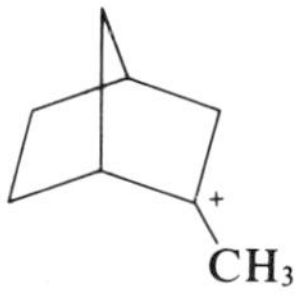 + CH_3	3.7 ± 0.2	1:7
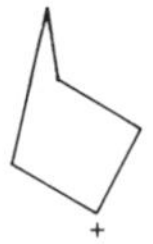 +	4.3 ± 0.5	1:4
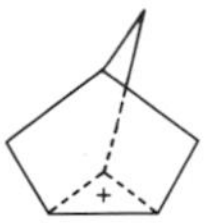 +	1.5 ± 0.2	2:5

secondary carbocation with no additional stabilization provided by σ-delocalization (such as the cyclopentyl cation). The facts, however, seem to be to the contrary. Direct experimental evidence for the unusual stability of the secondary 2-norbornyl cation comes from the low-temperature solution calorimetric studies by Arnett and co-workers.[45] In a series of investigations, they determined[45] the heats of ionization (ΔH_i) of secondary and tertiary chlorides in $SbF_5:SO_2ClF$ and more recently alcohols in $HSO_3F:SbF_5:SO_2ClF$ solutions.[46]

$$R{-}Cl + SbF_5 \xrightarrow{\Delta H_i} R^+ \ SbF_5Cl^-$$

$$R{-}OH + HSO_3F{:}SbF_5 \xrightarrow{\Delta H_i} R^+ \ SbF_5(FSO_3)^- + H_3O^+$$

However, it was found that whereas the difference observed in the heats of ionization of 2-methyl-2-exo-norbornyl chloride and 2-exo-norbornyl chloride in $SbF_5:SO_2ClF$ solution is 7.4 kcal · mol, the same difference between the corresponding alcohols in $HSO_3F:SbF_5:SO_2ClF$ solution is only 2.5 kcal · mol^{-1}. This indicates that the heats of ionization value (ΔH_i) seem to largely depend on the nature of the starting precursors (initial state effects). However, the observed differences are remarkably small for the corresponding secondary and tertiary cations, which generally is 10–15 kcal · mol^{-1}. In the case of norbornyl, there seems to be at least 7.5 kcal · mol^{-1} extra stabilization. A further compelling evidencc for the nonclassical stabilization of the 2-norbornyl cation also comes from Arnett's measured heats of isomerization of secondary cations to tertiary cations.[304] The measured heat of isomerization of 4-methyl-2-norbornyl cation **218** (secondary system) to 2-methyl-2-norbornyl cation **219** is 6.6 kcal · mol^{-1}. In contrast, the related isomerization of *sec*-butyl cation **18** and *t*-butyl cation **1** involved a difference in ΔH_i of 14.2 kcal · mol^{-1}. Taking this latter value as characteristic for isomerization

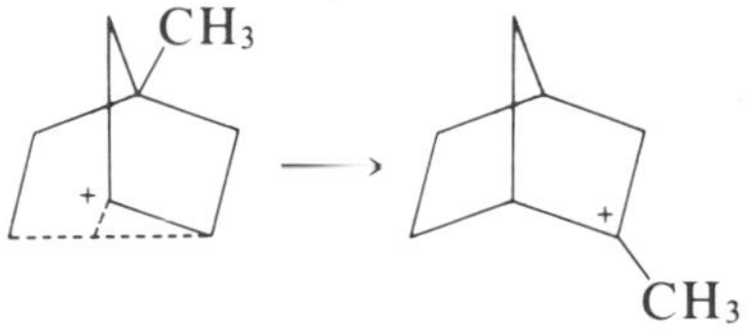

CH$_3$ **218** → **219** CH$_3$ $\Delta H_{isomerization} = -6.6$ kcal · mol^{-1}

18 → **1** $\Delta H_{isomerization} = -14.2$ kcal · mol^{-1}

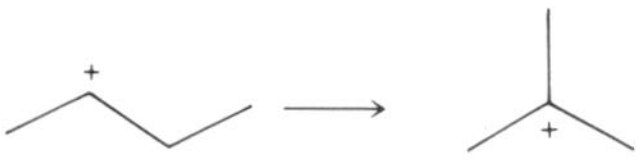

of secondary to tertiary ions, one must conclude that the secondary norbornyl ion has an extra stabilization of at least 7.6 kcal · mol^{-1}. Farcasiu[305] questioned these conclusions, arguing that they neglected to account for extra stabilization by bridgehead methyl substitution as indicated by his molecular force field calculations. Schleyer and Chandrasekhar[306] have subsequently pointed out that Farcasiu failed to include corrections for β-alkyl branching. Correcting for this effect, there is still

6 ± 1 kcal · mol^{-1} extra stabilization in the 2-norbornyl cation for which no other reasonable explanation other than σ-bridging was offered.

Gas-phase mass spectrometric studies[307,308] also indicate exceptional stability of the 2-norbornyl cation relative to other secondary cations.

Theoretical quantum mechanical calculations[309–313] also have been performed on the 2-norbornyl cation at various levels. These calculations reveal significant preference for σ-delocalized nonclassical structures. The most recent and extensive calculation by Schaefer and co-workers[312] using full geometry optimization for symmetrically and unsymmetrically bridged systems showed a difference of only 1.0 kcal · mol^{-1} between these structures. (Some confusion was introduced by Schaefer who called the unsymmetrically bridged ion ''classical.'') A similar high-level calculation including electron correlations (with a double zeta plus polarization basis set) by Ragavachari, Schleyer,[313a] Haddon and Schaefer, as well as Liu et al.[313b] indicates that the only minimum on the 2-norbornyl cation potential energy surface is the symmetrically bridged structure. The structure with ''classical'' 2-norbornyl-like geometry, however, did not correspond to a fixed point on the potential energy surface. The extra stabilization of the bridged structure was roughly estimated at 15 kcal · mol^{-1} at this high level of *ab initio* theory.[313]

In conclusion,[20] all the investigations reported on the structure of the 2-norbornyl cation clearly show that the ion has the bridged nonclassical structure and is not a classical equilibrating system.

3.5.2.3 Bicyclo[2.1.1]hexyl Cation

The bicyclo[2.1.1]hexyl cation **220** was first observed[314] by Seybold, Vogel, Saunders, and Wiberg in superacid media by ^{1}H-nmr spectroscopy. From the observed chemical shift data, they suggested a symmetrically bridged structure **221** for the ion, although they could not freeze out the degenerate equilibria. Similar conclusions were drawn from solvolytic studies.[315]

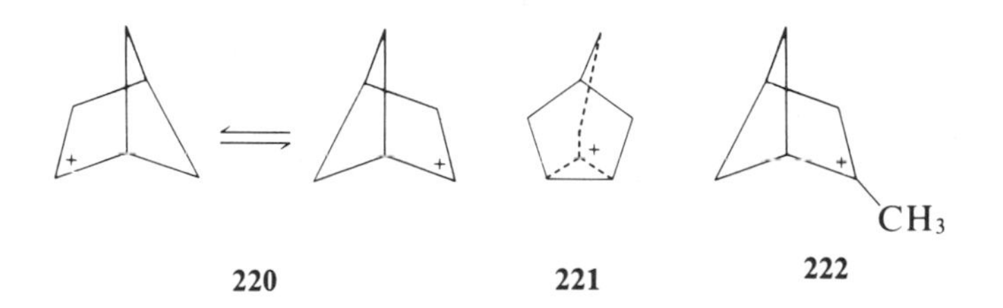

In a subsequent ^{13}C-nmr study,[316] Olah, Liang, and Jindal concluded that there is very little σ-bridging in the rapidly equilibrating ion.

The ^{1}H-nmr spectrum of the ion **220** in $SbF_5:SO_2ClF$ showed three resonances at δ ^{1}H 8.32 (two protons), 3.70 (six protons), and 2.95 (one proton) with no significant line broadening down to −140°C. The ^{13}C-nmr spectrum of the ion **220** also shows three resonances at δ 157.8 (doublet, J = 184.5 Hz; C-1 and C-2), δ 49.1 (triplet, J = 156.9 Hz; C-3, C-5, and C-6), and δ 43.4 (doublet, J = 164.6 Hz; C-4). Above −90°C, the ion irreversibly rearranges to cyclohexenyl cation.[317]

A recent study by Saunders and Wiberg[318] by deuterium labeling at the exchanging sites indicates that there is significant σ-bridging in the ion **220.** Sorensen and Schmitz[319] have shown that the free-energy difference between cations **220** and **222** is 7–9.8 kcal · mol^{-1} compared with 5.5 and 11.4 kcal · mol^{-1} for the analogous 2-norbornyl and cyclopentyl ions substantiating the intermittent (partially bridged) nature of the ion **220.**

Attempts to prepare[320] the analogous bicyclopentyl cation **223,** however, were unsuccessful and instead gave the rearranged cyclopentenyl cation **55.**

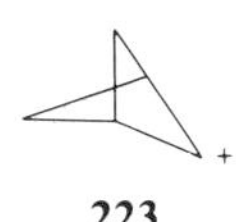

223

3.5.2.4 Degenerate Cyclopropylmethyl and Cyclobutyl Cations

In contrast to "classical" tertiary and secondary cyclopropylmethyl cations (showing substantial charge delocalization into cyclopropane ring but maintaining their identity), primary cyclopropylmethyl cations rearrange to cyclobutyl and homoallylic cations under both solvolytic and stable ion conditions.[9,111,321–322] The nonclassical nature of cyclopropylmethyl and 1-methylcyclopropylmethyl cations **224** and **225** is now firmly established.[97–101]

The cyclopropylmethyl cation **224** can be generated from both cyclobutyl and cyclopropylmethyl precursors (^{1}H-nmr spectrum in Fig. 3.18).

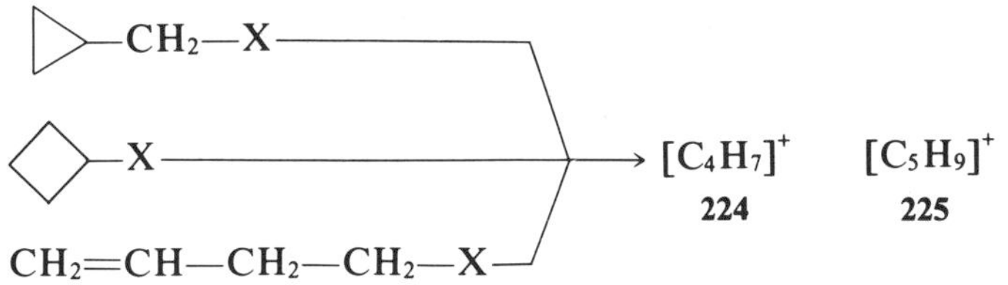

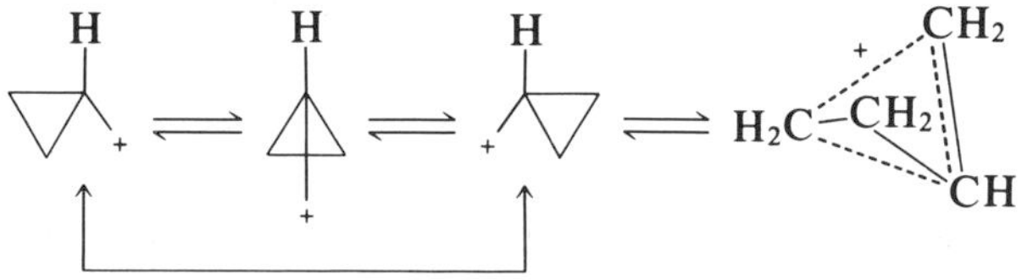

—CH_2—X

—X

CH_2=CH—CH_2—CH_2—X

→ $[C_4H_7]^+$ **224** $[C_5H_9]^+$ **225**

H H H

$\rightleftharpoons$ $\rightleftharpoons$ $\rightleftharpoons$ H_2C CH_2 CH_2 CH

226 **227**

At the lowest temperatures studied (ca. −140°C), ^{13}C-nmr spectroscopy indicates that **224** is still an equilibrating mixture of bisected σ-delocalized cyclopropylcarbinyl cations **226** and the bicyclobutonium ion **227.**[98] From the comparison of calculated nmr shifts, the low-lying species is considered to be the bicyclobutonium ion.[98] A similar conclusion has been reached by Saunders and Roberts based on

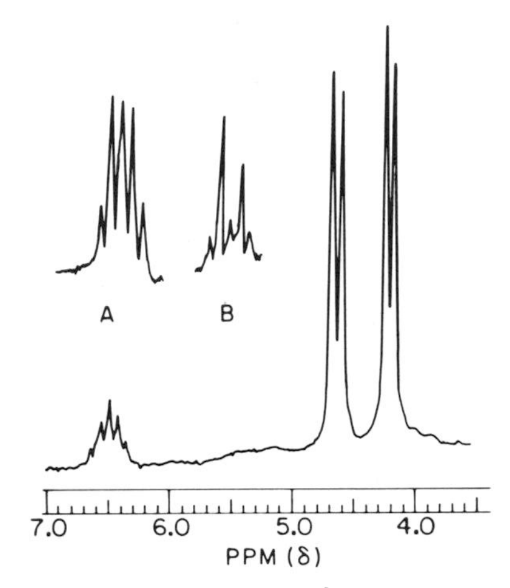

FIGURE 3.18 100-MHz pmr spectrum of the cyclopropylcarbinyl cation in SbF_5-SO_2ClF solution at −80°: (a) 60-MHz spectrum of H_2 region; (b) 60-MHz spectrum of H_2 region from the α,α-dideuter-iocyclopropylcarbinyl precursor.

isotopic perturbation studies.[323] However, in the case of $C_5H_9^+$ **225,** the low-lying species are the nonclassical methylbicyclobutonium ions **228** with no contribution from either the bisected 1-methyl-cyclopropylmethyl cation **229** or the 1-methyl-cyclobutyl cation **230.**

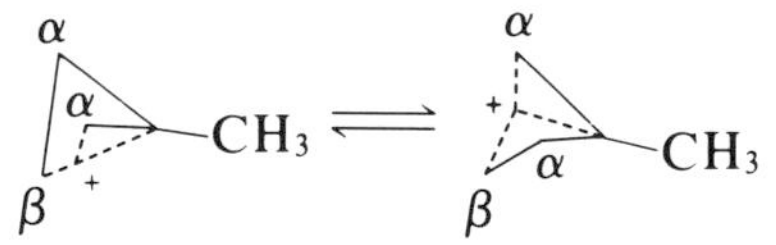

228 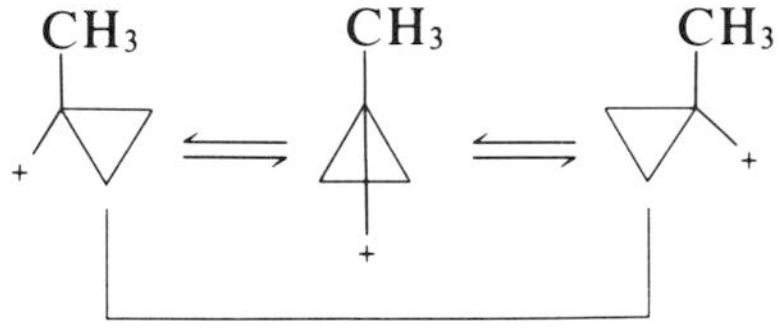 **229** 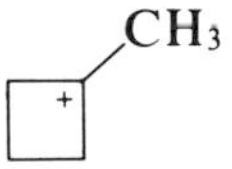 **230**

The highly shielded β-methylene resonance at δ −2.83 in the ^{13}C-nmr spectrum is particularly convincing evidence for the nonclassical bicyclobutonium structures.[101]

Recently Schmitz and Sorensen[324a] have prepared a primary cyclopropylmethyl cation **231** which shows static behavior. The nortricyclylmethyl cation **231** is regarded as vinyl bridged 2-norbornyl cation **231b.** The support for the structure comes not only from ^1H- and ^{13}C-nmr studies but also from molecular orbital calculations.[324]

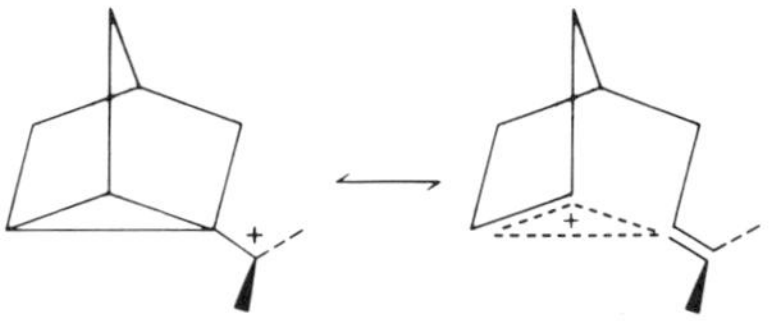

a b

231

Some examples of equilibrating cyclopropylmethyl-cyclobutyl systems that have been interpreted as partially bridged or unbridged species are the following.

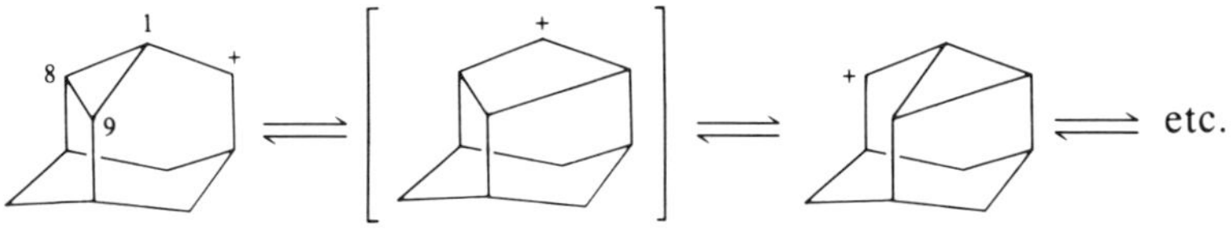

etc.

232[116]

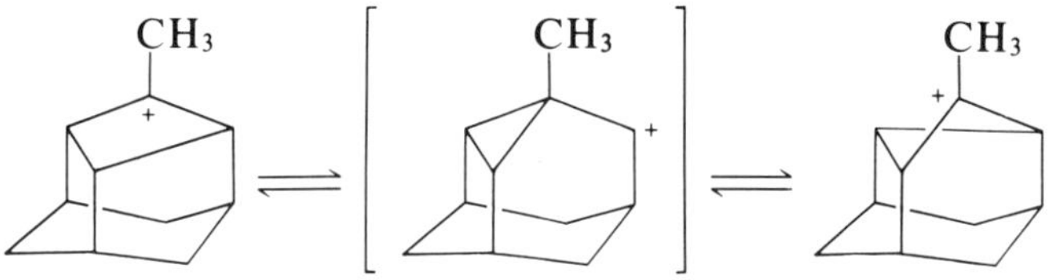

233[116]

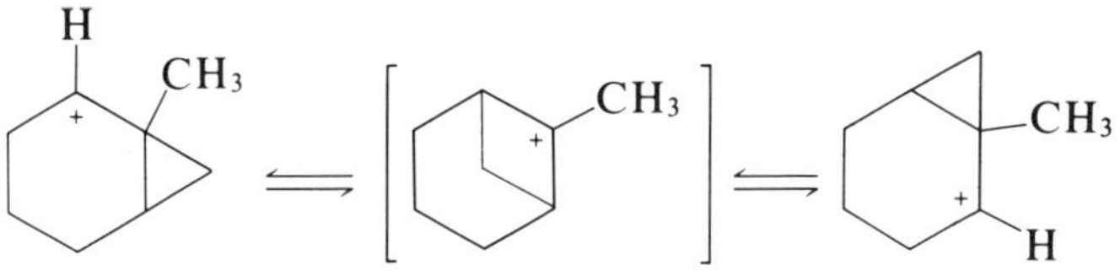

234[118]

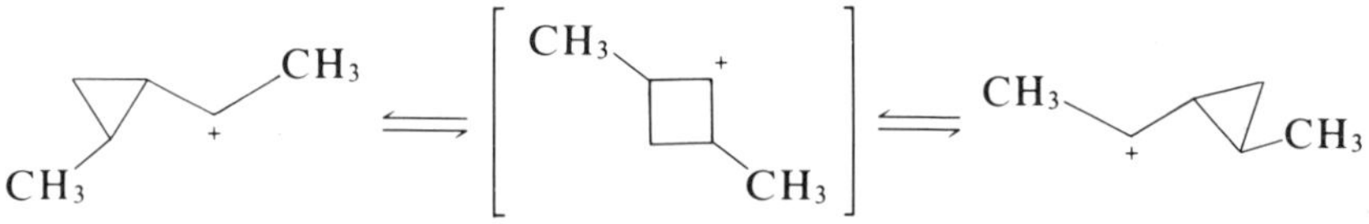

235[325]

DEGENERATE CYCLOPROPYLMETHYLCYCLOBUTYL CATION EQUILIBRIA (DOWN TO −140°C)

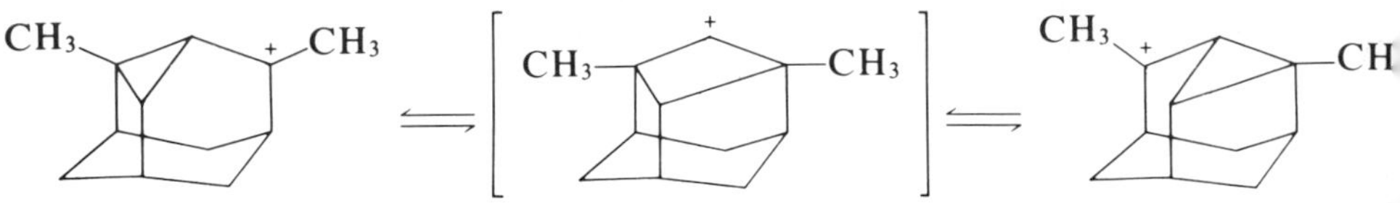

236[326] $\Delta G^{\ddagger}_{-112°C} = 7.4 \pm 0.5$ kcal/mol

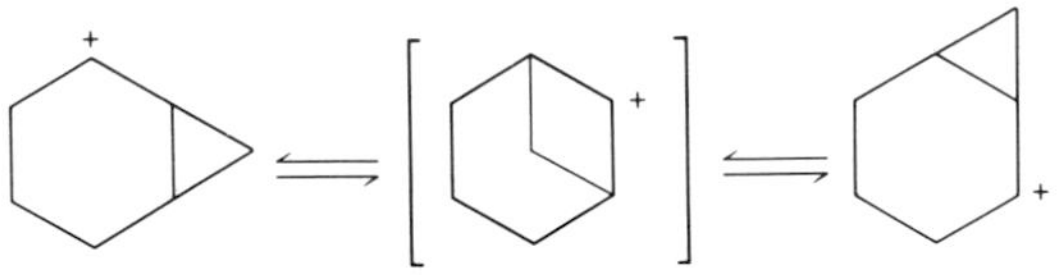

237[118] $\Delta G^{\ddagger}_{-85°C} = 8.5 \pm 0.5$ kcal/mol

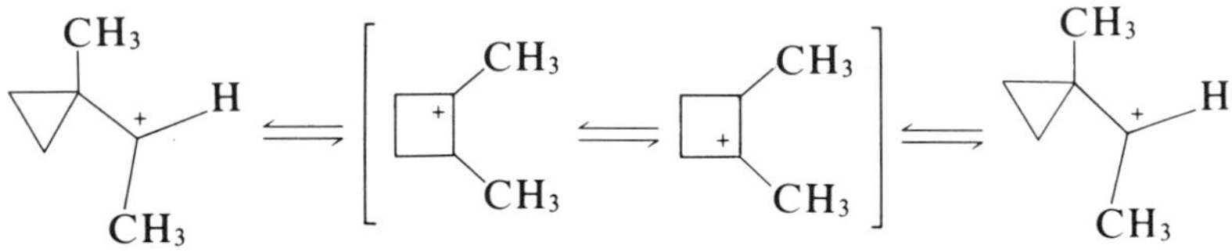

238[327] $\Delta G^{\ddagger}_{-80°C} = 10.9 \pm 0.5$ kcal/mol

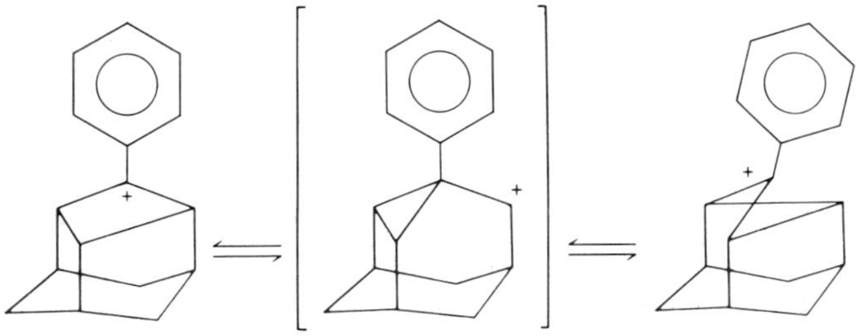

239[326] $\Delta G^{\ddagger}_{-110°C} = 6.8 \pm 0.5$ kcal/mol

CYCLOPROPYLMETHYLCYCLOBUTYL CATION EQUILIBRIA WITH BARRIERS THAT CAN BE FROZEN OUT ON THE NMR TIME SCALE

3.5.2.5 Shifts to Distant Carbons

Although there are many examples of 1,2-hydrogen and alkyl shifts, the occurrence of 1,3, 1,4, and 1,5 shifts must also be considered. A sequence of 1,2 shifts, however, can often yield the same result as a 1,3 or 1,4 shift and the unambiguous demonstration of such can be difficult.

The 2,4-dimethyl-2-pentyl cation **240** is able to undergo a degenerate 1,3-hydrogen shift[328,329] (E_a = 8.5 kcal · mol^{-1}). The alternative mechanism of the successive 1,2-hydrogen shift can be eliminated in this case, since line broadening of methyl peak but not methylene peak occurs (in the nmr spectrum) in the temperature range of −70°C to −100°C. A third possible mechanism involving a corner-protonated cyclopropane **241** is highly unlikely based on energy estimates.[63]

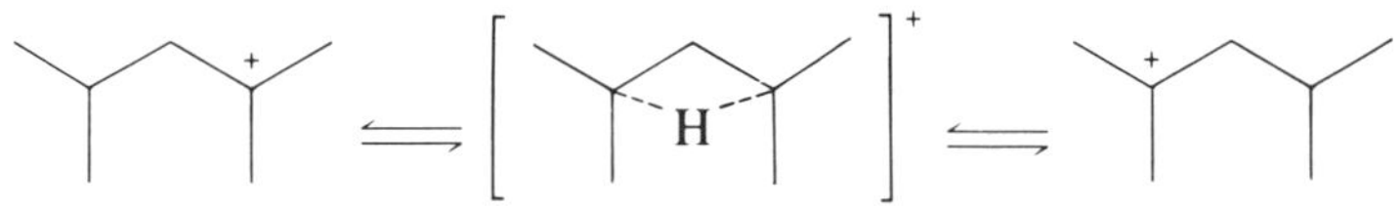

H

240

H

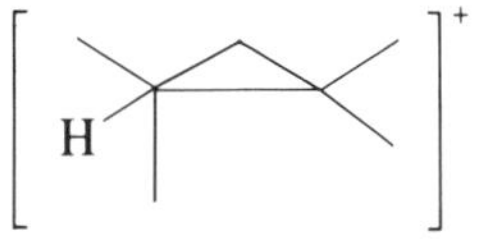

241

A similar activation energy barrier of 10.5 kcal · mol^{-1} is found for 1,3-hydrogen shift in 1,3-dimethylcyclohexyl cation **242**[330], incorporation of the six-membered ring constrains the transition state and raises the activation energy barrier.

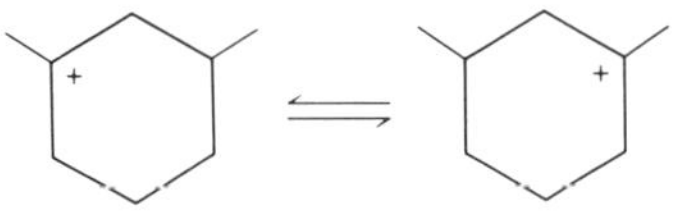

242

An activation energy barrier of 12–13 kcal · mol^{-1} was estimated for the 1,4-hydrogen shift in 2,5-dimethyl-2-hexyl cation **243** using magnetization transfer techniques.[329] The possibility of protonated cyclobutane intermediate similar to the previously considered protonated cyclopropane intermediate is highly unlikely. A similar degenerate 1,4-hydrogen shift is found to occur in the 1,4-dimethyl-1-cyclohexyl cation **244.** The occurrence of successive 1,2- or 1,3-hydrogen shifts was clearly ruled out from a variable temperature nmr study. The activation energy barrier for such a process was estimated at 13 kcal · mol^{-1}.

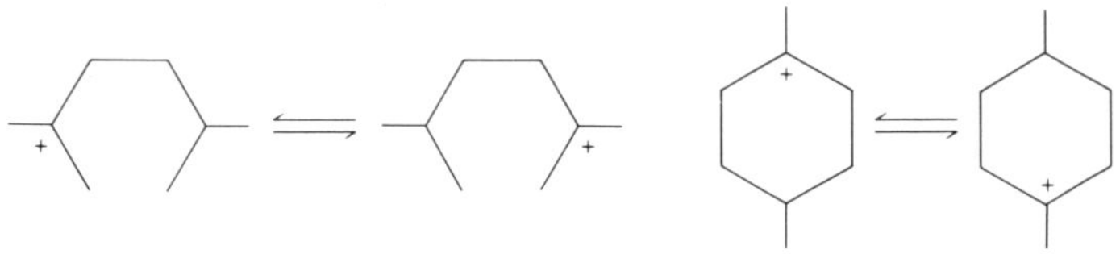

243 **244**

Intrigued by fascile transannular hydride transfers[331,332] in medium-sized rings, Saunders and co-workers[34] examined the 2,6-dimethylheptyl cation **245.** Even at the lowest temperature studied (ca. -100°C), the ion exhibits a single averaged peak for the four methyls, implying that the 1,5-hydrogen shift occurs with an activation energy barrier of 5 kcal · mol^{-1} or less or the ion could have a symmetrically bridged structure such as **246.**

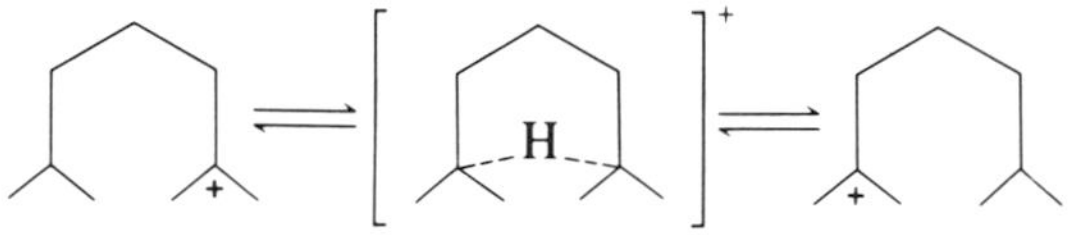

245 246

The studies of Prelog[331] and Cope[333] established that medium-sized cycloalkyl rings (C_8 to C_{11}) undergo direct transannular hydride shifts in reactions involving an electrophilic (i.e., carbocationic) intermediate. Sorensen and co-workers[334a] have shown that at very low (−130°C) temperature the cyclodecyl cation **247** exists as a static 1,6- or 1,5-hydrido structure **248c** or **248e.** Similar behavior[334b] was also observed for the 1,6-dimethyl analog **249.** The bridging hydrogen in ion **248c** is observed at an unusually high field of δ ^{1}H −6.85. Stable hydrido-bridged cy-

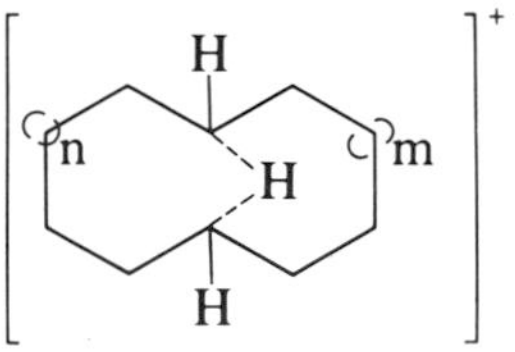

a m = 0, n = 0
b m = 1, n = 0
c m = 1, n = 1
d m = 2, n = 1 or m = 3, n = 0

248

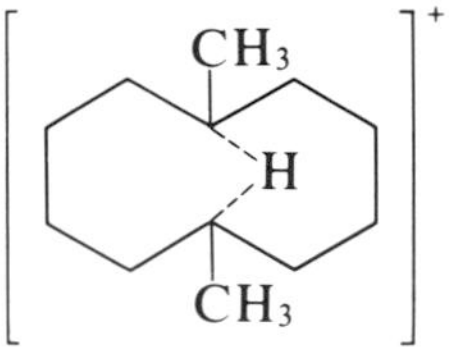

249 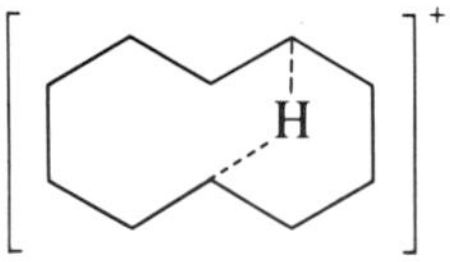 248e

cloalkyl cations **248 a, b, d** (8-, 9-, and 11-membered rings) have subsequently been observed.[335] The bridging hydrogen was found to be increasingly more shielded in the ^{1}H-nmr spectra as the ring size was increased. This indicates increased negative charge on the bridged hydrogen (conversely increased positive charge on the terminal hydrogens) as the separation between the two bridged carbons is increased. The

TABLE 3.6 ^{1}H-nmr Chemical Shifts of Bridged and Terminal Hydrogen Atoms in μ-Hydrido Bridged Ions

Cation	Terminal hydrogens	Bridged hydrogens
248a	+7.9	−7.7
b	+6.8	−6.6
c	+6.8	−6.85
d	+6.3	−6.0
249	—	−3.9

^{1}H-nmr chemical shifts of terminal and bridging hydrogens of various bridged carbocations are shown in the Table 3.6.

Sorensen[336] also obtained evidence for 1,5-μ-hydrido bridging between secondary and tertiary carbon sites [C---H---C(H)] in several substituted cyclooctyl cations. The μ-1,5-hydrido-bridged 1,5-dimethylcyclodecyl cation **249a** has also been obtained[337] and studied as a distinct stable species.

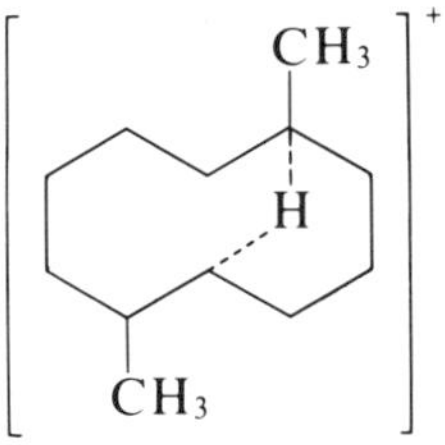

CH_3, H, CH_3

249a

Application of Saunders's isotopic perturbation technique[34] to **249** has confirmed the bridged structure. With one trideuteromethyl group, an isotopic splitting of only 0.5 ppm is observed in the ^{13}C resonance of bridged carbon and this clearly supports the assigned nonclassical hydrido-bridged structure.

3.5.2.6 9-Barbaralyl (tricyclo[3.3.1.0^{2,8}]nona-3,6-dien-9-yl) Cations and Bicyclo[3.2.2]nona-3,6,8-trien-2-yl Cations

The 9-barbaralyl cation **250** is the cationic counterpart of bullvalene **251.** The unique stereoelectronic composition of the structural elements of **250** suggests that it is very reactive in both degenerate (partial and total) as well as nondegenerate rearrangements. Bullvalene shows total degeneracy through a series of Cope rearrangements.[338] There are several intriguing structural and mechanistic questions connected with the barbaralyl cations; for example, what are their structures and how does the positive charge influence the degenerate Cope rearrangement in ion **250?** Ion **250** is closely related to the bicyclo[3.2.2]nona-3,6,8-trien-2-yl cation **252** which has been of interest in connection with the development of the concept of bicycloaromaticity.[339]

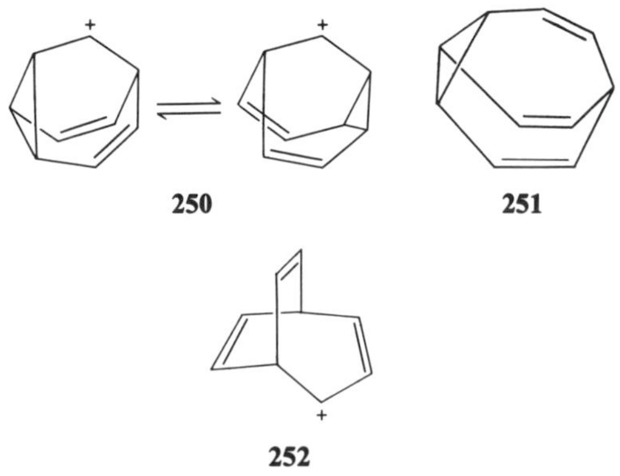

250 **251**

252

When bicyclo[3.2.2]nonatrien-2-ol treated with superacid at $-135°C$ and was observed at the same temperature by ^1H-nmr spectroscopy, a sharp singlet at δ 6.59 was obtained. A rapidly rearranging carbocation was inferred to be responsible for

HO H $\xrightarrow[-135°C]{HSO_3F:SO_2ClF\ 1:4\ v/v}$ **250** $\xrightarrow[-125°C]{\Delta F^{\ddagger} = 10.4\ kcal \cdot mol^{-1}}$ **253**

the observed singlet since there is no regular polyhedron with nine corners, i.e., nine equivalent positions. The chemical shift of the singlet compared with that estimated from appropriate reference compounds indicated that the ion that was rapidly exchanging all its nine CH-groups was the 9-barbaralyl cation **250** rather than the bicyclo[3.2.2]nonatrienyl cation **252.** A combination of mechanisms **VI** and **VII** or structure **254** itself was proposed for the total degeneracy, and the rearrangement barrier was estimated to be ~6 kcal · mol^{-1}. Even at $-125°C$, the singlet disappeared rapidly, and a novel type of ion, a 1,4-bishomotropylium ion, bicyclo[4.3.0]nonatrienyl cation **253** was quantitatively formed.[24,340–341]

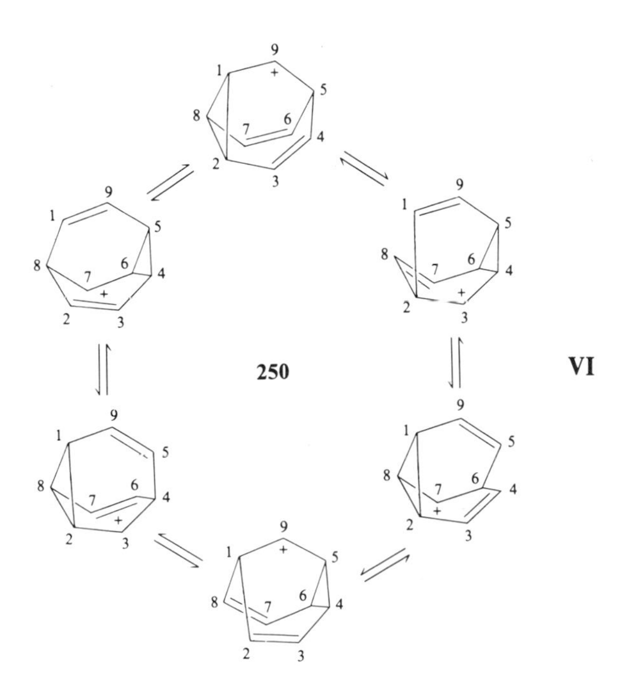

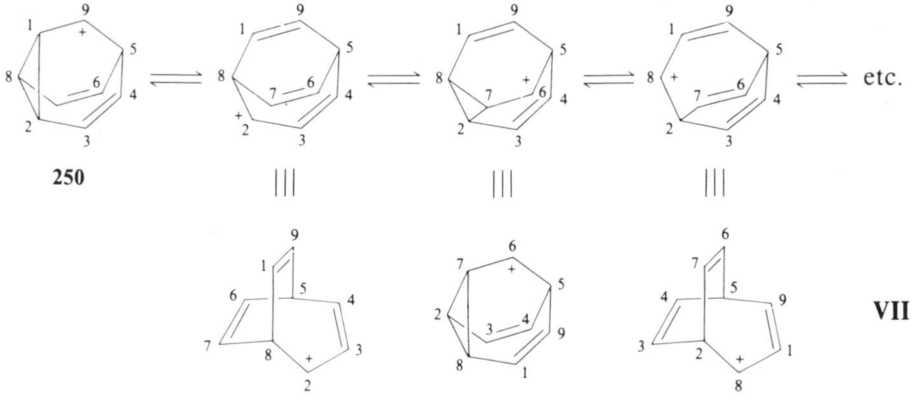

250 ⇌ ⇌ ⇌ ⇌ etc.

VII

252

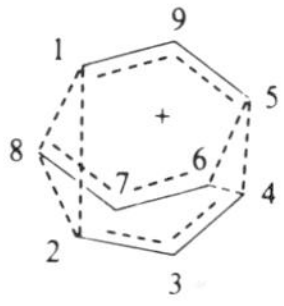

254

The total degeneracy was found to be slow on the ^{13}C-nmr time scale. No signal was detected above the noise level in the spectrum. Thus, the signals must be very broad as a consequence of slow rearrangements.[32] In the hope of being able to solve controversies concerning the structure and mechanisms of rearrangement of the barbaralyl cation, the ^{13}C-labeled precursor was synthesized and the ^{13}C-labeled barbaralyl cation **255** was prepared.[342] The ^{13}C-nmr spectrum at −135.5°C showed

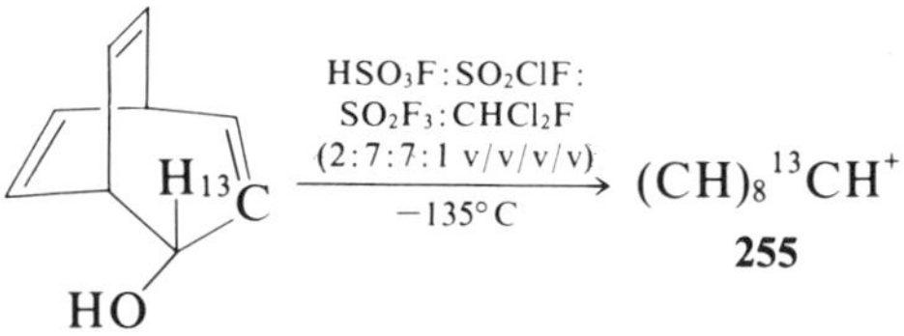

$$\xrightarrow[-135^\circ C]{HSO_3F:SO_2ClF:SO_2F_2:CHCl_2F\ (2:7:7:1\ v/v/v/v)} (CH)_8{}^{13}CH^+$$

255

a broad band at δ ^{13}C 118.5 confirming the fast scrambling of all the nine carbon atoms. The barrier was found to be 5.5 kcal · mol^{-1}. Upon lowering the temperature to −150°C, the signal broadened and split into two new signals at 101 ppm and 152 ppm with the area ratio 6:3. Further lowering of the temperature to −152°C sharpened the signals. These results exclude ion **252** as the observed ion. However, the data do not allow discrimination between the two proposed ions **250** and **254.** If rearrangement **(VI)** gives the area ratio 6:3, the barrier for such a rearrangement is estimated to be ≤4 kcal · mol^{-1}. The static ion **254** should also show the area ratio observed.

Recently, the controversy was solved by Ahlberg et al.[343,344] using a combination of ^{13}C labeling and isotopic perturbation.[34] The specifically octadeuterated and ^{13}C-labeled precursor was synthesized and reaction with superacid gave the $(CD)_8{}^{13}CH^+$ cation **256.** The ^{13}C-nmr spectrum of **256** was similar to that of **255** but had some important differences. The broad singlet observed from **256** at $-135°C$ appeared 4.5 ppm downfield of that of **255** and the two signals at $-151°C$ had also been shifted; the low-field one had been shifted upfield by ca. 1 ppm and the high-field

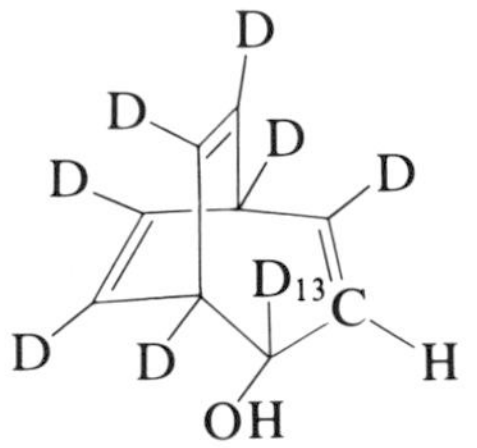

$\longrightarrow (CD)_8{}^{13}CH^+$

256

one downfield by 6 ppm. Furthermore, the area ratio had changed from 6:3 to 5:3. These changes caused by the isotopic perturbation are only consistent with the labeled perturbed ion being a 9-barbaralyl cation **250** and not the D_{3h} structure **254.**

If the ion **256** has structure **254** then the following spectral changes would have been expected. Due to equilibrium **(VIII),** which is likely to have an equilibrium constant smaller than 1 because of the difference in zero-point energy between a "cyclopropane" C—H and "olefinic" C—H, a downfield shift of the singlet might be observed. The two signals at $-151°C$ on the other hand are not expected to shift relative to those of **255,** since the six "cyclopropane" carbon-hydrogen bonds are equivalent and so also are the three "olefinic" carbon-hydrogen bonds. However, an area ratio different from 6:3 is expected.

If on the other hand **256** has the 9-barbaralyl cationic structure **250,** the observed shifts of the two signals at $-151°C$ are as expected: the carbon-hydrogen bonds involving C(1), C(2), C(4), C(5), C(6), and C(8) are not equivalent and therefore ^{13}C—H will preferentially be found in olefinic positions since olefinic C—H bonds

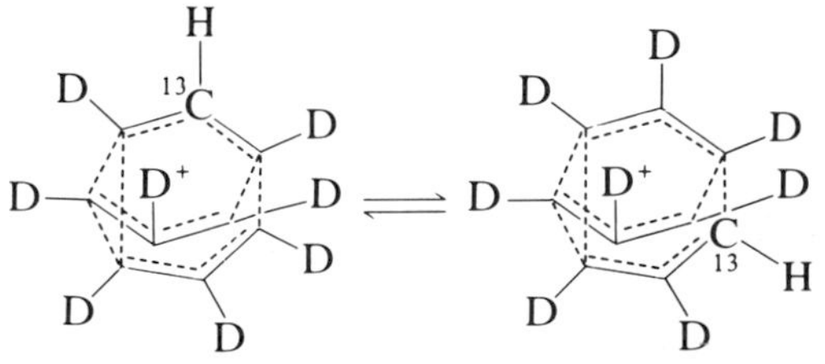

VIII

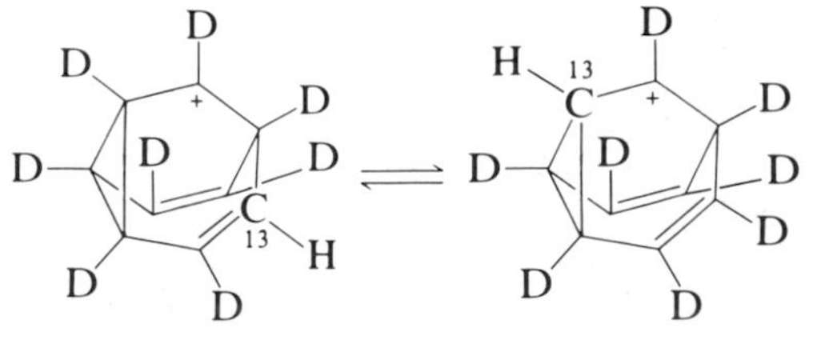

IX

have lower zero-point energy than saturated C—H bonds. Therefore, the ^{13}C average chemical shift will be shifted downfield (by the process **IX**). By the same reasoning, only a minor shift is expected for the C(3), C(7), and C(9) group involving olefinic carbon-hydrogen bonds. The observed shift at $-135°C$ of the broad singlet is also as expected. Thus, the barbaralyl cation has 9-barbaralyl cationic structure **250** and undergoes the six-fold degenerate rearrangement **(VI)** which has a barrier of only 4 kcal · mol^{-1}. Structure **254** has been excluded as either transition state or intermediate in this rearrangement.[343]

Complete degeneracy of ion **250** is probably achieved through mechanism (scheme **VII**) where ion **252** is either an intermediate or transition state.[345]

Similar rearrangements are also observed in 9-substituted 9-barbaralyl cations. The mechanisms of such rearrangements have been thoroughly investigated.[346,347]

3.5.2.7 1,3,5,7-Tetramethyl and 1,2,3,5,7-Pentamethyl-2-Adamantyl Cations

The nature of 2-adamantyl cation **257** has been difficult to study under stable ion conditions because it undergoes facile rearrangements to the more stable 1-adamantyl cation **36.**[106] This difficulty was circumvented by Lenoir, Schleyer, Saunders and co-workers[348] by blocking all four bridgehead positions by mcthyl groups in a study of 1,3,5,7-tetramethyl- and 1,2,3,5,7-pentamethyl-2-adamantyl cations **258**

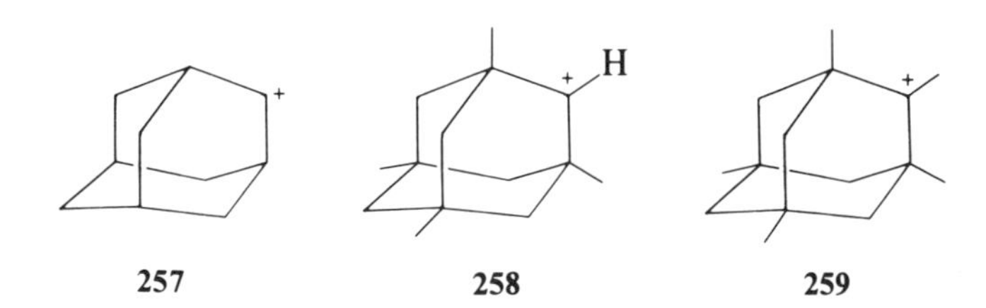

and **259.** The ^1H-nmr spectrum of **258** in superacid had the correct number of peaks to fit the symmetry of a static 2-adamantyl cation, but the chemical shift of the CH proton at the presumed carbocation center C(2) was δ 5.1. This is 8 ppm to higher field than expected for a typical static secondary carbenium ion such as the 2-propyl cation **2.** Since the symmetry of the spectrum was incompatible with either a static bridged 2-adamantyl cation **260** or a static tertiary protoadamantyl cation **261,** two mechanisms (**X** and **XI**) were postulated involving sets of **260** or **261** undergoing rapid degenerate rearrangements at $-47°C$.

Apart from one of these degenerate rearrangements, **258** also underwent a non-degenerate rearrangement **(XII)** to a more stable tertiary 2-adamantyl cation **263** with a half-life of 1 h at $-47°C$. The kinetics of this rearrangement, which involves protoadamantyl cations **262** as intermediates, was advantageously studied in the tertiary 2-adamantyl system **259** where it is degenerate. Line-shape analysis for the degenerate rearrangement of **259** gave $E_a = 12.1 \pm 0.4$ kcal · mol^{-1} in accord with molecular mechanics calculations.

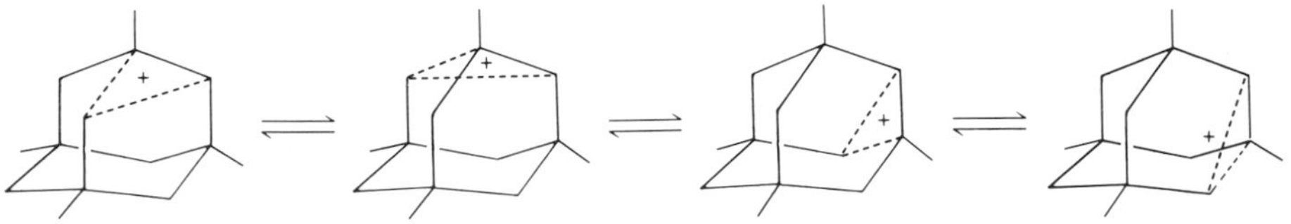

260

X

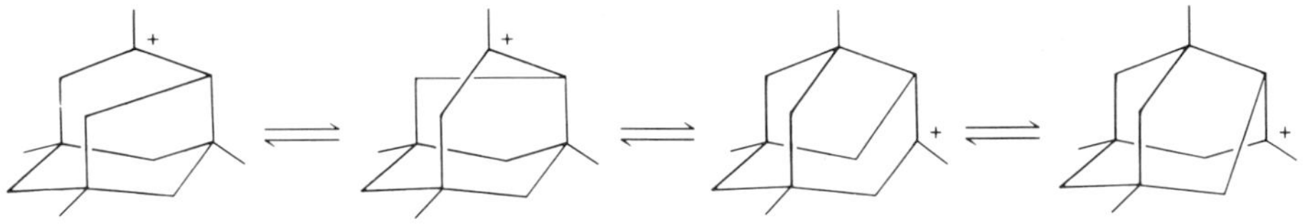

261

XI

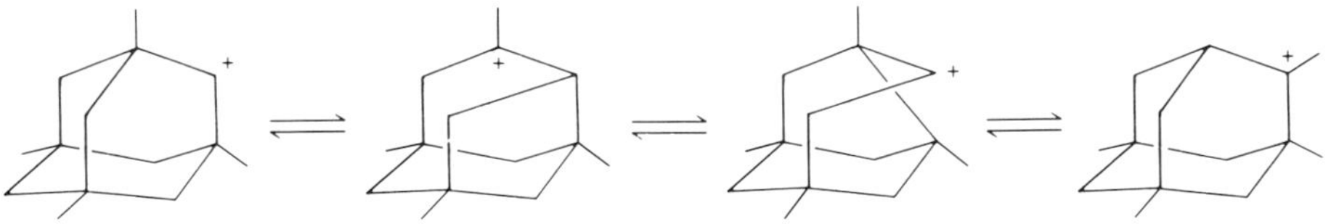

258 262 263

XII

Since it was difficult to make an exclusive choice between mechanisms **(X)** and **(XI)** for the degenerate rearrangement and average structure of **258** on the basis of ^{1}H-nmr data only, further arguments were taken from a solvolytic study, and mechanism **(X)** was preferred as explanation for the behavior of **258** in superacid.

Criticism of these conclusions by Farcasiu[349,350] led Schleyer, Olah, et al.,[33] to study **258** and **259** further by ^{13}C-nmr spectroscopy. The spectra of **259** confirmed its classical static carbenium ion structure at low temperature. At 30°C an average of the C(1), C(2), and C(3) signals and those of the CH_3 groups attached to these positions, respectively, were observed due to the degenerate rearrangement via mechanism **(XI).**

A totally different spectrum was obtained for **258** with the C(2) ^{13}C resonance at δ 92.3, more than 200 ppm removed from the position expected for a static classical cation. Since a static structure like **258** clearly was incompatible with the observed spectrum, a chemical shift estimate was made for the protoadamantyl cation **261.** However, the discrepancy between these estimated chemical shifts and those observed was too large to explain the behavior of the 1,3,5,7-tetramethyl-2-adamantyl cation within the properties of an equilibrating set of ions **261** even with the partial contribution of **258.** This left the set of σ-bridged ions **260** equilibrating according to mechanism **(X)** as the only possible structure for this ion.

3.5.3 Homoaromatic Cations

The concept of homoaromaticity was advanced by Winstein in 1960.[351] It represented a challenge to experimental and theoretical chemists alike.[352] The question of homoaromatic overlap has been mainly studied in 6-π electron Huckeloid systems,[351] although several 2-π electron systems have been discovered in recent years.[353–356]

3.5.3.1 *Monohomoaromatic Cations*

The simplest 2-π monohomoaromatic cation, the homocyclopropenyl cation **264** and its analogs, have been prepared and studied by Olah and his co-workers.[355] The experimental evidence for the existence of 1,3-overlap has been derived from ^{1}H- and ^{13}C-nmr data.[355] At lower temperature, the methylene protons of the ion

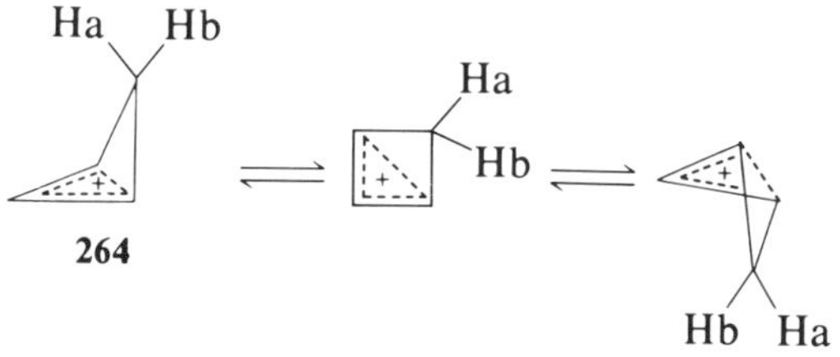

Ha Hb

Ha

Hb

264

Hb Ha

264 exhibit nonequivalence indicating ring puckering. The experimentally determined barrier[355] for ring flipping of 8.4 kcal · mol^{-1} compares closely with the theoretical estimates.[357] Other examples of monohomoaromatic cations are the 6-π homotropylium ion **109** of Pettit and Winstein[358] and the previously discussed 2-π homoaromatic 11-methyl-11-tricyclo[4.4.1.$0^{1,6}$]undecyl cation **32.**[94]

3.5.3.2 *Bishomoaromatic Cations*

The magnitude of any homoaromatic stabilization is expected to decrease with increasing interruption by methylene groups of the otherwise conjugated π-framework in neutral molecules. However, in an ionic species additional driving force is present for charge delocalization. Two of the most widely studied bishomoaromatic cations are the 7-norbornenyl and 7-norbornadienyl cations **265** and **266.**[354,359–361]

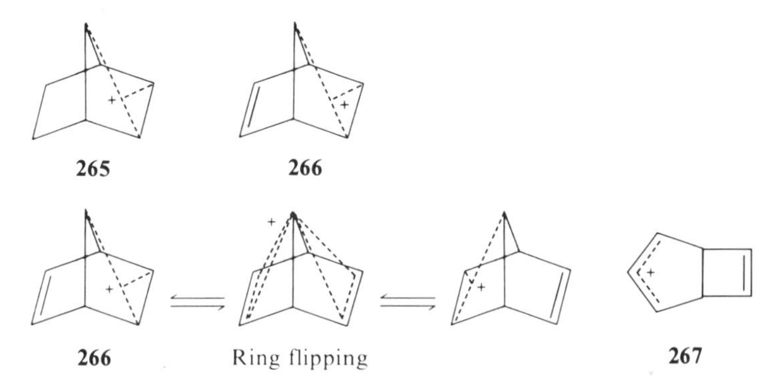

The ^{13}C-nmr spectrum of the cation **265** shows substantial shielding of both the C-7 cationic and vinylic carbon chemical shifts (δ ^{13}C, 34.0 and 125.9, respectively). A similar shielding phenomenon is observed for ion **266.** Interestingly, ion **266** undergoes bridge flipping rearrangement[360] as well as ring contraction-expansion through the intermediacy of bicyclo[3.2.0]heptadienyl cation **267.** These two processes can result in scrambling of all the seven carbon atoms.

Several studies including the application of the tool of increasing electron demand[295,361] best describes ion **265** as a symmetrical bridged π-bishomocyclopropenyl cation and not as a rapidly equilibrating pair of cyclopropylmethyl cations, such as **268.** The observed unusually large $^{13}C—H$ coupling constants at the C-7 position of **265** and **266** (218.9 Hz and 216.4 Hz, respectively) demonstrate the higher coordination of the carbocationic carbon.

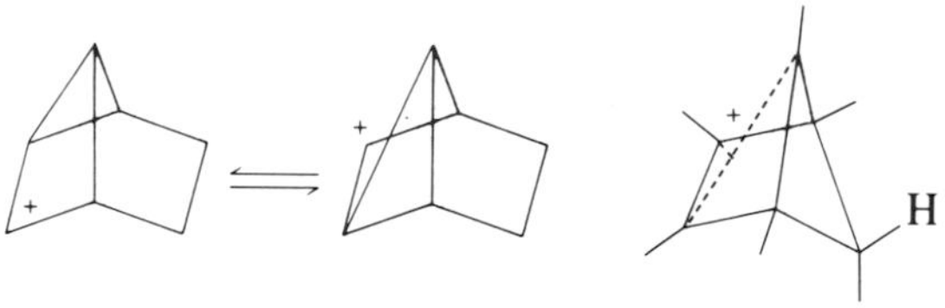

H

268 **269**

Several studies[21] on hexamethylbicyclo[2.1.1]hexenyl cation have shown that the ion is best represented as a bishomoaromatic species **269** analogous to **265** and **266.**

The extent of bishomoaromatic delocalization as expected is critically dependent upon structural geometry. Attempts to prepare the parent bishomoaromatic 4-cyclopentenyl cation **271** from 4-halocyclopentene **270** were unsuccessful.[362] They gave instead the cyclopentenyl cation. The lack of formation of bishomoaromatic ions from cyclopentenyl derivatives is mainly due to steric reasons. The planar cyclopentene skeleton has to bend into the "chair" conformation to achieve any significant overlap between empty *p*-orbital and π-*p* lobe of the olefinic bond, which is sterically unfavorable. However, such a conformation already exists in ions **265** and **266**

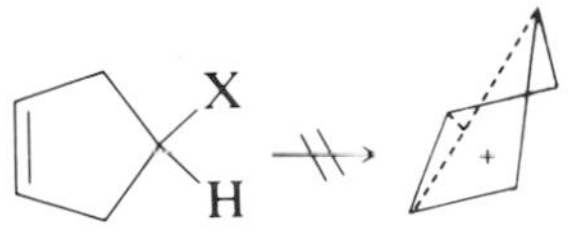

X
H
X = Cl, Br
270
271

Cations **272, 273,** and **274** are some of the 6-π bishomoaromatic cations that have been prepared and studied.[351d]

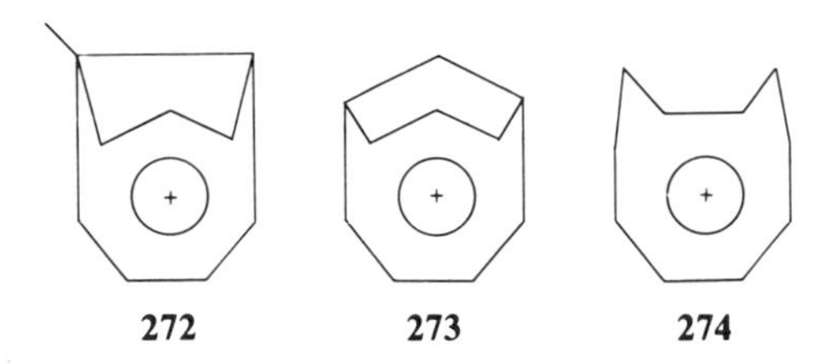

3.5.3.3 Trishomoaromatic Cations

Following Winstein's proposal[363] of the formation of a trishomoaromatic cation, in the solvolysis of *cis*-bicyclo[3.1.0]hexyl tosylate, extensive effort was directed toward its generation under stable ion conditions.[362a] Masamune and co-workers[353] were first able to prepare the ion **275** from *cis*-3-chlorobicyclo[3.1.0]hexane in the superacidic media. Subsequently, it has also been generated from the corresponding *cis*-bicyclo[3.1.0]hexan-3-ol in protic acid-free SbF_5:SO_2ClF solution. The observed nmr spectral data agree very well with the C_{3v} symmetry of the ion, **275.** The ^{13}C-nmr shifts are highly shielded for the three equivalent five-coordinated

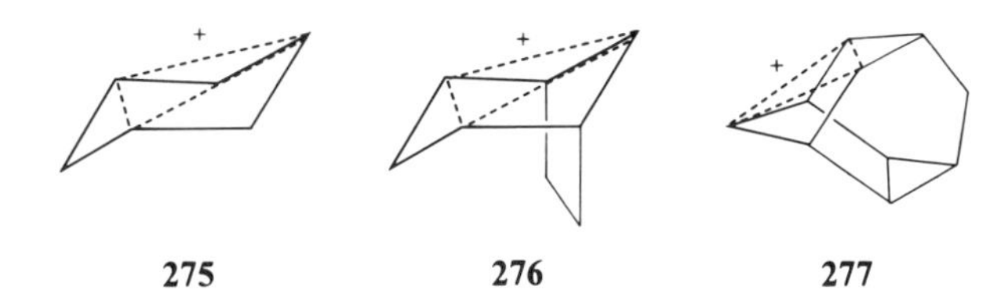

carbons (δ ^{13}C 4.9 with a ^{13}C—H coupling constant of 195.4 Hz), which is indicative of the nonclassical nature of the ion. Attempts to prepare methyl- and aryl-substituted trishomocyclopropenyl cations were, however, unsuccessful,[362b] which is consistent with Jorgensen's calculations.[365]

The ethanobridged analog **276** was also prepared[364] from the 8-chlorotricyclo[3.2.1.0$^{2.4}$]octane precursor. Ion **276** closely resembles the trishomoaromatic ion **275** in its spectral properties.

Another interesting trishomoaromatic system which has C_{3v} symmetry is the 9-pentacyclo[4.3.0.0$^{2.4}$0$^{2.8}$0$^{5.7}$]nonyl cation **277** prepared by Coates and co-workers.[356] Ion **277** has been thoroughly investigated in solvolysis.[366,367] A rate enhancement of 10^{10} to 10^{12} compared with ordinary systems uncovered its highly delocalized nature (strain relief is not the reason for the degenerate rearrangement). Also, analysis of remote and proximate substituent effects upon ionization[368] and application[296,369] of the tool of increasing electron demand fully reinforced its nonclassical nature. A recent deuterium labeling study also confirms this conclusion.[291]

3.5.4 Pyramidal Cations

3.5.4.1 $(CH)_5^+$ Type Cations

The close relationship between carbocations and boranes led Williams[370] to suggest the square pyramidal structure **278** for the $(CH)_5^+$ cation based on the square pyramidal structure of pentaborane. Stohrer and Hoffman[371] subsequently came to the same conclusion concerning the preferred square-pyramidal structure for the $(CH)_5^+$ cation using extended Hückel MO calculations.

Cation **278** can be viewed as square cyclobutadiene capped by CH^+.[371] Since then, several calculations of $(CH)_5^+$ at a more sophisticated level have appeared.[372] The MINDO/3 method[372c,373] indicated that the pyramidal cation **278** is less stable by 14.4 kcal · mol^{-1}, compared with the isomeric singlet cyclopentadienyl cation **279.** The triplet ion **280** was found to be more stable than the singlet ion **279** but only by 1.6 kcal · mol^{-1}.

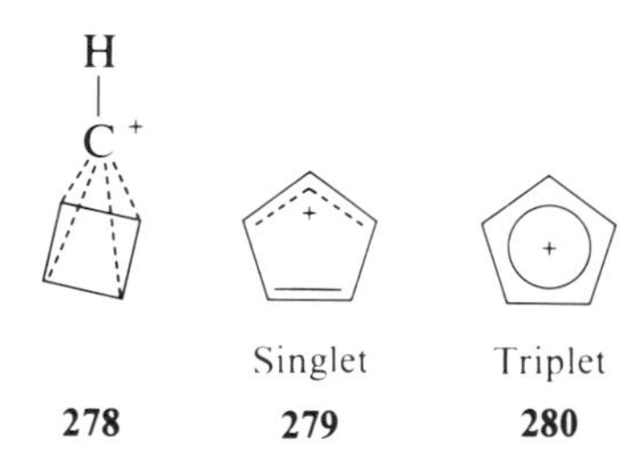

Although experimental work on the parent square pyramidal cation **278** has not been reported, the triplet ion **280** has been prepared and studied by electron spin resonance (esr) spectroscopy by Breslow, Saunders, and their co-workers.[374] A dimethyl-substituted derivative **281** of the pyramidal ion **278,** however, has been prepared by Masamune and studied by ^{1}H- and ^{13}C-nmr spectroscopy.[375]

Ionization of the dimethylhomotetrahedranol in $HSO_3F:SO_2ClF$ at $-78°C$ gave the pyramidal ion **281.** The alternative singlet cyclopentadienyl structure such as **279** for the species was eliminated based on ^{1}H- and ^{13}C-nmr data. The highly

281 **282**

shielded C-5 carbon chemical shift (δ ^{13}C -23.4) supports the structure **281** over a set of rapidly equilibrating structures such as **282.**[375b,c] The quenching of ion **281** at low temperature affords cyclopentenes.

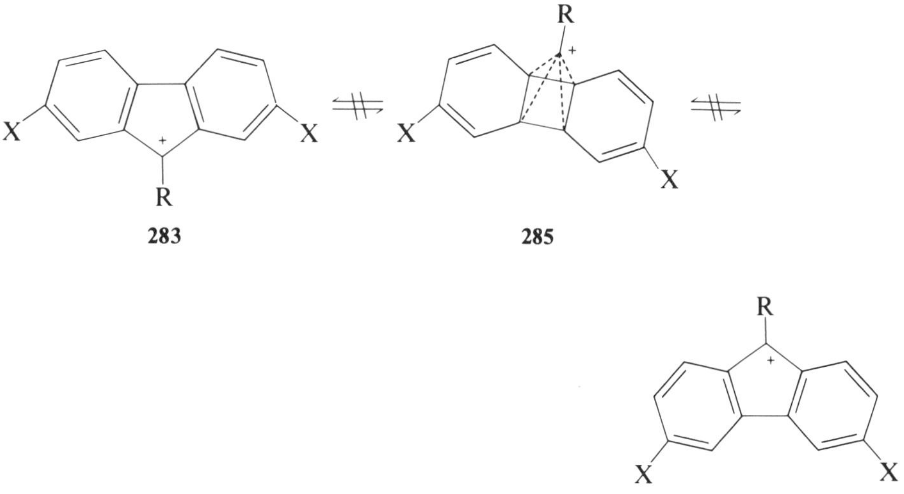

283 285

284

Attempts have been made to observe[373] the assumed interconversion of **283** to **284** in fluorenyl cations. Such intramolecular interconversion **(283–284)** through the capped pyramidal ion **285** was not observed. MINDO/3 calculations[373] on isomeric structures of cyclopentadienyl, indenyl, and fluorenyl cations indicated strongly decreasing relative stabilities of pyramidal forms due to benzoannulation.

Insertion of a methylene group into the four-membered ring of the pyramidal $(CH)_5^+$ cation **278** would give rise to the homo derivative **286.** As discussed earlier, hexamethylbicyclo[2.1.1]hexenyl cation[21,376–380] is best represented as a bishomocyclopropenyl cation **270** and not pyramidal type ion **287.**

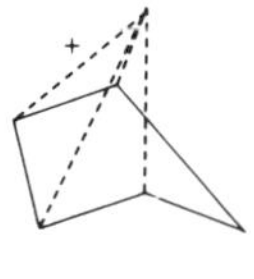

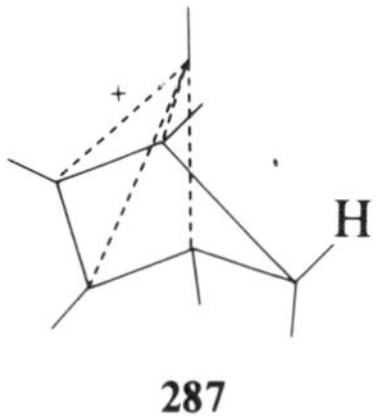

286 287

The trishomocyclopropenyl cation **288** has been investigated by both solvolytic[381] and stable ion studies.[375b] The obtained ^{1}H and ^{13}C-nmr data could be explained with the intermediacy of bishomopyramidal ion **289** although no conclusive distinction could be made between rapidly equilibrating systems of **288** and **289** structures. However, the ion **289** was preferred based on related MINDO/2 calculations.[382]

The other two substituted bishomo-$(CH)_5^+$ cations that have been investigated in the superacid media are the octamethylated ion **290** and its methanobridged analog **291**.[375c]

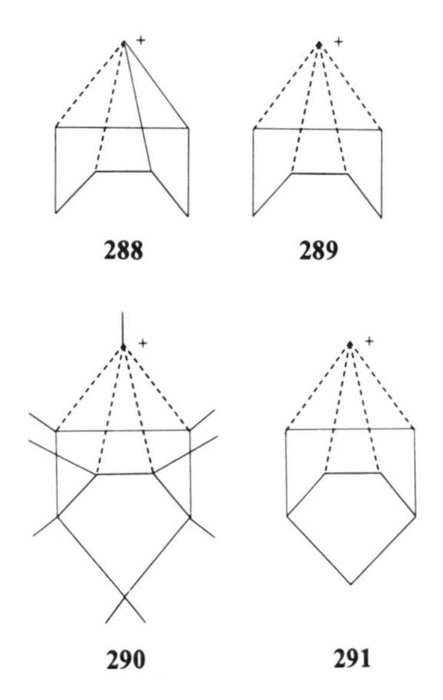

The observed[374c,383] ^{13}C-nmr data of both ions are consistent with their highly symmetrical structures **290** and **291**, but the data could also be explained with degenerate rapidly equilibrating systems of lesser symmetry. Hart and Willer[383e] have determined the apical ^{13}C—H coupling constant of ion **290** (J_{C-H} = 220 Hz), which is consistent with the single pyramidal structure **291** with nearly *sp* hybridization of the apical carbon atom. Surprisingly, the bishomoaromatic 7-norbornenyl cation, **266** has nearly identical ^{13}C—^{1}H coupling constant as the ion **290** at the C-7 carbon.[354]

3.5.4.2 $(CH)_6^{2+}$ Type Dications

The first known representative of the $(CH)_6^{2+}$-type hexacoordinate pyramidal dication is Hogeveen's hexamethyl cation **292**. The dication **292** can be prepared[21,384–385] from a variety of precursors in superacidic media (HSO_3F, $HSO_3F:SbF_5$) at low temperature. The observed five-fold symmetry in the ^{1}H- and ^{13}C-nmr spectra even at very low temperature (−150°C) with no line broadening leaves only two alternatives for the structure of the dication: the nonclassical five-fold symmetrical, static structure **292** or a set of rapidly equilibrating degenerate dications **293** with an activation energy barrier of less than 5 kcal · mol^{-1}.

^{1}H- and ^{13}C-nmr data, the rate of deuterium exchange, the rate of carbonylation, and the thermal stability of **292** provide strong evidence for the nonclassical nature of this carbodication.[385] However, the structure can also be considered in terms of the rapidly equilibrating set of structures **293** that have only one plane of symmetry during the degenerate process; the apical C-atom describes a circle.

Although the observed ^{1}H- and ^{13}C-nmr data strongly support the bridged nature of the ion **292,** further definite evidence comes from nmr isotope perturbation studies.[386] Very little isotopic perturbation (the degeneracy is not lifted due to CD_3 substitution) on the basal carbon signal is observed in the ^{13}C-nmr spectrum of the

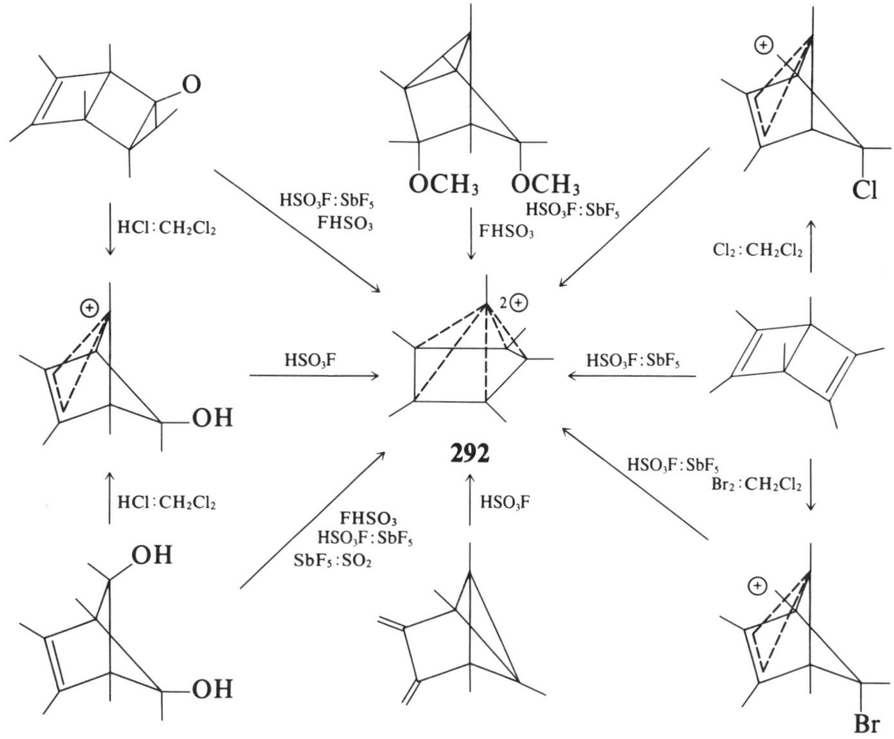

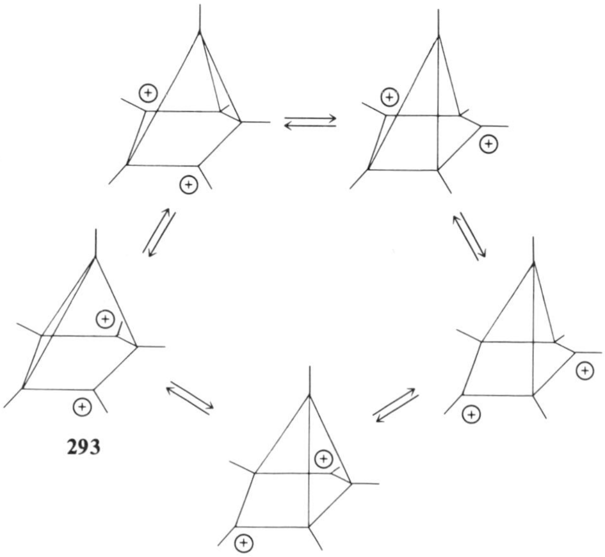

dication $[C_6(CH_3)_4(CD_3)_2]^{2+}$ indicating that the ion is symmetrically bridged. Other supporting evidence comes from a comparison of the ^{13}C- and ^{11}B-nmr shifts of the isoelectronic borane B_6H_{10} **294.** Williams and Field[387] have shown that the ^{13}C-nmr chemical shifts of a series of nonclassical cations may be compared with the ^{11}B-nmr chemical shifts of the isoelectronic polyhedral polyboranes.

Recently Hogeveen et al.[388] have also prepared the ethyl and isopropyl substituted analogs **295a–d** of the dipositive ion $(CH)_6^{2+}$.

294 **295**

a, $R^1 = CH_3$, $R^2 = C_2H_5$
b, $R^1 = CH_3$, $R^2 = i\text{—}C_3H_7$
c, $R^1 = R^2 = C_2H_5$
d, $R^1 = R^2 = i\text{-}C_3H_7$

REFERENCES

1a. J. F. Norris, W. W. Saunders, *J. Am. Chem. Soc. 23,* R85 (1901).
1b. F. Kehrmann, F. Wentzel, *Ber. Dtsch. Chem. Ges. 34* (1901); 3801 (1901).
1c. M. Gomberg, *Ibid. 35,* 1822 (1902).
1d. A. Baeyer, V. Villiger, *Ibid. 35,* 1189, 3013 (1902).
2. H. Meerwein, K. Van Emster, *Chem. Ber. 55,* 2500 (1922).
3. For a summary, see C. K. Ingold, *Structure and Mechanisms in Organic Chemistry,* Cornell Univ. Press, Ithaca, New York, 1953.
4. F. C. Whitmore, *J. Am. Chem. Soc. 54,* 3274, 3276 (1932); *Ann. Rep. Progr. Chem.* (*Chem. Soc. London*), 177 (1933).
5a. H. Meerwein, in *Methoden der organischen Chemie* (*Houben-Weyl*) E. Müller, Ed. 4th ed, Vol. VI/3, p. 329, Thieme, Stuttgart, 1965 and references therein.
5b. F. Z. Seel, *Anorg. Allg. Chem.* 250, 331 (1943); *Ibid.* 252, 24 (1943).
6. For earlier reviews, see G. A. Olah, *Chem. Eng. News,* 45, 76 (1967); *Science,* 168, 1298 (1970).
7. Magic Acid is a registered trade name.
8. G. A. Olah, *Angew. Chem. Int. Ed. Engl. 12,* 173 (1973).
9. For comprehensive reviews, see *Carbonium Ions,* G. A. Olah, P. v. R. Schleyer, Eds., Wiley-Interscience, New York, London, Vol. 1, 1968; Vol. II, 1970; Vol. III, 1972; Vol. IV, 1973; and Vol. V, 1976.
10. For very recent reviews, see G. A. Olah, *Top. Curr. Chem. 80,* 21 (1979); *Chem. Scr. 18,* 97 (1981).

11. G. A. Olah, *J. Am. Chem. Soc. 94,* 808 (1972).
12. "Existence" is defined by Webster as: "The state of fact of being especially as considered independently of human consciousness."
13. R. J. Gillespie, *J. Chem. Soc.* 1002 (1952) and for rebuttal M. J. S. Dewar, *Ibid.* 2885 (1953); H. H. Jaffe, *J. Chem. Phys. 21,* 1893 (1953).
14. R. J. Gillespie, *J. Chem. Phys. 21,* 1893 (1953).
15. E. L. Muetterties, R. A. Schunn, *Quart. Rev. 20,* 245 (1966).
16. For a review and critical discussion, see F. R. Jensen, B. Rickborn, *Electrophilic Substitution of Organomercurials,* McGraw-Hill, New York, 1968.
17. J. D. Roberts seems to have first used the term "nonclassical ion" when he proposed the tricyclobutonium structure for the cyclopropylcarbinyl cation. See J. D. Roberts, R. H. Mazur, *J. Am. Chem. Soc. 73,* 1542 (1951). Winstein referred to the nonclassical structure of norbornyl, cholesteryl, and 3-phenyl-2-butyl cations. See S. Winstein, D. Trifan, *Ibid. 74,* 1154 (1952). Bartlett's definition is widely used for nonclassical ions; it states "an ion is nonclassical if its ground state has delocalized bonding σ-electrons." Also see Ref. 9.

18a. For leading references, see P. D. Bartlett, *Nonclassical Ions,* reprints and commentary, W. A. Benjamin, New York, 1965.

18b. G. D. Sargent, *Quart. Rev. 20,* 301 (1966).

18c. S. Winstein, *Ibid. 23,* 141 (1969).

18d. For an opposing view, H. C. Brown, *Chem. Eng. News,* pp. 86–97 (February 13, 1967) and references given therein.

18e. H. C. Brown, *Acc. Chem. Res. 6,* 377 (1973).

18f. H. C. Brown, *Tetrahedron, 32,* 179 (1976).

18g. G. A. Olah, *Acc. Chem. Res. 41,* 9 (1976).

18h. H. C. Brown, *The Nonclassical Ion Problem,* with Commentary by P. v. R. Schleyer, Plenum Press, New York, N.Y. 1977.

18i. H. C. Brown, *Acc. Chem. Res. 16,* 432 (1983).

18j. C. A. Grob, *Ibid., 16,* 426 (1983).

18k. C. Walling, *Ibid. 16,* 448 (1983).

19a. G. A. Olah, A. M. White, J. R. DeMember, A. Commeyras, D. Y. Liu, *J. Am. Chem. Soc. 92,* 4627 (1970).

19b. G. A. Olah, G. Liang, G. D. Mateescu, J. L. Riemenschneider, *Ibid. 95,* 8698 (1973).

20. G. A. Olah, G. K. S. Prakash, M. Saunders, *Acc. Chem. Res. 16,* 440 (1983).
21. H. Hogeveen, P. W. Kwant, *Acc. Chem. Res. 8,* 413 (1975).

22a. For a review and discussion, see J. J. Jenner, *Chimia 20,* 309 (1966); however, for a contrasting view on general use of "carbonium ions", see G. Volz, *Unserer Zeit, 4,* 101 (1970).

22b. G. A. Olah, *Pure Appl. Chem. 51,* 1725 (1979).

22c. V. Gold, *Ibid. 55,* 1281 (1983).

23. This method has been employed in most of his work.
24. P. Ahlberg, *Chem. Scr. 2,* 81, 231 (1972).
25. M. Saunders, D. Cox, W. Ohlmstead, *J. Am. Chem. Soc. 95,* 3018 (1973).

26a. P. Ahlberg, C. Engdahl, *Chem. Scr. 11,* 95 (1977).

26b. P. Ahlberg, M. Ek, *Chem. Commun.* 624 (1979).

27. M. Saunders, D. Cox, J. R. Lloyd, *J. Am. Chem. Soc. 101,* 6656 (1979).
28. P. C. Myhre, C. S. Yannoni, *J. Am. Chem. Soc. 103,* 230 (1981).
29. D. P. Kelly, H. C. Brown, *Aust. J. Chem. 29,* 957 (1976).

30a. S. Forsen, R. A. Hoffman, *Acta Chem. Scand. 17,* 1787 (1963).

30b. S. Forsen, R. A. Hoffmann, *J. Chem. Phys. 39,* 2892 (1963).
31. S. Forsen, R. A. Hoffmann, *J. Chem. Phys. 40,* 1189 (1964).
32. C. Engdahl, P. Ahlberg, *J. Am. Chem. Soc. 101,* 3940 (1979).
33. P. v. R. Schleyer, D. Lenoir, P. Mison, G. Liang, G. K. S. Prakash, G. A. Olah, *J. Am. Chem. Soc. 102,* 683 (1980).
34. M. Saunders, J. Chandrasekhar, P. v. R. Schleyer in *Rearrangements in Ground and Excited States,* P. D. Mayo, Ed., Vol. 1, Academic Press, New York, 1980, p. 1.
35a. M. Saunders, M. H. Jaffe, P. Vogel, *J. Am. Chem. Soc. 93,* 2558 (1971).
35b. M. Saunders, P. Vogel, *Ibid. 93,* 2559, 2561 (1971).
36. M. Saunders, L. A. Talkowski, M. R. Kates, *J. Am. Chem. Soc. 99,* 8070 (1977).
37. M. Saunders, M. R. Kates, *J. Am. Chem. Soc, 99,* 8071 (1977).
38a. C. S. Yannoni, *Acc. Chem. Res. 15,* 201 (1982).
38b. J. Lyerla, C. S. Yannoni, C. A. Fyfe, *Ibid. 15,* 208 (1982).
39. For a detailed account of the method, see G. A. Olah, G. K. S. Prakash, D. G. Farnum, T. P. Clausen, *J. Org. Chem. 48,* 2146 (1983).
40a. J. M. Hollander, W. L. Jolly, *Acc. Chem. Res. 3,* 1931 (1970).
40b. D. Betteridge, A. D. Baker, *Anal. Chem. 42,* 42A (1970).
41. G. A. Olah, G. D. Mateescu, L. A. Wilson, M. H. Goss, *J. Am. Chem. Soc. 92,* 7231 (1970).
42a. G. A. Olah, E. B. Baker, J. C. Evans, W. S. Tolgyesi, J. S. McIntyre, J. I. Bastien, *J. Am. Chem. Soc. 86,* 1360 (1964).
42b. G. A. Olah, A. Commeyras, J. DeMember, J. L. Bribes, *Ibid. 93,* 459 (1971).
43. G. A. Olah, C. U. Pittmann, Jr., R. Waack, M. Doran, *J. Am. Chem. Soc. 88,* 1488 (1966).
44. G. A. Olah, C. U. Pittmann, Jr., M. C. R. Symons, in *Carbonium Ions,* Vol. I, Ref. 9.
45. E. M. Arnett, C. Petro, *J. Am. Chem. Soc. 100,* 2563 (1978).
46. E. M. Arnett, T. C. Hofelich, *J. Am. Chem. Soc. 105,* 2889 (1983).
47. For an extensive review, see *Mass Spectrometry of Organic Ions,* F. W. McLafferty, Ed., Academic Press, New York, 1963.
48. H. C. Brown, H. W. Pearsall, L. P. Eddy, *J. Am. Chem. Soc. 42,* 5347 (1950).
49. E. Wertyporoch, T. Firla, *Ann. Chim. 500,* 287 (1933).
50. G. A. Olah, S. J. Kuhn, J. A. Olah, *J. Chem. Soc.* 2174 (1957).
51. F. Fairbrother, *J. Chem. Soc.* 503 (1945).
52. J. Rosenbaum, M. C. R. Symons, *Proc. Chem. Soc. (London),* 92 (1959); J. Rosenbaum, M. Rosenbaum, M. C. R. Symons, *Mol. Phys. 3,* 205 (1960); J. Rosenbaum, M. C. R. Symons, *J. Chem. Soc. (London),* 1 (1961).
53. A. C. Finch, M. C. R. Symons, *J. Chem. Soc.* 378 (1965).
54. For a summary, see M. C. Deno, *Prog. Phys. Org. Chem. 2,* 129 (1964).
55. For preliminary communications and lectures, see G. A. Olah, Conference Lecture at the 9th Reaction Mechanism Conference, Brookhaven, New York, August 1962; Abstr. 142nd Nat. Meet. Am. Chem. Soc. (Atlantic City, N. J. 1962), p. 45; W. S. Tolgyesi, J. S. McIntyre, I. J. Bastien, M. W. Meyer, p. 121; G. A. Olah, *Angew. Chem. 75,* 800 (1963); *Rev. Chim. Acad. Rep. Populaire Roumaine 7,* 1139 (1962); preprints, *Div. Petr. Chem., ACS, 9,* (7), 21 (1964); Organic Reaction Mechanism Conference, Cork, Ireland, June 1964, Special Publications No. 19, Chemical Society, London, 1965.
56. C. U. Pittman, Jr., *Adv. Phys. Org. Chem. 4,* 305 (1960).
57. G. A. Olah, W. S. Tolgyesi, S. J. Kuhn, M. E. Moffatt, I. J. Bastien, E. B. Baker, *J. Am. Chem. Soc. 85,* 1378 (1963).
58a. J. Beacon, P. A. W. Dean, R. J. Gillespie, *Can. J. Chem. 47,* 1655 (1969).

58b. A. Commeyras, G. A. Olah, *J. Am. Chem. Soc. 90,* 2929 (1968).

59. For definition, see Chapter 1.

60. For a comprehensive report, see G. A. Olah, A. M. White, *J. Am. Chem. Soc. 91,* 5801 (1969).

61. G. A. Olah, D. J. Donovan, *J. Am. Chem. Soc. 99,* 5026 (1977).

62. M. Saunders, E. L. Hagen, *J. Am. Chem. Soc. 90,* 6881 (1968).

63. M. Saunders, P. Vogel, E. L. Hagen, J. Rosenfeld, *Acc. Chem. Res. 6,* 53 (1973).

64. C. C. Lee, D. J. Woodcock, *J. Am. Chem. Soc. 92,* 5992 (1970).

65. G. J. Karabatsos, C. Zioudrou, S. Meyerson, *J. Am. Chem. Soc. 92,* 5996 (1970).

66. L. Radom, J. A. Pople, V. Buss, P. v. R. Schleyer, *J. Am. Chem. Soc. 93,* 1813 (1971).

67. L. Radom, J. A. Pople, V. Buss, P. v. R. Schleyer, *J. Am. Chem. Soc. 94,* 311 (1972).

68. N. Bodor, M. J. S. Dewar, *J. Am. Chem. Soc. 93,* 6685 (1971).

69. N. Bodor, M. J. S. Dewar, D. H. Lo, *J. Am. Chem. Soc. 94,* 5303 (1972).

70. H. Lischka, H. J. Kohler, *J. Am. Chem. Soc. 100,* 5297 (1978).

71. H. Hogeveen, C. J. Gaasbeek, *Recl. Trav. Chim. Pays-Bas. 89,* 857 (1970).

72. M. Saunders, S. P. Budiansky, *Tetrahedron 35,* 929 (1979).

73. G. K. S. Prakash, A. Husain, G. A. Olah, *Angew. Chem. 95,* 51 (1983).

74. G. D. Mateescu, J. L. Riemenschneider, "Application of Electron Spectroscopy in Organic Chemistry," in *Electron Spectroscopy,* D. A. Shirley, Ed., North Holland, Amsterdam, London, 1972.

75. G. A. Olah, M. B. Comisarow, C. A. Cupas, C. V. Pittman, Jr., *J. Am. Chem. Soc. 87,* 2997 (1965).

76. G. A. Olah, E. Namanworth, *J. Am. Chem. Soc. 88,* 5327 (1966).

77. G. A. Olah, J. Sommer, E. Namanworth, *J. Am. Chem. Soc. 89,* 3576 (1967).

78. G. A. Olah, P. Schilling, J. M. Bollinger, J. Nishimura, *J. Am. Chem. Soc. 96,* 2221 (1974).

79. G. A. Olah, D. H. O'Brien, *J. Am. Chem. Soc. 89,* 1725 (1967).

80. G. A. Olah, C. U. Pittman, Jr., *J. Am. Chem. Soc. 89,* 2996 (1967).

81. G. A. Olah, J. Lukas, *J. Am. Chem. Soc. 89,* 2227, 4739 (1967).

82. H. Hogeveen, C. J. Gaasbeek, *Recl. Trav. Chim. Pays-Bas. 89,* 395 (1970) and references cited therein.

83. G. A. Olah, D. H. O'Brien, C. U. Pittman, Jr., *J. Am. Chem. Soc. 89,* 2996 (1967).

84. G. A. Olah, P. Schilling, *Liebigs Ann. Chem. 761,* 77 (1972).

85. G. A. Olah, N. Friedman, N. Bollinger, J. M. Lukas, *J. Am. Chem. Soc. 88,* 5328 (1966).

86. G. A. Olah, J. Lukas, *J. Am. Chem. Soc. 90,* 933 (1968).

87. M. Saunders, J. Rosenfeld, *J. Am. Chem. Soc. 91,* 7756 (1969).

88. D. M. Brouwer, E. L. Mackor, *Proc. Chem. Soc. (London),* 147 (1964); D. M. Brouwer, *Recl. Trav. Chim. Pays-Bas. 87,* 210 (1968).

89. G. A. Olah, A. P. Fung, T. N. Rawdah, G. K. S. Prakash, *J. Am. Chem. Soc. 103,* 4646 (1981).

90. R. P. Kirchen, T. S. Sorensen, K. E. Wagstaff, *J. Am. Chem. Soc. 100,* 5134 (1978).

91. G. A. Olah, A. L. Berrier, G. K. S. Prakash, *Proc. Natl. Acad. Sci. USA, 78,* 1998 (1981).

92. J. D. Roberts, V. Chambers, *J. Am. Chem. Soc. 73,* 3176, 5034 (1951).

93a. G. A. Olah, J. M. Bollinger, *J. Am. Chem. Soc. 90,* 6082 (1968).

93b. G. Biale, A. J. Parker, S. G. Smith, I. D. R. Stevens, S. Winstein, *Ibid. 92,* 115 (1970).

93c. P. v. R. Schleyer, T. M. Su, M. Saunders, J. C. Rosenfeld, *Ibid. 91,* 5174 (1969).

93d. J. M. Bollinger, J. M. Brinich, G. A. Olah, *Ibid. 92,* 4025 (1970).

94. G. A. Olah, G. Liang, D. B. Ledlie, M. G. Costopoulos, *J. Am. Chem. Soc. 99,* 4196 (1977).

95a. L. Radom, J. A. Pople, P. C. Hariharan, P. v. R. Schleyer, *J. Am. Chem. Soc. 95,* 6531 (1973), and references therein.

95b. V. Buss, P. v. R. Schleyer, L. C. Allen, *Top. Stereochem. 7,* 253 (1973).

95c. L. Radom, J. A. Pople, P. v. R. Schleyer, *J. Am. Chem. Soc. 95,* 8193 (1973).

95d. D. H. Aue, W. R. Davidson, M. T. Bowers, *Ibid. 98,* 6700 (1976). This reference gives the heats of formation of cyclopropyl cations in the gas phase.

96a. V. Schollkopf, K. Fellenberger, M. Potsch, P. v. R. Schleyer, T. Su, G. W. Van Dine, *Tetrahedron Lett.* 3639 (1967).

96b. W. Kutzeinigg, *Ibid.* 4965 (1967).

96c. V. Schollkopf, *Angew. Chem. Int. Ed. Engl. 7,* 588 (1968).

97. G. A. Olah, C. L. Jeuell, D. P. Kelly, R. D. Porter, *J. Am. Chem. Soc. 94,* 146 (1972).

98. J. S. Staral, I. Yavari, J. D. Roberts, G. K. S. Prakash, D. J. Donovan, G. A. Olah, *J. Am. Chem. Soc. 100,* 8016 (1978).

99a. M. Saunders, J. Rosenfeld, *J. Am. Chem. Soc. 92,* 2548 (1970).

99b. G. A. Olah, R. J. Spear, P. C. Hiberty, W. J. Hehre, *Ibid. 98,* 7470 (1976).

100. R. P. Kirchen, T. S. Sorensen, *J. Am. Chem. Soc. 99,* 6687 (1977).

101. G. A. Olah, G. K. S. Prakash, D. J. Donovan, I. Yavari, *J. Am. Chem. Soc. 100,* 7085 (1978).

102. J. Bredt, J. Houben, P. Levy, *Chem. Ber. 35,* 1286 (1902); J. Bredt, H. Thouet, J. Schmitz, *Ann. Chem.* 437 (1924).

103. P. D. Bartlett, L. H. Knox, *J. Am.Chem. Soc. 61,* 3184 (1939).

104. H. Koch, W. Haaf, *Angew. Chem. 72,* 628 (1960).

105. P. v. R. Schleyer, R. C. Fort, Jr., W. E. Watts, M. B. Comisarow, G. A. Olah, *J. Am. Chem. Soc. 86,* 4195 (1964).

106. G. A. Olah, G. K. S. Prakash, J. Shih, V. V. Krishnamurthy, G. D. Mateescu, G. Sipos, G. Liang, P. v. R. Schleyer, *J. Am. Chem. Soc.* (submitted).

107. G. A. Olah, G. Liang, *J. Am. Chem. Soc. 95,* 194 (1973).

108. G. A. Olah, G. Liang, P. W. Westerman, *J. Org. Chem. 39,* 367 (1974).

109. G. A. Olah, G. Liang, P. v. R. Schleyer, W. Parker, C. I. F. Watt, *J. Am. Chem. Soc. 99,* 966 (1977).

110. P. v. R. Schleyer, W. E. Watts, R. C. Fort, Jr., M. B. Comisarow, G. A. Olah, *J. Am. Chem. Soc. 86,* 5679 (1964).

111. J. D. Roberts, R. H. Mazur, *J. Am. Chem. Soc. 73,* 2509 (1951).

112. H. Hart, J. M. Sandri, *J. Am. Chem. Soc. 81,* 320 (1959).

113. N. C. Deno, H. G. Richey, Jr., J. S. Liu, J. D. Hodge, J. J. Houser, M. J. Wisotsky, *J. Am. Chem. Soc. 84,* 2016 (1962).

114. G. A. Olah, P. W. Westerman, J. Nishimura, *J. Am. Chem. Soc. 96,* 3548 (1974).

115. C. U. Pittman, Jr., G. A. Olah, *J. Am. Chem. Soc. 98,* 2998 (1965).

116. G. A. Olah, G. Liang, K. A. Babiak, T. M. Ford, D. L. Goff, T. K. Morgan, Jr., R. K. Murray, Jr., *J. Am. Chem. Soc. 100,* 1494 (1978) and references given therein.

117. D. S. Kabakoff, E. Namanworth, *J. Am. Chem. Soc. 92,* 3234 (1970).

118. G. A. Olah, G. K. S. Prakash, T. N. Rawdah, *J. Org. Chem. 45,* 965 (1980).

119. G. A. Olah, G. Liang, *J. Am. Chem. Soc. 97,* 1920 (1975).

120. G. A. Olah, G. Liang, *J. Am. Chem. Soc. 98,* 7026 (1976).

121. G. A. Olah, G. Liang, R. K. Murray, Jr., K. A. Babiak, *J. Am. Chem. Soc. 98,* 576 (1976).

122. G. A. Olah, G. Liang, *J. Org. Chem. 40,* 2108 (1975).

123. G. A. Olah, G. K. S. Prakash, G. Liang, *J. Org. Chem. 42,* 2666 (1977).

124. G. K. S. Prakash, Ph.D. dissertation, University of Southern California, 1978.

125. N. C. Deno, H. G. Richey, Jr., N. Friedman, J. D. Hodge, J. J. Houser, C. U. Pittman, Jr., *J. Am. Chem. Soc. 85,* 2991 (1963); N. C. Deno, J. M. Bollinger, N. Friedman, K. Hafer, J. D. Hodge, J. J. Houser, *Ibid. 85,* 2998 (1963).
126. T. S. Sorensen, *Can. J. Chem. 42,* 2768 (1969).
127. G. A. Olah, M. B. Comisarow, *J. Am. Chem. Soc. 86,* 5682 (1964).
128a. G. A. Olah, G. Liang, *J. Am. Chem. Soc. 94,* 6434 (1972) and references therein.
128b. G. A. Olah, G. Liang, Y. K. Mo, *Ibid. 94,* 3544 (1972).
128c. G. A. Olah, G. Liang, S. P. Jindal, *J. Org. Chem. 40,* 3249 (1975).
129. G. A. Olah, R. J. Spear, *J. Am. Chem. Soc. 97,* 1539 (1975).
130. G. A. Olah, G. Asensio, H. Mayr. *J. Org. Chem. 43,* 1518 (1978).
131. G. A. Olah, G. K. S. Prakash, G. Liang, *J. Org. Chem.* 41, 2820 (1976).
132. B. K. Carpenter, *J. Chem. Soc., Perkin. Trans. II,* 1 (1974).
133. Ref. 9, Vol. II.
134. G. A. Olah, J. S. Staral, R. J. Spear, G. Liang, *J. Am. Chem. Soc. 97,* 5489 (1975).
135. C. U. Pittman, Jr., *Chem. Commun.* 122 (1969).
136. T. S. Sorensen, *J. Am. Chem. Soc. 87,* 5075 (1965); *Can. J. Chem. 43,* 2744 (1965).
137. G. A. Olah, C. U. Pittman, Jr., T. S. Sorensen, *J. Am. Chem. Soc. 88,* 2331 (1966).
138. G. A. Olah, G. K. S. Prakash, G. Liang, *J. Org. Chem. 42,* 661 (1977).
139. D. M. Dytnerski, K. Ranganayakalu, B. P. Singh, T. S. Sorensen, *Can. J. Chem. 60,* 2993 (1982).
140. D. Dytnerski, K. Ranganayakalu, B. P. Singh, T. S. Sorensen, *J. Org. Chem. 48,* 309 (1983).
141a. A. A. Verrijn Stuart, E. L. Mackor, *J. Chem. Phys. 27,* 826 (1957); G. Dallinga, E. L. Mackor, A. A. Varrijn Stuart, *Mol. Phys. 1,* 123 (1958); G. Maclean, E. L. Mackor, *Discus. Faraday Soc. 34,* 165 (1962).
141b. T. Birchall, R. J. Gillespie, *Can. J. Chem. 42,* 502 (1964).
142. V. A. Koptyug, *Top. Curr. Chem.,* **122,** 1 (1984).
143. G. A. Olah, S. J. Kuhn, *Nature 178,* 693 (1956); *Naturwissenschaften 43,* 59 (1956); *J. Am. Chem. Soc. 80,* 6535 (1958); G. A. Olah, Abstr. 138th Natl. Meet. Am. Chem. Soc., New York, 1960, p. 3P; *Ibid. 87,* 1103 (1965).
144. G. A. Olah, R. H. Schlosberg, R. D. Porter, Y. K. Mo, D. P. Kelly, G. D. Mateescu, *J. Am. Chem. Soc. 94,* 2034 (1972).
145. G. A. Olah, J. S. Staral, G. Asensio, G. Liang, D. A. Forsyth, G. D. Mateescu, *J. Am. Chem. Soc. 100,* 6299 (1978).
146. G. A. Olah, G. D. Mateescu, Y. K. Mo, *J. Am. Chem. Soc. 95,* 1865 (1973).
147a. G. D. Mateescu, Ph.D. dissertation, Case Western Reserve University, 1971.
147b. G. A. Olah, J. M. Bollinger, C. A. Cupas, J. Lukas, *J. Am. Chem. Soc. 89,* 2692 (1967).
148. W. E. Doering, M. Saunders, H. G. Boyton, H. W. Earhart, E. F. Wadley, W. R. Edwards, G. Laber, *Tetrahedron 4,* 178 (1958).
149. C. A. Fyfe, D. Bruck, L. R. Lyerla, C. S. Yannoni, *J. Am. Chem. Soc. 101,* 4770 (1979).
150. P. Warner, S. Winstein, *J. Am. Chem. Soc. 91,* 7785 (1969).
151. D. J. Cram, J. M. Cram, *Acc. Chem. Res. 4,* 204 (1971).
152. D. J. Cram, *J. Am. Chem. Soc. 71,* 3863 (1949) and subsequent publications; for a critical review see D. J. Cram, *J. Am. Chem. Soc. 86,* 3767 (1964).
153a. G. A. Olah, M. B. Comisarow, E. Namanworth, B. Ramsay, *J. Am. Chem. Soc. 89,* 5259 (1967).
153b. G. A. Olah, M. B. Comisarow, C. J. Kim, *Ibid. 91,* 1458 (1969).
153c. G. A. Olah, R. D. Porter, *Ibid. 92,* 7267 (1970); 93, 6877 (1971).
153d. G. A. Olah, R. J. Spear, D. A. Forsyth, *Ibid. 99,* 2615 (1977).

154. G. A. Olah, G. Liang, *J. Am. Chem. Soc. 98,* 6304 (1976).
155a. G. A. Olah, B. P. Singh, *J. Am. Chem. Soc. 104,* 5168 (1982).
155b. G. A. Olah, B. P. Singh, G. Liang, *J. Org. Chem.* **48,** 4830 (1983).
155c. G. A. Olah, B. P. Singh, *J. Am. Chem. Soc.* **106,** 3265 (1984).
156. L. Eberson, S. Winstein, *J. Am. Chem. Soc. 87,* 3506 (1965).
157a. For reviews on vinyl cations, see H. G. Richey, Jr., J. M. Richey, *Carbonium Ions,* Vol. II, Ref. 9.
157b. M. Hanack, *Acc. Chem. Res. 3,* 209 (1970).
157c. G. Modena, U. Tonellato, *Adv. Phys. Org. Chem. 9,* 185 (1971).
157d. P. J. Stang, *Prog. Phys. Org. Chem. 10,* 276 (1973).
157e. P. J. Stang, Z. Rappoport, M. Hanack, L. R. Subramanian, *Vinyl Cations,* Academic Press, New York, New York, 1979.
158a. C. U. Pittmann, Jr., G. A. Olah, *J. Am. Chem. Soc. 87,* 5632 (1965).
158b. G. A. Olah, R. J. Spear, P. W. Westerman, J. H. Denis, *Ibid. 96,* 5855 (1974).
158c. G. A. Olah, A. L. Berrier, L. D. Field, G. K. S. Prakash, *Ibid. 104,* 1349 (1982).
159a. H-U. Siehl, H. Mayr, *J. Am. Chem. Soc. 104,* 909 (1982).
159b. H-U. Siehl, E-W. Koch, *J. Org. Chem. 49,* 575 (1984).
160. V. Gold, F. L. Tye, *J. Chem. Soc.* 2172 (1952).
161. D. G. Farnum, *J. Am. Chem. Soc. 86,* 934 (1964).
162. G. A. Olah, *J. Am. Chem. Soc. 86,* 932 (1964).
163. G. A. Olah, C. U. Pittman, Jr., *J. Am. Chem. Soc. 87,* 3507 (1965).
164. G. A. Olah, C. U. Pittman, Jr., E. Namanworth, M. B. Comisarow, *J. Am. Chem. Soc. 88,* 5571 (1966).
165. C. A. Cupas, M. B. Comisarow, G. A. Olah, *J. Am. Chem. Soc. 88,* 361 (1966).
166. J. M. Bollinger, M. B. Comisarow, C. A. Cupas, G. A. Olah, *J. Am. Chem. Soc. 89,* 5687 (1967).
167. G. A. Olah, R. D. Porter, D. P. Kelly, *J. Am. Chem. Soc. 93,* 464 (1971).
168. A. S. Siegel, *J. Am. Chem. Soc. 92,* 5277 (1970).
169. G. K. S. Prakash, T. N. Rawdah, G. A. Olah, *Angew. Chem. Int. Ed. Engl. 22,* 390 (1983).
170. H. Hart, R. W. Fish, *J. Am. Chem. Soc. 80,* 5894 (1958); 82, 5419 (1960); *83,* 4460 (1961).
171. R. J. Gillespie, E. A. Robinson, *J. Am. Chem. Soc. 86,* 5676 (1964).
172. N. C. Deno, N. Friedman, J. Mockus, *J. Am. Chem. Soc. 86,* 5676 (1964).
173. E. A. Robinson, J. A. Ciruna, *J. Am. Chem. Soc. 86,* 5677 (1964).
174. H. Hart, T. Sulzberg, R. R. Rafos, *J. Am. Chem. Soc. 85,* 1800 (1963).
175. H. Volz, M. J. Volz de Lecea, *Tetrahedron Lett.* 1871 (1964); *J. Am. Chem. Soc. 88,* 4863 (1966).
176. G. A. Olah, J. L. Grant, R. J. Spear, J. M. Bollinger, A. Serianz, G. Sipos, *J. Am. Chem. Soc. 98,* 2501 (1976).
177. J. M. Bollinger, C. A. Cupas, K. J. Friday, M. L. Woolfe, G. A. Olah, *J. Am. Chem. Soc. 89,* 156 (1967).
178. G. A. Olah, G. Liang, P. v. R. Schleyer, E. M. Engler, M. J. S. Dewar, R. C. Bingham, *J. Am. Chem. Soc. 95,* 6829 (1973).
179. G. K. S. Prakash, A. P. Fung, T. N. Rawdah, G. A. Olah, *J. Am. Chem. Soc.* Submitted.
180. G. A. Olah, et al., unpublished results.
181. G. A. Olah, G. K. S. Prakash, T. N. Rawdah, *J. Am. Chem. Soc. 102,* 6127 (1980).
182a. G. A. Olah, J. S. Staral, G. Liang, L. A. Paquette, W. P. Melega, M. J. Carmody, *J. Am. Chem. Soc. 99,* 3349 (1977).

182b. G. A. Olah, G. Liang, L. A. Paquette, W. P. Melega, *Ibid. 98,* 4327 (1976).

182c. G. A. Olah, J. S. Staral, L. A. Paquette, *Ibid. 98,* 1267 (1976).

183. G. A. Olah, M. Arvanaghi, G. K. S. Prakash, *Angew. Chem. Int. Ed. Engl. 22,* 712 (1983).

184. K. Lammertsma, H. Cerfontain, *J. Am. Chem. Soc. 102,* 3257 (1980).

185. K. Lammertsma, *J. Am. Chem. Soc. 103,* 2062 (1981).

186a. N. V. Bodoev, V. I. Mamatyuk, A. P. Krysin, V. A. Koptyug, *J. Org. Chem. USSR,* (Eng. Transl.) *14,* 1789 (1978).

186b. V. I. Mamatyuk, A. P. Krysin, N. V. Bodoev, V. A. Koptyug, *Izv. Akad. Nauk. SSSR. Ser Khim.* 2392 (1974).

187a. R. Breslow, C. Yuan, *J. Am. Chem. Soc. 80,* 5591 (1958).

187b. R. Breslow, J. T. Groves, G. Ryan, *Ibid. 89,* 5048 (1967).

187c. J. Ciabattoni, E. C. Nathan, *Ibid. 91,* 4766 (1969).

187d. J. Ciabattoni, E. C. Nathan, A. E. Feiring, P. J. Kocienski, *Org. Synth. 54,* 97 (1974).

188a. P. Müller, R. Etienne, J. Pfyffer, N. Pineda, M. Schipoff, *Helv. Chim. Acta. 61,* 2482 (1978).

188b. P. Müller, J. Pfyffer, E. W. Byrne, U. Burger, *Ibid., 61,* 2081 (1978).

188c. B. Halton, H. M. Hugel, D. P. Kelly, P. Müller, U. Burger, *J. Chem. Soc., Perk. II,* 258 (1976).

188d. U. Burger, P. Müller, L. Zuidema, *Helv. Chim. Acta, 57,* 1881 (1974).

189a. G. A. Olah, J. S. Staral, *J. Am. Chem. Soc. 98,* 6290 (1976).

189b. G. A. Olah, J. M. Bollinger, A. M. White, *Ibid. 91,* 3667 (1969).

189c. G. A. Olah, G. D. Mateescu, *Ibid. 92,* 1430 (1970).

189d. G. A. Olah, J. S. Staral, R. J. Spear, G. Liang, *Ibid. 97,* 5849 (1975).

190a. T. Nozoe, in *Non Benzonoid Aromatic Compounds,* D. Ginsburg, Ed., Chapter 7, Wiley-Interscience, New York, 1959; K. M. Harmon, in *Carbonium Ions,* Vol. 4, Ref. 9.

190b. R. W. Murray, M. L. Kaplan, *Tetrahedron Lett.* 2903 (1965).

190c. *Idem. Ibid.* 1307 (1967).

191. I. S. Akhrem, E. I. Fedin, B. A. Kvasov, M. E. Volpin, *Tetrahedron Lett.* 5265 (1967).

192. G. A. Olah, G. Liang, *J. Am. Chem. Soc. 98,* 3033 (1976).

193. G. A. Olah, G. Liang, *J. Am. Chem. Soc. 99,* 6045 (1977).

194. J. F. M. Oth, D. M. Smith, U. Prange, G. Schröder, *Angew. Chem. Int. Ed. Engl. 12,* 327 (1973).

195. D. H. O'Brien, A. J. Hart, C. R. Russel, *J. Am. Chem. Soc. 97,* 4410 (1975).

196. H. Spiesecke, W. G. Schneider, *J. Chem. Phys. 35,* 731 (1961); *Can. J. Chem. 41,* 966 (1963).

197a. H. Spiesecke, W. G. Schneider, *Tetrahedron Lett.* 468 (1961).

197b. J. B. Stothers *Carbon-13 NMR Spectroscopy,* Academic Press, New York, 1972.

197c. G. C. Levy, G. L. Nelson, *Carbon-13 Nuclear Magnetic Resonance for Organic Chemists,* Wiley-Interscience, New York, 1972.

198. F. Borchers, K. Levsen, *Org. Mass. Spect. 10,* 584 (1975) and references cited therein.

199. O. Hammerich, V. D. Parker, *Electrochim. Acta, 18,* 537 (1973).

200a. H. J. Shine, Y. Murata, *J. Am. Chem. Soc. 91,* 1872 (1969).

200b. Y. Murata, H. J. Shine, *J. Org. Chem. 34,* 3368 (1969).

201. L. S. Marcoux, *J. Am. Chem. Soc. 93,* 537 (1971).

202. I. C. Lewis, L. S. Singer, *J. Chem. Phys. 43,* 2712 (1965).

203. W. I. Aalbersberg, G. J. Hoijtink, E. L. Mackor, W. P. Weijland, *J. Chem. Soc.* 3055 (1959).

204. W. Th. A. M. Vander Lugt, H. M. Buck, L. J. Oosterhoff, *Tetrahedron 24,* 4941 (1968).

205. D. M. Brouwer, J. A. van Doorn, *Recl. Trav. Chim. Pays-Bas. 91,* 1110 (1972).

206a. S. I. Weisman, E. Boer, J. J. Conradi, *J. Chem. Phys. 26,* 963 (1957).

206b. G. J. Hoijtink, W. P. Weijland, *Recl. Trav. Chim. Pays-Bas. 76,* 936 (1957).

206c. D. M. Brouwer, *Chemistry and Industry (London),* 177 (1961).

206d. D. M. Brouwer, *J. Catalysis, 1,* 372 (1962).

206e. J. J. Pooney, R. C. Pink, *Proc. Chem. Soc.* 70 (1961).

207. G. A. Olah, D. A. Forsyth, *J. Am. Chem. Soc. 98,* 4086 (1976).

208. K. Lammertsma, G. A. Olah, C. M. Berke, A. Streitwieser, *J. Am. Chem. Soc. 101,* 6658 (1979).

209. G. A. Olah, B. P. Singh, *J. Org. Chem. 48,* 4830 (1983).

210a. I. Willner, M. Rabinowitz, *J. Am. Chem. Soc. 100,* 337 (1978).

210b. I. Willner, J. Y. Becker, M. Rabinowitz, *Ibid. 101,* 395 (1979).

211. G. A. Olah, G. K. S. Prakash, M. Arvanaghi, V. V. Krishnamurthy, S. C. Narang, *J. Am. Chem. Soc.* **106,** 2378 (1984).

212. G. A. Olah, M. B. Comisarow, C. A. Cupas, *J. Am. Chem. Soc. 88,* 362 (1966).

213a. For a summary see G. A. Olah, Y. K. Mo, in *Advances in Fluorine Chemistry,* Vol. 7, 69 (1972).

213b. G. A. Olah, Y. K. Mo, in *Carbonium Ions,* Vol. IV. Ref. 9.

214. G. A. Olah, R. D. Chambers, M. B. Comisarow, *J. Am. Chem. Soc. 89,* 1268 (1967).

215. G. A. Olah, C. U. Pittman, Jr., *J. Am. Chem. Soc. 88,* 3310 (1966).

216. G. A. Olah, M. B. Comisarow, *J. Am. Chem. Soc. 89,* 1027 (1967).

217. R. Filler, C. S. Wang, M. A. McKinney, F. N. Miller, *J. Am. Chem. Soc. 89,* 1026 (1967).

218. P. B. Sargent, C. G. Krespan, *J. Am. Chem. Soc. 91,* 415 (1969).

219. H. Volz, *Tetrahedron Lett.* 3413 (1963); *Ibid.* 5229 (1966).

220. M. Ballester, J. Riera-Figueras, A. Rodriguez-Siurana, *Tetrahedron Lett.* 3615 (1970).

221. R. West, P. T. Kwitowski, *J. Am. Chem. Soc. 88,* 5280 (1966).

222. G. A. Olah, M. B. Comisarow, *J. Am. Chem. Soc. 91,* 2955 (1969).

223. G. A. Olah, Y. Halpern, Y. K. Mo, *J. Am. Chem. Soc. 94,* 3551 (1972); G. A. Olah, G. Liang, Y. K. Mo, *J. Org. Chem. 39,* 2394 (1974).

224. A. Hantzsch, *Chem. Ber. 55,* 953 (1922); *Chem. Ber. 58,* 612, 941 (1925).

225. H. Meerwein, K. Bodenbeneer, P. Borner, F. Kunert, K. Wunderlich, *Ann. Chem. 632,* 38 (1960); H. Meerwein, V. Hederick, H. Morschel, K. Wunderlich, *Ibid. 635,* 1 (1960).

226. B. C. Ramsey, R. W. Taft, Jr., *J. Am. Chem. Soc. 88,* 3058 (1966).

227. G. A. Olah, J. M. Bollinger, *J. Am. Chem. Soc. 89,* 2993 (1967).

228. G. A. Olah, S. Yu, G. Liang, G. D. Mateescu, M. R. Bruce, D. J. Donovan, M. Arvanaghi, *J. Org. Chem. 46,* 571 (1981).

229. C. MacLean, E. L. Mackor, *J. Chem. Phys. 34,* 2207 (1961).

230. T. Birchall and R. J. Gillespie, *Can. J. Chem. 43,* 1045 (1965).

231. H. Hogeveen, *Recl. Trav. Chim. Pays-Bas. 86,* 696 (1967); D. M. Brouwer, *Ibid. 86,* 879 (1967).

232. M. Brookhart, G. C. Levy, S. Winstein, *J. Am. Chem. Soc. 89,* 1735 (1967).

233. G. A. Olah, D. H. O'Brien, M. Calin, *J. Am. Chem. Soc. 89,* 3582 (1967); G. A. Olah, M. Calin, D. H. O'Brien, *Ibid. 89,* 3586 (1967); G. A. Olah, M. Calin, *Ibid. 89,* 4736 (1967); G. Liang, Ph.D. Thesis, Case Western Reserve University, 1973.

234a. G. A. Olah and A. M. White, *J. Am. Chem. Soc. 91,* 5801 (1969).

234b. G. A. Olah and J. L. Grant, *J. Org. Chem. 42,* 2237 (1977).

235a. G. A. Olah, A. L. Berrier, G. K. S. Prakash, *J. Am. Chem. Soc. 104,* 2372 (1982).

235b. V. V. Krishnamurthy, G. K. S. Prakash, P. K. Iyer, G. A. Olah, *Ibid.* in press.

236. G. A. Olah, D. G. Parker, N. Yoneda, *Angew. Chem. Int. Ed. Engl. 17,* 909 (1978).

237. O. V. Lubinskaya, A. S. Shashkov, V. A. Chertkov, W. A. Smit, *Synthesis,* 742 (1976). Also see V. A. Smit, A. V. Semenovskii, O. V. Lubinskaya, V. F. Kucherov, *Doklady Akad. Nauk. SSSR, 203,* 604 (1972); *Chem. Abs. 77,* 87955 (1972).

238. G. A. Olah, A. M. White, *J. Am. Chem. Soc.* 3591, 4752 (1967); H. Hogeveen, *Rec. Trav. Chim. Pays-Bas. 86,* 289 (1967); H. Hogeveen, *Adv. Phys. Org. Chem. 10,* 29 (1973).

239. G. A. Olah, D. H. O'Brien, A. M. White, *J. Am. Chem. Soc. 89,* 5694 (1967); H. Hogeveen, *Recl.Trav. Chim. Pays-Bas. 86,* 816 (1967); H. Hogeveen, A. F. Bickel, C. W. Hilbers, E. L. Mackor, C. MacLean, *Chem. Commun.* 898 (1966); *Recl. Trav. Chim. Pays-Bas. 86,* 687 (1967).

240. G. A. Olah, A. M. White, *J. Am. Chem. Soc. 90,* 1884 (1968).

241. D. Cook, *Can. J. Chem. 37,* 48 (1959); for a review, also see D. Cook, in *Friedel-Crafts and Related Reactions,* G. A. Olah, Ed., Vol. 1, pp. 767–820, Interscience, New York, 1963.

242. B. P. Susz, J. J. Wuhrman, *Helv. Chim. Acta 40,* 971 (1957) and subsequent publications.

243. G. A. Olah, S. Kuhn, S. Beke, *Chem. Ber. 89,* 862 (1956); G. A. Olah, S. Kuhn, W. S. Tolgyesi, E. B. Baker, *J. Am. Chem. Soc. 84,* 2733 (1962); W. S. Tolgyesi, S. J. Kuhn, M. E. Moffatt, I. J. Bastien, E. B. Baker, *J. Am. Chem. Soc. 85,* 1328 (1963).

244. N. C. Deno, C. U. Pittman, Jr., and M. J. Wisotsky, *J. Am. Chem. Soc. 86,* 4370 (1964).

245. G. A. Olah, A. M. White, *J. Am. Chem. Soc. 89,* 7072 (1967).

246. C. U. Pittman, Jr., G. A. Olah, *J. Am. Chem. Soc. 87,* 5632 (1965); G. A. Olah, R. J. Spear, P. W. Westerman, J. H. Denis, *J. Am. Chem. Soc. 96,* 5855 (1974).

247. F. P. Boer, *J. Am. Chem. Soc. 88,* 1572 (1966).

248. G. A. Olah, M. B. Comisarow, *J. Am. Chem. Soc. 88,* 4442 (1966).

249. G. A. Olah, M. B. Comisarow, *J. Am. Chem. Soc. 88,* 3313 (1966).

250. G. A. Olah, M. B. Comisarow, *J. Am. Chem. Soc. 89,* 2694 (1967).

251a. G. A. Olah, A. M. White, D. H. O'Brien, *Chem. Rev. 70,* 561 (1970) and references given therein.

251b. Oxonium ions, carbonium ions, Vol. 5, Ref. 9.

252. G. A. Olah, T. Nakajima, G. K. S. Prakash, *Angew. Chem. Int. Ed. Engl. 19,* 811 (1980).

253. G. A. Olah, G. K. S. Prakash, T. Nakajima, *Angew. Chem. Int. Ed. Engl. 19,* 812 (1980).

254. R. J. Gillespie, T. Birchall, *Can. J. Chem. 41,* 148 (1963).

255. V. C. Armstrong, D. W. Farlow, R. B. Moodie, *Chem. Commun.* 1362 (1968).

256. G. A. Olah, A. M. White, A. T. Ku, *J. Org. Chem. 36,* 3585 (1971).

257. G. A. Olah, D. L. Brydon, R. D. Porter, *J. Org. Chem. 35,* 317 (1970).

258. J. L. Sudmeier, K. E. Schwartz, *Chem. Comm.* 1646 (1968).

259. G. A. Olah, A. M. White, *J. Am. Chem. Soc. 90,* 6087 (1968).

260a. G. A. Olah, T. E. Kiovsky, *J. Am. Chem. Soc., 90,* 4666 (1968).

260b. H. Meerwein, P. Laasch, R. Mersch, J. Spille, *Chem. Ber. 89,* 209 (1956).

261. J. F. McGarrity, D. P. Cox, *J. Am. Chem. Soc. 105,* 3961 (1983).

262. R. Petit, L. W. Haynes, in *Carbonium Ions,* Vol. V, pp. 2263, Ref. 9.

263. F. H. Field, M. S. B. Munson, *J. Am. Chem. Soc. 87,* 3289 and references given therein.

264a. D. Smith, N. G. Adams, E. Alge, *J. Chem. Phys. 77,* 1261 (1982).

264b. G. A. Olah et al., unpublished results.

265. G. A. Olah, G. Klopman, R. H. Schlosberg, *J. Am. Chem. Soc. 91,* 3261 (1969).

266a. V. Dyczmons, V. Staemmler, W. Kutzelnig, *Chem. Phys. Lett. 5,* 361 (1970).

266b. A. Dedieu, A. Veillard, Lecture at 21st Annual Meeting at the Societe Chimique Physique, Paris, Sept. 1970; A Veillard, private communication.

266c. W. A. Lathan, W. J. Hehre, J. A. Pople, *Tetrahedron Lett.* 2699 (1970).

266d. *Idem., J. Am. Chem. Soc. 93,* 808 (1971).

266e. E. L. Muetterties, *Ibid., 91,* 1636 (1969).
267. D. P. Stevenson, C. D. Wagner, O. Beek, J. W. Otvos, *J. Am. Chem. Soc.,* **74,** 3269 (1952).
268. G. A. Olah, Y. Halpern, J. Shen, Y. K. Mo, *J. Am. Chem. Soc. 93,* 1251 (1971).
269. R. H. Schlosberg, M. Siskin, W. P. Kocsi, F. J. Parker, *J. Am. Chem. Soc. 98,* 7723 (1976).
270. P. Ahlberg, G. Jonsall, C. Engdahl, *Adv. Phys. Org. Chem. 19,* 223 (1983).
271. M. Saunders, E. L. Hagen, *J. Am. Chem. Soc. 90,* 2436 (1968).
272. M. Saunders, M. R. Kates, *J. Am. Chem. Soc. 100,* 7082 (1978).
273. M. R. Kates, Ph.D. Thesis, Yale University, 1978.
274. G. A. Olah, A. M. White, *J. Am. Chem. Soc. 91,* 3954 (1969).
275. P. C. Myhre, J. D. Kruger, B. L. Hammond, S. M. Lok, C. S. Yannoni, V. Macho, H. H. Limbach, H. M. Vieth, *J. Am. Chem. Soc.,* **106,** 6079 (1984).
276. G. A. Olah, J. R. DeMember, A. Commeyras, J. L. Bribes, *J. Am. Chem. Soc. 93,* 459 (1971).
277. G. A. Olah, P. Schilling, P. W. Westerman, H. C. Lin, *J. Am. Chem. Soc. 96,* 3581 (1974).
278. G. A. Olah, G. Liang, *J. Am. Chem. Soc. 96,* 189 (1974); *Ibid. 96,* 195 (1974).
279. G. A. Olah, G. Liang, P. W. Westerman, *J. Org. Chem. 39,* 367 (1974).
280. Y. M. Jun, J. W. Timberlake, *Tetrahedron Lett.* 1761 (1982).
281. S. Winstein, D. S. Trifan, *J. Am. Chem. Soc. 71,* 2953 (1949), *74,* 1147 (1952).
282. G. D. Sargent, in *Carbonium Ions,* Vol. III, Ref. 9.
283. G. M. Kramer, *Adv. Phys. Org. Chem. 11,* 177 (1975).
284. H. C. Brown, *Top. Curr. Chem. 80,* 1 (1979).
285. C. A. Grob, *Angew. Chem. Int. Ed. Engl. 21,* 87 (1982).
286. G. A. Olah, G. K. S. Prakash, Preprints, *Div. Petr. Chem., ACS. 28(2),* 366 (1983).
287. M. Saunders, P. v. R. Schleyer, G. A. Olah, *J. Am. Chem. Soc. 86,* 5680 (1964).
288. G. A. Olah, G. K. S. Prakash, M. Arvanaghi, F. A. L. Anet, *J. Am. Chem. Soc. 104,* 7105 (1982).
289. C. S. Yannoni, V. Macho, P. C. Myhre, *J. Am. Chem. Soc. 104,* 907 (1982).
290. C. S. Yannoni, V. Macho, P. C. Myhre, *J. Am. Chem. Soc. 104,* 7380 (1982).
291. M. Saunders, M. R. Kates, *J. Am. Chem. Soc. 102,* 6867 (1980).
292a. M. Saunders, M. R. Kates, *J. Am. Chem. Soc. 105,* 3571 (1983).
292b. K. Ranganayakalu, T. S. Sorensen, *J. Org. Chem.,* **49,** 4310 (1984).
293. G. A. Olah, G. K. S. Prakash, G. Liang, *J. Am. Chem. Soc. 99,* 5683 (1977).
294a. D. G. Farnum, R. E. Botto, *Tetrahedron Lett. 46,* 4013 (1975).
294b. D. G. Farnum, R. E. Botto, W. T. Chambers, B. Lam, *J. Am. Chem. Soc. 100,* 3847 (1978).
294c. For preceding ^{1}H studies, see D. G. Farnum, H. D. Wolf, *J. Am. Chem. Soc. 96,* 5166 (1974).
295. G. A. Olah, A. L. Berrier, M. Arvanaghi, G. K. S. Prakash, *J. Am. Chem. Soc. 103,* 1122 (1981).
296. D. G. Farnum, T. P. Clausen, *Tetrahedron Lett. 22,* 549 (1981).
297. G. A. Olah, A. L. Berrier, G. K. S. Prakash, *J. Org. Chem. 47,* 3903 (1982).
298. G. A. Olah, R. D. Porter, C. L. Jeuell, A. M. White, *J. Am. Chem. Soc. 94,* 2044 (1972).
299. H. C. Brown, D. P. Kelly, M. Periasamy, *Proc. Natl. Acad. Sci. USA, 77,* 6956 (1980).
300. H. C. Brown, M. Periasamy, D. P. Kelly, J. J. Giansiracusa, *J. Org. Chem. 47,* 2089 (1982) and references cited therein.
301. The ESCA spectrum was obtained on an HP 7950A ESCA spectrometer at Jet Propulsion Labs in Pasadena, California.
302. D. W. Goetz, H. B. Schlegel, L. C. Allen, *J. Am. Chem. Soc. 99,* 8118 (1977).
303. D. T. Clark, B. J. Cromarty, L. Colling, *J. Am. Chem. Soc. 99,* 8121 (1977).
304. E. M. Arnett, N. Pienta, C. Petro, *J. Am. Chem. Soc. 102,* 398 (1980).
305. D. Farcasiu, *J. Org. Chem. 46,* 223 (1981).

306. P. v. R. Schleyer, J. Chandrasekhar, *J. Org. Chem. 46,* 225 (1981).
307a. F. Kaplan, P. Cross, R. Prinstein, *J. Am. Chem. Soc. 92,* 1445 (1970).
307b. J. J. Solomon, F. H. Field, *Ibid. 98,* 1567 (1976).
307c. R. H. Staley, R. D. Weiting, L. J. Beauchamp, *Ibid. 99,* 5964 (1977).
308. P. P. S. Saluja, P. Kebarle, *J. Am. Chem. Soc. 101,* 1084 (1979).
309. M. J. S. Dewar, R. C. Haddon, A. Komornicki, H. Rzepa, *J. Am. Chem. Soc. 99,* 377 (1977).
310. G. Wenke, D. Lenoir, *Tetrahedron 35,* 489 (1979).
311. H. J. Kohler, H. Lischka, *J. Am. Chem. Soc. 101,* 3479 (1979).
312. J. D. Goddard, Y. Osamura, H. Schaefer, III, *J. Am. Chem. Soc. 104,* 3258 (1982).
313a. K. Raghavachari, R. C. Haddon, P. v. R. Schleyer, H. F. Schaefer, III, *J. Am. Chem. Soc. 105,* 5915 (1983).
313b. M. Yoshimine, A. D. McLean, B. Liu, D. J. Defrees, J. S. Binkley, *Ibid. 105,* 6185 (1983).
314. G. Seybold, P. Vogel, M. Saunders, K. B. Wiberg, *J. Am. Chem. Soc. 95,* 2045 (1973).
315a. J. Meinwald, P. G. Gassman, *J. Am. Chem. Soc. 85,* 57 (1963).
315b. J. Meinwald, J. K. Crandall, *Ibid. 88,* 1292 (1966).
316. G. A. Olah, G. Liang, S. P. Jindal, *J. Am. Chem. Soc. 98,* 2508 (1976).
317. G. A. Olah, G. Liang, *J. Am. Chem. Soc. 97,* 1987 (1975).
318. M. Saunders, M. R. Kates, K. B. Wiberg, W. Pratt, *J. Am. Chem. Soc. 99,* 8072 (1977).
319. T. S. Sorensen, L. R. Schmitz, *J. Am. Chem. Soc. 102,* 1645 (1980).
320. G. A. Olah, G. K. S. Prakash, G. Liang, *J. Am. Chem. Soc. 101,* 3932 (1979).
321. R. H. Mazur, W. N. White, D. A. Semenow, C. C. Lee, M. S. Silver, J. D. Roberts, *J. Am. Chem. Soc. 81,* 4390 (1959).
322. K. B. Wiberg, *Tetrahedron, 24,* 1083 (1968); K. B. Wiberg, G. Szemies, *J. Am. Chem. Soc. 90,* 4195 (1968).
323a. M. Saunders, H-U. Siehl, *J. Am. Chem. Soc. 102,* 6868 (1980).
323b. W. J. Brittain, M. E. Squilliacote, J. D. Roberts, *J. Am. Chem. Soc.,* **106,** 7280 (1984).
324a. L. R. Schmitz, T. S. Sorensen, *J. Am. Chem. Soc. 104,* 2600 (1982).
324b. *Idem. Ibid. 104,* 2605 (1982).
325. G. A. Olah, G. K. S. Prakash, T. Nakajima, *J. Am. Chem. Soc. 104,* 1031 (1982).
326. R. K. Murray, Jr., T. M. Ford, G. K. S. Prakash, G. A. Olah, *J. Am. Chem. Soc. 102,* 1865 (1980).
327. G. A. Olah, D. J. Donovan, G. K. S. Prakash, *Tetrahedron Lett.* 4779 (1978).
328. D. M. Brouwer, J. A. Van Doorn, *Recl. Trav. Chim. Pays-Bas. 88,* 573 (1969).
329. M. Saunders, J. J. Stofko, Jr., *J. Am. Chem. Soc. 95,* 252 (1973).
330. A. P. W. Hewett, Ph.D. dissertation, Yale University, 1975.
331. V. Prelog, T. J. Traynham, in *Molecular Rearrangements,* Vol. 1, P. de Mayo, Ed., p. 593, Interscience, New York, 1963.
332a. W. Parker, C. I. F. Watt, *J. Chem. Soc., Perk. II,* 1642 (1975).
332b. L. Stehglin, L. Kannellias, G. Ourisson, *J. Org. Chem. 38,* 847, 851 (1973).
332c. L. Stehglin, J. Lhomme, G. Ourisson, *J. Am. Chem. Soc. 93,* 1650 (1971).
333. A. C. Cope, M. M. Martin, M. A. McKervey, *Quart. Rev. 20,* 119 (1966).
334a. R. P. Kirchen, T. S. Sorensen, K. Wagstaff, *J. Am. Chem. Soc. 100,* 6761 (1978).
334b. R. P. Kirchen, T. S. Sorensen, *Chem. Commun.* 769 (1978).
335. R. P. Kirchen, T. S. Sorensen, *J. Am. Chem. Soc. 101,* 3240 (1979).
336. R. P. Kirchen, N. Okazawa, K. Ranganayakalu, A. Rauk, T. S. Sorensen, *J. Am. Chem. Soc. 103,* 597 (1980).
337. R. P. Kirchen, K. Ranganayakalu, A. Rauk, B. P. Singh, T. S. Sorensen, *J. Am. Chem. Soc. 103,* 588 (1980).

338. G. Schröder, *Angew. Chem.*, *75,* 722 (1963).

339. M. J. Goldstein, R. Hoffman, *J. Am. Chem. Soc. 93,* 6193 and references cited therein.

340a. P. Ahlberg, D. L. Harris, S. Winstein, *J. Am. Chem. Soc. 92,* 4454 (1970).

340b. P. Ahlberg, D. L. Harris, M. Roberts, P. Warner, P. Seidl, M. Sakai, D. Cook, A. Diaz, J. P. Dirlam, H. Hamberger, S. Winstein, *Ibid. 94,* 7063 (1972).

341. C. Engdahl, P. Ahlberg, *J. Chem. Res(s).* 342 (1977).

342. C. Engdahl, G. Jonsall, P. Ahlberg, *Chem. Commun.* 626 (1979).

343. P. Ahlberg, C. Engdahl, G. Jonsall, *J. Am. Chem. Soc. 103,* 1583 (1981).

344. C. Engdahl, G. Jonsall, P. Ahlberg, *J. Am. Chem. Soc. 105,* (1983).

345. T. D. Bouman, C. Trindle, *Theor. Chim. Acta 37,* 217 (1975).

346. J. W. McIver, Jr., *Acc. Chem. Res. 7,* 72 (1974).

347. P. Ahlberg, J. B. Grutzner, D. L. Harris, S. Winstein, *J. Am. Chem. Soc. 92,* 3478 (1970).

348. D. Lenoir, P. Mison, E. Hyson, P. v. R. Schleyer, M. Saunders, P. Vogel, L. A. Telkowski, *J. Am. Chem. Soc. 96,* 2157 (1974).

349. D. Farcasiu, *J. Am. Chem. Soc. 98,* 5301 (1976).

350. D. Farcasiu, *J. Org. Chem. 43,* 3878 (1978).

351a. For reviews on homoaromaticity, see S. Winstein, *Quart. Rev. 23,* 141 (1969); *Chem. Soc. Spec. Publ.* No. 21, 5 (1967).

351b. P. R. Story, B. C. Clark, Jr., in *Carbonium Ions,* Vol. III, Ref. 9.

351c. P. J. Garatt, M. V. Sargent, in *Nonbenzenoid Aromatics,* Vol. II, J. F. Snyder, Ed., Academic Press, New York, 1971, p. 208.

351d. L. A. Paquett, *Angew. Chem. Int. Ed. Engl. 17,* 106 (1978).

352a. W. J. Hehre, *J. Am. Chem. Soc. 95,* 5807 (1973); *96,* 5207 (1974).

352b. R. Haddon, *Tetrahedron Lett.* 2797, 4303 (1974).

352c. R. S. Brown, T. G. Traylor, *J. Am. Chem. Soc. 95,* 8025 (1973).

353. S. Masamune, M. Sakai, A. V. K. Jones, T. Nakashima, *Can. J. Chem. 52,* 855 (1974).

354. G. A. Olah, G. Liang, *J. Am. Chem. Soc. 97,* 6803 (1975), and references cited therein.

355a. G. A. Olah, J. S. Staral, G. Liang, *J. Am. Chem. Soc. 96,* 6233 (1974).

355b. G. A. Olah, J. S. Staral, R. J. Spear, G. Liang, *Ibid. 97,* 5489 (1975).

356. R. M. Coates, E. R. Fretz, *J. Am. Chem. Soc. 97,* 2538 (1975).

357. W. L. Jorgensen, *J. Am. Chem. Soc. 97,* 3082 (1975); *98,* 6784 (1976).

358a. J. L. Rosenber, J. E. Mahler, R. Pettit, *J. Am. Chem. Soc. 84,* 2842 (1962); C. E. Keller, R. Pettit, *Ibid. 88,* 604 (1966); J. D. Holmes, R. Pettit, *Ibid. 85,* 2531 (1963).

358b. S. Winstein, H. D. Kaesz, C. G. Kreiter, E. C. Friearich, *J. Am. Chem. Soc. 87,* 3267 (1965); P. Warner, D. L. Harris, C. H. Bradley, S. Winstein, *Tetrahedron Lett.* 4013 (1970).

358c. S. Winstein, C. G. Kreiter, J. A. Brauman, *J. Am. Chem. Soc. 88,* 2047 (1966).

359. R. K. Richey, Jr., J. D. Nichols, *Tetrahedron Lett.* 1699 (1970).

360a. R. K. Lustagarten, M. Brookhart, S. Winstein, *J. Am. Chem. Soc. 89,* 6350 (1967).

360b. M. Brookhart, R. K. Lustgarten, S. Winstein, *Ibid. 89,* 6352 (1967).

360c. R. K. Lustgarten, M. Brookhart, S. Winstein, *Ibid. 94,* 2347 (1972).

361a. P. G. Gassman, J. Zeller, J. Lumb, *Chem. Commun. 69, (1968).*

361b. P. G. Gassman, A. F. Fentiman, Jr., *J. Am. Chem. Soc. 91,* 1545 (1969), *92,* 2549 (1970).

361c. H. G. Richey, Jr., D. Nichols, P. G. Gassman, A. F. Fentiman, Jr., S. Winstein, M. Brookhart, R. K. Lustgarten, *Ibid. 92,* 3783 (1970).

362a. G. A. Olah, G. Liang, Y. K. Mo, *J. Am. Chem. Soc. 94,* 3544 (1972).

362b. G. A. Olah, G. K. S. Prakash, T. N. Rawdah, D. Whittaker, J. C. Rees, *J. Am. Chem. Soc. 101,* 3935 (1979).

363. S. Winstein, *J. Am. Chem. Soc. 81,* 6524 (1959); S. Winstein, J. Sonnerberg, L. DeVries, *Ibid. 81,* 6523 (1959); S. Winstein, J. Sonnerberg, *Ibid. 83,* 3235 and 3244 (1961), S. Winstein, E. C. Frederick, R. Baker, Y. I. Liu, *Tetrahedron Lett. 22,* 5621 (1966).
364. S. Masamune, H. Sakai, A. V. K. Jones, *Can. J. Chem. 858* (1974).
365. W. L. Jorgensen, *Tetrahedron Lett.* 3029, 3033 (1976).
366. R. M. Coates, J. L. Kirkpatrick, *J. Am. Chem. Soc. 90,* 4162 (1968).
367. R. M. Coates, J. L. Kirkpatrick, *J. Am. Chem. Soc. 92,* 4883 (1970).
368. R. M. Coates, E. R. Fretz, *J. Am. Chem. Soc. 99,* 297 (1977).
369. H. C. Brown, M. Ravindranathan, *J. Am. Chem. Soc. 99,* 299 (1977).
370. R. E. Williams, *Inorg. Chem. 10,* 210 (1971).
371. W. D. Stohrer, R. Hoffomann, *J. Am. Chem. Soc. 94,* 1661 (1972).
372a. S. Yoneda, Z. Yoshida, *Chem. Lett.* 607 (1972).
372b. H. Kollmar, H. O. Smith, P. v. R. Schleyer, *J. Am. Chem. Soc. 95,* 5834 (1973).
372c. M. J. S. Dewar, R. C. Haddon, *Ibid. 95,* 583 (1973).
372d. W. J. Hehre, P. v. R. Schleyer, *Ibid. 95,* 5837 (1973).
372e. M. J. S. Dewar, R. C. Haddon, *Ibid. 96,* 255 (1974).
373. G. A. Olah, G. K. S. Prakash, G. Liang, P. W. Westerman, K. Kunde, J. Chandrasekhar, P. v. R. Schleyer, *J. Am. Chem. Soc. 102,* 4485 (1980).
374. R. Breslow, *Acc. Chem. Res. 6,* 393 (1973) and references given therein.
375a. S. Masamune, M. Sakai, H. Ona, A. L. Jones, *J. Am. Chem. Soc. 94,* 8956 (1972).
375b. S. Masamune, M. Sakai, A. V. Kemp-Jones, H. Ona, A. Venot, T. Nakashima, *Angew. Chem. 85,* 829 (1973).
375c. A. V. Kemp-Jones, N. Nakamura, S. Masamune, *Chem. Commun.* 109 (1974).
376. H. Hogeveen, H. C. Volger, *Recl. Trav. Chim. Pays-Bas, 87,* 385 (1968); *87,* 1042 (1968); *88,* 353 (1969).
377. L. A. Paquette, G. R. Krow, J. M. Bollinger, G. A. Olah, *J. Am. Chem. Soc. 90,* 7147 (1968).
378. H. Hogeveen, P. W. Kwant, *J. Am. Chem. Soc. 95,* 7315 (1973).
379. P. W. Kwant, dissertation, Universitat Groningen, 1974.
380. H. Hogeveen, P. W. Kwant, E. P. Schudde, P. A. Wade, *J. Am. Chem. Soc. 96,* 7518 (1974).
381. R. K. Lustgarten, *J. Am. Chem. Soc. 94,* 7602 (1972); S. Masamune, R. Vukok, M. J. Benett, J. T. Purdham, *Ibid. 94,* 8239 (1972); P. G. Gassman, X. Creary, *Ibid. 95,* 2729 (1973).
382. K. Morio, S. Masamune, *Chem. Lett.* 1107 (1974).
383a. H. Hart, M. Kuzuya, *J. Am. Chem. Soc. 94,* 8958 (1972).
383b. *Idem. Ibid., 95,* 4096 (1973).
383c. *Idem. Ibid. 96,* 6436 (1974).
383d. *Idem. Tetrahedron Lett.* 4123 (1973).
383e. H. Hart, R. Willer, *Ibid.* 4189 (1978).
384. H. Hogeveen, P. W. Kwant, *Tetrahedron Lett.* 1665 (1973).
385. H. Hogeveen, P. W. Kwant, *J. Am. Chem. Soc. 96,* 2208 (1974) and references cited therein.
386. H. Hogeveen, E. M. G. A. Van Kruchten, *J. Org. Chem. 46,* 1350 (1981).
387. R. E. Williams, L. D. Field in *Boron Chemistry,* R. W. Parry and G. Kodama, Eds., Pergamon Press, New York, 1980.
388a. C. Giordano, R. F. Heldeweg, H. Hogeveen, *J. Am. Chem. Soc. 99,* 5181 (1977).
388b. R. F. Heldeweg, dissertation, Universitat Groningen, 1977.

chapter 4

HETEROCATIONS IN SUPERACIDS

4.1 INTRODUCTION

This chapter deals with cations generated in superacid media wherein the positive charge is located on an atom other than carbon. The heteroatoms include oxygen, sulfur, selenium, tellurium, nitrogen, chlorine, bromine, iodine, hydrogen, xenon, and krypton. This discussion is divided for convenience into the following sections: (1) Higher valency onium ions such as oxonium, sulfonium, selenonium, telluronium, halonium, diazonium, and other species. The discussion of ammonium and phosphonium ions are excluded. (2) Enium ions of nitrogen (nitrenium), phosphorous (phosphenium), boron (borenium), and silicon (silicenium). (3) Homo and heteropolycations of halogens (interhalogen cations), polyhomoatomic cations of oxygen, sulfur, selenium, and tellurium, and other polyheteroatomic cations. (4) Miscellaneous cations of hydrogen, xenon, and krypton.

4.2 ONIUM IONS

4.2.1 Oxonium Ions

4.2.1.1 Hydronium Ion (H_3O^+)

The hydronium ion, H_3O^+ **1,** is the parent of saturated oxonium ions. It was first postulated in 1907[1] and gained wider acceptance with the acid-base theory of

Brønsted and Lowry.[2] The life time of H_3O^+ **1** in aqueous solution has been estimated to be about 10 times longer than the time scale of molecular vibrations.[3] The hydronium ion **1** has been studied both in gas phase as well as in solution (ir,[4] Raman,[5] nmr,[6] mass spectrometry,[7] and neutron diffraction[8]). In the mass spectrum, evidence has also been obtained for H_2O solvated hydronium ions such as $H_5O_2^+$, $H_7O_3^+$, etc. Extensive evidence for the long-lived hydronium ion has been obtained in superacid solution.[6] In fact, Christe and co-workers[9] have isolated hydronium salts with a variety of counter ions such as SbF_6^-, AsF_6^-, and BF_4^-. Complete vibrational[4] and neutron diffraction studies[8] support the pyramidal nature of the hydronium ion **1.**

The ^{1}H-nmr spectrum of H_3O^+ **1** in superacids such as $HSO_3F:SbF_5$ (Magic Acid) and $HF:SbF_5$ is observed at δ 11.0 (from tetramethylsilane).[6] Recent ^{17}O-nmr spectroscopic studies[10,11] in $HF:SbF_5$ with ^{17}O-enriched hydronium ion have indicated strong ^{17}O—^{1}H coupling (Fig. 4.1). In the proton-coupled spectrum, a quartet is observed at δ ^{17}O 9 ± 0.2 (with reference to SO_2 at 505 ppm) with $J_{^{17}O\text{-}^1H} = 106 \pm 1.5$ Hz.

Mateescu and Benedikt[10] suggested that H_3O^+ **1** is planar and not pyramidal, on the basis of a 33% increase in ^{17}O—^{1}H coupling constant upon going from H_2O to H_3O^+. They assumed a linear relationship between $J_{^{17}O\text{-}^1H}$ and the hybridization state of oxygen. Symons[12] subsequently showed that the assumed sp^3 hybridization for H_2O by Mateescu and Benedikt was not true (H_2O has a bond angle of 104.5°[13]). Redoing the calculation, Symons[12] showed that the hydronium ion **1** is pyramidal with an H—O—H bond angle of 111.3°. Similar conclusions have been reached on the basis of recent quantum mechanical calculations.[14]

4.2.1.2 Primary Oxonium Ions

It has been shown that methyl and ethyl alcohol in sulfuric acid give stable solutions of the corresponding alkyl hydrogen sulfates.[15] Many other alcohols show similar initial behavior, but the solutions are not stable at room temperature.

$$\text{R-OH} + 2H_2SO_4 \longrightarrow RHSO_4 + H_3O^+ + HSO_4^-$$

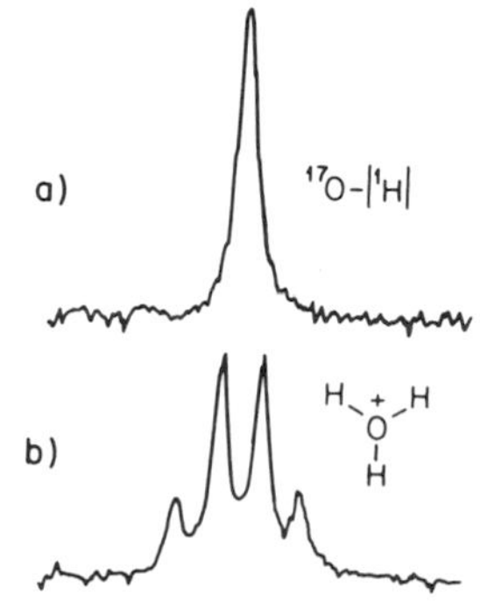

FIGURE 4.1 ^{17}O NMR spectrum of H_3O^+ in $HF:SbF_5/SO_2$ at −20°, (a) proton decoupled (b) proton coupled.

The first direct nmr spectroscopic evidence for the existence of primary alkyloxonium ions (protonated alcohols) in superacid solutions was found in 1961 by Mackor. The nmr spectrum of ethanol in BF_3-HF solution at $-70°$ gave a well-resolved triplet at about δ 9.90 for the protons on oxygen coupled to the methylene protons.[16] In HSO_3F this fine structure is not observed, even at $-95°$, due to fast exchange.[17]

The nmr spectra of a series of aliphatic alcohols have been investigated in the stronger acid system, $HSO_3F:SbF_5$ using sulfur dioxide as diluent.[18,19] Methyl, ethyl, *n*-propyl, isopropyl, *n*-butyl, isobutyl, *sec*-butyl, *n*-amyl, neopentyl, *n*-hexyl, and neohexyl alcohol all give well-resolved nmr spectra at $-60°$, under these conditions.

$$\mathrm{ROH} \xrightarrow[-60°]{\mathrm{HSO_3F:SbF_5:SO_2}} \mathrm{ROH_2}^+$$

The strength of this acid system is reflected by the fact that even at 25°C solutions of primary alcohols in $HSO_3F:SbF_5$ show fine structure for the proton on oxygen (Fig. 4.2). This indicates that at this relatively high temperature the exchange rate is still slow on the nmr time scale. In ^{1}H-nmr, the OH_2^+ protons of the primary alcohols appear at a lower field (δ 9.3–9.5) than the isomeric secondary alcohols (δ 9.1). This is due to a different charge distribution, as confirmed by the C_1 protons of the primary alcohols appearing at higher field (δ 5.0–4.7) compared with the C_1 methine proton of the secondary alcohols (δ 5.4 and 5.5). The ^{17}O chemical shift[11] of protonated methyl alcohol **2** is deshielded by approximately 25 ppm from the neutral methyl alcohol. The ^{17}O-H coupling constant of 107.6 Hz is similar in magnitude to that of the parent oxonium ion OH_3^+ **1.**

Aliphatic glycols in $HSO_3F:SbF_5:SO_2$ solution give diprotonated species at low temperatures.[20] In diprotonated diols, the protons on oxygen are found at lower fields than in protonated alcohols reflecting the presence of two positive charges. This is especially true for ethylene glycol (δ 11.2) where the positive charges are adjacent. As the separation of the positive charges becomes greater with increasing chain length, the chemical shift of the protons on oxygen of protonated diols approaches that of protonated alcohols.

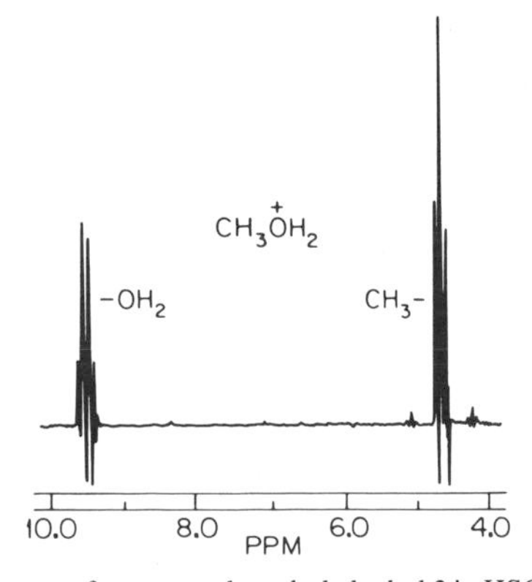

FIGURE 4.2 ^{1}H NMR spectrum of protonated methyl alcohol *2* in HSO_3F-SbF_5-SO_2 solution at $-60°$.

The reactivity of protonated alcohols[19] and protonated diols[20] in strong acids has been studied by nmr spectroscopy. Protonated methyl alcohol shows surprising stability in $HSO_3F:SbF_5$ and can be heated to 50°C without significant decomposition, although surprising condensation to *t*-butyl cation was occasionally observed. Protonated ethyl alcohol **3** is somewhat less stable and begins to decompose at 30°C. The cleavage of protonated *n*-propyl alcohol **4** has been followed in the temperature range 5–25°C, giving a mixture of *t*-butyl and *t*-hexyl cations.

$$\underset{\mathbf{4}}{CH_3CH_2CH_2\overset{+}{O}H_2} \xrightarrow[HSO_3F:SbF_5]{5\text{–}25°C} (CH_3)_3C^+ + (CH_3)_2CH\overset{+}{C}(CH_3)_2 + H_3O^+$$

Higher protonated alcohols cleave to stable tertiary alkyl cations. For protonated primary and secondary alcohols, the initially formed primary and secondary carbocations rapidly rearrange to the more stable tertiary carbenium ions under the conditions of the reaction. For example, protonated *n*-butyl alcohol **5** cleaves to *n*-butyl cation which rapidly rearranges to *t*-butyl cation.

$$\underset{\mathbf{5}}{CH_3CH_2CH_2CH_2\overset{+}{O}H_2} \xrightarrow{k_1} [CH_3CH_2CH_2CH_2{}^+] \xrightarrow{k_2} (CH_3)_3C^+ \qquad k_2 \gg k_1$$

The cleavage to carbocations, shown to be first order, is enhanced by branching of the chain: protonated 1-pentanol is stable up to 0°C, isopentyl alcohol is stable up to −30°C, and neopentyl alcohol cleaves at −50°C. The stability of the protonated primary alcohols also decreases as the chain length is increased.

When the $HSO_3F:SbF_5:SO_2$ solutions of diprotonated glycols are allowed to warm up, pinacolone rearrangements, formation of allylic carbenium ions, and cyclization reactions of diprotonated glycols can be directly observed by nmr spectroscopy. Diprotonated ethylene glycol **6** rearranges to protonated acetaldehyde in about 24 h at room temperature. Protonated 1,2-propanediol undergoes a pinacolone rearrangement to protonated propionaldehyde, probably through the initial cleavage of water from the secondary position. Diprotonated 2,3-butanediol **7** rearranges to

$$\underset{\mathbf{6}}{CH_3\underset{\underset{{}^+OH_2}{|}}{CH}-\underset{\underset{{}^+OH_2}{|}}{CH_2}} \xrightarrow{-H_2O} CH_3-\overset{+}{C}H-CH_2-\overset{+}{O}H_2 \longrightarrow CH_3CH_2C(=\overset{+}{O}H)H$$

protonated methyl ethyl ketone either through a direct hydride shift (Equation 1) or through a bridged intermediate (Equation 2).

$$\underset{\mathbf{7}}{CH_3-\underset{\underset{H}{|}}{\overset{\overset{{}^+OH_2}{|}}{C}}-\underset{\underset{{}^+OH_2}{|}}{CH}-CH_3} \longrightarrow CH_3-\underset{\underset{H}{|}}{\overset{\overset{H\overset{+}{O}H}{|}}{C}}-\overset{+}{C}HCH_3 \longrightarrow CH_3-\overset{\overset{{}^+OH}{\|}}{C}-CH_2CH_3 \qquad (1)$$

$$\underset{\mathbf{7}}{CH_3-\underset{H}{\overset{\overset{+}{O}H_2}{C}}-\underset{H}{\overset{\overset{+}{O}H_2}{C}}-CH_3} \longrightarrow CH_3-\underset{H}{C}\overset{\overset{H}{O^+}}{\text{—}}\underset{H}{C}-CH_3 \longrightarrow CH_3-\overset{\overset{+}{O}H}{\overset{\|}{C}}-CH_2-CH_3 \quad (2)$$

Diprotonated 2,4-pentanediol **8** loses water and rearranges to form 1,3-dimethyl allyl cation. Diprotonated 2,5-hexanediol, **9** above ~ – 30°C, rearranges to a mixture of protonated *cis*- and *trans*-δ, δ′-dimethyltetrahydrofurans. This would seem to indicate that there is either a significant amount of the monoprotonated form present or that the carbocation formed can easily lose a proton before ring formation occurs.

$$\underset{\mathbf{8}}{CH_3\underset{\overset{+}{O}H_2}{C}HCH_2\underset{\overset{+}{O}H_2}{C}HCH_3} \xrightarrow{-H_2O} \left[CH_3\overset{+}{C}H-\overset{H}{C}H-\overset{+}{C}HCH_3 \right] \xrightarrow{-H^+}$$

$$CH_3CH{=}CH-\overset{+}{C}HCH_3 \longleftrightarrow CH_3\overset{+}{C}H-CH{=}CHCH_3$$

$$\underset{\mathbf{9}}{CH_3\underset{\overset{+}{O}H_2}{C}HCH_2CH_2\underset{\overset{+}{O}H_2}{C}HCH_3} \xrightarrow{-H_2O} CH_3\underset{\overset{+}{O}H_2}{C}HCH_2-CH_2\overset{+}{C}HCH_3 \underset{+H^+}{\overset{-H^+}{\rightleftarrows}}$$

$$CH_3\underset{OH}{C}HCH_2CH_2\overset{+}{C}HCH_3 \longrightarrow \text{2,5-dimethyltetrahydrofuran, O-protonated } (H_2C\text{—}CH_2,\ HC(H_3C)\text{—}\overset{+}{O}H\text{—}CH(CH_3))$$

In the case of phenols, either C or O-protonated ions **10** or **11** are formed, depending upon acidity of the medium. In relatively weak acid medium such as H_2SO_4[22] only O-protonation is observed. However, in HSO_3F[23] only the C-protonated species is observed. In a much stronger acid system such as $HSO_3F:SbF_5$ (Magic Acid), both C and O-protonated dication **12** can be observed.

10: R-substituted phenyl–$\overset{+}{O}H_2$ **11**: R-substituted cyclohexadienyl cation (+), OH, CH_2 (H H) **12**: R-substituted cyclohexadienyl cation (+), $\overset{+}{O}H$, CH_2 (H H)

4.2.1.3 Secondary Oxonium Ions

^{1}H-nmr and ir investigation[23] of protonated ether salts (hexachloroantimonates) in dichloromethane solutions showed the formation of both dialkyloxonium ions in which the proton is bound to only one oxygen, **13** and bidentate complexes in which the proton is shared between two ether molecules, **14.**

$$(R)_2\overset{+}{O}H\ SbCl_6^- \qquad (R)_2\overset{+}{O}H \longleftarrow O(R)_2SbCl_6^-$$

13 **14**

In the ^{1}H-nmr, the proton on oxygen showed up as a singlet around δ 7–9 with no coupling with adjacent alkyl protons, indicating rapid proton exchange under the reaction conditions. Investigations using the superacid system, $HSO_3F:SbF_5:SO_2$ at low temperatures leads to comparable values for the chemical shift of the proton on oxygen for a variety of aliphatic ethers of δ 7.88–9.03. Because of the stronger acid system and the low temperature, however, the exchange rate is slowed down sufficiently so that the expected splitting of the proton resonance on oxygen by the adjacent hydrogens is observed[24] (Fig. 4.3).

The cleavage of protonated ethers in strong acid systems has not been studied extensively. Kinetic investigation of the cleavage of ethers in 99.6% sulfuric acid using cryoscopic methods showed that cleavage takes place by unimolecular fission of the conjugate acid of the ether to form the most stable carbocation and an alcohol. The carbocation and alcohol formed rapidly unite with the hydrogen sulfate anion. The overall rate in sulfuric acid, however, appears to be dependent upon the concentration of sulfur trioxide. To rationalize these observations, the following mechanism has been proposed.[25]

$$R{-}O{-}R' + H_2SO_4 \rightleftharpoons R{-}\overset{H}{\underset{+}{O}}{-}R' + HSO_4^-$$

$$R{-}\overset{H}{\underset{+}{O}}{-}R' + SO_3 \rightleftharpoons R{-}\overset{+}{\underset{SO_3H}{O}}{-}R'$$

$$R{-}\overset{+}{\underset{SO_3H}{O}}{-}R' \xrightarrow{RDS} R^+ + R'HSO_4$$

$$R^+ \xrightarrow{HSO_4^-} RHSO_4$$

In a solution of $HSO_3F:SbF_5$, protonated *n*-butyl methyl ether **15** does not show any significant change, neither cleavage nor rapid exchange, as indicated by the nmr spectrum up to +40°C. Above +40°C, however, it cleaves and a sharp singlet appears at δ 4.0. This can be attributed to the rearrangement of the *n*-butyl cation, formed in the cleavage, to *t*-butyl cation.

$$CH_3\overset{+}{O}CH_2CH_2CH_2CH_3 \longrightarrow CH_3\overset{+}{O}H_2 + [CH_3CH_2CH_2CH_2]^+ \longrightarrow (CH_3)_2C^+$$

15

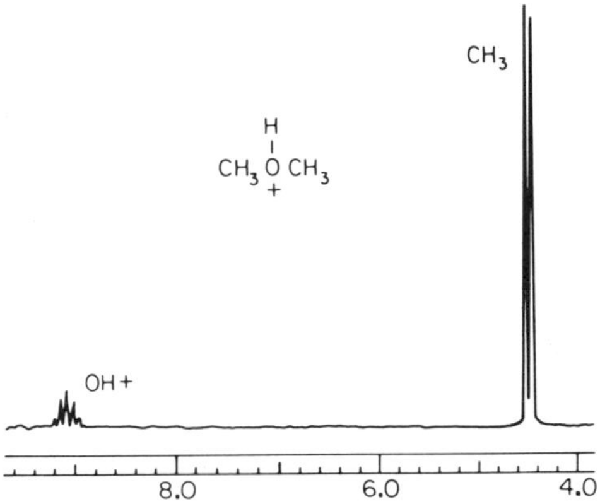

FIGURE 4.3 ^{1}H NMR spectrum of protonated dimethyl ether in HSO_3F-SbF_5-SO_2 solution.

Ethers in which one of the groups is secondary begin to show appreciable cleavage at $-30°C$. Protonated *s*-butyl methyl ether **16** cleaves cleanly at $-30°C$ to protonated methanol and *t*-butyl cation. Ethers in which one of the alkyl groups is tertiary cleave rapidly even at $-70°C$.

It was found possible to measure the kinetics of cleavage of protonated *s*-butyl methyl ether **16** by following the disappearance of the methoxy doublet in the nmr spectrum with simultaneous formation of protonated methanol and *t*-butyl cation. The cleavage shows pseudo-first-order kinetics. Presumably the rate-determining step is the formation of *s*-butyl cation followed by rapid rearrangement to the more stable *t*-butyl cation ($k_1 \ll k_2$).

$$\underset{\mathbf{16}}{CH_3CH_2\underset{CH_3}{\underset{|}{C}}H\overset{H}{\overset{|}{\underset{+}{O}}}CH_3} \xrightarrow{k_1} \begin{array}{c} CH_3\overset{+}{O}H_2 + \\ CH_2CH_2\underset{+}{C}HCH_3 \end{array} \xrightarrow{k_2} (CH_3)_3C^+$$

Alkylaryl ethers and diaryl ethers undergo protonation either on oxygen or carbon depending upon the acidity of the medium.[26]

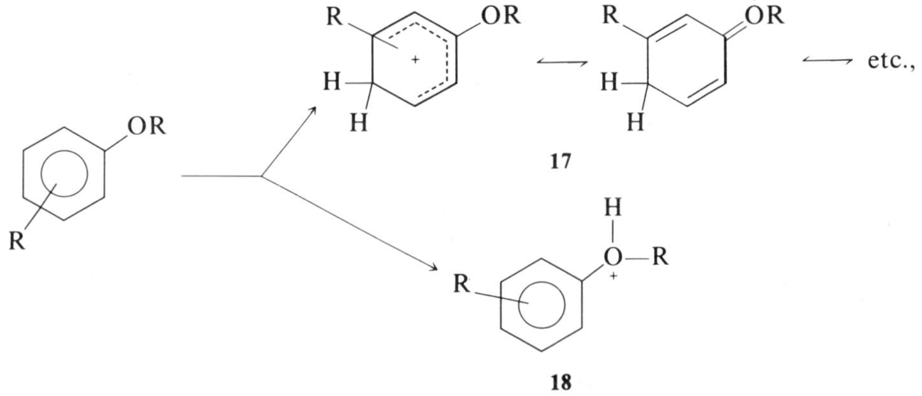

Both species **17** and **18** have been observed. The evidence mainly comes from nmr and uv data. Sommer and co-workers[27] have used *p*-anisaldehyde as an indicator in acidity measurements in the superacidity range. The barrier of rotation around C_{ipso}-O bond upon O-protonation has been used as a criteria in such studies. Also recently, the torsional barrier around phenyl-alkoxy bond in the C-protonated forms of aromatic ethers has been measured by spin-saturation transfer measurements.[28]

4.2.1.4 Tertiary Oxonium Ions

The synthesis, structural study, and chemistry of compounds containing trivalent, positively charged oxygen (R_3O^+) has been extensively investigated and reviewed since Meerwein's pioneering work on tertiary oxonium ions in 1937.[29,30] Trialkyloxonium salts are generally prepared by the alkylation of dialkyl ethers.[30] Alkyl halides can be used as alkylating agents in the presence of strong halide acceptors such as BF_3, PF_5, $SbCl_5$, and SbF_5.

$$R_2O + SbCl_5 + R{-}Cl \longrightarrow R_3O^+SbCl_6^-$$

A metathetic silver salt reaction has been employed[31] to prepare bicyclo[2.2.2]octyl-1-oxonium hexafluoroantimonate **19.**

$$\text{(iodoalkyl cyclic ether)}\ \ddot{O} + AgSbF_6 \longrightarrow \text{bicyclo[2.2.2]octyl-1-}O{:}\oplus\ SbF_6^{\ominus} + AgI$$

19

Trialkyloxonium salts can also be prepared by the reaction of secondary oxonium ion salts with diazoalkanes.[32]

$$R_2\overset{+}{O}H\ X^- + CH_2N_2 \longrightarrow R_2\overset{+}{O}-CH_3\ X^- + N_2$$

Disproportionation of dialkyl ether–PF_5 adducts also give trialkyl oxonium ions.[33]

$$3R_2O \rightarrow PF_5 \rightarrow 2R_3O^+PF_6^- + F_3P{=}O$$

Another important method is by transalkylation reactions[34] with other oxonium ions. However, these reactions are reversible to a certain extent. The equilibrium between two oxonium salts and the corresponding ethers in solution is established by both differences in solubility and stability of the oxonium salts. By an exchange reaction of this type, the trimethyloxonium salt **20** can be prepared in excellent yield (92%) from the readily available triethyloxonium salt.[34,35]

$$\text{C}_5\text{H}_{10}\text{:O:} + R_3O^{\oplus}\,BF_4^{\ominus} \longrightarrow \text{C}_5\text{H}_{10}\,{\oplus}\text{:O—R}\;BF_4^{\ominus} + R_2O$$

$$R_3O^{\oplus}\,BF_4^{\ominus} + R'_2O \rightleftharpoons R_2O + R_2'OR^{\oplus}\,BF_4^{\ominus} \rightleftharpoons R_2OR'^{\oplus}BF_4^{\ominus} + R'OR$$

$$R_2'OR^{\oplus}\,BF_4^{\ominus} + R_2'O \rightleftharpoons R_3'O^{\oplus}\,BF_4^{\ominus} + R'OR$$

In many cases, the main step in the syntheses of trialkyloxonium salts is the alkylation of a dialkyl ether with a reactive intermediate oxonium ion formed in situ. Thus, the most widely used method for preparing trialkyloxonium tetrafluoroborates by the reaction of epichlorohydrin and BF_3 etherate is based on the intermediacy of the inner oxonium salt **21**.[34,36,37]

$$\text{epichlorohydrin} + R_2O{\rightarrow}BF_3 \longrightarrow \underset{\mathbf{21}}{\text{H}_2\text{C—O}^{\oplus}R_2,\ \text{ClH}_2\text{C—C(H)—OBF}_3^{\ominus}} \xrightarrow{:OR_2} \text{H}_2\text{C(OR)—C(H)(CH}_2\text{Cl)—OBF}_3^{\ominus} + R_3O^{\oplus}$$

The method has been extended by the use of reactive intermediate oxonium ions such as sulfonyl cations–ether adducts, dialkoxycarbenium salts with ether, etc. The more reactive dialkyl halonium ions also readily alkylate dialkyl ethers to the corresponding oxonium ions.[38]

In contrast to trialkyloxonium salts, the triphenyloxonium salt **22** can be obtained by the reaction of benzenediazonium tetrafluoroborate with diphenyl ether in poor (2%) yield and is extremely inert toward nucleophiles. This fact is clearly demonstrated by the remarkable stability of triphenyloxonium halides (Cl^-, Br^-, I^-).[39,40]

The surprising stability of triaryloxonium ions has been demonstrated by the fact that reduction of the tris(*p*-nitrophenyl)oxonium salt **23** can be performed to give the tris(*p*-aminophenyl)oxonium salt **24** without cleavage of aryl oxygen bonds. Diazotization of this amino derivative and reaction with iodide leads to the tris(*p*-iodophenyl)oxonium salt **25** in excellent yields.[41]

Dialkylaryl or alkyldiaryl oxonium ions are less stable and are prepared by alkylation of alkylaryl ethers or diaryl ethers with either methyl or ethyl hexafluoroantimonate in SO_2ClF at low temperatures.[42,43] Dialkylaryloxonium ions are stable below $-70°C$, above this temperature they are transformed into ring-alkylated alkoxybenzenes. These onium ions are stronger alkylating agents than trialkyloxonium ions, as shown by the immediate formation of trimethyloxonium ion from dimethylphenyloxonium ion and dimethyl ether (the reverse alkylation of anisole fails with trimethyloxonium salt).

The protons of methyl or methylene groups in the α-position of the oxonium center of trialkyloxonium ions exhibit a deshielding effect of 1 ppm in the ^{1}H-nmr spectra compared with protons of the corresponding dialkyl ethers.[44] A deshielding effect of 15–20 ppm is also observed in the ^{13}C-nmr spectra.[45] The spectra generally reflect the expected result of replacing a neutral center by a positively charged one.

The geometry of trialkyloxonium salts has been presumed to be pyramidal rather than planar. This assumption has been proved by an nmr experiment in which pyramidal oxygen inversion in the case of the O-alkyloxiranium ion **26** was unambiguously demonstrated:[46]

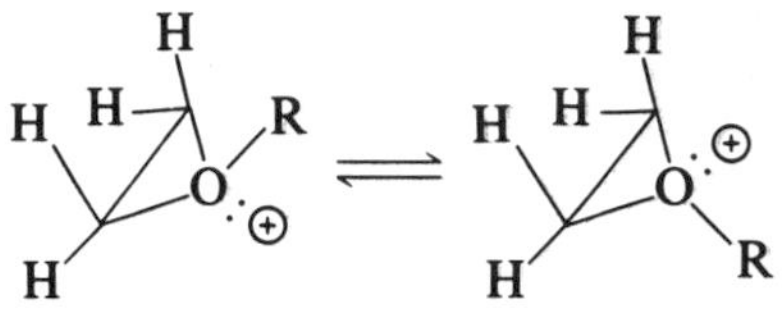

26 $R = i\text{-}C_3H_7$

At +40°C, the ^{1}H-nmr spectrum of **26** contains a very sharp singlet due to the ring protons. As the temperature is lowered, the ring proton resonance broadens to coalescence at −50°C. At lower temperatures (−70°C) the signal sharpens to a closely coupled ($\nu_{AB} \sim 3$ Hz) AA′BB′ spectrum. Thus, at the lowest temperature, the slowing of the rate of oxygen inversion results in the nonequivalence of the ring protons. The activation energy of the pyramidal inversion has been determined to be 10 ± 2 kcal · mol^{-1}.[46] The nmr spectra of six-membered cyclic oxonium ions are temperature independent; therefore, even at −70°C, rapid oxygen inversion must be assumed.[46] Further evidence for the pyramidal nature of trialkyloxonium salts comes from recent X-ray crystallographic studies on triethyloxonium ion.[47] On the contrary, the triphenyloxonium ion has a planar structure.[47]

Trialkyloxonium ions have been extensively used to carry out *O*-alkylation of lactams, amides, sulfoxides, oxo-sulfonium ylids, *N*-alkylation of *N*-heterocycles, *S*-alkylation of thioethers, thioacetals, thioamides, and thiolactams.[30] Although trialkyloxonium ions do not alkylate benzene or toluene directly, alkylations readily occur in combination with highly ionizing superacids such as $HSO_3F:SbF_5$. This may indicate protosolvation of nonbonded electron pair on oxygen in the superacid medium.

Recently, intermediacy of trimethyloxonium ion has been proposed in the first step of the acid-base-catalyzed conversion of methyl alcohol into the gasoline range hydrocarbons.[49] The crucial step is the base-catalyzed subsequent deprotonation of the trimethyloxonium ion to the very reactive surface-bound methylene-dimethyloxonium ylide. The ylide subsequently undergoes alkylation at its negative pole and the resulting ethyldimethyloxonium ion then cleaves to give ethylene and dimethyl ether. The key to the success of such reactions is the use of catalysts that have both acidic as well as basic sites.

4.2.1.5 Hydrogen Peroxonium Ion ($H_3O_2^+$)

The use of hydrogen peroxide as a source of electrophilic oxygen under acidic conditions is gaining increasing importance.[50] The hydrogen peroxonium ion $H_3O_2^+$ **27** has been invoked as an intermediate in the electrophilic hydroxylation of aromatics, oxygenation of alkanes, etc.[51] The hydrogen peroxonium ion **27** has been isolated recently under superacidic conditions.[52] H_2O_2 is also protonated in $HF:MF_5$(M = Sb,As) solutions.

$$H_2O_2 + HF + MF_5 \longrightarrow \underset{\mathbf{27}}{H_3O_2^+MF_6^-} \qquad M = As,Sb$$

The $H_3O_2^+MF_6^-$ **27** salts are white crystalline solids with marginal stability. They exothermically decompose at room temperature to hydronium ion salt and molecular oxygen.

$$\underset{\mathbf{27}}{H_3O_2^+MF_6^-} \longrightarrow \underset{\mathbf{1}}{H_3O^+MF_6^-} + \tfrac{1}{2}O_2$$

A detailed infrared analysis[52] of $H_3O_2^+SbF_6^-$ **27** shows that the ion possesses C_s symmetry similar to its isoelectronic analog NH_2OH. The ^{1}H-nmr spectrum of **27** in $HF:AsF_5$ solution at $-80°C$ shows only one broad signal at δ 11.06 indicating rapid proton exchange between **27** and the HF solution. This has been recently confirmed by ^{17}O-nmr spectroscopy.[11] The $H_3O_2^+$ ion **27** has a ^{17}O-nmr chemical shift of δ 151 (with respect to SO_2 at 505 ppm) and is shielded by 36 ppm with respect to hydrogen peroxide. The synthetic utility of hydrogen peroxonium ion reagent is discussed in Chapter 5.

Similarly, protonation and cleavage reactions of hydroperoxides and peroxides have been investigated extensively[50] and are also discussed in Chapter 5.

4.2.1.6 Ozonium Ion (O_3^+H)

Ozone is a resonance hybrid of canonical structure **28a–d.**[53] Ozone does in fact act

(a) ⟷ (b) ⟷ (c) ⟷ (d)

28

as a 1,3-dipole, i.e., either as an electrophile or a nucleophile. The electrophilic nature of ozone has been recognized for a long time in its reactions with alkenes, alkynes, arenes, amines, phosphines, sulfides, etc.;[54–58] However, its reactivity as a nucleophile has not been well recognized.[59]

Recently, ozone has been shown to be protonated in the superacid media to ozonium ion O_3H^+ **29** which reacts with alkanes[50] as powerful electrophilic oxygenating agent. Similarly, ozone reacts with carbocations giving alkylated ozonium ion which undergoes further cleavage reactions. These reactions are well covered in Chapter 5.

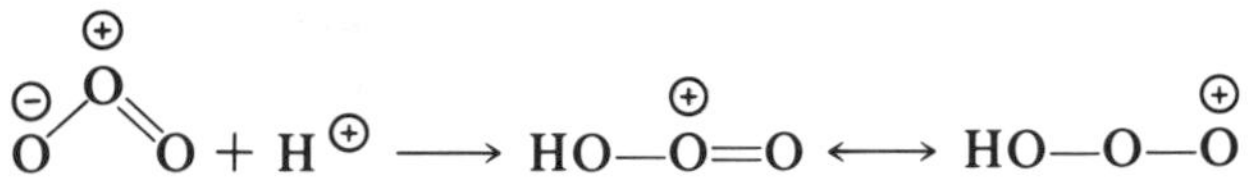

$$O_3 + H^{\oplus} \longrightarrow HO{-}\overset{\oplus}{O}{=}O \longleftrightarrow HO{-}O{-}\overset{\oplus}{O}$$

29

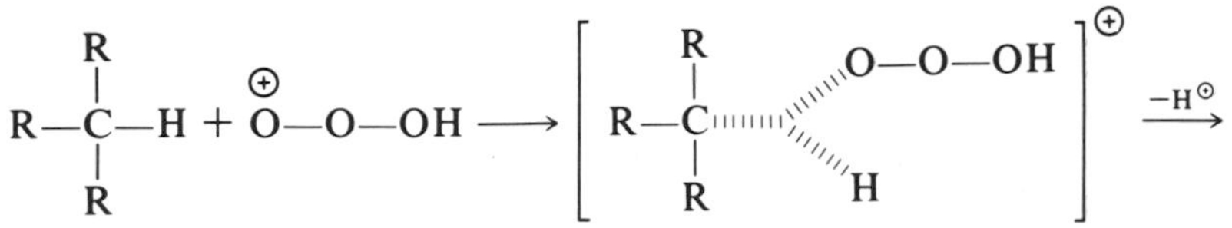

$$R_3C{-}H + \overset{\oplus}{O}{-}O{-}OH \longrightarrow \left[R_3C \cdots (O{-}O{-}OH)(H)\right]^{\oplus} \xrightarrow{-H^{\oplus}}$$

29

$$R_3C{-}O{-}O{-}OH \xrightarrow{+H^{\oplus}} R_3C{-}O{-}\overset{\oplus}{O}H{-}OH \longrightarrow R_2C{=}\overset{\oplus}{O}{-}R + H_2O_2$$

$$R^1R^2R^3CX \xrightarrow[SbF_5:SO_2ClF]{} R^1R^2R^3C^{\oplus} \xrightarrow{O_3} \left[R^1R^2R^3C{-}O{-}O{-}O\right]^{\oplus}$$

$X = F, Cl, OH$

$$\xrightarrow{-O_2} \left[R^1R^2R^3C{-}O^{\oplus}\right] \longrightarrow R^2R^3C{=}\overset{\oplus}{O}{-}R^1$$

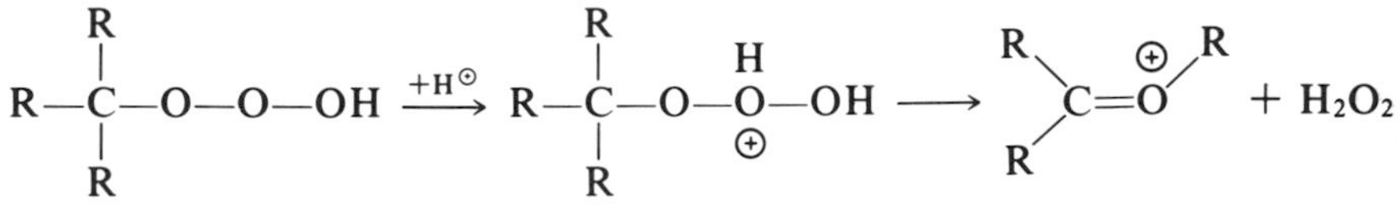

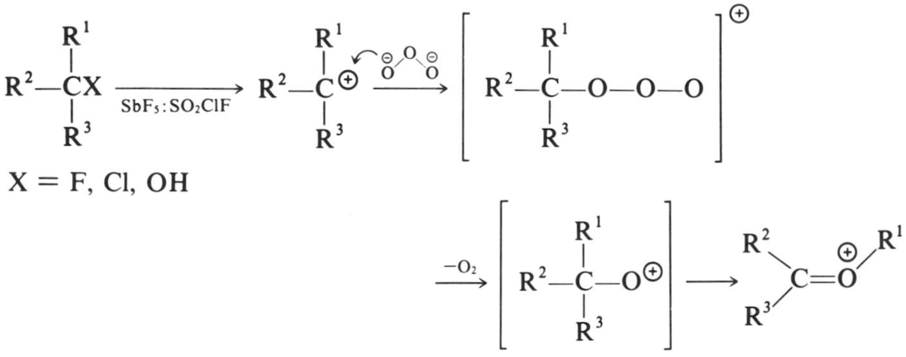

4.2.2 Sulfonium Ions

4.2.2.1 Hydrosulfonium Ion (H_3S^+)

The parent of sulfonium ions is the protonated H_2S, the hydrosulfonium ion H_3S^+ **30.** Olah and co-workers observed[60] H_3S^+ **30** for the first time in $HSO_3F:SbF_5:SO_2$ media at low temperature by ^1H-nmr spectroscopy. H_3S^+ appeared as a singlet at δ ^1H 6.6 from tetramethylsilane. H_3S^+ ion has also been studied in the gas phase by mass spectrometry.[61] Christe[62] has isolated $H_3S^+SbF_6^-$ salt by treating H_2S with $HF:SbF_5$.

$$H_2S + HF + SbF_5 \longrightarrow H_3S^+SbF_6^-$$

30

The hexafluoroantimonate salt of **30** is a stable white solid that reacts with water to produce H_2S. This reaction can be conveniently used to generate H_2S. The ion **30** has been thoroughly characterized by vibrational spectroscopy and normal coordinate analysis.[62]

4.2.2.2 Primary Sulfonium Ions

Aliphatic thiols are completely protonated in $HSO_3F:SbF_5$ diluted with SO_2 at $-60°C$.[60]

$$\mathrm{RSH} \xrightarrow[-60°\mathrm{C}]{\mathrm{HSO_3F:SbF_5:SO_2}} \mathrm{RSH_2{}^+SbF_5FSO_3{}^-}$$

The proton on sulfur is at considerably higher field in the ^{1}H-nmr (δ 5.9–6.6) than the corresponding proton on oxygen in protonated alcohols (δ 9.1–9.5), reflecting the larger size of the sulfur atom compared with oxygen. The protonated thiols are considerably more stable than protonated alcohols. Protonated *t*-butyl thiol **31** shows no appreciable decomposition at $-60°C$ in $HSO_3F:SbF_5:SO_2$, whereas protonated tertiary alcohols could not be observed under the same conditions and even protonated secondary alcohols cleave at a significant rate. Protonated thiols cleave at higher temperatures to give protonated hydrogen sulfide (singlet, δ 6.60) and stable alkyl cations. For example, protonated *t*-butyl thiol **31** slowly cleaves to *t*-butyl cation and protonated hydrogen sulfide when the temperature is increased to $-30°$ ($t_{\frac{1}{2}} \sim 15$ min). Protonated *t*-amyl thiol **32** also cleaves at this temperature to the *t*-amyl cation.

$$\underset{\mathbf{31}}{\mathrm{(CH_3)_3CSH_2{}^+}} \xrightarrow{-30°\mathrm{C}} \mathrm{(CH_3)_3C^+} + \mathrm{H_3S^+}$$

$$\underset{\mathbf{32}}{\mathrm{CH_3CH_2{-}C(CH_3)_2{-}SH_2{}^+}} \xrightarrow[\mathrm{H+}]{-30°\mathrm{C}} \mathrm{CH_3CH_2{-}C^+(CH_3)_2} + \mathrm{H_3S^+}$$

Protonated secondary thiols are stable even at higher temperatures. Protonated isopropyl thiol **33** cleaves slowly at 0°C in $HSO_3F:SbF_5$ (1:1 *M*) solution. No well-identified carbocations were found in the nmr spectra due to the instability of the isopropyl cation under these conditions. Protonated *s*-butyl thiol **34** cleaves to *t*-butyl cation at this temperature.

$$\underset{\mathbf{34}}{\mathrm{CH_3CH_2CH(CH_3)SH_2{}^+}} \xrightarrow{0°\mathrm{C}} [\mathrm{CH_3CH_2\overset{+}{C}HCH_3}] \longrightarrow \mathrm{(CH_3)_3C^+}$$

Protonated primary thiols are stable at much higher temperatures. Protonated *n*-butyl thiol **35** slowly cleaves to *t*-butyl cation only at +25°C.

$$\underset{\mathbf{35}}{CH_3CH_2CH_2CH_2SH_2^+} \xrightarrow{+25°C} [CH_3CH_2CH_2CH_2^+] \longrightarrow (CH_3)_3C^+$$

4.2.2.3 Secondary Sulfonium Ions

Protonated aliphatic sulfides have been studied at low temperatures by nmr spectroscopy in strong acid systems.[60] They show well-resolved nmr spectra, the proton on sulfur being observed at about δ 6.0.

$$RSR \xrightarrow[-60°C]{HSO_3F:SbF_5:SO_2} R\text{—}\overset{H}{\underset{+}{S}}\text{—}R \quad SbF_5FSO_3^-$$

The nmr spectrum of protonated cyclohexathiane-3,3,5,5-d_4 **36** has also been studied in $HSO_3F:SO_2$ to determine the conformational position of the proton on sulfur in this six-membered ring and to study the ring inversion process.[63] The proton on sulfur resides exclusively in the axial position.

The protonated sulfides are less susceptible to cleavage than the corresponding protonated ethers and also more stable than the protonated thiols. Protonated methyl *t*-butyl ether is completely cleaved to *t*-butyl cation and protonated methanol even at −70°C. On the other hand, protonated methyl *t*-butyl sulfide **37** is stable even at −60°C. When the temperature is increased to −15°C, protonated methyl *t*-butyl sulfide very slowly cleaves to *t*-butyl cation and protonated methyl thiol.

$$\underset{\mathbf{37}}{(CH_3)_3C\text{—}\overset{H}{\underset{+}{S}}\text{—}CH_3} \xrightarrow[H^+]{-15°C} (CH_3)_3C^+ + CH_3SH_2^+$$

Protonated di-*t*-butyl sulfide **38** shows very little cleavage at −60°C. At 35°C it cleaves slowly ($t\frac{1}{2}$ ~ 1 h) to *t*-butyl cation and protonated hydrogen sulfide, the latter showing the ^{1}H-nmr peak at δ 6.60.

$$\underset{\mathbf{38}}{(CH_3)_3C\text{—}\overset{H}{\underset{+}{S}}\text{—}C(CH_3)_3} \xrightarrow[H^+]{-35°C} (CH_3)_3C^+ + (CH_3)_3CSH_2^+$$

$$(CH_3)_3CSH_2^+ \xrightarrow{H^+} (CH_3)_3C^+ + H_3S^+$$

Protonated secondary sulfides show extraordinary stability toward the strongly acidic medium. Protonated isopropyl sulfide **39** shows no appreciable cleavage up to +70°C in a solution of $HSO_3F:SbF_5$ (1:1).

4.2.2.4 Tertiary Sulfonium Ions

In contrast to tertiary oxonium ions, the tertiary sulfonium ions are stable and are prepared rather readily.[64] They are even stable in aqueous solutions. Trialkylsulfonium ions are obtained by alkylation of dialkyl sulfides with alkyl halides. Trialkyl oxonium and dialkylhalonium ions readily transalkylate basic dialkylsulfides.

In contrast, alkylation of diaryl sulfides and thiophene requires rather drastic conditions due to poor nucleophilicity of sulfur. The alkylations have been achieved using alkyltriflates[65] or with alkyl halide and silver tetrafluoroborate.[66]

$$(C_6H_5)_2S + C_4H_9OSO_2CF_3 \longrightarrow (C_6H_5)_2\overset{+}{S}\,C_4H_9\;CF_3SO_3^-$$

40

$$\text{thiophene} + CH_3I \xrightarrow[ClCH_2CH_2Cl]{AgBF_4} \text{S-methylthiophenium } (\overset{+}{S}-CH_3)\;\overset{-}{B}F_4$$

41

The most important application of trisubstituted sulfonium ions is in the generation of sulfur ylides.[67]

4.2.3 Selenonium and Telluronium Ions

A great number of trialkyl(aryl) selenonium and telluronium ions are known, and their synthesis does not require the use of strong electrophilic alkylating or arylating agents.[68,69] However, acidic selenonium and telluronium ions were obtained only under superacidic conditions.[70] Similarly, some trialkylselenonium and telluronium ions have been prepared as fluorosulfate salts.[70]

Hydrogen selenide is very easily oxidized to elemental selenium. As a result, fluoroantimonic acid ($HF:SbF_5$) and Magic Acid ($HSO_3F:SbF_5$), generally used in the preparation of acidic onium ions, cannot be used, because they both oxidize hydrogen selenide. It has been found, however, that hydrogen selenide can be protonated without oxidation by $HF:BF_3$, in excess HF solution. The hydroselenonium ion **42** formed in this way at −70°C shows a singlet ^{1}H-nmr resonance at δ 5.8, deshielded by 6.1 ppm from the ^{1}H-nmr resonance of parent H_2Se. In the ^{77}Se-nmr spectrum, **42** is observed at δ −142 (from dimethyl selenide) with no resolvable ^{77}Se–^{1}H coupling even at −90°C.[71] Methyl selenide also undergoes protonation in $HF:BF_3$ media to methylselenonium ion **43.** The methylselenonium ion **43,** however, shows ^{77}Se–^{1}H coupling in ^{77}Se-nmr spectrum ($J_{^{77}Se-^1H}$ = 129.5

Hz). A long-range ^{77}Se-methyl proton coupling of 8.9 Hz is also observed.[71]

$$H_2Se + HF + BF_3 \longrightarrow H_3\overset{+}{Se}\ BF_4^- \quad \mathbf{42}$$

Alkyl selenides are much more stable to oxidation than hydrogen selenide, and can be protonated in $HSO_3F:SbF_5:SO_2$ solution. The dimethylselenonium ion **44** (protonated dimethyl selenide) shows in its ^{1}H-nmr spectrum the methyl doublet at δ 2.90 (J = 7.0 Hz) and the SeH septet at δ 4.50 (J = 7.0 Hz). A double irradiation experiment showed that the doublet and septet are coupled. The ^{1}H-nmr spectrum also shows an unidentified small doublet at δ 3.50 and a singlet at δ 3.80 for the $(CH_3)_2Se:SbF_5$ complex. The diethylselenonium ion **45** shows the methyl triplet at δ 2.00, the methylene quintet at δ 3.77, and the SeH quintet at δ 4.40. The ^{13}C- and ^{77}Se-nmr data on **44** and **45** agree very well with ^{1}H-nmr results.

$$R_2Se \xrightarrow[SO_2,\ -60^\circ C]{HSO_3F:SbF_5} R_2SeH^+$$

The acidic, secondary alkylselenonium ions are remarkably stable. The ^{1}H-nmr spectra show no significant change from -60° to 65°C.

Trialkylselenonium fluorosulfates are conveniently prepared by the reaction of dialkyl selenide and alkyl fluorosulfate, using 1,1,2-trichlorotrifluoroethane as the reaction solvent. Trimethyl selenonium fluorosulfate **46** thus prepared is a stable,

$$R_2Se + ROSO_2F \longrightarrow R_3Se^+FSO_3^- \ (R = CH_3, C_2H_5)$$

white solid, mp 83–85°C, which, then dissolved in liquid sulfur dioxide, exhibits a singlet proton nmr absorption at δ 2.7. Triethylselenonium fluorosulfate **47** has also been prepared in the same way. It is also a stable, white solid, mp 25–28°C. When dissolved in liquid sulfur dioxide, triethylselenonium fluorosulfate **47** shows the methylene protons at δ 3.2 (quartet) and the methyl protons at δ 1.4 (triplet). The parent hydrotelluronium ion, H_3Te^+ **48,** could not be observed in superacid solution of hydrogen telluride, under conditions where the selenonium ion is observed. Alkyl tellurides in $HSO_3F:SbF_5:SO_2$ solution at -60°C show deshielded alkyl proton chemical shifts (with no long-range proton coupling) as compared with the corresponding dialkyl tellurides themselves in SO_2. This indicates that in this medium the tellurides are probably oxidized. However, using $HF:BF_3$ in excess HF solution, both the TeH^+ proton and its coupling to secondary alkyl groups can be observed.

$$R_2Te + HF + BF_3 \longrightarrow R_2TeH^+BF_4^-$$

The dimethyltelluronium ion **49** (protonated dimethyl telluride) shows the methyl doublet at δ 2.7 (J = 7 Hz) and the TeH septet at δ 1.6. Similarly, the diethyltelluronium ion **50** (protonated diethyl telluride) shows the methyl triplet at δ 1.9, the methylene quintet at δ 3.4, and the TeH multiplet, partially overlapping the methyl triplet, at δ 1.6.

Trialkyltelluronium fluorosulfates were prepared similarly to the trialkylselenonium salts from dialkyl telluride and alkyl fluorosulfate. Trimethyltelluronium

$$(CH_3)_2Te + CH_3OSO_2F \longrightarrow \underset{\mathbf{51}}{(CH_3)_3Te^+FSO_3^-}$$

fluorosulfate **51** prepared in this way is a stable, light-yellow salt, mp 130°, which, when dissolved in liquid sulfur dioxide, exhibits a singlet ^{1}H-nmr signal at δ 2.3. The triethyltelluronium salt **52** could not be isolated, although prepared in solution it is also quite stable. No cleavage of the ions in solution is observed up to 65°C.

The proton on selenium in selenonium ions and on tellurium in telluronium ions is considerably more shielded than the proton on oxygen in the related oxonium ions (δ 7.88–9.21) and the proton on sulfur in the corresponding sulfonium ions (δ 5.80–6.52). There is a consistent trend of increasing shielding going from related oxonium to sulfonium to selenonium to telluronium ions (which is particularly significant when considering the directly observed protons on heteroatoms). Charge delocalization and shielding by increasingly heavier atoms is thus indicated.

4.2.4 Halonium Ions

Organic halogen cations, i.e., halonium ions of acyclic ($R\overset{+}{X}R$) or cyclic [$\overset{\oplus}{X}$]nature have gained increasing significance, both as reaction intermediates and preparative reagents. They are related to oxonium ions in reactivity but offer much more selectivity. Halogen atoms that form organic halonium ions are chlorine, bromine, and iodine. To date, no stable organic fluoronium ion has been prepared or characterized. A monograph[38] pertaining to the chemistry of halonium ions has been published which covers more than 200 research publications.

In 1894, Hartmann and Meyer[72] were the first to prepare a diphenyliodonium ion salt when they reacted iodosobenzene in concentrated H_2SO_4 and obtained *p*-iodophenylphenyliodonium bisulfate. Diphenyliodonium ions have since been studied by various research groups, most notably by Nesmeyanov and Beringer since 1950.[40] Diarylbromonium and -chloronium ions, although considerably less stable and much less investigated, were also prepared in the 1950s by Nesmeyanov and co-workers.[40]

Open-chain dialkylhalonium ions of the type R_2X^+ (X = I, Br, Cl) were unknown until recently, as were alkylarylhalonium ions ($Ar\overset{+}{X}R$). Realization of their possible role as intermediates in alkylation reactions of haloalkanes and -arenes has followed their preparation and study.[73]

One of the most daring proposals of a organic reaction mechanism of its time was made in 1937 by Roberts and Kimball[74] who suggested that the observed *trans* stereospecificity of the bromine addition to alkenes is a consequence of intermediate bridged bromonium ion formation. The brief original publication suggested the actual structure of the ion is undoubtedly intermediate between **53** and **54.** Structure

53 was not intended to represent a conventional free carbocation, however. Since the two carbons in either structure are joined by a single bond and by a halogen bridge, free rotation is not to be expected. A clear description of the difference in bonding between carbon and bromine in **53** and **54** was not given.

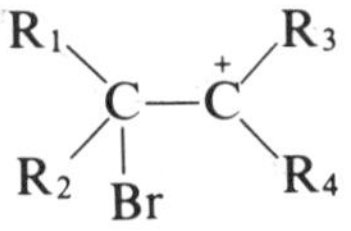

53 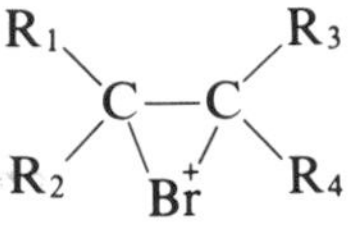 **54**

The bromonium ion concept was quickly used by other investigators to account for stereospecific transformation of olefins,[75] notably by Winstein and Lucas,[76] but was not unanimously accepted.[77] For example, in discussing the mechanistic concepts of bromonium ion formation, Gould,[78] in his still popular text, wrote in 1959, "Although a number of additions are discussed in terms of the halogenonium-ion mechanism, the reader should bear in mind that few organic mechanisms have been accepted so widely while supported with such limited data."

In 1967, Olah and Bollinger[79] reported the first preparation and spectroscopic characterization of stable, long-lived bridged alkylenehalonium ions. This was followed by Olah and DeMember's[74] preparation in 1969 of the first dialkylhalonium ions. Since then the field of organic halonium ions has undergone rapid development through substantial contributions from an increasing number of investigators, notably by Peterson.[80]

The halogen atom in organic halonium ions is generally bound to two carbon atoms, although in the case of acidic halonium ions, that is, protonated alkyl and aryl halides one ligand is hydrogen.

Of the dihydridohalonium ions, i.e., acidic halonium ions only chloronium ion is characterized. HCl has been protonated to dihydridochloronium ion **55** in $HSO_3F:SbF_5$ media and studied by ^1H-nmr spectroscopy.[81a] Christe has managed to isolate $H_2Cl^+SbF_6^-$ salt at low temperatures.[62] Under these strongly acidic conditions, HBr and HI are readily oxidized. H_2F^+ in the condensed state cannot be considered as dihydrodofluoronium ion,[81b] because the very electronegative fluorine atom resists acquiring positive charge. Instead the HF solvated proton can have linear or more probably $2e$–$3c$ bonding.

Organic halonium ions can be divided into two categories: acyclic (open-chain) halonium ions and cyclic halonium ions.

4.2.4.1 Acyclic (Open-Chain) Halonium Ions

Alkyl and Arylhydridohalonium Ions. The self-condensation of alkyl halides in strongly acidic media represents a convenient preparative route to symmetric dialkylhalonium ions ($R\overset{+}{X}R$). This reaction involves hydridohalonium ions

($R\overset{+}{X}H$) as intermediates that subsequently undergo nucleophilic attack by excess alkyl halide.

$$RX \xrightarrow{H^+} R\overset{+}{X}H \longrightarrow R\overset{+}{X}R + HX$$

$$X = Cl, Br, I \quad \ddot{X}R$$

Methyl bromide and methyl iodide in $HSO_3F:SbF_5:SO_2ClF$ at $-78°C$ give both dimethylhalonium as well methylhydridohalonium ions[82] (**56,** X = Br; **57,** X = I). These hydrido halonium ions **56** and **57** have been characterized by ^{13}C-nmr spectroscopy.[82] Methyl chloride does not give the methylhydridochloronium ion under similar conditions.

Halobenzenes do not form diarylhalonium ions under superacidic conditions. The protonation occurs either on the halogen or on the aromatic ring.[83] Indeed, chloro- and bromobenzenes quantitatively yield the corresponding 4-halobenzenium ions on protonation with $HSO_3F:SbF_5:SO_2ClF$ at $-78°C$. Iodobenzene under the same conditions give the hydridoiodonium ion **58.** The ion **58** is rather stable and does not rearrange to ring-protonated $C_6H_6I^+$ even when the temperature is raised to $-20°C$.

$$C_6H_5I \xrightarrow[SO_2ClF,\ -78°C]{HSO_3F:SbF_5} C_6H_5\overset{+}{I}H$$

58

Dialkylhalonium ions. Dialkylhalonium ions were first obtained as stable fluoroantimonates and characterized by Olah and DeMember in 1969.[73] Since then, a large number of unsymmetrical and symmetrical halonium ions have been prepared. The alkylating ability as well as intermolecular exchange reactions of dialkylhalonium ions were also studied.[84,85]

There are two general methods for the preparation of dialkylhalonium ions: (1) the reaction of excess primary and secondary alkyl halides with $SbF_5:SO_2$, anhydrous fluoroantimonic acid ($HF:SbF_5$), or anhydrous silver hexafluoroantimonate (or related complex fluoro silver salts) in SO_2 solution, and (2) the alkylation of alkyl halides with methyl or ethyl fluoroantimonate (or alkylcarbenium fluoroantimonates) in SO_2 solution.[73,84] The first method is only suitable for the preparation of symmetrical dialkylhalonium ions, whereas the second method can be used for both symmetrical and unsymmetrical dialkylhalonium ions. Additional methods for the preparation of dialkylhalonium ions are also available, but these methods generally have less practical value.

$$2RX \xrightarrow[-60°C]{SbF_5:SO_2} R\overset{+}{X}R\ SbF_5X^-$$

$$\text{or } 2RX + \overset{+}{H}SbF_6^- \longrightarrow R\overset{+}{X}R\ SbF_6^- + HX$$

$$\text{or } 2RX + \overset{+}{Ag}SbF_6^- \longrightarrow R\overset{+}{X}R\ SbF_6^- + AgX$$

$$R'X + R^+SbF_6^- \xrightarrow[-60°C]{SO_2} R'\overset{+}{X}R\ SbF_6^-$$

The following are some of the representative dialkylhalonium ions **59–63** that have been prepared and characterized.[38]

$CH_3\overset{+}{X}CH_3$ **59** X = Cl,Br,I

$C_2H_5\overset{+}{X}C_2H_5$ **60**

$i\text{-}C_3H_7\overset{+}{X}i\text{-}C_3H_7$ **61**

$CH_3\overset{+}{X}C_2H_5$ **62**

$CH_3\overset{+}{X}i\text{-}C_3H_7$ **63**

Even dicyclopropylbromonium ion **64** has been prepared[86] by the ionization of cyclopropyl bromide in $SbF_5:SO_2ClF$ at low temperature. Also a series of alkylcyclopropylhalonium ions **65, 66,** and **67** has been studied.[86]

$c\text{-}C_3H_5\text{—}\overset{+}{Br}\text{—}c\text{-}C_3H_5$ **64**

$c\text{-}C_3H_5\text{—}\overset{+}{Cl}\text{—}R$ **65** R = CH_3, i—C_3H_7

$c\text{-}C_3H_5\text{—}\overset{+}{Br}\text{—}R$ **66** R = CH_3, C_2H_5, i—C_3H_7

$c\text{-}C_3H_5\text{—}\overset{+}{I}\text{—}R$ **67** R = CH_3, C_2H_5, i—C_3H_7

Recently, alkylvinylhalonium ions **68** and **69,** which are stable only below −90°C, have been investigated.[87]

$CH_2{=}CH\text{—}\overset{+}{Br}\text{—}R$ **68** R = CH_3, C_2H_5

$CH_2{=}CH\text{—}\overset{+}{Cl}\text{—}R$ **69** R = CH_3, C_2H_5

In search for nonvolatile and, therefore, safer chloromethylating agents, even chloromethylhalonium ions have been synthesized.[88] Bis(chloromethyl)chloronium ion is formed by the ionization of dichloromethane in $SbF_5:SO_2ClF$ at −130°C.

Generally, the preparation of symmetrical dialkylhalonium ions is simpler and the reactions are clean. Unsymmetrical dialkylhalonium ions undergo disproportionation and alkylation (self-condensation) reactions, even at low temperatures (ca. −30°C).

A qualitative measure of the decreasing stability of dialkyl halonium ions is in the order $R\overset{+}{I}R > R\overset{+}{Br}R > R\overset{+}{Cl}R$ and sequence of reactivity is in the opposing order $R\overset{+}{Cl}R > R\overset{+}{Br}R > R\overset{+}{I}R$.

Dimethylhalonium fluoroantimonates such as dimethylbromonium and -iodonium fluoroantimonates can be isolated as crystalline salts. They are stable in a dry atmosphere at room temperature, and some are now commercially available. Dimethylhalonium fluoroantimonate salts are very hygroscopic, and exposure to atmospheric moisture leads to immediate hydrolysis.

The ^{1}H- and ^{13}C-nmr data[84] on dialkylhalonium ions seem to indicate the following neighboring group deshielding order $\diagup\overset{+}{Cl}\diagdown > \diagup\overset{+}{Br}\diagdown > \diagup\overset{+}{I}\diagdown$ indicating inductive effect of the halogen atoms. Chlorine, being the smallest halogen atom in halonium ions (fluoronium ions are not known in solution) can accommodate the least amount of positive charge, whereas iodine, the largest of the halogen atoms, can accept essentially most of the charge. However, the β-protons in the ^{1}H-nmr spectra of related homologous dialkylhalonium ions show an opposite deshielding order $\diagup\overset{+}{I}\diagdown > \diagup\overset{+}{Br}\diagdown > \diagup\overset{+}{Cl}\diagdown$, indicating that the inductive effect of the positively charged halogen atoms diminishes and the anisotropy effect of the halogen atoms causes an opposite trend.

The Raman and ir spectroscopic studies of dimethylhalonium ions **59** seem to indicate that these species exist in a bent conformation.

Dialkylhalonium ions are reactive alkylating reagents. The alkylation of π-donor (aromatic and olefins) and *n*-donor bases with dialkylhalonium ions has been studied.[84c] Alkylation of aromatics with dialkylhalonium ions was found to be not significantly different from conventional Friedel-Crafts alkylations, showing particular similarities in the case of alkylation with alkyl iodides. Alkylation of *n*-donor bases with dialkylhalonium salts provides a simple synthetic route to a wide variety of onium ions.

Alkylation of alkylene dihalides with methyl and ethyl fluoroantimonate ($CH_3F:SbF_5:SO_2$ and $C_2H_5F:SbF_5:SO_2$) gives monoalkylated halonium ions and/or dialkylated dihalonium ions, depending on the reaction conditions.[89] Iodine shows an unusual ability to stabilize positive charge, as demonstrated by the formation of dialkyl alkylenediiodonium ions [$RI^+(CH_2)_nI^+R$, $n = 1$ to 6, R = CH_3, C_2H_5]. In the extreme case ($n = 1$), the two positive iodonium cation sites are separated by only a single methylene group. However, dialkyl alkylenedibromonium ions were formed only when the two positive bromines were separated by three methylene groups. In the case with four methylene groups, rearrangement to the more stable five-membered-ring tetramethylenebromonium ion takes place. Dialkyl alkylenedichloronium ions have not yet been directly observed. Consequently, the ease of formation of dialkyl alkylenedihalonium ions is similar to that of dialkylhalonium ions, that is, $\diagup\overset{+}{I}\diagdown > \diagup\overset{+}{Br}\diagdown > \diagup\overset{+}{Cl}\diagdown$.

Alkylarylhalonium ions. Dence and Roberts[90] attempted to prepare the cyclopropylphenyliodonium ion from phenyliodoso chloride and cyclopropyllithium. However, they were unable to obtain the corresponding iodonium ion or any cyclopropylbenzene from the reaction mixture. Thus, the iodonium ion was not formed, even as an unstable reaction intermediate.

$$C_6H_5ICl_2 + RLi \longrightarrow [C_6H_5\overset{+}{I}R]\ C\bar{l} \longrightarrow C_6H_5I + RI + C_6H_5Cl + RCl$$
$$R = \text{Cyclopropyl}$$

Perfluoroalkyliodoso trifluoroacetate reacts with aromatic compounds to give perfluoroalkylaryliodonium ions **70.**[91]

$$RI(OCOCF_3)_2 + C_6H_5R' \longrightarrow R\text{—}\overset{+}{I}\text{—}C_6H_4\text{—}R' \quad CF_3COO^-$$

$R = n\text{-}C_3H_7,\ n\text{-}C_6F_{13},\ C_6F_5$

70

Alkylarylhalonium ions (other than perfluorinated derivatives) were first prepared by Olah and Melby.[92] When a SO_2 solution of iodobenzene was added to a SO_2 solution of the $CH_3F:SbF_5$ complex (methyl fluoroantimonate) at $-78°C$, a clear, slightly colored solution resulted. The ^1H-nmr spectrum of this solution at $-80°C$ showed in addition to the excess methyl fluoroantimonate a methyl singlet at δ 3.80 and a multiplet aromatic region (7.7–8.3) with a peak area ratio of 3:5. The aromatic signals showed the same coupling pattern as that of iodobenzene in SO_2 but were deshielded by approximately 0.5 ppm. The species that accounts for the nmr data is the methylphenyliodonium ion ($CH_3I^+C_6H_5$) **71**-I. When bromobenzene and other aryl bromides or iodides were added in the same manner to methyl fluoroantimonate in SO_2, analogous spectra were obtained, indicating the formation of the corresponding methylarylbromonium ions.

$$C_6H_5\text{—}X + RF \rightarrow SbF_5 \xrightarrow{SO_2} C_6H_5\text{—}\overset{+}{X}\text{—}R \quad SbF_6^-$$

X = Br, I

71, $R = CH_3$
72, $R = C_2H_5$

Likewise, the reaction of bromo and iodoarenes with ethyl fluoroantimonate in SO_2 gave the corresponding ethylarylhalonium ions. The structures of all alkylarylhalonium ions prepared have been characterized by ^1H- and ^{13}C-nmr spectroscopy.

Similarly a series of dialkylphenylenedihalonium ions such as **73–77** have been prepared and characterized.[92] Even trihalonium ions such as **78** have been prepared.

Many of the above mentioned halonium ions are stable only below $-20°C$ and above which they undergo ring alkylations.

$$CH_3\text{—}\overset{+}{Br}\text{—}C_6H_4\text{—}\overset{+}{Br}\text{—}CH_3 \qquad CH_3\text{—}\overset{+}{I}\text{—}C_6H_4\text{—}\overset{+}{I}\text{—}CH_3 \qquad C_6H_4(\overset{+}{I}\text{—}CH_3)_2$$

73 **74** **75**

Diarylhalonium Ions. In contrast to dialkylhalonium ions and alkylarylhalonium ions, diarylhalonium ions are considerably more stable. This is particularly the case for diaryliodonium ions, which have been known for 80 years.[72] It is interesting, however, to compare the discovery and assumed significance of these ions with that of triarylmethyl cations. The latter were also discovered at about the turn of the century, but were considered only as a specific class of organic cations limited exclusively to the highly stabilized triarylmethyl systems. The general significance of carbocations as intermediates in electrophilic reactions was not recognized until many years later, when it became evident that they are intermediates in all electrophilic organic reactions. Subsequently, methods were developed to prepare practically any conceivable type of carbocations under stable ion conditions. Halonium ions represent a somewhat similar case. Diaryliodonium (or to a lesser extent diarylbromonium and -chloronium ions) were also for long considered a

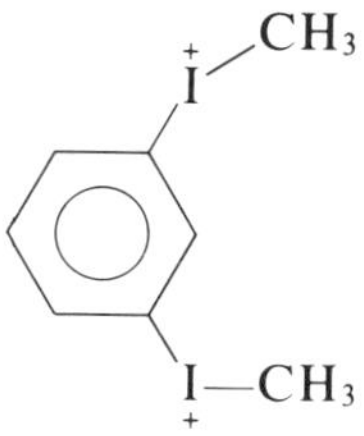

$CH_3-\overset{+}{Br}-C_6H_4-\overset{+}{I}-CH_3$

76 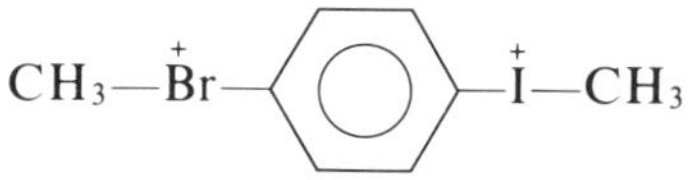 **77** 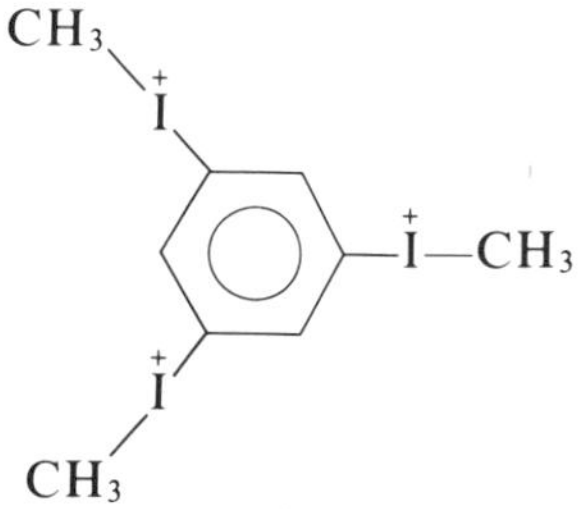 **78**

specific class of highly stabilized halonium ions. No relationship or significance was attached to these ions until many years later, when it was realized that many other types of halonium ions (such as dialkyl-, alkylaryl-, and alkylenehalonium ions) can exist, and the significance of dialkylhalonium ions in electrophilic alkylation with alkyl halides was pointed out.

Some of the representative diarylhalonium ions are shown below.[38] The synthesis of these diarylhalonium ions does not require strongly acidic conditions and thus will not be discussed here.

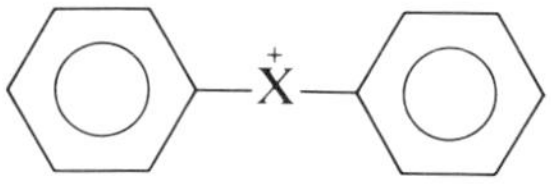

BF_4^- **79** HSO_4^- **80**

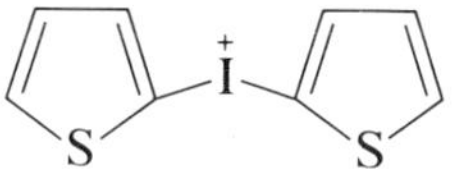

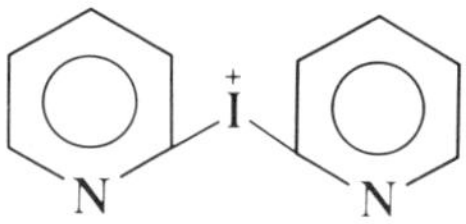

Cl^- **81** X^- X^- **82**

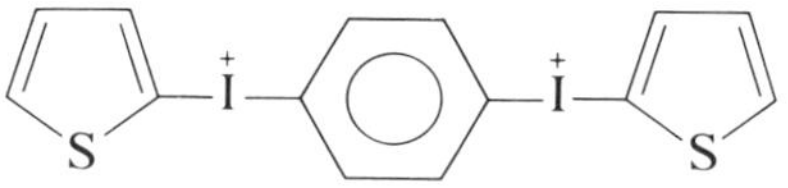

Olah and co-workers have studied[93,94] the ^{13}C-nmr spectra of a series of diarylhalonium ions.

4.2.4.2 Cyclic Halonium Ions

Ethylenehalonium Ions. The parent ethylenehalonium ions (X = Br and I) **83** were obtained when 1-halo-2-fluoroethanes were ionized in $SbF_5:SO_2$ solution at $-60°$.[95] The ^{1}H-nmr spectra of the ethylenebromonium and iodonium ions show a singlet at δ 5.53 and 5.77, respectively. Under similar experimental conditions,

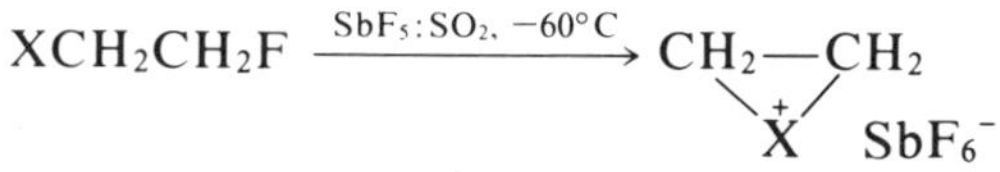

$$XCH_2CH_2F \xrightarrow{SbF_5:SO_2,\ -60°C} \underset{\overset{+}{X}}{CH_2—CH_2}\ SbF_6^-$$

X = Br, I

83

when 1,2-dichloroethane and 1-chloro-2-fluoroethane were treated with $SbF_5:SO_2$ solution, only donor-acceptor complexes were formed instead of the ethylenechloronium ions.

$$SbF_5 \leftarrow ClCH_2CH_2Cl \qquad SbF_5 \leftarrow FCH_2CH_2Cl$$

Subsequently, the ethylenechloronium ion **83**-Cl was obtained.[96] When, instead of SO_2, SO_2ClF was used as the solvent in the reaction of antimony pentafluoride with 1-chloro-2-fluoroethane at $-80°C$ a solution was obtained whose ^{1}H-nmr spectrum consisted of these absorptions: a doublet (δ 4.6, 3H, J = 6 Hz), a quartet (δ 13.3 1H, J = 6 Hz), and a singlet (δ 5.9), consistent with the formation of ethylenechloronium ion **83**-Cl and methylchlorocarbenium ion **84**, respectively.

$$ClCH_2CH_2F \xrightarrow[1,2\text{-H shift}]{SbF_5:SO_2ClF,\ -80°C} \underset{\overset{+}{Cl}}{CH_2—CH_2}\ SbF_6^- \ (\textbf{83}\text{-Cl})$$

$$CH_3\overset{+}{C}H(Cl)\ SbF_6^- \longleftrightarrow CH_3CH{=}Cl^+\ SbF_6^- \quad (\textbf{84})$$

The ethyleneiodonium ion **83**-I has also been prepared by the direct iodination of ethylene using $ICN:SbF_5:SO_2$ solution.[97] However, similar reaction with either BrCN or ClCN did not give the corresponding halonium ions **83**-Br and **83**-Cl.

$$CH_2{=}CH_2 \xrightarrow{ICN:SbF_5:SO_2} \textbf{83}\text{-I}$$

Propyleneiodonium and propylenebromonium ions **85** (X = Br, I) were also obtained by the ionization of 2-fluoro-1-iodo and 2-fluoro-1-bromopropanes in

$SbF_5:SO_2$ solution at $-78°C$.[95] The propylenechloronium ion **85**-Cl is also known.

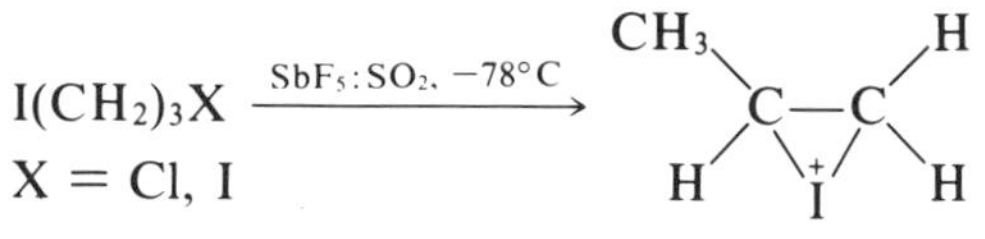

$$I(CH_2)_3X \xrightarrow{SbF_5:SO_2,\ -78°C} \text{85-I}$$

X = Cl, I

85-I

Similarly, a variety of dimethyl, trimethyl, and tetramethylethylene halonium ions have been prepared and studied by ^1H- and ^{13}C-nmr spectroscopy.[38,95] Some of the representative examples are the following.

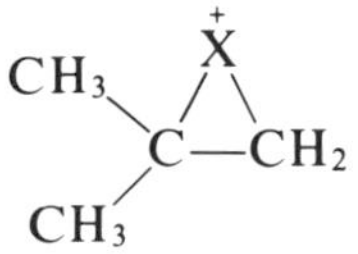

86 X = Br,Cl 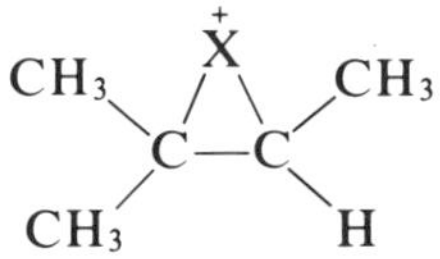 **87** X = Br,Cl,I 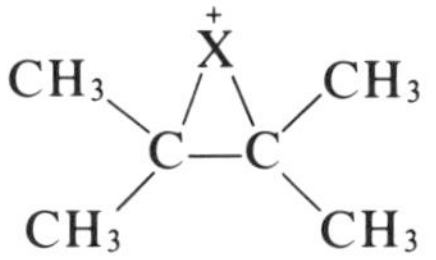 **88** X = Br,Cl,I

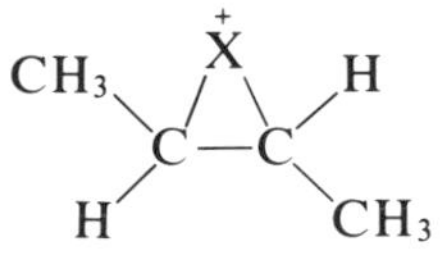

89 X = Br,I 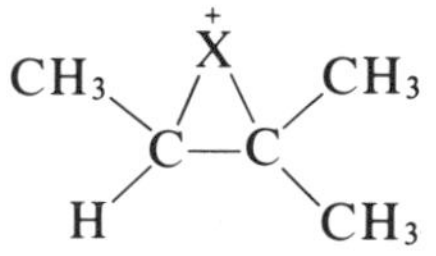 **90** X = Br,Cl,I

The unsymmetrical halonium ions such as **87**-Br is found in equilibrium with open-chain β-bromocarbenium ion. Many of the above-mentioned halonium ions have been prepared by the protonation of the appropriate cyclopropyl halides in superacids.[98]

Attempts to obtain the tetramethylfluoronium ion **88**-F has, however, been unsuccessful.[79] Ionization of 2,3-difluoro-2,3-dimethylbutane in $SbF_5:SO_2$ at $-60°C$ gave an equilibrating β-fluorocarbenium ion **91.** The equilibration was shown to occur through thc intermediacy of 1-*t*-butyl-1-fluoroethyl cation **92** by a recent ^{13}C-nmr spectroscopic study.[99]

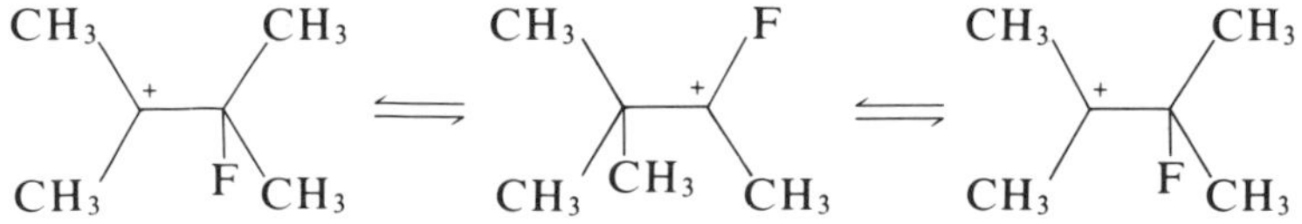

91 ⇌ **92** ⇌

Strating, Wieringa, and Wynberg[100,101] reported that adamantylideneadamantane, a highly sterically hindered olefin, reacted with chlorine in hexane and bromine in CCl_4 solution to give the corresponding chloronium and bromonium salts **93.** They proposed the structure of these rather insoluble salts (on which no spectroscopic study was conducted) to be that of three-membered-ring ethylenehalonium ions (σ-complexes). Olah and co-workers[97] subsequently investigated these stable salts

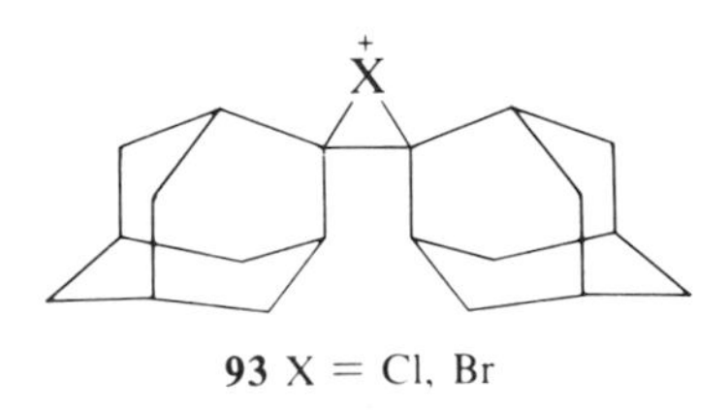

93 X = Cl, Br

by ^{1}H and ^{13}C-nmr spectroscopy, and concluded that they are not three-membered-ring halonium ions but molecularly bound π-complexes **94**.

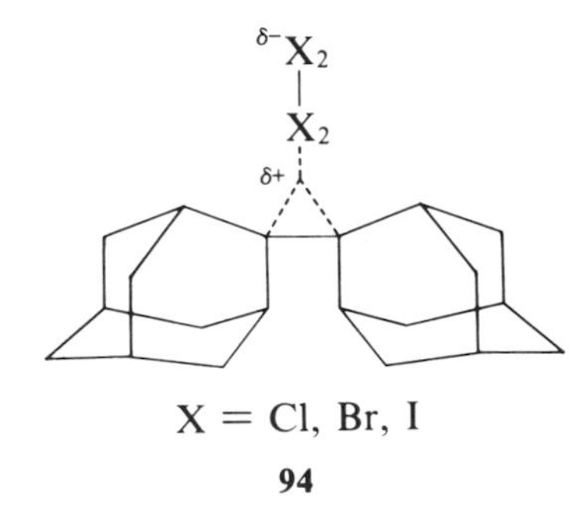

X = Cl, Br, I

94

Trimethylenehalonium Ions. Attempts to prepare trimethylenehalonium ions by ionizing the appropriate 1,3-dihalopropanes with $SbF_5:SO_2$ or with 1:1 $HSO_3F:SbF_5:SO_2$ solution at low temperature have been unsuccessful.[102] The ions obtained were either three- or five-membered-ring halonium ions, formed through ring contraction or expansion, respectively. For example, when 1-halo-3-iodopropanes were reacted with $SbF_5:SO_2ClF$ solution at −78°C, the propyleneiodonium ion **85**-I is formed. This ion could be formed through the trimethyleneiodonium ion which was, however, not observed as an intermediate. Alternatively, the non-assisted primary ion could undergo rapid 1,2-hydrogen shift to the secondary ion, which then would form the propylene-iodonium ion via iodine participation. 1-Halo-3-bromopropanes behave similarly yielding **85**-Br.

$I(CH_2)_3X$, X = F, Cl, I $\xrightarrow{SbF_5:SO_2ClF,\ -78°C}$ [four-membered ring $\overset{+}{I}$] ⟶ CH_3-substituted three-membered ring $\overset{+}{I}$ (**85**-I)

$I(CH_2)_3X$ ⟶ $(ICH_2CH_2CH_2^+)$ ⟶ $ICH_2C^+HCH_3$ ⟶ **85**-I

The ionization of 1,3-dihalo-2-methylpropanes with $SbF_5:SO_2$ gave both three- and five-membered ring halonium ions and open-chain halocarbenium ions.

The only reported preparation of long-lived four-membered-ring halonium ions is that reported by Exner et al.[103] 3,3-Bis(halomethyl)trimethylenebromonium ions **95** were prepared by treating tetrahaloneopentanes with $SbF_5:SO_2ClF$ solution at low temperature. Seemingly, halogen substitution stabilizes the four-membered ring. The ^1H-nmr spectrum of fluorinated ions shows a broad singlet at δ 5.28 and a doublet at δ 4.68 (J_{HF} = 47 Hz). In contrast, the brominated ion displays a

$$RCH_2C(CH_2R')_2CH_2Br \xrightarrow{SbF_5:SO_2ClF,\ -55°C} (RCH_2)_2C(CH_2)_2Br^+$$

R = R′ = F
R = R′ = Br
R = F, R′ = Br

R = F
R = Br

95

temperature-dependent proton absorption at δ 5.17 (from internal $Me_4\overset{+}{N}\overset{-}{B}F_4$), indicating the occurrence of the following exchange process.

$$(BrCH_2)_2C(CH_2^+)(CH_2Br) \rightleftharpoons (BrCH_2)_2C(CH_2)_2Br^+ \rightleftharpoons (BrCH_2)(CH_2Br)C(CH_2)_2Br^+ \rightleftharpoons \text{, etc.}$$

Tetramethylenehalonium Ions. 1,4-Halogen participation was first postulated to occur in the acetolysis of 4-iodo- and 4-bromo-1-butyl tosylates.[104] In subsequent studies, Peterson and co-workers found anomalous rates in the addition of trifluoroacetic acid to 5-halo-1-hexenes[105,106] and 5-halo-1-pentynes.[107] Such observations were recognized as due to 1,4-halogen participation.[80] Also, a study of the solvolysis of δ-chloroalkyl tosylates indicated rate-accelerating 1,4-chlorine participation effects up to 99-fold.[108]

Direct experimental evidence for 1,4-halogen participation comes from the direct observation (by nmr spectroscopy) of five-membered-ring tetramethylenehalonium ions by Olah and Peterson[109] and by Olah et al.[102]

The ionization of 1,4-dihalobutanes in $SbF_5:SO_2$ solution gave the parent tetra-

$$XCH_2CH_2CH_2CH_2X \xrightarrow{SbF_5:SO_2,\ -60°C} \text{(CH}_2)_4X^+ \quad SbF_5X^-$$

X = Cl,Br,I

96

methylenehalonium ions **96.**[109] The ^1H-nmr spectra of these ions are similar to each other. Subsequently it was found that even 1,2- and 1,3-dihalobutanes when reacted with $SbF_5:SO_2$ solution[102] give the same tetramethylenehalonium ions.[89]

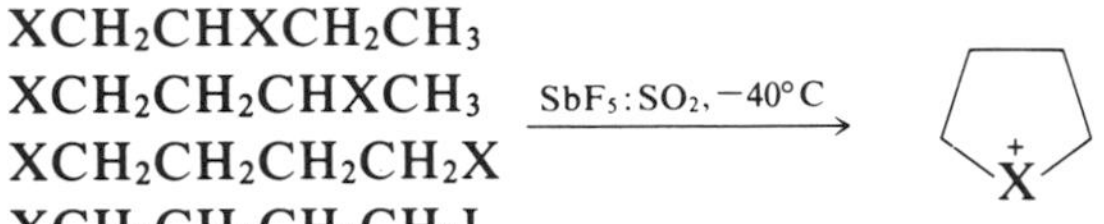

$$\begin{array}{l} XCH_2CHXCH_2CH_3 \\ XCH_2CH_2CHXCH_3 \\ XCH_2CH_2CH_2CH_2X \\ XCH_2CH_2CH_2CH_2I \end{array} \xrightarrow{SbF_5:SO_2,\ -40^\circ C} \mathbf{96}$$

X = Br,Cl

96 X = Cl,Br,I

Similarly several 2- and 5-substituted tetramethylenehalonium ions have been prepared and studied by both ^{1}H- and ^{13}C-nmr spectroscopy.[38,108,110,111]

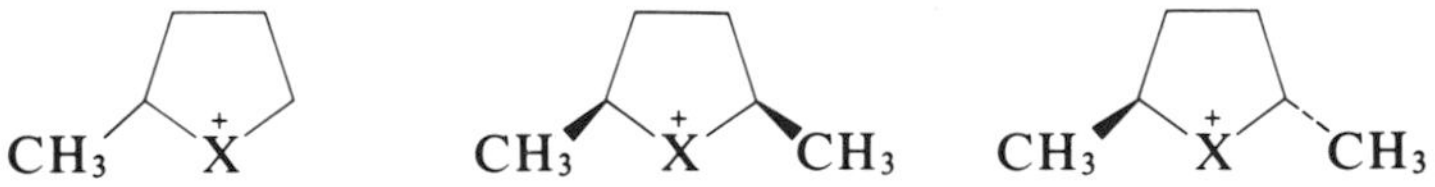

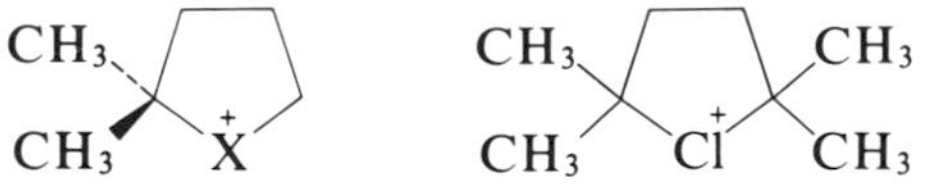

97 X = Cl,Br,I

98 X = Cl,Br,I

99 X = Cl,Br,I

100 X = Cl,Br,I

101

The 2,2,5,5-tetramethylchloronium ion **101** shows only one ^{1}H- and ^{13}C-nmr signal for the nonequivalent methyl groups indicating that some kind of equilibrium exists with the open-chain carbenium ion.[111]

Even 2-methylenetetramethyleneiodonium ion **102** has been prepared[107] by the protonation of 5-iodopentyne.

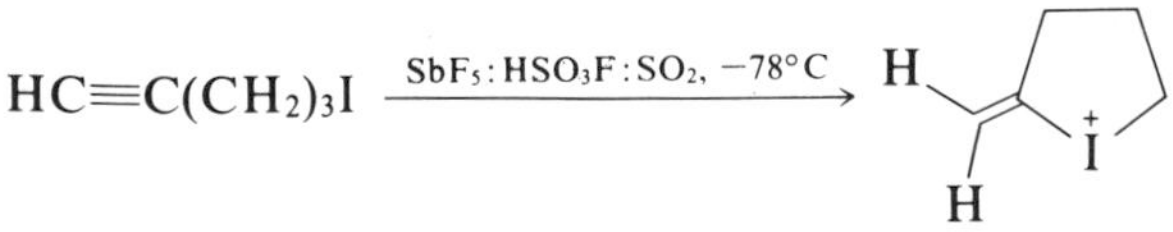

$$HC{\equiv}C(CH_2)_3I \xrightarrow{SbF_5:HSO_3F:SO_2,\ -78^\circ C} \mathbf{102}$$

Pentamethylenehalonium Ions. Attempts to prepare six-membered-ring halonium ions by treating 1,5-dihalopentanes with $SbF_5:SO_2$ gave exclusive rearrangement to five-membered-ring halonium ions.[110] Peterson and co-workers, however, were able to prepare[112] six-membered-ring halonium ions by the methylation of 1,5-dihalopentanes with methyl fluoroantimonate (CH_3F-SbF_5) in SO_2 solution. However, some rearrangement to five-membered-ring halonium ions was also observed.

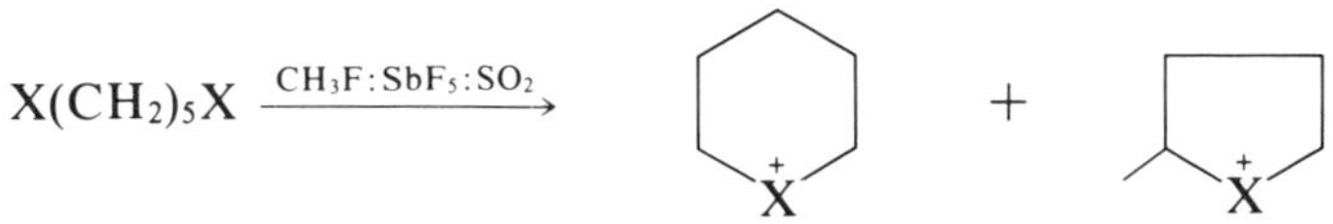

$$X(CH_2)_5X \xrightarrow{CH_3F:SbF_5:SO_2}$$ **103** + **97**

103: X = Br (72-55%), X = I (95%)

97: X = Br (28-45%), X = I (5%)

Alternatively, six-membered-ring halonium ions were also formed when equimolar amount 1,5-dihalopentane was added to dihalonium ions.

The dihalonium ions were prepared from 1,5-dihalopentane and 2 mol of methyl

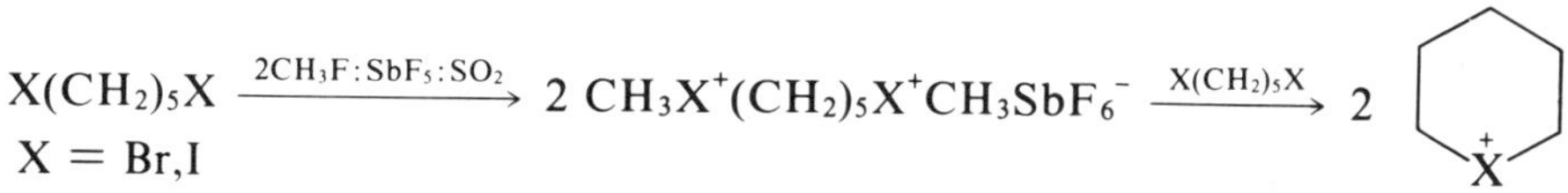

$$X(CH_2)_5X \xrightarrow{2CH_3F:SbF_5:SO_2} 2\ CH_3X^+(CH_2)_5X^+CH_3SbF_6^- \xrightarrow{X(CH_2)_5X} 2$$ **103**

X = Br,I (reactant); X = Br,I (**103**)

fluoroantimonate. Furthermore, the dimethylbromonium ion **59**-Br is also a sufficiently active methylating agent to form cyclic pentamethylenebromonium ion from 1,5-dibromopentane.

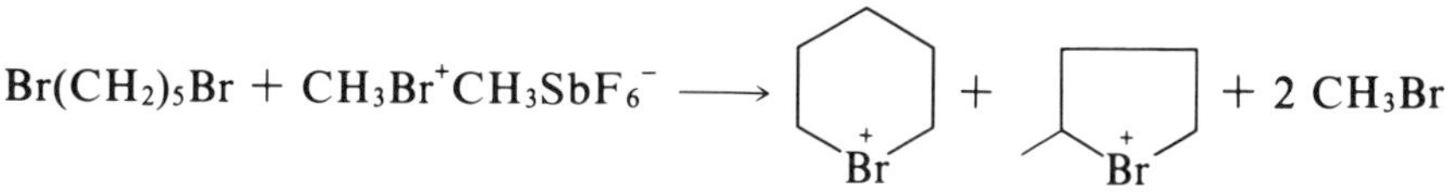

$$Br(CH_2)_5Br + CH_3Br^+CH_3SbF_6^- \longrightarrow$$ **103**-Br 73% + 27% + $2\ CH_3Br$

Bicyclic Halonium Ions. Although halogen additions to cycloalkenes are assumed to proceed through the corresponding bicyclic halonium ions, these ions are quite elusive. Olah, Liang, and Staral[113] were able to prepare the cyclopentenebromonium ions **104** by the ionization of *trans*-1,2-dibromocyclopentane in $SbF_5:SO_2ClF$ solution at $-120°C$. The ^{1}H-nmr (60 MHz) spectrum of the ion solution showed

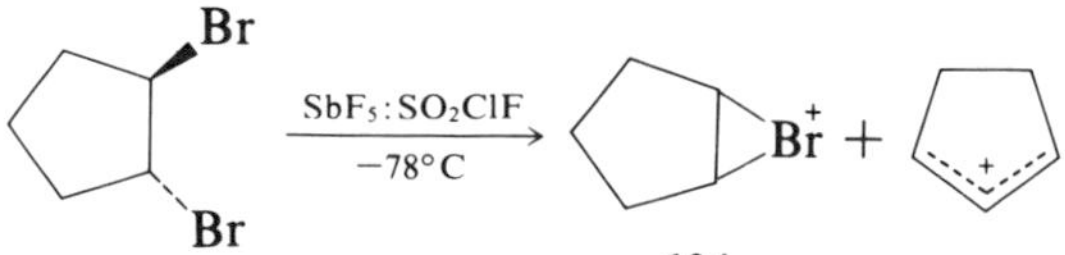

$$\xrightarrow[-78°C]{SbF_5:SO_2ClF}$$ **104** +

a broadened peak at δ 7.32 (two protons) and two broad peaks centered at δ 3.14 (four protons) and δ 2.50 (two protons). The ^{1}H-nmr spectrum of the solution also showed the presence of the related cyclopentenyl cation. When the solution was slowly warmed to $-80°C$, the cyclopentenebromonium ion gradually and cleanly transformed into the allylic ion. The cyclopentenyl ion present initially in solution might be formed as a result of local overheating during preparation.

The cyclopentenebromonium ion **104** was also obtained via protonation of 4-bromocyclopentene in $HSO_3F:SbF_5:SO_2ClF$ solution at $-120°C$, through the following reaction sequence.

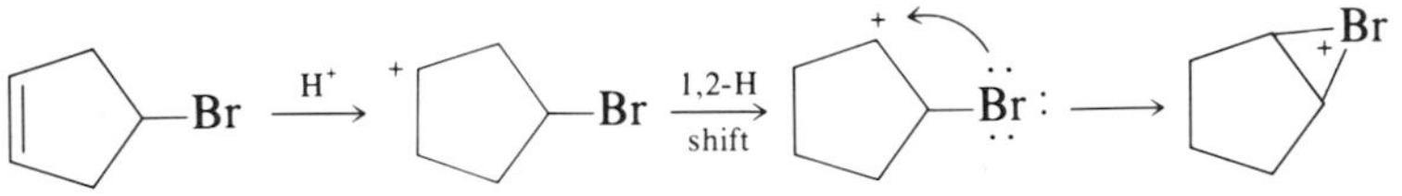

Br $\xrightarrow{H^+}$ Br $\xrightarrow[\text{shift}]{1,2\text{-H}}$ Br: $\longrightarrow$ Br

104

The proton noise-decoupled ^{13}C-nmr spectrum of the cyclopentenebromonium ion **104** shows three carbon resonances at $\delta_{^{13}C}$ 114.6 (doublet, J_{CH} 190.6 Hz), 31.8 (triplet, J_{CH} 137.6 Hz), and 18.7 (triplet, J_{CH} 134.0 Hz).

Attempts were also made to prepare the cyclopentenechloronium ion via ionization of *trans*-1,2-dichlorocyclopentane in $SbF_5:SO_2ClF$ at $-120°C$. However, instead of the cyclopentenechloronium ion **105** only the 1-chloro-1-cyclopentyl cation **106** was obtained. Apparently, the participation of the smaller chlorine atom could not effectively compete with the fast 1,2-hydride shift forming the tertiary ion. The larger bromine atom, however, preferentially participates with the neighboring electron-deficient center, forming the bicyclic bridged ion.

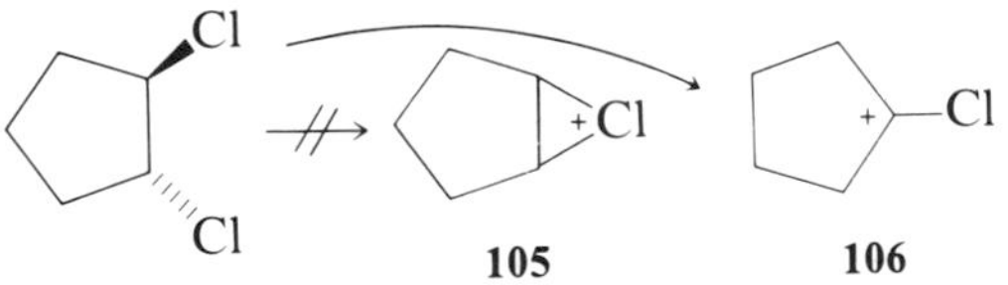

Cl, Cl ⇸ +Cl — Cl

105 **106**

Peterson and Bonazza have[114] reported that ionization of *cis*-1,2-bis(chloromethyl)cyclohexane in $SbF_5:SO_2$ solution at $-78°C$ gives the bicyclic five-membered-ring chloronium ion **107** along with smaller amounts of other species. Warming the solution containing the ion to $-10°C$ leads to the formation of the open-chain tertiary carbenium ion **108**.

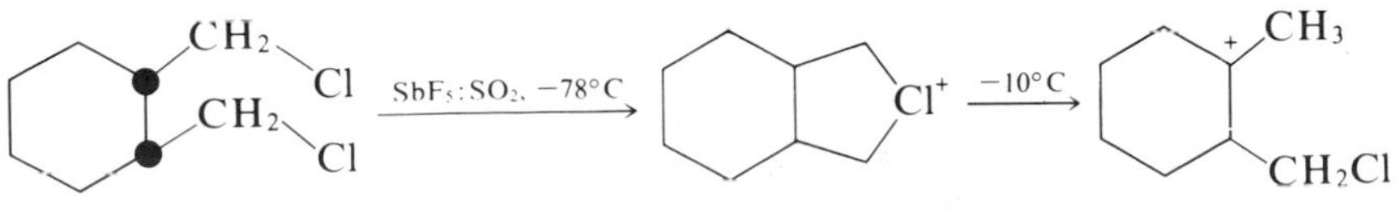

CH_2Cl, CH_2Cl $\xrightarrow{SbF_5:SO_2,\ -78°C}$ Cl^+ $\xrightarrow{-10°C}$ CH_3, CH_2Cl

107 **108**

Heteroaromatic Halophenium Ions. Halophenium ions are a class of halonium ions analogous to thiophene, furan, and pyrrole. Many stabilized tetraphenyl, benzo-, and dibenzoiodophenium ions have been prepared by Beringer and co-workers.[115,116] Some of them have been analyzed by X-ray crystal structure investigations. Representative members are

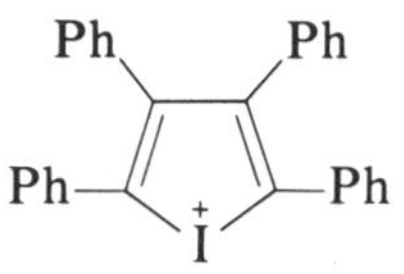

I^- **109** Cl^- **110**

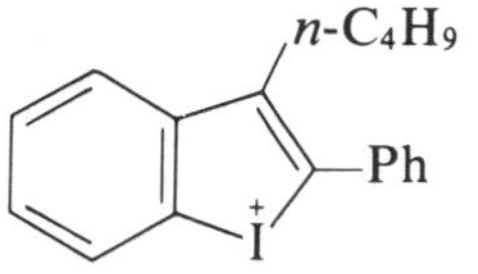

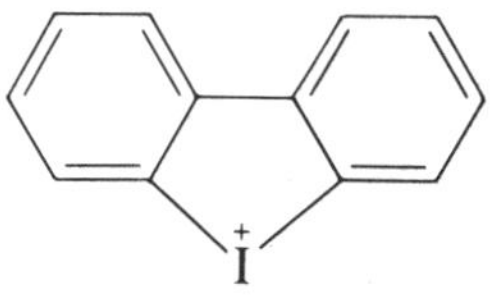

111 Cl^- 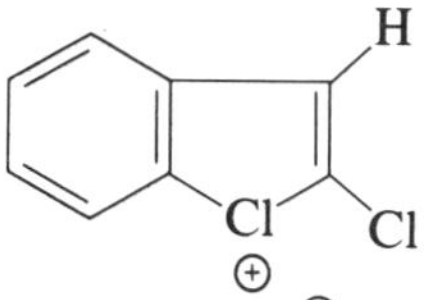 $PF_6^\ominus$ **112**

Olah and Yamada[117] have shown that thermal decomposition of *o*-(β,β-dichloroethenyl)phenyl diazoniumfluorophosphate yielding *o*-(β-chloroethynyl)-

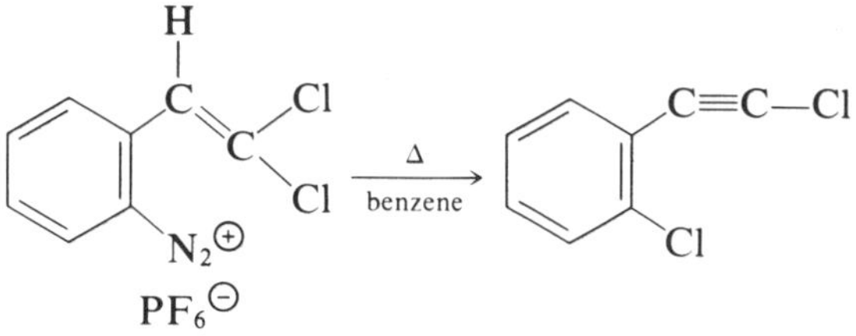

$\xrightarrow[\text{benzene}]{\Delta}$

chlorobenzene involves benzochloronium ion **112** as the intermediate. To date, however, no parent halophenium ion **113** is known.

113 X = Cl, Br, I

4.2.5 Azonium Ions

4.2.5.1 Diazonium Ions

The simplest diazonium ion, protonated dinitrogen **114** is still elusive. In fact, molecular nitrogen is a very weak base. An analog of **114,** fluorodiazonium ion **115,** is, however, known.[118] Ionization of *cis*-difluorodiazine with arsenic pentafluoride below ambient temperature gives fluorodiazonium hexafluoroarsenate **115** as a white solid. The same reaction does not occur with *trans*-difluorodiazine.

$$H{-}\overset{+}{N}{\equiv}N \qquad F{-}\overset{+}{N}{\equiv}N\ AsF_6^-$$

114 **115**

The ion **115** has been studied by X-ray diffraction, ^{19}F- and ^{14}N-nmr spectroscopy.[119,120] The $J_{^{14}N\text{-}F}$ value measured for **115** at 339.0 Hz is the largest known ^{14}N–^{19}F coupling constant, indicating the high *s*-character of the nitrogen–nitrogen bond.

The intermediacy of alkyldiazonium ions in a variety of organic reactions is well established.[121] They are common intermediates in the acid-catalyzed decomposition of diazo compounds and the nitrous acid-induced deamination of aliphatic primary amines. The evidence for RN_2^+ (R = alkyl) intermediates come from both rate data and product analysis studies. However, direct investigation of alkyldiazonium ions have been difficult due to their instability. The first direct observation[122] of an aliphatic diazonium ion was achieved by protonation of trifluoromethyldiazomethane in HSO_3F at $-60°C$. The 2,2,2-trifluoroethyldiazonium ion **116** is stable for 1 h at $-60°C$. Similarly bis(trifluoromethyl)methane diazonium ion **117** has been prepared and characterized.[123] These ions were studied by ^1H-nmr spectroscopy. Similar diazonium structures have been assigned to protonated 2-diazo-5α-cholestan-3-one.[124] None of these studies, however, showed nitrogen protonation. Recently, McGarrity and co-workers[125] have succeeded in protonating diazomethane in $HSO_3F:SO_2ClF$ at $-120°C$. In this acid media, exclusive formation of methyldiazonium ion **118** is observed. With a more acidic $HSO_3F:SbF_5$ system, both methyldiazonium ion **118** as well as methylenediazenium ion are observed.

$$CF_3CH_2\overset{+}{N}_2 \qquad (CF_3)_2CH\overset{+}{N}_2$$

116 **117**

$$CH_3\overset{+}{N}_2 \qquad CH_2{=}\overset{+}{N}{=}NH$$

118 **119**

The two ions **118** and **119** have been characterized by ^1H-, ^{13}C-, and ^{15}N-nmr spectroscopy.

In contrast to alkyldiazonium ions, aryldiazonium ions are well studied.[126] They were known as early as 1894. They are isolable as ionic salts with variety of counter ions such as BF_4^-, PF_6^-, $SbCl_6^-$, SbF_6^-, AsF_6^-, ClO_4^-, etc. They undergo variety of nucleophilic reactions and an excellent review is available on the subject.[127]

The ambident reactivity of aryldiazonium ions has also been established.[128] The diazonium group is an interesting substituent on the aryl ring and by far the most strongly electron-withdrawing substituent known ($\sigma_p = 1.8$).[129] A recent ^{13}C-nmr spectroscopic investigation on a series of aryldiazonium ions seems to support the above fact and also indicates their ambident electrophilic character. Aryldiazonium ions undergo interesting reaction of N_α–N_β inversion catalyzed by metals that complex molecular nitrogen, probably through the intermediacy of a phenyl cation.[130,131] Such inversions have been observed by Zollinger and co-workers in dediazotization reaction of β-^{15}N-labeled diazonium ions with nitrogen gas under pressure (300 atmospheres).[132,133] Such exchange reactions have been further studied with sterically hindered diazonium ions.[134,135]

$$Ar{-}\overset{+}{N}{\equiv}{}^{15}N \rightleftharpoons Ar{-}{}^{15}\overset{+}{N}{\equiv}N$$

$$Ar{-}{*}N_2^+ + N_2 \rightleftharpoons Ar{-}N_2^+ + {*}N_2$$

A series of aminodiazonium ions have been prepared under superacid conditions. Schmidt[136] described the preparation and ir spectra of protonated hydrazoic acid **120** and methylazide **121**-CH_3 as their hexachloroantimonate salts. Recently Olah and co-workers have carried out a comprehensive study[137] on aminodiazonium ions (protonated azides) by ^1H-, ^{13}C-, and ^{15}N-nmr spectroscopy. Even the electrophilic aminating ability of aromatics of **120** has been explored.[137] The tetrachloroaluminate salt of **120** has also been prepared.[137]

$$HN_3 + HSO_3F{:}SbF_5 \text{ or } HF{:}SbF_5 \text{ or } HF{:}BF_3 \longrightarrow \underset{\mathbf{120}}{H_2N{-}\overset{+}{N}{\equiv}N\ X^-}$$

$$NaN_3 + AlCl_3 + 2HCl \longrightarrow \underset{\mathbf{120}}{NH_2{-}\overset{+}{N}{\equiv}N\ AlCl_4^-} + NaCl$$

The evidence for the aminodiazonium structure **120** for the protonated hydrazoic acid comes from ^{15}N-nmr spectroscopy. The iminodiazenium structure **122** is not observed either in the case of hydrazoic acid or the alkyl azides. Alkyl azides are also protonated to alkylaminodiazonium ions **121**.

$$\underset{\substack{\mathbf{122} \\ R\,=\,H,\ CH_3,\ C_2H_5}}{R{-}N{=}\overset{+}{N}{=}N{-}H} \qquad \underset{\substack{\mathbf{121} \\ R\,=\,CH_3,\ C_2H_5}}{RNH{-}\overset{+}{N}{\equiv}N}$$

4.2.5.2 Nitronium Ion (NO_2^+)

Nitration is one of the most studied and best understood of organic reactions.[138] The species responsible for electrophilic aromatic nitration was shown to be the nitronium ion NO_2^+ **123.** Since the turn of the century extensive efforts have been directed toward the identification of this ion, whose existence was first shown by Hantzsch and later firmly established by Ingold and Hughes.[138b]

Raman spectroscopic studies in the 1930s on $HNO_3:H_2SO_4$ mixtures by Médard[139] showed the presence of nitronium ion **123** in the media. A sharp band at 1400 cm^{-1} was assigned to the symmetric stretching mode of the species. Since then the nitronium ion has been characterized in a variety of Brønsted and Lewis acid mixtures of nitric acid.[140]

There are more than 15 crystalline nitronium ion salts that have been isolated and characterized with a variety of counterions. The most important salts are of BF_4^-, PF_6^-, SbF_6^-, ClO_4^-, FSO_3^- counter ions. The X-ray crystal structure of nitronium ion is known with hydrosulfate anion.[141] The most widely used nitronium tetrafluoroborate salt ($NO_2^+BF_4^-$) is prepared by treating a mixture of nitric acid and anhydrous hydrogen fluoride with boron trifluoride.[142]

$$HNO_3 + HF + 2BF_3 \longrightarrow \underset{\mathbf{123}}{NO_2^+\bar{B}F_4} + BF_3:H_2O$$

In the ^{15}N-nmr spectrum, the nitronium ion[143] is about 130 ppm deshielded from NO_3^- of aqueous sodium nitrate solution.[120,143] Recently it has been shown by ^{14}N-nmr spectroscopy that a mixture of nitric acid in 88% sulfuric acid contains both nitronium ion as well as free nitric acid.[144]

The applications of nitronium salts as a synthetic reagent[145] are discussed in Chapter 5. Until recently, the nitronium ion was recognized only as a nitrating agent. However, it has been found that it possesses significant ambident reactivity. This has been recently shown in the oxidation of sulfides, selenides, and phosphines. In fact, the sulfide reaction has been monitored by ^{15}N-nmr spectroscopy wherein both nitrosulfonium and nitritosulfonium ions **124** and **125** were detected as distinct intermediates.[143]

$$R{-}S{-}R + NO_2^+ \rightleftharpoons \underset{\mathbf{124}}{\left[R{-}\overset{\overset{NO_2}{|}}{S}{-}R \right]^+} \longrightarrow \underset{\mathbf{125}}{\left[R{-}\overset{\overset{\overset{NO}{|}}{O}}{\underset{}{\overset{|}{S}}}{-}R \right]^+}$$

$$R{-}\overset{\overset{O}{\|}}{S}{-}R + NO^+$$

4.2.5.3 *Nitrosonium Ion (NO^+)*

Nitrosonium ion (NO^+) **126** is an important electrophilic species that is generally present in nitrous acid media. It acts as a powerful nitrosating agent of amines (both aromatic and aliphatic) resulting in the diazotization reaction.

$$R{-}NH_2 + NO^+ \longrightarrow RN_2^+ + H_2O$$

126

The first isolation of nitrosonium ion **126** as a distinct species was in the reaction of dinitrogen trioxide and dinitrogen tetroxide with boron trifluoride.[146]

$$3N_2O_3 + 8BF_3 \longrightarrow 6NO^+BF_4^- + B_2O_3$$

126

$$3N_2O_4 + 8BF_3 \longrightarrow 3NO^+BF_4^- + 3\ NO_2^+BF_4^- + B_2O_3$$

Since then a variety of nitrosonium salts have been isolated. The important ones are with the following counter ions: BF_4^-, PF_6^-, FSO_3^-, HSO_4^-, BCl_4^-, and $SbCl_6^-$. The ion has been characterized by ^{15}N-nmr, ir, and X-ray analysis.[120,140,143,146,147]

The nitrosonium ion does not react toward aromatics except in activated systems. It forms a π-complex with aromatics with deep color.[148,149] However, it is a powerful hydride-abstracting agent in the case of activated benzylic or allylic positions. Olah and Friedman have demonstrated[150] that isopropyl benzene undergoes hydride abstraction to a cumyl cation which further reacts to give various condensation products. The reaction has been employed to prepare a variety of stable carbocations.[151]

$$R{-}C_6H_4{-}CH(CH_3)_2 \xrightarrow{\overset{+}{N}OPF_6^-} R{-}C_6H_4{-}\overset{+}{C}(CH_3)_2 + HNO$$

The unique hydride abstracting property has been gainfully employed in developing novel synthetic reactions.[145] Reactive hydrocarbons such as triphenylmethane, adamantane, and diamantane are readily fluorinated in presence of nitrosonium ion in HF/pyridine media.[152] In the presence of a suitable oxygen donor such as dimethylsulfoxide, the nitrosonium ion can act as a nitrating agent.[153] The initially formed nitrito onium ion **125**[143] transfer nitrates aromatics rather readily. The NO^+-induced reactions are further reviewed in Chapter 5.

$$NO^+ + CH_3{-}\overset{\overset{O}{\|}}{S}{-}CH_3 \longrightarrow \left[CH_3{-}\underset{+}{\overset{\overset{NO}{\diagup}\ O}{\overset{|}{S}}}{-}CH_3 \right]$$

125

$$\xrightarrow{ArH} ArNO_2 + H^+ + CH_3{-}S{-}CH_3$$

4.3 ENIUM IONS

4.3.1 Nitrenium Ions

Nitrenium ions containing divalent positive nitrogen have been postulated as intermediates in rearrangement, synthesis, and cleavage of nitrogen containing organic compounds.[154,155] In contrast to a trivalent carbocation, the nitrenium ion **127** is unusual in that it has both a positive charge and a nonbonding pair of electrons. Hence, the nitrenium could exist both as a singlet and a triplet. The singlet would resemble a carbocation and triplet a radical cation with great tendency for hydride abstraction.[156]

X | R—N̈—R ⟶ R—N⁺—R ⟶ R—N⁺·—R

Singlet **127a** Triplet **127b**

Attempted generation of nitrenium ions as distinct species under long-lived stable ion conditions has thus far been unsuccessful[157] except in the case of the trifluorodiazenium ion $N_2F_3^+$ **128**,[158] which could be considered as a potential nitrenium ion. Protonation of nitrosobenzenes in superacidic media has led only to benzeniumiminium dications **129.**[157,158] Christe and Schack[158] have obtained $N_2F_3^+$ **128** by

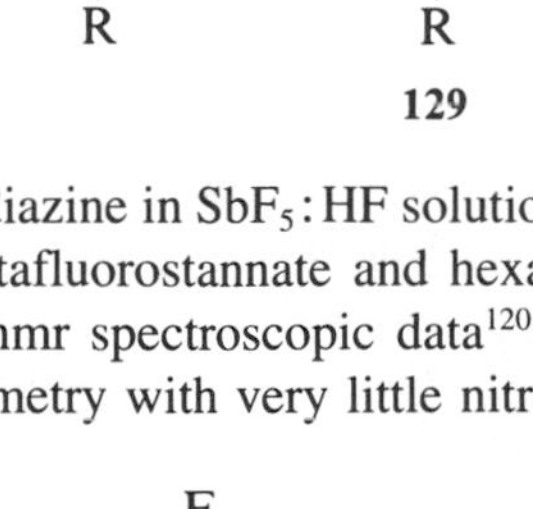

129

the ionization of tetrafluorodiazine in SbF_5:HF solutions. Similarly, they have been successful in preparing pentafluorostannate and hexafluoroarsenate salts. The vibrational and ^{19}F- and ^{15}N-nmr spectroscopic data[120,158] are consistent with planar structure **128a** with C_s symmetry with very little nitrenium character **128b.**

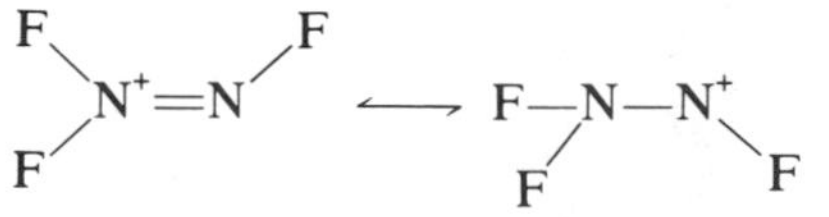

$F_2N^+{=}NF \longleftrightarrow F{-}N(F){-}N^+{-}F$

128a **128b**

The ambivalent nature of a α-cyano group on a carbocationic center has been demonstrated by the solvolytic work of Gassman and coworkers.[159,160] Inductively, the cyano group is strongly destabilizing. However, the major portion of this effect is offset by the mesomeric nitrenium ion structure **130a-b.**

$$R_2C^+{-}C{\equiv}N \longleftrightarrow R_2C{=}C{=}N^+$$

130a **130b**

Olah and co-workers have prepared a series of α-cyanodiarylcarbenium ions **130** (R = Ar) under superacid conditions and have evaluated their mesomeric nitrenium ion character by ^{1}H-, ^{13}C-, and ^{15}N-nmr spectroscopy.[162] A recent one-bond ^{13}C—^{13}C coupling constant measurements[163] also indicate significant mesomeric nitrenium character of **130** (R = Ar). Protonated aroyl cyanides **131,** however, exist predominantly in the carboxonium ion form **131a** over the nitrenium ion form **131b.**[164]

$$Ar{-}C({=}\overset{+}{O}H){-}C{\equiv}N \longleftrightarrow Ar{-}C(OH){=}C{=}N^+$$

131a **131b**

4.3.2 Borenium Ions

Boron cations with coordination number four such as $[BH_2(NH_3)_2]^+$ and many others of the type $[H_2BL_2]^+$, $[HXBL_2]^+$, and $[X_2BL_2]^+$ are well known,[165–168] X being halogen and L an electron-donating ligand. Even doubly and triply charged tetracoordinate boron cations are known.

However, dicoordinate borenium ions are much less known. It is well recognized that electron deficiency of boron compounds can be considerably compensated by π-back bonding. Exploiting this principle, Nöeth and Staudigl[169] have succeeded in obtaining borenium ions.

Reaction of anhydrous aluminum bromide with a series of 2,2,6,6-tetramethylpiperidinoaminoboron bromides in dichloromethane leads to specific displacement of bromide, which is trapped as tetrabromoaluminate. By formation of this less nucleophilic anion, and owing to the steric and electronic shielding of the β atom by the bulky 2,2,6,6-tetramethylpiperidino moiety, dicoordinated borenium ions **132** are generated.

$$\text{(TMP)N{-}B(Br)Y} \xrightarrow[CH_2Cl_2]{AlBr_3} [\text{(TMP)N{=}B{=}Y}]^+ \; AlBr_4^-$$

$Y = N(CH_3)_2,\ N(C_2H_5)_2,\ C_6H_5, CH_3$

132

^{27}Al-nmr spectra confirm exclusive formation of the $AlBr_4^-$ anion, which, compared to Al_2Br_6, is characterized by its substantially sharper signal; the linewidth of $\nu_{1/2} \sim 20$ Hz observed in **132**-N$(CH_3)_2$ and **132**-N$(C_2H_5)_2$ corresponds to an undistorted tetrahedral $AlBr_4^-$. The ^{11}B-nmr signals of **132** are shifted 6–8 ppm downfield relative to those of the starting compounds tetramethylpiperidinoboron bromide; their linewidths, which are greater by a factor of about 5, are consistent with a linear heteroallene structure. Also consistent with such a structure are the isotopically split ir bands at 1850–1900 cm^{-1}, which are assigned to an antisymmetric BN_2 vibration. The heteroallene structure has been confirmed by X-ray analysis of **132**-N-$(CH_3)_2$.

Such studies have been extended to a variety of amidoborenium ions.[170] Recently Parry and co-workers have isolated and characterized bis(diisopropylamino)borenium ion **133** at low temperature as tetrachloroaluminate salt.[171] Attempts to prepare the analogous bis(dimethylamino)borenium ion **134** was, however, unsuccessful.

$$[(i\text{-}C_3H_7)_2N{=}B{=}N(i\text{-}C_3H_7)_2]^+ \quad [(CH_3)_2N{=}B{=}N(CH_3)_2]^+$$

133 **134**

To date, no dicoordinate borenium ion with either only alkyl and/or aryl substituent is known.

4.3.3 Oxenium Ions

Oxenium ions similar to nitrenium ions are in general too reactive to be observed. The parent ion, that is, the hydroxyl cation "$\overset{+}{O}H$", is elusive, and it is improbable that it can be observed in its "free" form in the condensed state. However, the incipient hydroxyl cation is involved in acid-catalyzed electrophilic hydroxylation with protonated (or Lewis acid complexed) hydrogen peroxide ($HO{-}\overset{+}{O}H_2$) or ozone ($\overset{+}{O}{-}O{-}OH$).[50] Nitrous oxide is also a potential precursor for the hydroxyl cation (in its protonated form). The hydroxy diazonium ion $\overset{+}{N}_2OH$ has not yet been observed.

Alkyl and aryloxenium ions (RO^+) are similarly too reactive to be observed. The methyloxenium ion, CH_3O^+, may be involved in the oxidation of methane under superacidic conditions with ozone (see Chapter 5). Of all the possible alkyloxenium ions, the *t*-butyloxenium ion is the most significant. Superacid cleavage of *t*-butyl hydroperoxide in a Hock-type reaction gives acetone and methyl alcohol, indicative of the intermediacy of the *t*-butyl oxenium ion.

$$(CH_3)_3C{-}OOH \xrightleftharpoons{H^+} [(CH_3)_3C{-}O^+] + H_2O$$

$$\longrightarrow (CH_3)_2C{=}\overset{+}{O}{-}CH_3 \xrightarrow{H_2O} CH_3CCH_3 (C{=}O) + CH_3OH$$

Under stable ion conditions even at low temperatures, only the rearranged carboxonium ion could, however, be observed.[50]

Similarly the cumyloxenium ion [$C_6H_5(CH_3)_2CO^+$] is involved in the acid-catalyzed cleavage rearrangement reaction of cumene hydroperoxide to phenol and acetone.

$$C_6H_5C(CH_3)_2OOH \xrightleftharpoons{H^+} [C_6H_5C(CH_3)_2O^+]$$

$$C_6H_5OH + (CH_3)_2C{=}O \xleftarrow{H_2O} C_6H_5\overset{+}{O}{=}C(CH_3)_2$$

4.3.4 Phosphenium Ions

In contrast to widely studied phosphonium ions (PR_4^+), the chemistry of the dicordinate phosphenium ion is little known.[172] It has been recognized that phosphenium ions can only be generated if one of the substituent is a dialkylamino group.[173]

$$R_2N(R')P{-}Cl + \tfrac{1}{2}Al_2Cl_6 \longrightarrow R_2N(R')\overset{+}{P}\ \overset{-}{A}lCl_4 \longleftrightarrow R_2\overset{+}{N}{=}P(R')\ \overset{-}{A}lCl_4$$

R = Alkyl, $(CH_3)_3Si$

R′ = $N(CH_3)_2$, $[(CH_3)_3Si]_2N$, Cl, $(CH_3)_3C$

Parry and co-workers have carried out detailed ^{31}P-nmr spectroscopic study of a series of phosphenium ions.[174,175] Their study indicates that the chemical shift of P^+ center depends upon the steric crowding around and the extent of back-donation from the nitrogen lone pair. In the case of *t*-butyldimethylaminophosphenium ion **135** the ^{31}P chemical shift of the P^+ center is at δ 510 (from 85% H_3PO_4) and it appears to be largest ever ^{31}P downfield shift to be measured. On the other hand bis(dimethylamino)phosphenium ion **136** shows a chemical shift of δ 264.

$$(CH_3)_3C{-}\overset{+}{P}{-}N(CH_3)_2 \quad \textbf{135} \qquad (CH_3)_2N{-}\overset{+}{P}{-}N(CH_3)_2 \quad \textbf{136}$$

Attempts have been made to observe bis and tris phosphenium ions of the following type.[173] In all the cases, however, only monoionization takes place.

Two-coordinate geometry of phosphenium ions has been confirmed by X-ray diffraction studies on $(i\text{-}Pr_2N)_2P^+AlCl_4^-$ **139.**[176] Recently, an unique ferrocenyl-stabilized two-coordinate phosphenium ion **140** has been prepared and analyzed.[177]

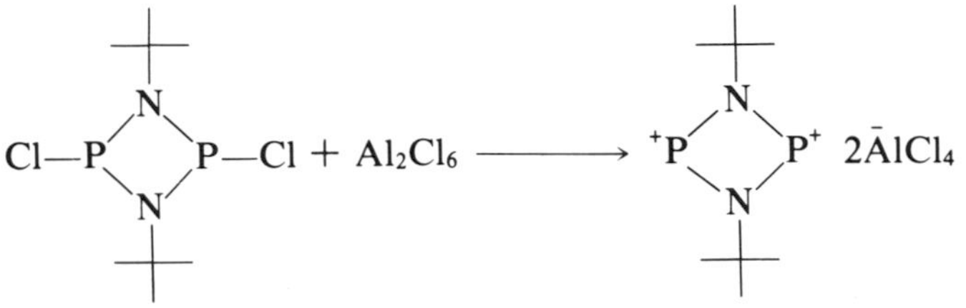

137

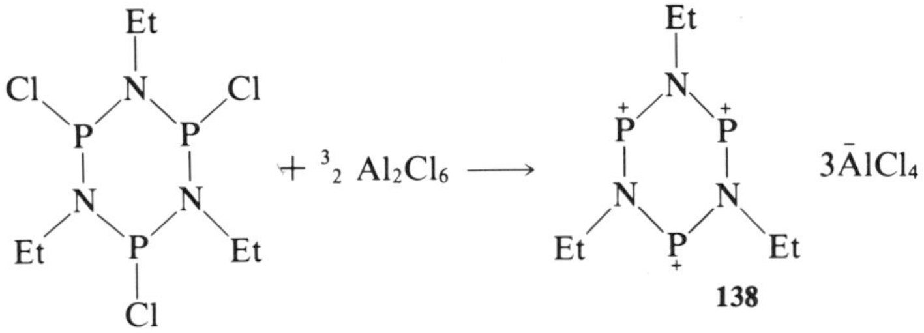

138

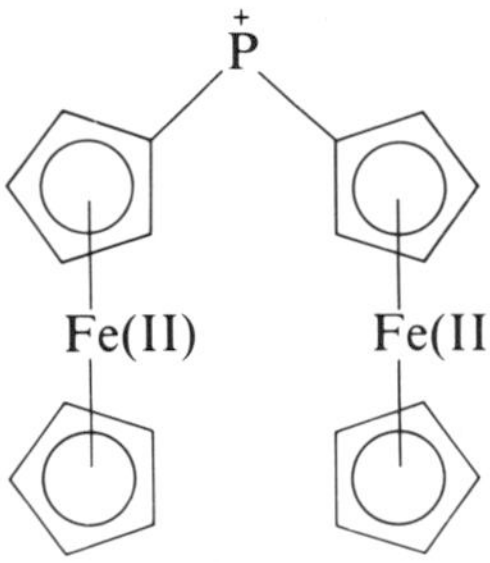

140

4.3.5 Silicenium Ions

One of the key organic intermediates that has not been observed in solution is trivalent positively charged silicon, the silicenium ion (R_3Si^+), the analog of a carbocation.[178,179] On the contrary, silicenium ions are well known in the gas phase as high-abundance fragments in the mass spectra of organosilicon compounds.[180] The failure to observe them in solution is due to the poor ability of silicon to undergo $p\pi$–$p\pi$ bonding. Whereas, carbocations are readily stabilized by $2p$–$2p$ resonance, the silicenium ion is more weakly stabilized through $2p$–$3p$ overlap over longer bonds with lone pairs or π-electrons on carbon, nitrogen, or oxygen.[181] Moreover, the very large bond strength of silicon-oxygen, nitrogen, and most halogens make common leaving groups unavailable. This is the main reason, why, attempts to prepare silicenium ions under superacid conditions have failed (due to nucleophilic fluorosulfate or fluoride ions which strongly bond electrophilic silicon).[182]

Olah and Field[183] were able to obtain only a polarized complex **141** from trimethylsilyl bromide and aluminum bromide in methylene bromide solution. However, they were able to correlate ^{29}Si-nmr chemical shifts with ^{13}C-nmr chemical

$$(CH_3)_3Si^{\delta+}\text{—}BR\longrightarrow{}^{\delta-}AlBr_3$$

141

shifts of analogous compounds. Based on such an empirical relationship, they have been able to predict the ^{29}Si chemical shift of trivalent silicenium ion. More recently, Lambert and Schulz[184] have prepared triisopropylthiosilicenium ion **142** by hydride abstraction from triisopropylthiosilane using trityl perchlorate in dichloromethane solution.

$$(i\text{-}PrS)_3Si\text{—}H + Ph_3C^+ \; ClO_4^- \longrightarrow Ph_3CH + (i\text{-}PrS)_3Si^+ClO_4^-$$

142

The evidence for **142** comes from both electrical conductivity measurements and ^{1}H- and ^{13}C-nmr spectra. The ir spectrum of **142** clearly showed the presence of perchlorate anion. The ^{1}H- and ^{13}C-nmr data on **142** were interpreted to indicate that there is not much positive charge delocalization from silicon to sulfur. Unfortunately, the authors[184] were not able to obtain a satisfactory ^{29}Si-nmr spectrum of **142.**

4.4 HOMO- AND HETEROPOLYCATIONS

In this section, the polyatomic cations of group VII elements (halogen and interhalogen cations and polycations) along with cations and polycations of group VI elements obtained in the superacid media will be discussed (O, S, Se, and Te).

4.4.1 Halogen Cations

The existence of many well-known compounds in which chlorine, bromine, and iodine are found in +1 oxidation state lead to the assumption that the cations Cl^+, Br^+, and I^+ are important as stable entities or at best as reaction intermediates. However, no evidence exists for monoatomic Cl^+, Br^+, and I^+ as stable species.[185,186] In contrast, a whole series of polyatomic halogen cations are known.

4.4.1.1 Iodine Cations

The existence of I_3^+ and I_5^+ ions, deduced by Masson[187] 30 years ago in aromatic iodination reactions, has been confirmed now by physical measurements. The controversy over the nature of blue solutions of iodine in strong acid media has now been settled. It has been shown conclusively that these solutions contain I_2^+[188–190] and not I^+ as suggested earlier.[191]

I_3^+ and I_5^+ Ions. The first evidence for stable iodine cations was obtained by Masson.[187] He postulated the presence of I_3^+ and I_5^+ ions **143** and **144** in solutions of iodine and iodic acid in sulfuric acid to explain the stoichiometry of the reaction of such solutions with chlorobenzene to form both iodo and iodoso derivatives. Later, Symons and co-workers[191] gave conductometric evidence for I_3^+ formed from iodic acid and iodine in 100% sulfuric acid and suggested that I_5^+ may be formed on the basis of changes in the uv and visible spectra when iodine is added to I_3^+ solutions. Gillespie and co-workers,[192] on the basis of detailed conductometric and cryoscopic measurements, confirmed that I_3^+ is formed from HIO_3 and I_2 in 100% sulfuric acid. The I_3^+ cation may also be prepared in fluorosulfuric acid.[188] Solutions

$$HIO_3 + 7I_2 + 8H_2SO_4 \longrightarrow \underset{\textbf{143}}{5I_3^+} + 3H_2O + 8HSO_4^-$$

of red brown I_3^+ **143** in H_2SO_4 or HSO_3F have characteristic absorption maxima at 305 and 470 nm, with a molar extinction coefficient of 5200 at 305 nm.

Solutions of I_3^+ in 100% H_2SO_4[192] or in fluorosulfuric acid,[188] dissolve at least 1 mol of iodine per mol of I_3^+, and a new absorption spectrum is obtained which has bands at 270, 340, and 470 nm. At the same time, there is no change in either the conductivity or the freezing point of the solutions. This leads to the conclusion that I_5^+ is formed. Some further iodine also dissolves in solutions of I_5^+, indicating possible formation of I_7^+.

$$I_3^+ + I_2 \longrightarrow \underset{\textbf{144}}{I_5^+}$$

Corbett et al.[193] have prepared the compounds $I_3^+AlCl_4^-$ and $I_5^+AlCl_4^-$, which they characterized by phase equilibria studies and nuclear quadrupole resonance spectroscopy.

Solution of I_3^+ **143** in H_2SO_4 give Raman spectra[194] that have three bands, in addition to the solvent peaks, at 114, 207, and 233 cm^{-1} which may be assigned as the ν_2, ν_1, and ν_3 vibrations of an angular molecule. The average stretching frequency of 220 cm^{-1} in the I_3^+ molecule is appreciably lower than the stretching frequency of 238 cm^{-1} for the I_2^+ molecule and, in fact, closer to the frequency of 213 cm^{-1} for the neutral molecule I_2. This is consistent with I_3^+ having a formal I—I bond order of 1.0 as in the simple valence bond formulation, whereas that in I_2^+ it is 1.5.

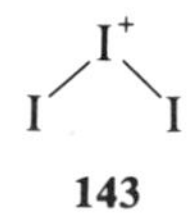

143

Corbett et al.[193] on the basis of ^{127}I nuclear quadrupole resonance (nqr) studies of $I_3^+AlCl_4^-$ have predicted a bond angle of 97° between the two bonding orbitals on the central atom. X-ray crystal structures of $I_3^+AsF_6^-$ and $I_5^+SbF_6^-$ have been obtained.[195,196]

I_2^+ Ion. Gillespie and Milne[188] have shown, by conductometric, spectrophotometric, and magnetic susceptibility measurements in fluorosulfuric acid, that the blue iodine species observed in strong acids is I_2^+ **145.** When iodine was oxidized by peroxodisulfuryl difluoride in fluorosulfuric acid, the concentration of the blue iodine species reached a maximum at the 2:1 I_2:$S_2O_6F_2$ mole ratio and not at the 1:1 mole ratio as would be anticipated for the formation of I^+. The conductivities

$$2I_2 + S_2O_6F_2 \longrightarrow \underset{\mathbf{145}}{2I_2^+} + 2SO_3F^-$$

$$I_2 + S_2O_6F_2 \longrightarrow 2I^+ + 2SO_3F^-$$

of 2:1 solutions of iodine:$S_2O_6F_2$ at low concentrations were found to be very similar to solutions of KSO_3F at the same concentration, showing that 1 mol of SO_3F^- had been formed per mole of iodine. The magnetic moment of the blue species in fluorosulfuric acid was found to be 2.0 ± 0.1 D, which agreed with the value expected for the $^3\pi_{3/2}$ ground state of the I_2^+ cation **145.** The I_2^+ has characteristic peaks in its uv absorption spectrum at 640, 490, and 410 nm and has a molar extinction coefficient of 2560 at 640 nm.

The I_2^+ cation is not completely stable in fluorosulfuric acid and undergoes some disproportionation to the more stable I_3^+ ion **143** and $I(SO_3F)_3$. This dispropor-

$$8I_2^+ + 3SO_3F^- \rightleftharpoons I(SO_3F)_3 + 5I_3^+$$

tionation is largely prevented in a 1:1 I_2:$S_2O_6F_2$ solution in which $I(SO_3F)_3$ is also formed.

$$5I_2 + 5S_2O_6F_2 \longrightarrow 4I_2^+ + 4SO_3F^- + 2I(SO_3F)_3^-$$

The disproportionation can also be prevented if the fluorosulfate ion concentration in fluorosulfuric acid is lowered by addition of antimony pentafluoride or by using the less basic solvent, 65% oleum.

$$SbF_5 + SO_3F^- \longrightarrow (SbF_5SO_3F)^-$$

In 100% H_2SO_4, the disproportionation of I_2^+ to I_3^+ and an iodine(III) species, probably $I(SO_4H)_3$, is essentially complete, and only traces of I_2^+ can be detected by means of its resonance Raman spectrum.

Solution of the blue iodine cation in oleum have been reinvestigated[189] by conductometric, spectrophotometric, and cryoscopic methods confirming the formation of I_2^+. In 65% oleum, iodine is oxidized to I_2^+.

$$2I_2 + 5SO_3 + H_2S_4O_{13} \longrightarrow 2I_2^+ + 2HS_4O_{13}^- + SO_2$$

Adhami and Herlem[195] have carried out a coulometric titration at controlled potential of iodine in fluorosulfuric acid and have shown that iodine is quantitatively oxidized to I_2^+ by removal of one electron per mole of iodine.

Pure crystalline $I_2^+Sb_2F_{11}^-$ has recently been prepared by the reaction of iodine with antimony pentafluoride in liquid sulfur dioxide as solvent.[197] After removal of insoluble SbF_3, deep blue crystals of $I_2^+Sb_2F_{11}^-$ were obtained from the solution. An X-ray crystallographic structure determination showed the presence of the discrete ions I_2^+ and $Sb_2F_{11}^-$.[198] Crystalline solids that can be formulated as $I_2^+Sb_2F_{11}$ have also been prepared by Kemmitt et al.[190] by the reaction of iodine with antimony or tantalum pentafluorides in iodine pentafluoride solutions.

I_4^{2+} Dication. Recently, Gillespie and co-workers were able to prepare[199] the I_4^{2+} dication **146** by the reaction of iodine with either AsF_5 or SbF_5 in SO_2 solution.

$$2I_2 + 4SbF_5 \longrightarrow \underset{\mathbf{146}}{I_4^{2+}(Sb_3F_{14}^-),\ SbF_6^-}$$

The X-ray crystal structure of salts $I_4^{2+}(AsF_6)_2^{2-}$ and $I_4^{2+}(Sb_3F_{14})^-\ SbF_6^-$ have been determined. The I_4^{2+} dication described as two I_2^+ cations bonded together by two relatively weak bonds.

The interaction between the two I_2^+ ions may be described as a four-center two-electron bond so that each of the long I—I bonds has a bond order of 0.5. The four-center orbital can be considered to be formed from the half-filled π antibonding orbitals on each I_2^+. This model is consistent with the long I—I distances and the diamagnetism of the dication.

4.4.1.2 Bromine Cations

Br_3^+. Ruff,[200] in 1906, prepared a compound by the reaction of Br_2 and SbF_5, which he formulated as SbF_5Br. Later McRae[201] showed that Br_3^+ **147** is formed in the system. Subsequently, Gillespie and Morton showed[202,203] that Br_3^+ **147** is formed quantitatively in the superacid medium $HSO_3F:SbF_5:SO_3$ (mainly by the reaction of $S_2O_6F_2$).

$$3Br_2 + S_2O_6F_2 \longrightarrow \underset{\mathbf{147}}{2Br_3^+FSO_3^-}$$

These solutions are brown in color and have a strong absorption at 300 nm with a shoulder at 375 nm. Solutions of Br_3^+ can also be obtained in a similar way in fluorosulfuric acid; however, they are not completely stable in this solvent and undergo some disproportionation.

$$Br_3^+ + SO_3F^- \rightleftharpoons Br_2 + BrOSO_2F$$

Glemser and Smalc[204] have prepared the compound $Br_3^+AsF_6^-$ by the displacement of oxygen in dioxygenyl hexafluoroarsenate by bromine and by the reaction of bromine pentafluoride, bromine, and arsenic pentafluoride. The compound is chocolate-brown and in solution has absorption bands at 300 nm and 375 nm; it has fair thermal stability and can be sublimed at 30–50°C under nitrogen atmosphere.

$$2O_2^+AsF_6^- + 3Br_2 \longrightarrow 2Br_3^+AsF_6^- + 2O_2$$

$$7Br_2 + BrF_5 + 5AsF_5 \longrightarrow 5Br_3^+AsF_6^-$$

Br_2^+. The Br_2^+ cation **148** can be prepared[203] by oxidation of bromine by $S_2O_6F_2$ in the superacid $HSO_3F:SbF_5:3SO_3$; however, even in this very weakly basic medium, the Br_2^+ ion is not completely stable as it undergoes appreciable disproportionation.

$$\underset{\textbf{148}}{2Br_2^+} + 2HSO_3F \rightleftharpoons Br_3^+ + BrOSO_2F + H_2SO_3F^+$$

Moreover, the $BrOSO_2F$ that is formed also undergoes some disproportionation by itself to Br_2^+, Br_3^+, and $Br(OSO_2F_3)_3$, and the equilibria in these solutions are quite complex.

$$5BrOSO_2F + 2H_2SO_3F \rightleftharpoons \underset{\textbf{148}}{2Br_2^+} + Br(OSO_2F)_3 + 4HSO_3F$$

$$4BrOSO_2F + H_2SO_3F^+ \rightleftharpoons Br_3^+ + Br(OSO_2F)_3 + 2HSO_3F$$

A solution of Br_2^+ **148** in superacid has a characteristic cherry red color with maximum absorption at 510 nm and a single band in the Raman spectrum at 360 cm^{-1}.

The paramagnetic scarlet crystalline compound $Br_2^+Sb_3F_{16}^-$[205,206] has been prepared by the following reaction.

$$9Br_2 + 2BrF_5 + 30SbF_5 \longrightarrow \underset{\textbf{148}}{10\ Br_2^+Sb_3F_{16}^-}$$

It is a stable salt and can be sublimed at 200°C. The X-ray crystal structure of Br_2^+–$Sb_3F_{16}^-$ **148** salt shows a bromine–bromine bond distance of 2.13Å. The shorter bond distance of **148** compared to neutral bromine is in accord with increase in bond order resulting from the loss of an antibonding electron from the neutral molecule.[198,206]

4.4.1.3 Chlorine Cations

There is no evidence for either Cl^+ or Cl_2^+ in superacid media.[207] However, Cl_2, ClF, and AsF_5 react at −70°C to form $Cl_3^+AsF_6^-$ **149.**

$$Cl_2 + ClF + AsF_5 \longrightarrow \underset{\textbf{149}}{Cl_3^+AsF_6^-}$$

The Cl_3^+ cation **149** has been identified by its Raman spectrum in the yellow solid which precipitates from a solution of Cl_2 and ClF in HF:SbF_5 at −76°C. At room temperature the Cl_3^+ cation completely disproportionates in the HF:SbF_5 media to chlorine and ClF_2^+ salts. There is no evidence for the formation of $Cl_3^+BF_4^-$ salt from mixtures of Cl_2, ClF, and BF_3 at temperatures ranging from ambient to −120°C.

The Raman spectrum[208] of $Cl_3^+AsF_6^-$ **149** shows bands due to the AsF_6^- ion, together with three relatively intense bands at 490 (split to 485 and 492), 225, and 508 cm^{-1}. These frequencies are very close to the vibrational frequencies of neutral SCl_2 molecule.[209]

4.4.2 Interhalogen Cations

Interhalogen cations form a class of polycations containing at least two different halogen atoms.[185] Cations containing one or more halogens and another element such as oxygen, nitrogen, or xenon will not be considered here. The class of interhalogen cations include, triatomic, pentaatomic and heptaatomic systems. Many of these interhalogen cation salts, which are strong oxidants, have been found useful for collecting radioactive noble gases such as ^{222}Rn ^{133}Xe.[210,211]

4.4.2.1 Triatomic Interhalogen Cations

Of all the possible 16 triatomic interhalogen cations ClF_2^+, BrF_2^+, IF_2^+, Cl_2F^+, Br_2F^+, I_2F^+, $ClBrF^+$, $ClIF^+$, $BrIF^+$, $BrCl_2^+$, ICl_2^+, Br_2Cl^+, I_2Cl^+, $BrICl^+$, IBr_2^+, and I_2Br^+, only five are known. They are ClF_2^+, BrF_2^+, IF_2^+, Cl_2F^+, and ICl_2^+. It seems reasonable to predict that the least electronegative halogen occupies the central position, where it carries a formal positive charge.

ClF_2^+. Adducts of ClF_3 with Lewis acids such as AsF_5, SbF_5, and BF_3 have been known for some time,[212] and it has been established by infrared, Raman, and ^{19}F-nmr spectroscopic studies that these compounds are best formulated as salts of ClF_2^+ cation **150,** e.g., ClF_2^+ AsF_6^-.[213,214] The spectroscopic data indicate a bent structure. Additional support for the bent structure of **150** comes from X-ray crystallographic studies[215] on ClF_2^+ SbF_6^- salt. The ClF_2^+ ion **150** has a bond angle of 95.9° and a bond length of 1.58 Å. There is a good evidence for the fluorine bridging between the anion and the cation and the two fluorine bridges formed by each ClF_2^+ give rise to a very approximately square coordination of fluorine around chlorine, which is the geometry predicted by the valence shell electron pair-repulsion theory for AX_4E_2 coordination (where X is a ligand and E a lone pair). It is interesting to note that the SbF_6^- ion in this structure forms *trans* bridges rather than the *cis* bridges that have been observed in other related structures.

BrF_2^+. The 1:1 adduct of BrF_3 with SbF_5 has been shown by X-ray crystallography[216] to contain BrF_2^+ **151** and SbF_6^- ions held together by fluorine bridging in such a way that bromine acquires a very approximately square planar configuration. Each bromine atom has two fluorine atoms at 1.69 Å, making an angle of 93.5° at bromine, and two other neighboring fluorine atoms at 2.29 Å which form part of the distorted octahedral coordination of the antimony atoms. The two fluorine bridges formed by SbF_6^- are *cis* rather than *trans,* as in the unusual $ClF_2^+SbF_6^-$ structure.

The ir and Raman spectra of $BrF_3—SbF_5$, $BrF_3—AsF_5$, and $(BrF_3)_2GeF_4$ have been reported.[217,218] The electrical conductivity of liquid bromine trifluoride[219] (specific conductance = 8×10^{-3} $ohm^{-1} \cdot cm^{-1}$) may be attributed to the following self-ionization reaction.

$$2\ BrF_3 \rightleftharpoons BrF_2^+ \cdot BrF_4^-$$

IF_2^+. The salts of IF_2^+ **152** with AsF_6^- and SbF_6^- anions have been prepared from IF_3 and AsF_5, and from IF_3 and SbF_5 in AsF_5 as solvent at −70°C.[220] The compound $IF_2^+SbF_6^-$ is stable up to 45°C and the solid gives two broad overlapping ^{19}F-nmr signals whose relative intensities were estimated to be 1:2.6, and which were assumed to arise, therefore, from fluorine on iodine and fluorine on antimony, respectively. $IF_2^+AsF_6^-$ was found to be stable only up to −20°C.

Cl_2F^+. Raman spectra of the adducts $AsF_5—2ClF$ and $BF_3—2ClF$ have established that these compounds contain the unsymmetrical $ClClF^+$ cation[213] and not the symmetrical $ClFCl^+$ cation previously reported on the basis of the ir spectrum alone.[221] The observed vibrational frequencies indicate that there is strong fluorine bridging between the cation and the anion in $Cl_2F^+AsF_6^-$ **153** salt.

The Cl_2F^+ ion **153** appears to be unstable in solution and was found to be completely disproportionated in $SbF_5—HF$ even at −76°C. The Cl_3^+ in the media disproportionate further at room temperature to give chlorine and ClF_2^+ **150.**[207]

$$\underset{\mathbf{153}}{2Cl_2F^+} \rightleftharpoons \underset{\mathbf{150}}{ClF_2^+} + Cl_3^+$$

ICl_2^+. X-ray crystallographic investigations[222] of the adducts of ICl_3 with $SbCl_5$ and $AlCl_3$ have shown that these may be regarded as ionic compounds, i.e., $ICl_2^+SbCl_6^-$ and $ICl_2^+AlCl_4^-$ **154,** although there is considerable interaction between the two ions via two bridging chlorines, which give an approximately square planar arrangement of four chlorines around the iodine atom, similar to the arrangement of fluorines around bromine and chlorine in $BrF_2^+SbF_6^-$ **151** and $ClF_2^+SbF_6^-$ **150,** respectively. The bond angle and bond length for ICl_2^+ **154** were found to be 92.5° and 2.31 Å in $ICl_2^+SbCl_6^-$ and 96.7° and 2.28 Å in $ICl_2^+AlCl_4^-$.

The electrical conductivity of liquid ICl_3 (specific conductance = 9.85×10^{-2} $ohm^{-1} \cdot cm^{-1}$) can be attributed to the self-ionization.[222]

I_2Cl^+. There is no certain evidence for the I_2Cl^+ cation **155** but presumably the electrical conductivity of liquid ICl (specific conductance = 4.52×10^{-2} $ohm^{-1} \cdot cm^{-1}$ at 31°C) which has previously been ascribed to the self-ionization,[223]

$$2ICl \rightleftharpoons I^+ + ICl_2^-$$

is in fact due to a self-ionization that produces the I_2Cl^+ ion **155** according to the equation:

$$3ICl \rightleftharpoons I_2Cl^+ + ICl_2^-$$

155

The I_2Cl^+ cation **155,** however, is possibly extensively disproportionated to give the known I_3^+ and ICl_2^+ cations

$$2I_2Cl^+ \rightleftharpoons I_3^+ + ICl_2^+$$

4.4.2.2 Pentaatomic Interhalogen Cations

Chlorine pentafluoride forms 1:1 adducts with AsF_5 and SbF_5. The interpretation of Raman spectra of these adducts indicate the formation of ClF_4^+ cation **156.**[224] Bromine pentafluoride forms the adducts $BrF_5 \cdot 2SbF_5$ and $BrF_5 \cdot SO_3$.[225] These may, presumably be formulated as BrF_4^+ ion **157** salts, although the latter compound might be the covalent BrF_4SO_3F. Iodine pentafluoride also forms adducts with SbF_5[226] and PtF_5.[227] The X-ray crystal structure analysis[228] of $IF_5 \cdot SbF_5$ adduct, has been shown to be $IF_4^+SbF_6^-$ salt. The IF_4^+ ion **158** has a structure like SF_4 with two fluorines occupying the axial positions of a trigonal bipyramid and two fluorines and a lone pair occupying the equatorial positions. The electrical conductivity of liquid IF_5 (specific conductance $= 2.30 \times 10^{-5}\ ohm^{-1} \cdot cm^{-1}$) has been attributed to the self-ionization.

$$2IF_5 \rightleftharpoons IF_4^+IF_6^-$$

In the Raman spectrum of IF_4^-, the observed nine lines have been assigned to IF_4^+ cation **158,**[229] which is consistent with its C_{2v} structure found by X-ray crystallography.

4.2.2.3 Heptaatomic Interhalogen Cations

ClF_6^+. The $ClF_6^+PtF_6^-$ salt has been prepared by the reaction of PtF_6 with chlorine fluorides or oxyfluorides.[230,231]

$$2ClF_5 + 2PtF_6 \xrightarrow[UV]{\text{Pyrex filter}} ClF_4^+PtF_6^- + ClF_6^+PtF_6^-$$

159

$$6FClO_2 + 6PtF_6 \longrightarrow 5ClO_2^+PtF_6^- + ClF_6^+PtF_6^- + O_2$$

159

The structure of ClF_6^+ cation **159** has been established beyond any reasonable doubt by ^{19}F-nmr spectroscopy.[214,231] The ClF_6^+ **159,** except for the $ClO_2F_4^+$ cation[232] is the only known heptacoordinate chlorine cation. Besides the well-known NF_4^+ cation,[233,234] it is the only known example of a fluorocation derived from the hitherto unknown compounds (i.e., NF_5 and ClF_7). Complete vibrational analysis of ClF_6^+ cation **159** has been reported[235] and it indicates the octahedral symmetry of the ion.

The $ClF_6^+PtF_6^-$ salt is canary yellow in color and is quite stable at 25°C when stored in Teflon-FEP containers. The ClF_6^+ salts are very powerful oxidizers and react explosively with organic compounds and water.

BrF_6^+. In 1974 Gillespie and Schrobilgen reported the direct oxidation of bromine pentafluoride to BrF_6^+ cation **160** by $Kr_2F_3^+$ cation.[236,237]

$$BrF_5 + \underset{(SbF_6^-)}{Kr_2F_3^+AsF_6^-} \longrightarrow \underset{\mathbf{160}}{\underset{(SbF_6^-)}{BrF_6^+AsF_6^-}} + KrF_2 + Kr$$

The ^{19}F-nmr and Raman[237,238] spectroscopic studies on BrF_6^+ **160** indicates the octahedral symmetry of the species. The ion **160** is a powerful oxidizing agent and readily oxidizes oxygen and xenon to O_2^+ and XeF^+ cations, respectively, under ambient conditions.

IF_6^+. Iodine heptafluoride has been shown to form the adduct IF_7AsF_5 and $IF_7 \cdot 3SbF_5$.[239] The latter complex was postulated to have the ionic structure $IF_4^{3+}(SbF_6^-)_3$. These findings were questioned by Christe and Sawodny,[240] who indicated that the adduct may contain IF_6^+ cation **161.** Indeed, they showed that IF_7AsF_5 adduct is actually $IF_6^+AsF_6^-$ salt. Hohorst, Stein, and Gebert have been successful in preparing $IF_6^+ \cdot SbF_6^-$ salt.[241]

The Raman and infrared spectral analyses of the salts indicate the octahedral nature of the IF_6^+ cation **161.**[241,242] The $IF_6^+SbF_6^-$ salt rapidly reacts with radon gas at room temperature forming a nonvolatile radon compound. The salt is claimed to have potential application in purifying radon-contaminated air and in the analysis of radon in air.[241]

4.4.3 Polyatomic Cations of Group-VI Elements

4.4.3.1 The O_2^+ Cation

The existence of O_2^+ **162** in the gas phase at low pressures has been well established.[243] However, it was not until 1962 that a compound containing O_2^+ was identified.[244] It was discovered as a reaction product during the fluorination of platinum in a silica apparatus. The product was first thought to be $PtOF_4$,[245] but later it was shown to be $O_2^+PtF_6^-$.[244] It was subsequently prepared by direct oxidation of molecular oxygen using platinum hexafluoride at room temperature. It now appears that the dioxygenyl salt $O_2^+BF_4^-$ may have been prepared prior to 1962.[246] There are at least nine O_2^+ salts known with a variety of anions. The anions are PtF_6^-, AsF_6^-, SbF_6^-, $Sb_2F_{11}^-$, PF_6^-, BF_4^-, VF_6^-, BiF_6^-, and SnF_6^-. The most convenient route to O_2^+ salts appears to be the photochemical synthesis of $O_2^+AsF_6^-(SbF_6)^-$ from oxygen, fluorine, and arsenic (antimony) pentafluoride.[247] Most O_2^+ preparations involve the reaction of fluoride ion acceptors with

O_2F_2 or O_4F_2 at low temperatures or with O_2 and F_2 mixtures under conditions favoring synthesis of the long-lived O_2F radical, e.g.,

$$O_2 + F_2 \xrightarrow{h\nu} O_2F + F$$

$$O_2F + AsF_5 \longrightarrow O_2{}^+AsF_6{}^-$$

Compounds containing $O_2{}^+$ **162** are colorless with the exception of $O_2{}^+PtF_6{}^-$, which is red due to the $PtF_6{}^-$ ion. The compound $O_2{}^+PF_6{}^-$ decomposes slowly[248] at $-80°C$, and rapidly at room temperature, giving oxygen, fluorine, and phosphorouspentafluoride. ^{18}F tracer studies on $O_2{}^+BF_4{}^-$ have led to the conclusion that the mechanism of the decomposition involves the equilibrium followed by a bimolecular decomposition of O_2F.[249]

$$O_2{}^+BF_4{}^- \rightleftharpoons O_2F(g) + BF_3(g)$$

Dioxygenyl hexafluoroantimonate has been studied by differential thermal analysis.[250] Decomposition of $O_2{}^+SbF_6{}^-$ proceeds in two stages, according to the mechanism

$$2O_2{}^+SbF_6{}^- \xrightarrow{\sim 240°C} O_2 + \tfrac{1}{2}F_2 + O_2{}^+Sb_2F_{11}{}^-$$

$$O_2{}^+Sb_2F_{11}{}^- \xrightarrow{\sim 280°C} O_2 + \tfrac{1}{2}F_2 + 2SbF_5$$

The $O_2{}^+Sb_2F_{11}{}^-$ was converted to $O_2{}^+SbF_6{}^-$ by heating at 130°C in vacuo, and conversely, $O_2{}^+Sb_2F_{11}{}^-$ was prepared by reaction of $O_2{}^+SbF_6{}^-$ and SbF_5 at 180–200°C. Dioxygenyl hexafluoroarsenate is markedly less stable than the fluoroantimonate salts; it decomposes rapidly at 130–180°C.[248] $O_2{}^+PtF_6^-$ can be sublimed above 90°C in vacuo and melts with some decomposition at 219°C in a sealed tube.[249]

X-ray powder data obtained from the cubic form of O_2PtF_6 were consistent with the presence of $O_2{}^+$ and $PtF_6{}^-$ ions.[249] The structure was refined using neutron diffraction powder data. The $O_2{}^+$ cation **162** resembles nitrosonium ion.[250] The Raman and esr spectra of $O_2{}^+$ salts have been studied in detail.[251–254]

The $O_2{}^+AsF_6{}^-$ salt readily oxidizes molecular bromine to $Br_3{}^+$ ion **147.**[204] The $O_2{}^+BF_4{}^-$ salt also reacts with xenon.[255] The major application of $O_2{}^+$ salts is as an oxidant for collecting ^{222}Rn in uranium mines, since it has negligible dissociation pressure at ambient temperature and releases oxygen as a gaseous product. Reactions of the salt with radon and components of diesel exhausts (CO, CO_2, CH_4, SO_2, NO, and NO_2) have therefore been studied in some detail.[210,211]

4.4.3.2 Polysulfur Cations

The nature of colored solutions obtained by dissolving elemental sulfur in oleum[256] remained a mystery for a long time, since their discovery by Bucholz[257] in 1804. Red, yellow, and blue solutions have been prepared, however, it appears that particular attention has been given to the blue solutions. The species responsible for blue color has been identified by various workers as S_2O_3, S_2, radical ion $(X_2S—SX_2)^+$ and the species designated as S_n.[258] The various colors have now been shown to be due to the cations S_{16}^{2+}, S_8^{2+}, and S_4^{2+}, **163, 164,** and **165,** respectively.[259–261] Recently, it has been shown by X-ray crystallography[262] that the originally assigned S_{16}^{2+} cation **163** is indeed S_{19}^{2+} **166.**

Sulfur can be oxidized by a variety of oxidizing agents including AsF_5, SbF_5, and $S_2O_6F_2$ and by protic acids such as $H_2S_2O_7$ and HSO_3F. Along with previously mentioned doubly charged species, several singly charged radical cations have been claimed of which only $S_5^{\dot{+}}$ has been positively identified.[263] So far these singly charged species have been observed only in solution. They do not appear to be as stable as the doubly charged species and hence are not well characterized. In contrast all the dipositive ions have been obtained in the form of relatively stable salts such as $S_4^{2+}(\bar{S}O_3F)_2$ and $S_8^{2+}(A\bar{s}F_6)_2$.[259]

In the case of $S_8^{2+}(AsF_6^-)_2$ a single crystal X-ray crystallographic analysis has shown unequivocally that this cation has an *exo-endo* cyclic structure with a long transannular bond.[261] No crystallographic studies have been reported on salts of S_4^{2+} cation **165.** By comparison with analogous selenium and tellurium ions, the S_4^{2+} ion is proposed to have a square planar structure.[264,265] Additional support comes from infrared spectra,[266] Raman and electronic spectra,[267,268] magnetic circular dichroism,[269] and molecular orbital calculations.[270]

The evidence for the third of the dipositive ion, S_{16}^{2+} **163,** has not been so conclusive. The existence of **163** was first proposed as a result of isolation of a red solid from the reaction of sulfur and AsF_5 in anhydrous HF solution.[271] From the analytical data, the solid appeared to have a composition corresponding to $S_{16}(AsF_6)_2$. In a study of the progressive oxidation of sulfur by $S_2O_6F_2$ in HSO_3F, it was observed that some unreacted sulfur always remained until sufficient oxidant had been added to give a ratio of sulfur to $S_2O_6F_2$ of 16 : 1. The cryoscopic measurements were also consistent with the formulation of this ion as S_{16}^{2+} **163.**[259] However, attempts to obtain a single crystal of the ion from various acid solutions was unsuccessful. More recently, Gillespie and co-workers[262] obtained both needle- and plate-like crystals originally believed to be $S_{16}(AsF_6)_2$ in 2 : 1 mixture of SO_2 and SO_2 ClF at $-25°C$. An X-ray crystallographic study of these crystals led to the surprising and the unexpected result that the compound had the composition $S_{19}(AsF_6)_2$, i.e., that it is a salt of S_{19}^{2+} **166** rather than S_{16}^{2+} **163.**

The structure of $S_{19}(AsF_6)_2$ contains discrete AsF_6^- anions and S_{19}^{2+} cations, the latter consisting of two seven-membered rings joined by a five-atom chain.

4.4.3.3 Polyselenium Cations

The colored solutions produced by dissolving elemental selenium in sulfuric acid were first observed by Magnus in 1827.[272] Since then a number of workers have investigated the nature of selenium solutions in sulfuric acid, oleum, and sulfur trioxide, providing [273] a substantial amount of data but with little understanding of the system. Recently, it has been shown that these solutions contain the yellow Se_4^{2+} **167** and green Se_8^{2+} **168** polyatomic cations.[274]

Polyselenium cations are less electrophilic than their sulfur analogs and give stable solutions in various strong acids.[274] In fluorosulfuric acid, selenium can be oxidized quantitatively by $S_2O_6F_2$ to give yellow Se_4^{2+} **167.**

$$4Se + S_2O_6F_2 \longrightarrow \underset{\mathbf{167}}{Se_4^{2+}} + 2SO_3F^-$$

The addition of selenium to the yellow solution up to a 8:1 ratio of $Se:S_2O_6F_2$ did not appreciably affect the conductivity. This indicated that the SO_3F^- ion concentration remained unchanged and that the Se_4^{2+} ion is reduced by selenium to Se_8^{2+} **168.**

$$Se_4^{2+} + 4Se \longrightarrow \underset{\mathbf{168}}{Se_8^{2+}}$$

Conductivity measurements of selenium in pure fluorosulfuric acid were also consistent with the formation of Se_8^{2+} **168.**

Solutions of Se_8^{2+} in 100% H_2SO_4 may be prepared by heating selenium in the acid at 50–60°C; the element is oxidized by sulfuric acid. The Se_4^{2+} **167** can also be oxidized to Se_8^{2+} **168** by selenium dioxide.

$$8Se + 5H_2SO_4 \longrightarrow Se_8^{2+} + 2H_3O^+ + 4HSO_4^- + SO_2$$

Various Se_4^{2+}- and Se_8^{2+}-containing compounds have been prepared by oxidizing selenium with $SeCl_4$ plus any one of the following acids; $AlCl_3$, SO_3, oleum, SbF_5, or AsF_5.[186] All polyatomic selenium cations are diamagnetic, and so far no evidence has been reported for radicals analogous to $S_5^{+\cdot}$. The crystal structure of $Se_4(HS_2O_7)_2$[264,275] has shown Se_4^{2+} **167** to be square planar with a Se—Se bond distance of 2.283(4) Å, significantly less than that of 2.34(2) Å found in the Se_8 molecule,[276] indicating some degree of multiple bonding. Such a result is consistent with a valence bond description of the molecule involving four-membered ring structures of the type **167a.** Alternatively, the structure can be understood in terms of molecular orbital theory. The circle in structure **167b** denotes a closed-shell (aromatic) 6π-electron system. Of the four π molecular orbitals, the two almost nonbonding (e_g) orbitals and the lower-energy (b_{2u}) bonding orbital are occupied by the six π electrons, leaving the upper antibonding (a_{1g}) orbitals empty. The intense yellow-orange color of Se_4^{2+} has been attributed to the dipole-allowed excitation of an electron from an e_g orbital to the lowest empty π orbital (b_{2u}). Stephens[269] has shown that the magnetic circular dichroism results are consistent with such a model. The structure is also consistent with the recent vibrational spectroscopic study and molecular orbital calculations.[266,270]

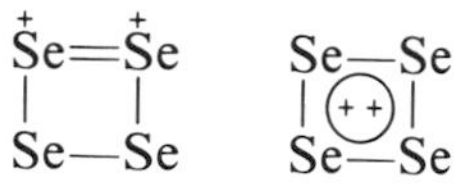

167a **167b**

The structure of Se_8^{2+} **168** in $Se_8(AlCl_4)_2$[277,278] is similar to that of S_8^{2+} **164** except that the cross-ring distance $Se_{(3)}$—$Se_{(7)}$ is relatively shorter than that found in the sulfur cation, and the other cross-ring distances, $Se_{(4)}$—$Se_{(6)}$ and $Se_{(2)}$—$Se_{(8)}$, are relatively long. The cation Se_8^{2+} is, therefore, reasonably well described by valence bond structure **168a.**

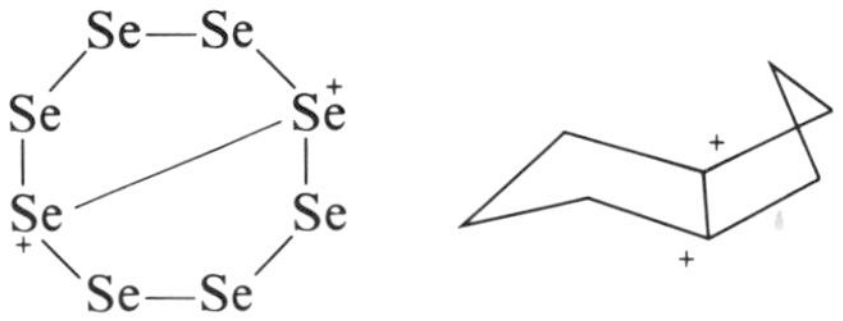

168a

Recently Se_{10}^{2+} cation **169** has been prepared and characterized.[278] The salts include $Se_{10}^{2+}(\bar{A}sF_6)_2$, $Se_{10}^{2+}(\bar{A}lCl_4)_2$, and $Se_{10}^{2+}(\bar{S}bF_6)_2$. The X-ray crystal structure of $Se_{10}^{2+}(\bar{S}bF_6)_2$ indicates that the Se_{10}^{2+} cation **169** can be described as a six-membered boat-shaped ring linked across the middle by a chain of four selenium atoms, i.e., a bicyclo[4.2.2]decane-type structure. The selenium–selenium bonds in the cation vary greatly in length from 2.24 to 2.44 Å.

4.4.3.4 Polytellurium Cations

The red color produced when tellurium dissolves in concentrated sulfuric acid was first observed as long ago as 1798,[279] but the origin of this color for long remained a mystery. Recently, Bjerrum and Smith[280] and Bjerrum[281] have studied the reaction of tellurium tetrachloride with tellurium in molten $AlCl_3$—NaCl. They obtained a purple melt which they concluded contained the species Te_{2n}^{n+} (probably Te_4^{2+}) formed by the following reaction.

$$7Te + Te^{4+} \longrightarrow \underset{\textbf{170}}{2Te_4^{2+}}$$

At about the same time solutions of tellurium in various acids were investigated in detail.[282,283] It was found that red solutions are produced when tellurium is dissolved in sulfuric acid, fluorosulfuric acid, or oleum with the simultaneous production of SO_2, indicating that tellurium is oxidized. The electronic spectra of the solutions were found to be identical with those obtained by Bjerrum and Smith from their melts. Conductometric and cryoscopic measurements of the acid solutions led to the conclusion that they contain a species Te_{2n}^{n+}, which was certainly not Te_2^+ but probably Te_4^{2+} **170.**

Reaction of tellurium with $S_2O_6F_2$,[283] SbF_5, and AsF_5 in SO_2 gave the compounds $Te_4^{2+}(SO_3F^-)_2$, $Te_4^{2+}(Sb_2F_{11}^-)_2$, and $Te_4^{2+}(AsF_6^-)_2$ and, from Te-($TeCl_4$-$AlCl_3$) melts, compounds $Te_4^{2+}(AlCl_4^-)_2$ and $Te_4^{2+}(Al_2Cl_7^-)_2$[284] were obtained.[285,286] The formulation of the red species as Te_4^{2+} **170** was finally confirmed by the determination of the crystal structures of these latter two compounds.[265] In both cases, the Te_4^{2+} **170** ion lies on a center of symmetry and is almost exactly square planar. The tellurium–tellurium distance of 2.66 Å is significantly shorter than the tellurium–tellurium distance of 2.864 Å within the spiral chain in elemental tellurium.[287] This is consistent with a structure exactly analogous to that for Se_4^{2+} in which each bond has 25% double-bond character. The Raman spectra of Te_4^{2+} in solution and the solid state are analogous to those of Se_4^{2+} and S_4^{2+} but shifted to lower frequency.[266] The magnetic circular dichroism[269] and visible and uv spectrum of solutions of Te_4^{2+} **170** are also similar to those of Se_4^{2+} **167** as expected on the basis of their structural similarity and a recent molecular orbital study.[270]

The acid solutions of tellurium described earlier when warmed or if the oleum is sufficiently strong, then the color of the solutions change from red to orange-yellow.[281] The same change may also be produced by addition of an oxidizing agent such as peroxydisulfate to the sulfuric acid solutions or $S_2O_6F_2$ to the HSO_3F solutions. Absorption spectra and conductometric, cryoscopic, and magnetic measurements on the solutions in HSO_3F suggested that the yellow species was tellurium in +1 oxidation state and it was formulated as "Te_n^{n+}" where n is even as the cation was found to be diamagnetic. Furthermore, these studies also established that "Te_n^{n+}" could not be Te_2^{2+} and is probably Te_4^{4+}, although higher-molecular-weight species such as Te_6^{6+} and Te_8^{8+} could not be ruled out.[283,285]

On the contrary, Paul and co-workers[286] concluded, from spectrophotometric measurements on solutions formed by the reduction of $TeCl_4$ with tellurium metal in $KAlCl_4$ melts buffered with KCl—$ZnCl_2$, that the species was Te_2^{2+}. Recently, Gillespie and co-workers[289,290] have been able to show that the Te_n^{n+} species is indeed a cluster cation Te_6^{4+} **171.** The single crystal X-ray crystallographic study on $Te_6(AsF_6)_4 \cdot AsF_3$ and $Te_6(AsF_6)_4 \cdot 2SO_2$ indicate that Te_6^{4+} **171** cation has a trigonal prismatic arrangement **171a.**

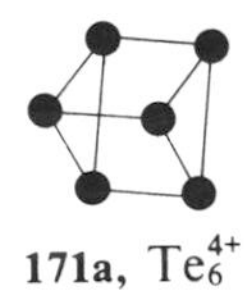

171a, Te_6^{4+}

4.4.3.5 Polyheteroatom Cations

So far only studies on homopolyatomic cations of oxygen, sulfur, selenium, and tellurium have been discussed. Most of them were of the type X_n^{2+}. There is a new class of polyatomic cations that comprise two or more heteroatoms.

When powdered tellurium metal (0.002 mol) was allowed to react with $Se_8(AsF_6)_2$ (0.001 mol)[291,292] in SO_2 at $-78°C$ and the mixture was allowed to warm up to room temperature under stirring, the dark green color of the Se_8^{2+} **168** slowly diminished (over a 12 h period) resulting in the formation of $Te_2Se_8(AsF_6)_2SO_2$.[293] The X-ray crystal structure of $Te_2Se_8(AsF_6)_2SO_2$ indicates that the adduct has $Te_2Se_8^{2+}$ cation **172** with a bicyclo[4.2.2]-type structure similar to that of Se_{10}^{2+} cation **169.** By changing the ratio of tellurium metal to Se_8^{2+} **168,** Gillespie and co-workers have managed to prepare $Te_{3.7}Se_{6.3}(AsF_6)_2$. The structure of $Te_{3.7}Se_{6.3}^{2+}$ is similar to that of **172.**[293]

Based on the same principle adducts $Te_3S_3(AsF_6)_2$, $Te_2Se_4(SbF_6)_2$, and $Te_2Se_4(AsF_6)_2$ have been prepared.[294] The $Te_3S_3^{2+}$ and $Te_2Se_4^{2+}$ cations **173** and **174** have novel structures that can be described as consisting of a three-membered ring and a five-membered ring fused together or as a boat-shaped six-membered ring with a cross-ring bond.

Recently the *trans*-$Te_2Se_2^{2+}$ **175** cation[266] has been prepared and studied by Raman and infrared spectroscopy. Similarly several cations containing both sulfur and nitrogen atoms are known. They are NS_2^+ **176**[295] (thionitronium), $S_3N_2^+$ **177,**[296] $S_6N_4^{2+}$ **178,**[296] and $S_4N_4^{2+}$ **179.**[297]

Lewis acid adducts of chlorine trifluoride oxide (ClF_3O),[298,299] chlorine trifluoride dioxide (ClF_3O_2),[300–302] and nitrogen trifluoride oxide (NF_3O)[303] are all known. They are best described as ClF_2O^+ **180,** $ClF_2O_2^+$ **181,** and NF_2O^+ **182,** respectively. The spectroscopic properties of these cations have been investigated extensively.[214,298–303]

4.5 MISCELLANEOUS CATIONS

4.5.1 Hydrogen Cations

4.5.1.1 H^+

The naked proton "H^+" **183** exists only in the gas phase. In the condensed state, the proton is always solvated, thus no free proton is capable of existence. It is customary, however, as a short hand notation to depict "H^+" as the solvated proton.

4.5.1.2 H_3^+

The H_3^+ ion **184** was first discovered by Thompson[304] in 1912 in hydrogen discharge studies. Actually, it was the first observed gaseous-ion molecule reaction product and the reaction sequence was established in 1925 by Hogness and Lunn.[305] Since then, extensive mass spectrometric studies of H_2, D_2, and their mixtures have been carried out in an effort to study thermodynamic and kinetic aspects of ion-molecule reactions of $(H, D)_3^+$ cations.[306]

$$H_2^+ + H_2 \longrightarrow \underset{\textbf{184}}{H_3^+} + H$$

Despite numerous studies in the gas phase on H_3^+ **184,** not much solution chemistry has been reported till recently. Gillespie and Pez[307] reported that according to their solubility, cryoscopic, and ^1H nmr spectroscopic measurements, $HSO_3F:SbF_5$:(Magic Acid) SO_2 is not sufficiently strong to protonate a series of weak bases, including molecular hydrogen. Their investigations pertained, however, to observe H_3^+ **184** as stable, detectable intermediate with a long life.

Olah, Shen, and Schlosberg[308] were subsequently able to observe the hydrogen–deuterium exchange of molecular H_2 and D_2, respectively, with 1:1 $HF:SbF_5$ and $HSO_3F:SbF_5$ at room temperature. The facile formation of HD does indicate that protonation or deuteration occurs involving **184** at least as transition states in the kinetic exchange process. The H_3^+ **184** is the simplest two-electron three-center-bonded entity. The H_3^+ ion **184** has been recently observed by ir spectroscopy.[309]

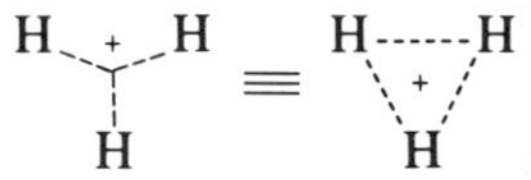

H + H ≡ H------H, +, H, H

184

4.5.2 Cations of Xenon and Krypton

Both xenon and krypton are known to undergo ion–molecule reaction with H^+ to give the corresponding onium ions. The XeH^+ **185** and KrH^+ **186** are well recognized in mass spectroscopic studies.[310–313] Even methylated xenon and krypton, CH_3Xe^+ **187** and CH_3Kr^+ **188,** have been observed by Beauchamp[314] in the gas phase. The carbon-inert gas atom bond strengths in **187** and **188** are estimated to be 43 ± 8 and 21 ± 15 kcal · mol^{-1}, respectively.

In solution, chemistry too attempts have been made to protonate xenon to **185** in the superacid media.[62,307] Evidence[315] for the protonation comes from suppression of proton-deuterium exchange rates of deuterium gas in the presence of xenon in strong acid medium.

The fluorides and oxyfluorides of xenon are well recognized.[316] Most of the xenon fluorides and oxyfluorides react with Lewis acids to give the corresponding cations. The following cations have been prepared and characterized spectroscopically: Xe^+F **189,**[317,318] XeF_3^+ **190,**[319] $Xe_2F_3^+$ **191,**[318] XeF_5^+ **192,**[318] $Xe_2F_{11}^+$ **193,**[320] XeO^+F_3 **194,**[319] and XeO_2^+F **195.**[319] The X-ray structure of **189, 190,** and **191** have been determined.[318] Similarly, krypton fluorides give the following cations[321,322] with Lewis acids: KrF^+ **196,** $Kr_3F_3^+$ **197.** As mentioned earlier, cation salts $Kr_2F_3^+SbF_6^-$ and $Kr_2F_3^+AsF_6^-$ are capable of oxidizing bromine pentafluoride to BrF_6^+ cation **160.**[236]

REFERENCES

1. A. Hantzsch, K. S. Caldwell, Z. *Phys. Chem. 58,* 575 (1907); H. Goldschmidt, O. Ubby, *Ibid. 60,* 728 (1907).

2. J. N. Brønsted, *Recl. Trav. Chim. Pays-Bas. 42,* 718 (1923); *J. Phys. Chem. 30,* 777 (1926); T. M. Lowry, *Trans. Faraday Soc. 20,* 13 (1924); *Chem. Ind.* (London), 1048 (1923).
3. B. E. Conway, J. O'M. Bockris, H. J. Linton, *J. Chem. Phys. 24,* 834 (1956); M. Eigen, L. DeMaeyer, *Proc. R. Soc. London, Ser. A, 247,* 505 (1958).
4. D. E. Bethell, N. Sheppard, *Chem. Phys. 21,* 1421 (1953); C. C. Ferisso, D. F. Horning, *Ibid. 23,* 1464 (1955); P. A. Giguere, *Rev. Chim. Miner. 3,* 627 (1966); M. Fournier, J. Roziere, *C. R. Acad. Sci. Ser. C, 270,* 729 (1970).
5a. R. C. Taylor, G. L. Vidale, *J. Am. Chem. Soc. 78,* 5999 (1956).
5b. P. A. Giguere, C. Madec, *Chem. Phys. Lett. 37,* 569 (1976) and references cited therein.
6a. R. E. Richards, J. A. S. Smith, *Trans. Faraday Soc. 47,* 1261 (1951).
6b. Y. Kakiuchi, H. Shono, H. Komatsu, K. Kigoshi, *J. Chem. Phys. 47,* 1261 (1951).
6c. M. Brookhart, G. C. Levy, S. Winstein, *J. Am. Chem. Soc. 89,* 1735 (1967).
6d. A. Commeyras, G. A. Olah, *Ibid. 91,* 2929 (1969).
6e. V. Gold, J. L. Grant, K. P. Morris, *Chem. Commun.* 397 (1976) and references cited therein.
7a. P. Kebarle, R.M. Haynes, J. G. Collins, *J. Am. Chem. Soc. 89,* 5753 (1967).
7b. A. W. Castelman, Jr., N. I. Tang, H. R. Munkelwitz, *Science, 173,* 1025 (1971).
8. J. O. Lundgren, J. M. Williams, *J. Chem. Phys. 58,* 788 (1973); J. O. Lundgren, R. Tellgren, I. Olovsson, *Acta. Crystallogr. Sec. B, B34,* 2945 (1978).
9. K. O. Christe, C. J. Schack, R. D. Wilson, *Inorg. Chem. 14,* 2224 (1975).
10. G. D. Mateescu, G. M. Benedikt, *J. Am. Chem. Soc. 101,* 3959 (1979).
11. G. A. Olah, A. L. Berrier, G. K. S. Prakash, *J. Am. Chem. Soc. 104,* 2373 (1982).
12. M. C. R. Symons, *J. Am. Chem. Soc. 102,* 3982 (1980).
13. C. A. Coulson, Volume Commemoratif Victor Henri, Contributions a l'Étude de la Structure Moleculaire, Liège, 1948, pp. 15.
14. W. R. Rodwell, L. Radom, *J. Am. Chem. Soc. 103,* 2865 (1981).
15. R. J. Gillespie, J. A. Leisten, *Quart. Rev. Chem. Soc. 8,* 40 (1954).
16. C. MacLean, E. L. Mackor, *J. Chem. Phys. 34,* 2207 (1961).
17. T. Birchall, R. J. Gillespie, *Can. J. Chem. 43,* 1045 (1965).
18. G. A. Olah, E. Namanworth, *J. Am. Chem. Soc. 88,* 5327 (1966).
19. G. A. Olah, J. Sommer, E. Namanworth, *J. Am. Chem. Soc. 89,* 3576 (1967).
20. G. A. Olah, J. Sommer, *J. Am. Chem. Soc. 90,* 927 (1968).
21. E. M. Arnett, C. Y. Wu, *J. Am. Chem. Soc. 82,* 5660 (1960).
22. T. Birchall, A. N. Bourns, R. J. Gillespie, P. J. Smith, *Can. J. Chem. 42,* 1483 (1964).
23. F. Klages, J. E. Gordon, H. A. Jung, *Chem. Ber. 98,* 3748 (1965).
24. G. A. Olah, D. H. O'Brien, *J. Am. Chem. Soc. 89,* 1725 (1967).
25. D. Jaques, J. A. Leisten, *J. Chem. Soc.* 4963 (1961).
26. G. A. Olah, A. M. White, D. H. O'Brien, *Chem. Rev. 70,* 561 (1970).
27. J. Sommer, P. Canivet, S. Schwartz, P. Rimmelin, *Nouv. J. Chim. 5,* 45 (1981).
28. R. Jost, J. Sommer, C. Engdahl, unpublished results.
29. H. Meerwein, G. Hinz, P. Hoffmann, E. Kronig, E. Pfeil, *J. Prakt. Chem.* [2], *147,* 257 (1937).
30. For a review concerning tertiary oxonium ions see H. Perst, in *Carbonium Ions,* G. A. Olah, P. v. R. Schleyer, Eds., John Wiley & Sons, New York, 1976, Vol. 5, pp. 1961–2047.
31. F. Klages, H. A. Jung, *Chem. Ber. 98,* 3747 (1965).
32. F. Klages, H. Meuresch, *Chem. Ber. 85,* 863 (1952); F. Klages, H. Meuresch, W. Steppich, *Ann. Chem. Liebigs, 592,* 116 (1955).
33. R. A. Goodrich, P. M. Treichel, *J. Am. Chem. Soc. 88,* 3509 (1966).

34. H. Meerwein, E. Battenberg, H. Gold, E. Pfeil, G. Willfang, *J. Prakt. Chem. 154,* 83 (1939).
35. H. Meerwein, *Org. Synth. 46,* 120 (1966).
36. H. Meerwein, G. Hinz, P. Hoffmann, E. Kronig, E. Pfeil, *J. Prakt. Chem. 147,* 257 (1937).
37. H. Meerwein, *Org. Synth. 46,* 113 (1966).
38. G. A. Olah, *Halonium Ions,* Wiley-Interscience, New York, 1975.
39. A. N. Nesmeyanov, L. G. Makarova, T. P. Tolstaya, *Tetrahedron 1,* 145 (1957); A. N. Nesmeyanov, T. P. Tolstaya, *Izv. Akad. Nauk. SSR Otd. Khim. Nauk* 647 (1959); *Chem. Abstr. 53,* 21796 (1957).
40. A. N. Nesmeyanov, *Selected Works in Organic Chemistry,* Pergamon Press, Oxford, 1963, p. 772.
41. A. N. Nesmeyanov, T. P. Tolstaya, A. V. Grib, *Dokl. Akad. Nauk. SSR 139,* 114 (1961); *Chem. Abstr. 56,* 1374 (1962).
42. G. A. Olah, E. G. Melby, *J. Am. Chem. Soc. 95,* 4971 (1973).
43. G. A. Olah, J. A. Olah, T. Ohyama, *J. Am. Chem. Soc.* **106**, 5284 (1984).
44a. G. A. Olah, J. R. DeMember, *J. Am. Chem. Soc. 92,* 2562 (1970).
44b. G. A. Olah, J. J. Svoboda, *Synthesis,* 203 (1973).
44c. B. G. Ramsey, R. W. Taft, *J. Am. Chem. Soc. 88,* 3058 (1966).
45. G. A. Olah et al., unpublished results.
46. J. B. Lambert, D. H. Johnson, *J. Am. Chem. Soc. 90,* 1349 (1968).
47. M. I. Watkins, W. M. Ip, G. A. Olah, R. Bau, *J. Am. Chem. Soc. 104,* 2365 (1982).
48. G. A. Olah, J. R. DeMember, Y. K. Mo, J. J. Svoboda, P. Schilling, J. A. Olah, *J. Am. Chem. Soc. 96,* 884 (1974).
49. G. A. Olah, H. Doggweiler, J. D. Felberg, S. Frohlich, M. J. Grdina, R, Karpeles, T. Keumi, S. Inaba, W. M. Ip, K. Lammertsma, G. Salem, D. C. Tabor, *J. Am. Chem. Soc. 106,* 2143 (1984).
50. For a review, see G. A. Olah, D. G. Parker, N. Yoneda, *Angew. Chem. Int. Ed. Engl. 17,* 909 (1978).
51. R. W. Alder, M. C. Whiting, *J. Chem. Soc.* 4707 (1964) and references given therein.
52. K. O. Christie, W. W. Wilson, E. C. Curtis, *Inorg. Chem. 18,* 2578 (1979).
53. R. Trambarulo, S. N. Ghosh, C. A. Barrus, W. Gordy, *J. Chem. Phys. 24,* 851 (1953).
54. P. D. Bartlett, M. Stiles, *J. Am. Chem. Soc. 77,* 2806 (1955).
55. J. P. Wibault, E. L. J. Sixma, L. W. E. Kampschidt, H. Boer, *Recl. Trav. Chim. Pays-Bas. 69,* 1355 (1950).
56. J. P. Wibault, E. L. J. Sixma, *Recl. Trav. Chim. Pays-Bas. 71,* 76 (1951).
57. P. S. Bailey, *Chem. Rev. 58,* 925 (1958).
58. L. Horner, H. Schaefer, W. Ludwig, *Chem. Ber. 91,* 75 (1958).
59. P. S. Bailey, J. W. Ward, R. E. Hornish, F. E. Potts, *Adv. Chem. Ser. 112,* 1 (1972).
60. G. A. Olah, D. H. O'Brien, C. U. Pittman, Jr., *J. Am. Chem. Soc. 89,* 2996 (1967).
61. D. Holtz, J. L. Beauchamp, *Science 173,* 1237 (1971).
62. K. O. Christe, *Inorg. Chem. 14,* 2230 (1975).
63. J. B. Lambert, R. G. Keske, D. K. Weary, *J. Am. Chem. Soc. 89,* 5921 (1967).
64. H. Meerwein, D. Delfs, H. Morschel, *Angew. Chem. 72,* 927 (1960).
65. C. S. F. Tang, H. Rapoport, *J. Org. Chem. 38,* 2809 (1973).
66. R. M. Acheson, D. R. Harrison, *Chem. Commun.* 724 (1969).
67. B. M. Trost, L. S. Melvin, Jr., "Sulfur Ylides, Emerging Synthetic Intermediates", in *Organic Chemistry, A series of Monographs,* Vol. 31, A. T. Blomquist, H. H. Wasserman, Eds., Academic Press, New York, 1975.

68. For a summary, see H. Reinboldt in *Houben-Weyl Methoden der Organischen Chemie,* Vol. 9, Georg Thieme Verlag, Stuttgart, 1955, pp. 1034, 1075.

69. R. Shine, in *Organic Selenium Compounds, Their Chemistry and Biology,* D. L. Klayman, W. H. H. Gunther, Eds., Wiley-Interscience, New York, 1973, pp. 223.

70. G. A. Olah, J. J. Svoboda, A. T. Ku, *J. Org. Chem. 38,* 4447 (1973).

71. G. A. Olah, A. Howard, G. K. S. Prakash, unpublished results.

72. C. Hartman, V. Meyer, *Chem. Ber. 27,* 426 (1894).

73. G. A. Olah, J. R. DeMember, *J. Am. Chem. Soc. 91,* 2113 (1969).

74. I. Roberts, G. E. Kimball, *J. Am. Chem. Soc. 59,* 947 (1937).

75a. B. Capon, *Quart. Rev.* (London), *18,* 45 (1964).

75b. R. C. Fay, in *Topics in Stereochemistry,* E. L. Eliel, N. L. Allinger, Eds., Wiley-Interscience, New York, Vol. 3, 1968, p. 237.

76. S. Winstein, H. J. Lucas, *J. Am. Chem. Soc. 61,* 1576, 2845 (1939).

77. J. C. Traynham, *J. Chem. Ed. 40,* 392 (1963).

78. E. Gold, *Mechanism and Structure in Organic Chemistry,* Holt, Rinehart & Winston, New York, 1959, p. 523.

79. G. A. Olah, J. M. Bollinger, *J. Am. Chem. Soc. 89,* 4744 (1967).

80. P. E. Peterson, *Acc. Chem. Res. 4,* 407 (1971) and references therein.

81a. G. A. Olah, J. Shen, *J. Am. Chem. Soc. 95,* 3582 (1973).

81b. K. O. Christe, *J. Fluo. Chem.* 269 (1984).

82. G. A. Olah, Y. Yamada, R. J. Spear, *J. Am. Chem. Soc. 97,* 680 (1975).

83. G. A. Olah, T. E. Kiovsky, *J. Am. Chem. Soc. 89,* 5692 (1967).

84a. G. A. Olah, J. R. DeMember, *J. Am. Chem. Soc. 92,* 718 (1970).

84b. G. A. Olah, J. R. DeMember, *Ibid. 92,* 2562 (1970).

84c. G. A. Olah, J. R. DeMember, Y. K. Mo, J. J. Svoboda, P. Schilling, J. A. Olah, *Ibid. 96,* 884 (1974).

85. G. A. Olah, Y. K. Mo, *J. Am. Chem. Soc. 96,* 3560 (1974).

86. G. A. Olah, G. K. S. Prakash, M. R. Bruce, *J. Am. Chem. Soc. 101,* 6463 (1979).

87. G. A. Olah, G. K. S. Prakash, M. R. Bruce, unpublished results.

88. G. A. Olah, M. R. Bruce, *J. Am. Chem. Soc. 101,* 4765 (1979).

89. G. A. Olah, Y. K. Mo, E. G. Melby, H. C. Lin, *J. Org. Chem. 38,* 367 (1973).

90. T. B. Dence, J. D. Roberts, *J. Org. Chem. 33,* 1251 (1968).

91. V. V. Lyalin, V. V. Orda, L. A. Alekseeva, L. M. Yagupoliskii, *Z. Org. Khim. 7,* 1473 (1971); *Chem. Abstr. 75,* 140390a (1971).

92. G. A. Olah, E. G. Melby, *J. Am. Chem. Soc. 94,* 6220 (1972).

93. G. A. Olah, P. W. Westerman, E. G. Melby, Y. K. Mo, *J. Am. Chem. Soc. 96,* 3565 (1974).

94. G. A. Olah, T. Sakakibara, G. Asensio, *J. Org. Chem. 43,* 463 (1978).

95a. G. A. Olah, J. M. Bollinger, *J. Am. Chem. Soc.* 90, 947 (1968).

95b. G. A. Olah, J. M. Bollinger, J. M. Brinich, *Ibid. 90,* 2587 (1968).

96. G. A. Olah, D. A. Beal, P. W. Westerman, *J. Am. Chem. Soc. 95,* 3387 (1973).

97. G. A. Olah, P. Schilling, P. W. Westerman, H. C. Lin, *J. Am. Chem. Soc. 96,* 3581 (1974).

98. G. A. Olah, J. M. Bollinger, *J. Am. Chem. Soc. 90,* 6082 (1968).

99. G. A. Olah, G. K. S. Prakash, V. V. Krishnamurthy, *J. Org. Chem. 48,* 5116 (1983).

100. J. Strating, J. H. Wieringa, H. Wynberg, *Chem. Commun.* 907 (1969).

101. J. H. Wieringa, J. Strating, H. Wynberg, *Tetrahedron Lett.* 4579 (1970).

102. G. A. Olah, J. M. Bollinger, J. M. Brinich, *J. Am. Chem. Soc. 94,* 1164, (1972).

103. J. H. Exner, L. D. Kershner, T. E. Evans, *Chem. Commun.* 361 (1973).
104. R. E. Glick, Ph.D. Thesis, University of California at Los Angeles, Los Angeles, Calif., 1954.
105. P. E. Peterson, G. Allen, *J. Am. Chem. Soc. 85,* 3608 (1963).
106. P. E. Peterson, E. V. P. Tao, *J. Am. Chem. Soc. 86,* 4503 (1964).
107a. P. E. Peterson, J. E. Duddey, *J. Am. Chem. Soc. 88,* 4990 (1966).
107b. P. E. Peterson, R. J. Bopp, Abstracts, 152nd National Meeting of the American Chemical Society, New York, Sept. 12–16, 1966, Paper S3.
108. P. E. Peterson, H. J. Bopp, D. M. Chevli, E. Curran, D. Dillard, R. J. Kamat, *J. Am. Chem. Soc. 89,* 5902 (1967).
109. G. A. Olah, P. E. Peterson, *J. Am. Chem. Soc. 90,* 4675 (1968).
110. P. E. Peterson, P. R. Clifford, F. J. Slama, *J. Am. Chem. Soc. 92,* 2840 (1970).
111. P. M. Henrichs, P. E. Peterson, *J. Am. Chem. Soc. 95,* 7449 (1973).
112. P. E. Peterson, B. R. Bonazza, P. M. Henrichs, *J. Am. Chem. Soc. 95,* 2222 (1973).
113. G. A. Olah, G. Liang, J. Staral, *J. Am. Chem. Soc. 96,* 8112 (1974).
114. P. E. Peterson, B. R. Bonazza, *J. Am. Chem. Soc. 94,* 5017 (1972).
115. F. M. Beringer, P. Ganis, G. Avitabile, H. Jaffe, *J. Org. Chem. 37,* 879 (1972).
116. F. M. Beringer, R. A. Nathan, *J. Org. Chem. 34,* 685 (1969).
117. G. A. Olah, Y. Yamada, *J. Org. Chem. 40,* 1107 (1975).
118. D. Moy, A. R. Young, II, *J. Am. Chem. Soc. 87,* 1889 (1965).
119. K. O. Christe, R. D. Wilson, W. Sawodny, *J. Mol. Struct. 8,* 245 (1971).
120. J. Mason, K. O. Christe, *Inorg. Chem. 22,* 1849 (1983).
121. W. Kirmse, *Angew. Chem. Int. Ed. Engl. 15,* 251 (1976).
122. J. R. Mohrig, K. Keegstra, *J. Am. Chem. Soc. 89,* 5492 (1967).
123. J. Mohrig, K. Keegstra, A. Maverick, R. Roberts, S. Wells, *Chem. Commun.* 780 (1974).
124. S. Farid, K.-H. Scholz, *Chem. Commun.* 412 (1968).
125. J. F. McGarrity, D. P. Cox, *J. Am. Chem. Soc. 105,* 3961 (1983).
126a. H. Zollinger, *Azo and Diazo Chemistry,* Interscience, New York, 1961.
126b. C. G. Overberger, J. P. Anselme, J. G. Lombardino, *Organic Compounds with Nitrogen–Nitrogen Bonds,* Ronald Press, New York, 1966.
126c. P. A. S. Smith, *Open Chain Nitrogen Compounds,* Vol. 1 and 2, W. A. Benjamin, New York, 1965, 1966.
126d. R. Huisgen, *Angew. Chem. 67,* 439 (1955).
126e. B. A. Porai-Koshits, *Russ. Chem. Rev. 39(4),* 2831 (1970).
126f. O. P. Studzinskii, K. Korobitsyna, *Ibid. 39(10),* 8341 (1970).
126g. J. H. Ridd, *Quart. Rev. Chem. Soc. 15,* 418 (1961).
126h. W. E. Bachmann, R. A. Hoffman, *Org. React. 2,* 224 (1944).
126i. C. S. Rondestvedt, Jr., *Ibid. 11,* 189 (1960).
126j. D. F. DeTar, *Ibid. 9,* 409 (1957).
126k. A. Roe, *Ibid. 5,* 193 (1949).
127. H. Zollinger, *Acc. Chem. Res. 6,* 335 (1973).
128. G. A. Olah, J. Welch, *J. Am. Chem. Soc. 97,* 208 (1975).
129. E. S. Lewis, M. D. Johnson, *J. Am. Chem. Soc. 81,* 2070 (1959).
130. J. M. Insole, F. S. Lewis, *J. Am. Chem. Soc. 85,* 122 (1963); E. S. Lewis, J. M. Insole, *Ibid. 86,* 32, 34 (1964); E. S. Lewis, R. E. Holliday, *Ibid. 88,* 5043 (1966); *91,* 426 (1969); E. S. Lewis, P. G. Kotcher, *Tetrahedron 25,* 4873 (1969).
131a. Reviews, G. Henrici-Olive, S. Olive, *Angew. Chem. Int. Ed. Engl. 8,* 650 (1969).
131b. M. E. Volpin, *Pure Appl. Chem. 30,* 607 (1972).

131c. W. Preetz, *Angew. Chem. Int. Ed. Engl. 11,* 243 (1972).

132. Y. Hashida, R. G. M. Landells, G. H. Wahl, Jr., H. Zollinger, *J. Am. Chem. Soc. 98,* 3301 (1976).

133. I. Szele, H. Zollinger, *J. Am. Chem. Soc. 100,* 2811 (1978).

134. W. Maurer, I. Szele, H. Zollinger, *Helv. Chim. Acta, 62,* 1079 (1979).

135. K. Laali, I. Szele, H. Zollinger, unpublished results.

136. A. Schmidt, *Chem. Ber. 99,* 2976 (1966).

137. A. Mertens, K. Lammertsma, M. Arvanaghi, G. A. Olah, *J. Am. Chem. Soc. 105,* 5657 (1983).

138a. For a current review, see G. A. Olah, S. C. Narang, J. A. Olah, K. Lammertsma, *Proc. Natl. Acad. Sci. USA, 79,* 4487 (1982).

138b. K. Schofield, *Aromatic Nitration,* Cambridge Univ. Press, New York, 1971.

139. L. Medard, *Compt. Rend. 199,* 1615 (1934).

140. D. Cook, in *Friedel-Crafts and Related Reactions,* Vol. I, G. A. Olah, Ed., Wiley-Interscience, New York, 1963, p. 769.

141. J. W. M. Steeman, C. H. MacGillavry, *Acta Crystal. 7,* 402 (1954).

142. G. A. Olah, S. J. Kuhn, *Org. Synth. 47,* 56 (1967).

143. G. A. Olah, B. G. B. Gupta, S. C. Narang, *J. Am. Chem. Soc. 101,* 5317 (1979).

144. D. S. Ross, K. F. Kuhlmann, R. Malhotra, *J. Am. Chem. Soc. 105,* 4299 (1983).

145. G. A. Olah, *Acc. Chem. Res. 13,* 330 (1980) and references given therein.

146. G. A. Olah, M. W. Meyer, in *Friedel-Crafts and Related Reactions,* Vol. I, G. A. Olah, Ed., Wiley-Interscience, New York, 1963, p. 684.

147. G. E. Toogood, C. Chieh, *Can. J. Chem. 53,* 831 (1975).

148. C. K. Ingold, *Structure and Mechanism in Organic Chemistry,* Cornell Univ. Press, Ithaca, New York, 1953.

149. J. Allan, J. Podstata, D. Snobl, J. Jarkovsky, *Tetrahedron Lett. 40,* 3565 (1965).

150. G. A. Olah, N. Friedman, *J. Am. Chem. Soc. 88,* 5330 (1966).

151. G. A. Olah, G. Salem, J. S. Staral, T. L. Ho, *J. Org. Chem. 43,* 173 (1978).

152. G. A. Olah, J. G. Shih, B. P. Singh, B. G. B. Gupta, *J. Org. Chem. 48,* 3356 (1983).

153. G. A. Olah, H. C. Lin, J. Olah, S. C. Narang, *Proc. Natl. Acad. Sci. USA, 75,* 1045 (1978).

154. P. G. Gassman, *Acc. Chem. Res. 3,* 36 (1970).

155. P. G. Gassman, G. D. Hartman, *J. Am. Chem. Soc. 95,* 449 (1973).

156. M. Wolff, *Chem. Rev. 63,* 55 (1963).

157a. G. A. Olah, D. J. Donovan, *J. Org. Chem. 43,* 1743 (1978).

157b. T. Okamoto, K. Shudo, T. Ohta, *J. Am. Chem. Soc. 97,* 7184 (1975).

157c. T. Ohta, K. Shudo, T. Okamoto, *Tetrahedron Lett.* 101 (1977).

157d. K. Shudo, T. Ohta, Y. Endo, T. Okamoto, *Ibid.* 105 (1977).

158. K. O. Christie, C. J. Schack, *Inorg. Chem. 17,* 2749 (1978).

159. P. G. Gassman, J. J. Talley, *J. Am. Chem. Soc. 102,* 1214 (1980).

160. P. G. Gassman, J. J. Talley, *J. Am. Chem. Soc. 102,* 4138 (1980).

161. P. G. Gassman, K. Saito, J. J. Talley, *J. Am. Chem. Soc. 102,* 7613 (1980).

162a. G. A. Olah, G. K. S. Prakash, M. Arvanaghi, *J. Am. Chem. Soc. 102,* 6640 (1980).

162b. G. A. Olah, M. Arvanaghi, G. K. S. Prakash, *Ibid. 104,* 1628 (1982).

163. V. V. Krishnamurthy, G. K. S. Prakash, P. S. Iyer, G. A. Olah, unpublished results.

164. A. Mertens, G. A. Olah, *Chem. Ber. 116,* 103 (1983).

165. O. P. Shitov, S. L. Ioffe, V. A. Tortakovskii, S. S. Novikov, *Russ. Chem. Rev.* (Engl. Transl.), *39,* 906 (1970)

166a. S. G. Shore, R. W. Parry, *J. Am. Chem. Soc. 77,* 6085 (1977).

166b. D. R. Schultz, R. W. Parry, *J. Am. Chem. Soc. 80,* 4 (1958).

167a. H. Nöeth, H. Beyer, H. Vetter, *Chem. Ber. 97,* 110 (1964).

167b. N. E. Miller, E. L. Muetterties, *J. Am. Chem. Soc. 88,* 1033 (1964).

168. H. Nöeth, S. Lukas, *Chem. Ber. 95,* 1505 (1962).

169. H. Nöeth, R. Staudigl, *Angew. Chem. Int. Ed. Engl. 20,* 794 (1981).

170. H. Nöeth, R. Staudigl, H-U. Wagner, *Inorg. Chem. 21,* 706 (1982).

171. J. Higashi, A. D. Eastman, R. W. Parry, *Inorg. Chem. 21,* 716 (1982).

172. P. Beck, in *Organic Phosphorus Compounds,* Vol. 2, Chapter 4, G. M. Kosolapoff, L. Maier, Eds., John Wiley & Sons, New York, 1972, p. 189.

173. A. H. Cowley, M. C. Cushner, M. Lattan, M. L. McKee, J. S. Szobota, J. C. Wilburn, *Pure Appl. Chem. 52,* 789 (1980).

174. C. W. Schultz, R. W. Parry, *Inorg. Chem. 15,* 3046 (1976).

175. M. G. Thomas, C. W. Schultz, R. W. Parry, *Inorg. Chem. 16,* 994 (1977).

176. A. H. Cowley, M. C. Cushner, J. S. Szobota, *J. Am. Chem. Soc. 100,* 7784 (1978).

177. S. G. Baxter, R. L. Collins, A. H. Cowley, S. F. Sena, *J. Am. Chem. Soc. 103,* 715 (1981).

178a. R. J. P. Corriu, M. J. Henner, *J. Organomet. Chem. 74,* 1 (1974).

178b. B. Boe, *J. Organomet. Chem. 107,* 139 (1976).

179. R. West, T. J. Barton, *J. Chem. Ed. 57,* 334 (1980).

180. W. P. Weber, R. A. Felix, A. K. Willare, *Tetrahedron Lett.* 907 (1970).

181. W. J. Pietro, W. J. Hehre, *J. Am. Chem. Soc. 104,* 4329 (1982).

182a. G. A. Olah, D. H. O'Brien, C. Y. Lin, *J. Am. Chem. Soc. 91,* 701 (1969).

182b. A. G. Brook, A. K. Pannel, *Can. J. Chem. 48,* 3679 (1973).

182c. G. A. Olah, Y. K. Mo, *J. Am. Chem. Soc. 93,* 4942 (1971).

182d. A. H. Cowley, M. C. Cushner, P. E. Riley, *J. Am. Chem. Soc. 102,* 624 (1980).

183. G. A. Olah, L. D. Field, *Organometallics 1,* 1485 (1982).

184. J. B. Lambert, W. J. Schultz, Jr., *J. Am. Chem. Soc. 105,* 1671 (1983).

185. R. J. Gillespie, M. J. Morton, *Quart. Rev. Chem. Soc.* 553 (1971).

186. R. J. Gillespie, J. Passmore, in *Advances in Inorganic Chemistry and Radiochemistry,* Vol. 17, H. J. Emeleus, A. J. Sharpe, Eds., Academic Press, New York, 1975.

187. I. Masson, *J. Chem. Soc.* 1708 (1938).

188. R. J. Gillespie, J. B. Milne, *Inorg. Chem. 5,* 1577 (1966).

189. R. J. Gillespie, K. C. Malhotra, *Inorg. Chem. 8,* 1751 (1969).

190. R. D. W. Kemmitt, M. Murray, V. M. McRae, R. D. Peacook, M. C. R. Symons, *J. Chem. Soc.* 862 (1968).

191. J. Arotsky, H. C. Mishra, M. C. R. Symons, *J. Chem. Soc.* 2582 (1962).

192. R. A. Garrett, R. J. Gillespie, J. B. Senior, *Inorg. Chem. 4,* 563 (1965).

193. D. J. Merryman, P. A. Edwards, J. D. Corbett, R. E. McCarley, *Chem. Commun.*, 779 (1972).

194. R. J. Gillespie, M. Morton, J. M. Sowa, *Adv. Raman Spectrosc. 1,* 530 (1972).

195. J. Passmore, G. Sutherland, P. S. White, *Inorg. Chem. 20,* 2169 (1981).

196. J. Passmore, P. Taylor, T. Whidden, P. S. White, *Can. J. Chem. 57,* 968 (1979).

197. G. Adhami, M. Herlem, *J. Electroanal. Chem. 26,* 363 (1970).

198. C. Davies, R. J. Gillespie, J. M. Sowa, *Can. J. Chem. 52,* 791 (1974).

199. R. J. Gillespie, R. Kapoor, R. Faggiani, C. J. L. Lock, M. Murchie, J. Passmore, *Chem. Commun.* 8 (1983).

200. O. Ruff, H. Gray, W. Heller, Knock, *Ber. 39,* 4310 (1906).

201. V. M. McRae, Ph.D. Thesis, University of Melbourne, 1966.

202. R. J. Gillespie, M. J. Morton, *Chem. Commun.* 1565 (1968).
203. R. J. Gillespie, M. J. Morton, *Inorg. Chem. 11,* 586 (1972).
204. O. Glemser, A. Smalc, *Angew. Chem. Int. Ed. Engl. 8,* 517 (1969).
205. A. J. Edwards, G. R. Jones, R. J. C. Sills, *Chem. Commun.* 1527 (1968).
206. A. J. Edwards, G. R. Jones, *J. Chem. Soc.* 2318 (1971).
207. R. J. Gillespie, M. J. Morton, *Inorg. Chem. 11,* 591 (1972).
208. R. J. Gillespie, M. J. Morton, *Inorg. Chem. 9,* 811 (1970).
209. H. Siebert, *Anwendungen der Schwingungs Spektroscopic in der Anorganischen Chemie,* Springer-Verlag, Berlin and New York, 1966.
210. L. Stein, *Chemistry 47,* 15 (1974); *Science 175,* 1463 (1972).
211. L. Stein, *Nature (London) 30,* 243 (1973).
212. *Halogen Chemistry,* V. Gutman, Ed., Academic Press, London and New York, 1967.
213. K. O. Christe, W. Sawodny, *Inorg. Chem. 6,* 313 (1967).
214. K. O. Christe, J. F. Hon, D. Pillipovich, *Inorg. Chem. 12,* 84 (1973).
215. A. J. Edwards, R. J. C. Sills, *J. Chem. Soc.* 2697 (1970).
216. A. J. Edwards, G. R. Jones, *J. Chem. Soc.* 1467 (1969).
217. K. O. Christe, C. J. Schack, *Inorg. Chem. 9,* 2296 (1970).
218. T. Surles, H. H. Hyman, L. A. Quarterman, A. I. Popov, *Inorg. Chem. 9,* 2726 (1970).
219. A. A. Banks, H. J. Emeleus, A. A. Wolf, *J. Chem. Soc.* 2861 (1949).
220. M. Schmeisser, W. Ludovici, D. Naumann, P. Sartori, E. Scharf, *Ber. 101,* 4214 (1968).
221. K. O. Christe, W. Sawodny, *Inorg. Chem. 8,* 212 (1969).
222. C. G. Vonk, E. H. Wiebenga, *Acta. Cryst. 12,* 859 (1959).
223. N. N. Greenwood, H. J. Emeleus, *J. Chem. Soc.* 987 (1950).
224. K. O. Christe, D. Pilipovich, *Inorg. Chem. 11,* 2205 (1969).
225. M. Schmeisser, E. Pammer, *Angew. Chem. 69,* 281 (1957).
226. A. A. Woolf, *J. Chem. Soc.* 3678 (1950).
227. N. Bartlett, D. H. Lohman, *J. Chem. Soc.* 8253 (1962); 619 (1964).
228. H. W. Baird, H. F. Giles, *Acta Cryst. A25,* S115 (1969).
229. J. Shamir, I. Yaroslavsky, *Israel. J. Chem. 7,* 495 (1969).
230. F. Q. Roberto, *Inorg. Nucl. Chem. Lett. 8,* 737 (1972).
231. K. O. Christe, *Inorg. Nucl. Chem. Lett. 8,* 741 (1972).
232. K. O. Christe, *Inorg. Nucl. Chem. Lett. 8,* 453 (1972).
233. K. O. Christe, W. W. Wilson, R. D. Wilson, *Inorg. Chem. 19,* 1494 (1980).
234. K. O. Christe, R. D. Wilson, C. J. Schak, *Inorg. Chem. 19,* 3046 (1980) and references given therein.
235. K. O. Christe, *Inorg. Chem. 12,* 1580 (1973).
236. R. J. Gillespie, G. J. Schrobilgen, *Chem. Commun.* 90 (1974).
237. R. J. Gillespie, G. J. Schrobilgen, *Inorg. Chem. 13,* 1230 (1974).
238. K. O. Christe, R. D. Wilson, *Inorg. Chem. 14,* 694 (1975).
239. F. Seel, O. Detmer, *Angew. Chem. 70,* 163 (1958); *Z. Anorg. Chem. 301,* 113 (1959).
240. K. O. Christe, W. Sawodny, *Inorg. Chem. 6,* 1783 (1967).
241. F. A. Hohorst, L. Stein, E. Gebert, *Inorg. Chem. 14,* 2233 (1975).
242. K. O. Christe, *Inorg. Chem. 9,* 2801 (1970).
243. G. Herzberg, *The Spectra of Diatomic Molecules,* Van Nostrand Reinhold, Princeton, New Jersey, 1950.
244. N. Bartlett, D. H. Lohmann, *Proc. Chem. Soc., London,* 115 (1962).

245. N. Bartlett, D. H. Lohmann, *Proc. Chem. Soc., London,* 14 (1960).
246. E. W. Lawless, I. C. Smith, *Inorganic High Energy Oxidizers,* Dekker, New York, 1968, and references therein.
247. D. E. McKee, N. Bartlett, *Inorg. Chem. 12,* 2738 (1973).
248. A. R. Young, T. Hirata, S. I. Morrow, *J. Am. Chem. Soc. 86,* 20 (1964).
249. J. N. Keith, I. J. Solomon, I. Sheft, H. H. Hyman, *Inorg. Chem. 7,* 230 (1968).
250. Z. K. Nikitina, V. Ya. Rosolovski, *Izv. Akad. Nauk. SSSR, Ser. Khim.* 2173 (1970).
251. N. Bartlett, *Angew. Chem. Int. Ed. Engl. 7,* 433 (1968).
252. J. Shamir, J. Bineboyn, H. H. Claasen, *J. Am. Chem. Soc. 90,* 6223 (1968).
253. K. R. Loos, V. A. Campanile, C. T. Goetschel, *Spectrochim. Acta A26,* 365 (1970).
254. I. B. Goldberg, K. O. Christe, R. D. Wilson, *Inorg. Chem. 14,* 152 (1975).
255. C. T. Goetschel, K. R. Loos, *J. Am. Chem. Soc. 94,* 3018 (1972).
256. J. W. Mellor, *Comprehensive Treatise on Inorganic and Theoretical Chemistry,* Vol. 10, Longmans, Green, New York, 1930, pp. 184 and 992.
257. C. F. Bucholz, *Gehlen's Neues. J. Chem. 3,* 7 (1804).
258. For a review, see *Inorganic Sulfur Chemistry,* G. Nickless, Ed., Elsevier, Amsterdam, 1968, p. 412.
259. R. J. Gillespie, J. Passmore, P. K. Ummat, O. C. Vaidya, *Inorg. Chem. 10,* 1327 (1971).
260. R. J. Gillespie, P. K. Ummat, *Inorg. Chem. 11,* 1674 (1972).
261. C. Davies, R. J. Gillespie, J. J. Park, J. Passmore, *Inorg. Chem. 10,* 2781 (1971).
262. R. C. Burns, R. J. Gillespie, J. F. Sawyer, *Inorg. Chem. 19,* 1423 (1980).
263. H. S. Low, R. A. Beaudet, *J. Am. Chem. Soc. 98,* 3849 (1976).
264. I. D. Brown, D. B. Crump, R. J. Gillespie, *Inorg. Chem. 10,* 2319 (1971).
265. T. W. Couch, D. A. Lokken, J. D. Corbett, *Inorg. Chem. 11,* 357 (1972).
266. R. C. Burns, R. J. Gillespie, *Inorg. Chem. 21,* 3877 (1982).
267. R. J. Gillespie, G. P. Pez, *Inorg. Chem. 8,* 1229 (1969).
268. J. Barr, R. J. Gillespie, P. K. Ummatt, *Chem. Commun.* 264 (1970).
269. P. J. Stephens, *Chem. Commun.* 1496 (1969).
270. R. C. Burns, R. J. Gillespie, J. A. Barnes, M. J. McGlinchey, *Inorg. Chem. 21,* 799 (1982).
271. R. J. Gillespie, J. Passmore, *Chem. Commun.* 1333 (1969).
272. G. Magnus, *Ann. Phys.* (Leipzig) [2], *10,* 491 (1827); *14*, 328 (1828).
273. Reference 256, pp. 922–923.
274. J. Barr, R. J. Gillespie, R. Kapoor, K. C. Malhotra, *Can. J. Chem. 46,* 149 (1968).
275. I. D. Brown, D. B. Crump, R. J. Gillespie, D. P. Santry, *Chem. Commun.* 853 (1968).
276. R. E. March, L. Pauling, J. D. McCullough, *Acta Cryst. Sec. B. 6,* 71 (1953).
277a. R. K. Mullen, D. J. Prince, J. D. Corbett, *Inorg. Chem. 10,* 1749 (1971).
277b. R. K. Mullen, D. J. Prince, J. D. Corbett, *Chem. Commun.* 1438 (1969).
278. R. C. Burns, W-L Chan, R. J. Gillespie, W-C Luk, J. F. Sawyer, D. R. Slim, *Inorg. Chem. 19,* 1432 (1980).
279. M. H. Klaproth, *Phil. Mag. 1,* 78 (1798).
280. N. J. Bjerrum, G. P. Smith, *J. Am. Chem. Soc. 90,* 4472 (1968).
281. N. J. Bjerrum, *Inorg. Chem. 9,* 1965 (1970).
282. J. Barr, R. J. Gillespie, R. Kapoor, G. P. Pez, *J. Am. Chem. Soc. 90,* 6855 (1968).
283. G. Barr, R. J. Gillespie, G. P. Pez, P. K. Ummat, O. C. Vaidya, *Inorg. Chem. 10,* 362 (1971).
284. D. J. Prince, J. D. Corbett, B. Garbisch, *Inorg. Chem. 9,* 2731 (1970).

285. J. Barr, R. J. Gillespie, G. P. Pez, P. K. Ummat, O. C. Vaidya, *J. Am. Chem. Soc. 92,* 1081 (1970).

286. R. C. Paul, C. L. Arora, J. K. Puri, R. N. Virmani, K. C. Malhotra, *J. Chem. Soc., Dalton Trans.* 781 (1972).

287. M. Z. Straumanis, *Z. Kristallogr. Kristallogeometrie., Kristallophys., Kristallochem. 102,* 432 (1946).

288. N. J. Bjerrum, *Inorg. Chem. 11,* 2648 (1972).

289. R. J. Gillespie, W. Luk, D. R. Slim, *Chem. Commun.* 791 (1976).

290. R. C. Burns, R. J. Gillespie, W-C Luk, D. R. Slim, *Inorg. Chem. 18,* 3086 (1979).

291. R. J. Gillespie, P. K. Ummat, *Can. J. Chem. 48,* 1239 (1970).

292. P. A. W. Dean, R. J. Gillespie, P. K. Ummat, *Inorg. Synth. 15,* 213 (1974).

293. P. Boldrini, I. D. Brown, R. J. Gillespie, P. R. Ireland, W. Luk, D. R. Slim, J. E. Vekris, *Inorg. Chem. 15,* 765 (1976).

294. R. J. Gillespie, W. Luk, E. Maharajah, D. R. Slim, *Inorg. Chem. 16,* 892 (1977).

295. R. Faggiani, R. J. Gillespie, C. J. L. Lock, J. D. Tyrer, *Inorg. Chem. 17,* 2975 (1975).

296. R. J. Gillespie, J. P. Kent, J. F. Sawyer, *Inorg. Chem. 20,* 3784 (1981).

297. R. J. Gillespie, J. P. Kent, J. F. Sawyer, D. R. Slim, J. D. Tyrer, *Inorg. Chem. 20,* 3799 (1981).

298. K. O. Christe, C. J. Schack, D. Pilipovich, *Inorg. Chem. 11,* 2205 (1972).

299. C. J. Schack, C. B. Lindahl, D. Pilipovich, K. O. Christe, *Inorg. Chem. 11,* 220 (1972).

300. K. O. Christe, *Inorg. Nucl. Chem. Lett. 8,* 453 (1972).

301. K. O. Christe, R. D. Wilson, *Inorg. Chem. 12,* 1356 (1973).

302. K. O. Christe, R. D. Wilson, E. C. Curtis, *Inorg. Chem. 12,* 1358 (1973).

303. K. O. Christe, W. Maya, *Inorg. Chem. 8,* 1253 (1968).

304. J. J. Thompson, *Phil. Mag. 24,* 209 (1912).

305. T. R. Hogness, E. G. Lunn, *Phys. Rev. 26,* 44 (1925); also see: H. D. Smith, *Phys. Rev. 25,* 452 (1925).

306. For a review see: L. Friedman, B. G. Reuben, *Adv. Chem. Phys. 19,* 59 (1971).

307. R. J. Gillespie, G. P. Pez, *Inorg. Chem. 8,* 1233 (1969).

308. G. A. Olah, J. Shen, R. H. Schlosberg, *J. Am. Chem. Soc. 95,* 4957 (1973).

309. T. Oka, *Phys. Rev. Lett. 43,* 531 (1980).

310. G. A. Sinnot, *Univ. Colo. Joint Inst. Lab. Astrophys. Inform. Center Rep.* No. 9 (1969) and references cited therein.

311. F. H. Field, H. H. Head, J. L. Franklin, *J. Am. Chem. Soc. 84,* 1118 (1962).

312a. T. O. Tiernan, P. S. Gill, *J. Chem. Phys. 50,* 5042 (1969).

312b. G. R. Hertel, W. S. Koski, *J. Am. Chem. Soc. 87,* 1686 (1965).

312c. S. Wexler, *J. Am. Chem. Soc. 85,* 272 (1963).

313a. M. S. B. Munson, F. H. Field, *J. Am. Chem. Soc. 87,* 4242 (1965).

313b. F. S. Klein, L. Friedman, *J. Chem. Phys. 41,* 1789 (1964).

313c. V. Aquilanti, A. Galli, A. Giardini-Guidoni, G. G. Volpi, *J. Chem. Phys. 43,* 1969 (1965).

313d. W. A. Chupka, M. E. Russell, *J. Chem. Phys. 49,* 5426 (1968).

314. D. Holtz, J. L. Beauchamp, *Science 173,* 1237 (1971).

315. G. A. Olah, J. Shen, *J. Am. Chem. Soc. 95,* 3582 (1973).

316. F. A. Cotton, G. Wilkinson, *Advanced Inorganic Chemistry,* John Wiley & Sons, New York, 1980, pp. 580.

317. V. M. McRae, R. D. Peacock, D. R. Russell, *Chem. Commun.* 62 (1969).

318. N. Bartlett, F. O. Sladky, *Comprehensive Inorganic Chemistry,* Vol. 1, J. C. Bailar, A. F. Trotman-Dickenson, Ed., Pergamon Press, Oxford, 1973, Chapter 6.
319. R. J. Gillespie, B. Landa, G. J. Schrobilgen, *Inorg. Chem. 15,* 1256 (1976) and references therein.
320. K. Leary, A. Zalkin, N. Bartlett, *Inorg. Chem. 13,* 775 (1974).
321. B. Frlec, J. H. Holloway, *Chem. Commun.* 89 (1974).
322. R. J. Gillespie, G. J. Schrobilgen, *Inorg. Chem. 15,* 22 (1976).

chapter 5

SUPERACID CATALYZED REACTIONS

In discussing superacids as catalysts for chemical reactions, we will review both liquid (Magic Acid, fluoroantimonic acid, etc.) and solid (Nafion-H, etc.) acid-catalyzed reactions, but not those of conventional Friedel-Crafts-type catalysts. The latter reactions have been extensively reviewed elsewhere (see G. A. Olah, *Friedel-Crafts Chemistry,* Wiley Interscience, 1973; *Friedel-Crafts and Related Reactions,* Vols. I–IV, G. A. Olah, Ed., Wiley Interscience, 1963–1965).

5.1 CONVERSION OF SATURATED HYDROCARBONS

Before the turn of the century, saturated hydrocarbons (paraffins) played only a minor role in industrial chemistry. They were mainly used as a source of paraffin wax as well as for heating and lighting oils. Aromatic compounds such as benzene, toluene, phenol, and naphthalene obtained from destructive distillation of coal were the main source of organic materials used in the preparation of dyestuffs, pharmaceutical products, etc. Calcium carbide-based acetylene was the key starting material for the emerging synthetic organic industry. It was the ever increasing demand for gasoline after the first world war that initiated study of isomerization and cracking reactions of petroleum fractions. After the second world war, rapid economic expansion necessitated more and more abundant and cheap sources for chemicals and this resulted in the industry switching over to petroleum based ethylene as the main source of chemical raw material. One of the major difficulties that had to be overcome is the low reactivity of some of the major components of the petroleum. The lower boiling components (up to 250°C) are mainly straight-chain saturated hydrocarbons or paraffins, which, as their name indicates (*parum affinis:* slight reactivity), have very little reactivity. Consequently the lower paraffins were cracked to give olefins (mainly ethylene, propylene, and butylenes). The straight-chain liquid hydrocarbons have also very low octane numbers which make them less desirable as gasoline components. To transform these paraffins into useful components for gasoline and other chemical applications, they have to undergo diverse reactions such as isomerization, cracking, or alkylation. These reactions, which are used on a large scale in industrial processes, necessitate acidic catalysts (at temperature around 100°C) or noble metal catalysts (at higher temperature, 200–500°C) capable of activating the strong covalent C—H or C—C bonds.[1] Since the early 1960s, superacids are known to react with saturated hydrocarbons, even at temperatures much below 0°C. This discovery initiated extensive studies devoted to hydrocarbon conversions.

5.1.1 C—H and C—C Bond Protolysis

The fundamental step in the acid-catalyzed hydrocarbon conversion processes is the formation of the intermediate carbocations. Whereas all studies involving iso-

$$\text{R—H} \xrightarrow{\text{Acid}} \text{R}^+ \begin{cases} \to \text{Isomerization} \\ \to \text{Cracking} \\ \to \text{Alkylation-homologation} \end{cases}$$

merization, cracking, and alkylation reactions under acidic conditions agree that a trivalent carbocation (carbenium ion) is the key intermediate, the mode of formation of this reactive species from the neutral hydrocarbon remained controversial for many years.

In 1946 Bloch, Pines, and Schmerling[2] observed that *n*-butane **1** would isomerize to isobutane **2** under the influence of pure aluminum chloride only in the presence of HCl. They proposed that the ionization step takes place through initial protolysis of the alkane as evidenced by formation of minor amounts of hydrogen in the initial stage of the reaction

$$\underset{\mathbf{1}}{n\text{-}C_4H_{10}} + HCl \xrightarrow[AlCl_3]{} \underset{\mathbf{3}}{sec\text{-}C_4H_9^+}\ AlCl_4^- + H_2$$

The first direct evidence of protonation of alkanes under superacid conditions has been reported independently by Olah and Lukas[3] as well as Hogeveen and coworkers.[4]

When *n*-butane **1** or isobutane **2** was reacted with $HSO_3F:SbF_5$, (Magic Acid), *t*-butyl cation **4** was formed exclusively as evidenced by a sharp singlet at 4.5 ppm (from TMS) in the ^1H-nmr spectrum. In excess Magic Acid, the stability of the ion is remarkable and the nmr spectrum of the solution remains unchanged even after having been heated to 110°C.

$$\underset{\mathbf{2}}{(CH_3)_3CH} \xrightarrow[\text{room temperature}]{HSO_3F:SbF_5} \underset{\mathbf{4}}{(CH_3)_3C^+\ SbF_5FSO_3^-} + H_2 \xleftarrow{HSO_3F:SbF_5} \underset{\mathbf{1}}{n\text{-}C_4H_{10}}$$

Recently, it was shown[5] that the *t*-butyl cation **4** undergoes degenerate carbon scrambling at higher temperatures (see Chapter 3). A lower limit of $E_a \sim 30$ kcal · mol^{-1} was estimated for the scrambling process which could correspond to the energy difference between *t*-butyl **4** and primary isobutyl cation **5** (the latter being partially delocalized).

$$\underset{\mathbf{2}}{(CH_3)_3CH} \xrightarrow[\text{room temperature}]{HSO_3F:SbF_5} \underset{\mathbf{4}}{(CH_3)_3C^+\ SbF_5FSO_3^-} + H_2 \xleftarrow{FSO_3H:SbF_5} \underset{\mathbf{1}}{n\text{-}C_4H_{10}}$$

$$\underset{\mathbf{5}}{(CH_3)_2CH\text{—}CH_2^+}$$

n-Pentane **6** and isopentane **7** are ionized under the same conditions to the *t*-amyl cation **8.** *n*-Hexane **9** and the branched C_6 isomers ionize in the same way to yield a mixture of the three tertiary hexyl ions as shown by their ^1H-nmr spectra.

$$\underset{\mathbf{9}}{n\text{-}C_6H_{14}} \xrightarrow{HSO_3F:SbF_5} (CH_3)_2C^+\text{—}CH(CH_3)_2 \rightleftharpoons (CH_3)_2CH\text{—}C^+(CH_3)_2 + CH_3CH_2C^+(CH_3)CH_2CH_3 + (CH_3)_2C^+CH_2CH_2CH_3$$

Both methylcyclopentane **10** and cyclohexane **11** were found to give the methylcyclopentyl ion **12,** which is stable at low temperature, in excess superacid.[6] When alkanes with seven or more carbon atoms were used, cleavage was observed with formation of the stable *t*-butyl cation **4.** Even paraffin wax (see Section 2.3.3 on Magic Acid) and polyethylene ultimately gave the *t*-butyl ion **4** after complex fragmentation and ionization processes.

In compounds containing only primary hydrogen atoms such as neopentane **13** and 2,2,3,3-tetramethylbutane **14,** a carbon–carbon bond is broken rather than a carbon–hydrogen bond.[7]

$$\underset{\mathbf{13}}{C(CH_3)_4} \xrightarrow{H^+} CH_4 + \underset{\mathbf{4}}{(CH_3)_3C^+}$$

Hogeveen and co-workers suggested a linear transition state for the protolytic ionization of hydrocarbons. This, however, may be the case only in sterically hindered systems. Results of protolytic reactions of hydrocarbons in superacidic media were interpreted by Olah as indication for the general electrophilic reactivity of covalent C—H and C—C single bonds of alkanes and cycloalkanes. The reactivity is due to the σ-donor ability of a shared electron pair (of σ-bond) via two-electron, three-center bond formation. The transition state of the reactions consequently are of three-center bound pentacoordinate carbonium ion nature. Strong indication for the mode of protolytic attack was obtained from deuterium–hydrogen exchange studies. Monodeuteromethane was reported to undergo C–D exchange without detectable side reactions in the HF:SbF_5 system.[8] d_{12}-Neopentane when treated with Magic Acid was also reported[9] to undergo H–D exchange before cleavage.

$$R_3C{-}H \xrightarrow{H^+} \left[R_3C\cdots\langle^{H}_{H} \right]^+ \longrightarrow (R_3)C^+ + H_2$$

$$HF{:}SbF_5 + CH_3D \longrightarrow \left[H_3C\cdots\langle^{H}_{D} \right]^+ \longrightarrow CH_4 + DF{:}SbF_5$$

Based on the demonstration of H–D exchange of molecular hydrogen (and deuterium) in superacid solutions,[10a] Olah suggested that these reactions go through trigonal isotopic $H_3{}^+$ ions in accordance with theoretical calculations and recent ir studies.[10b]

$$H_2D^+ \text{ (trigonal H, H, D)}, \quad HD_2^+ \text{ (trigonal H, D, D)}$$

Consequently, the reverse reaction of protolytic ionization of hydrocarbons to carbenium ions, i.e., the reduction of carbenium ion by molecular hydrogen[11,12] can be considered as alkylation of H_2 by the electrophilic carbenium ion through the pentacoordinate carbonium ion. Indeed Hogeveen has experimentally proved this point by reacting stable alkyl cations in superacids with molecular hydrogen.

$$R_3C^+ + H{-}H \longrightarrow \left[R_3C\cdots\langle^{H}_{H} \right]^+ \longrightarrow R_3CH + H^+$$

Further evidence for the pentacoordinate carbonium ion mechanism of alkane protolysis was obtained in the H–D exchange reaction observed with isobutane.

When isobutane **2** is treated with deuterated superacids ($DSO_3F:SbF_5$ or $DF:SbF_5$) at low temperature (−78°C) and atmospheric pressure, the initial hydrogen deuterium exchange is observed only at the tertiary carbon. Ionization yields only deuterium-free *t*-butyl cation **4** and HD.[13] Recovered isobutane from the reaction mixture shows at low temperature only methine hydrogen–deuterium exchange. This result is best explained as proceeding through a two-electron, three-center bound pentacoordinate carbonium ion **15.**

$$\underset{\mathbf{2}}{(CH_3)_3CH} \xrightarrow{D^+} \underset{\mathbf{15}}{\left[(CH_3)_3C\cdots\langle^{H}_{D} \right]^+} \longrightarrow (CH_3)_3CD$$

$$\mathbf{15} \xrightarrow{-HD} \mathbf{4}$$

The H–D exchange in isobutane in superacid media is fundamentally different from the H–D exchange observed by Otvos et al.[14] in D_2SO_4, who found the eventual exchange of all the nine methyl hydrogens but not the methine hydrogen.

$$\underset{\mathbf{2}}{CH_3-\overset{\displaystyle CH_3}{\underset{\displaystyle CH_3}{C}}-H} \xrightarrow[\text{Excess}]{D_2SO_4} \underset{\mathbf{4}}{(CH_3)_3C^+} \xrightarrow{-H^+} \underset{\mathbf{16}}{CH_3-\overset{\displaystyle CH_2}{\overset{\|}{C}}-CH_3}$$

$$\mathbf{16} \xrightarrow{D^+} (CH_3)_2C^+-CH_2D \underset{+D^+}{\overset{-H^+}{\rightleftarrows}} \longleftarrow \longleftarrow (CD_3)_3C^+$$

$$(CD_3)_3C^+ \xrightarrow{(CH_3)_3CH} (CD_3)_3CH$$

Otvos suggested[14] that under the reaction conditions a small amount of *t*-butyl cation **4** is formed in an oxidative step which deprotonates to isobutylene **16.** The reversible protonation (deuteration) of **16** was responsible for the H–D exchange on the methyl hydrogens, whereas tertiary hydrogen is involved in intermolecular hydride transfer from unlabeled isobutane (at the CH position). Under superacidic conditions, where no olefin formation occurs, the reversible isobutylene protonation cannot be involved in the exchange reaction. On the other hand, a kinetic study of hydrogen deuterium exchange in deuteroisobutane[15] showed that the exchange of the tertiary hydrogen was appreciably faster than the hydride abstraction by C—H protolysis.

The nucleophilic nature of the alkanes is also shown by the influence of the acidity level on their solubility. Torck and co-workers[16] have investigated the composition of the catalytic phase obtained when *n*-pentane or *n*-hexane is thoroughly mixed with HF:SbF_5 in an autoclave under hydrogen pressure. The total amount of hydrocarbon in the catalytic phase (dissolved ions and neutrals) was obtained by extraction with excess of methylcyclopentane. The amount of physically dissolved hydrocarbons was obtained by extracting the catalytic phase with Freon 113. The amount of cations is calculated by difference. The results are shown in Fig. 5.1.

The total amount of hydrocarbons increases from 1.6% to 14.6% in weight when the SbF_5 concentration varies from 0 to 6.8 mol/liter. The amount of carbenium ions increases linearly with the SbF_5 concentration, and the solubility of the hydrocarbon itself reaches a maximum for 5 mol/liter of SbF_5. The apparent decrease in solubility of the hydrocarbon at higher SbF_5 concentration may be due to the rapid rate of hydrocarbon protolysis as well as the change in the composition of the acid: SbF_6^- anions being transformed to $Sb_2F_{11}^-$ anions.

$$RH + H^+Sb_2F_{11}^- \longrightarrow R^+Sb_2F_{11}^- + H_2$$

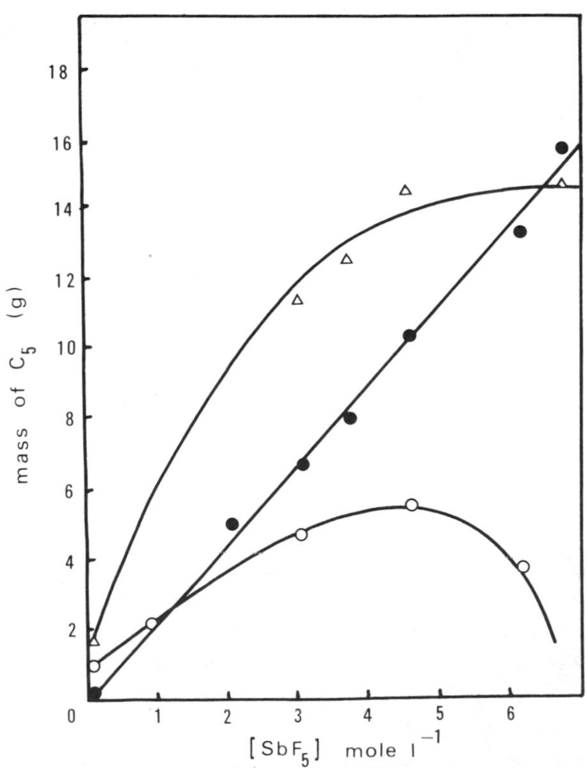

FIGURE 5.1 Variation of the composition of the catalytic phase as a function of the SbF_5 concentration in the pentane isomerization in HF—SbF_5 (ref. 16). T = 15°C, P_{H_2} = 5 bars, volume of the catalytic phase: 57 ml.

(●): mass C_5^+

(○): mass C_5H (Freon 113 extract)

(△): % weight $C_5^+ + C_5H$ (methylcylopentane extract)

One of the difficulties in understanding the carbocationic nature of acid-catalyzed transformations of alkanes via the hydride abstraction mechanism was that no stoichiometric amount of hydrogen gas evolution was observed from the reaction mixture. For this reason, a complementary mechanism was proposed involving direct hydride abstraction by the Lewis acid.[17]

$$RH + 2SbF_5 \longrightarrow R^+SbF_6^- + SbF_3 + HF$$

Olah has pointed out that if SbF_5 would abstract H^-, it would need to form SbF_5H^- ion involving an extremely weak Sb—H bond compared to the strong C—H bond being broken.[13] Calculations based on thermodynamics[18] also show that the direct oxidation of alkanes by SbF_5 is not feasible. Hydrogen is generally assumed to be partially consumed in the reduction of one of the superacid components.

$$2HSO_3F + H_2 \longrightarrow SO_2 + H_3O^+ + HF + SO_3F^- \qquad \Delta H = -33 \text{ kcal/mol}$$
$$SbF_5 + H_2 \longrightarrow SbF_3 + 2HF \qquad \Delta H = -49 \text{ kcal/mol}$$

The direct reduction of SbF_5 in the absence of hydrocarbons by molecular hydrogen necessitates, however, more forcing conditions (50 atm, high temperature) which suggests that the protolytic ionization of alkanes proceeds probably via solvation of protonated alkane by SbF_5 and concurrent ionization-reduction.[13]

$$\left[R_3C^{+}\cdots\begin{matrix} H\text{---}F \\ H\text{---}F \end{matrix} SbF_3 \right] SbF_6^- \longrightarrow (R)_3C^+ \; SbF_6^- + 2HF + SbF_3$$

The oxidation of alkanes consequently takes place much more readily in the more acidic solutions and can lead to some useful synthetic applications as for example in the synthesis of enone **18** starting from cyclohexane **11** (*vide infra*).

HSO₃F / AcOH:SbF₅(2:3)

11 → **17** ⇌ **12** ⇌ (1-methylcyclopentene) —CH_3CO^+→ (cation) → **18**

5.1.2 Electrochemical Oxidation in Strong Acids

The protolytic oxidation of alkanes is also strongly supported by electrochemical studies. In 1973, Fleischman, Plechter and co-workers[19] showed that the anodic oxidation potential of several alkanes in HSO_3F was dependent on the proton donor ability of the medium. This acidity dependence shows that there is a rapid protonation equilibrium before the electron transfer step and it is the protonated alkane that undergoes oxidation.

$$\mathrm{RH} \xrightarrow{\mathrm{HSO_3F}} \mathrm{RH_2^+} \xrightarrow[\text{Pt anode}]{} \mathrm{R^+} \longrightarrow \begin{cases} \xrightarrow{\mathrm{RH}} \text{Oligomers} \\ \xrightarrow{\mathrm{CO}} \text{Acids} \\ \xrightarrow{\mathrm{CH_3COO^-}} \text{Esters} \end{cases}$$

More recently, the electrochemical oxidation of lower alkanes in the HF solvent system has been investigated by Devynck and co-workers over the entire pH range.[20] Classical and cyclic voltammetry show that the oxidation process depends largely on the acidity level. Isopentane **7** (2-methylbutane, M2BH) for example undergoes two-electron oxidation in $HF:SbF_5$ and $HF:TaF_5$ solutions.[21]

$$\underset{\mathbf{7}}{\mathrm{M2BH}} - 2e^- \longrightarrow \mathrm{M2B^+} + \mathrm{H^+}$$

In the higher-acidity region, the intensity-potential curve shows two peaks (at 0.9 V and 1.7 V, respectively, vs the Ag/Ag^+ system). The first peak corresponds to the oxidation of the protonated alkane and the second the oxidation of the alkane itself.

The chemical oxidation process in the acidic solution (Reaction 1) can be considered as a sum of two electrochemical reactions (Reactions 2 and 3).

$$\mathrm{RH} + \mathrm{H^+} \rightleftharpoons \mathrm{R^+} + \mathrm{H_2} \tag{1}$$

$$2\mathrm{H^+} + 2e^- \rightleftharpoons \mathrm{H_2}(E^0_{\mathrm{H^+/H_2}}) \tag{2}$$

$$\mathrm{RH} \rightleftharpoons \mathrm{R^+} + \mathrm{H^+} + 2e^-(E^0_{\mathrm{R^+/RH}}) \tag{3}$$

The chemical equilibrium (1) is characterized by the constant K_1:

$$K_1 = \frac{[\mathrm{R^+}]\, p_{\mathrm{H_2}}}{[\mathrm{RH}]\,[\mathrm{H^+}]} \quad \text{and} \quad R_A = \frac{[\mathrm{R^+}]}{[\mathrm{RH}]} = \frac{K_1}{p_{\mathrm{H_2}}} \cdot [\mathrm{H^+}]$$

With the experimental conditions set to control p_{H_2}, the acidity level, and R^+ concentration, [RH] can be evaluated from voltammetric results and K_1 can be determined.

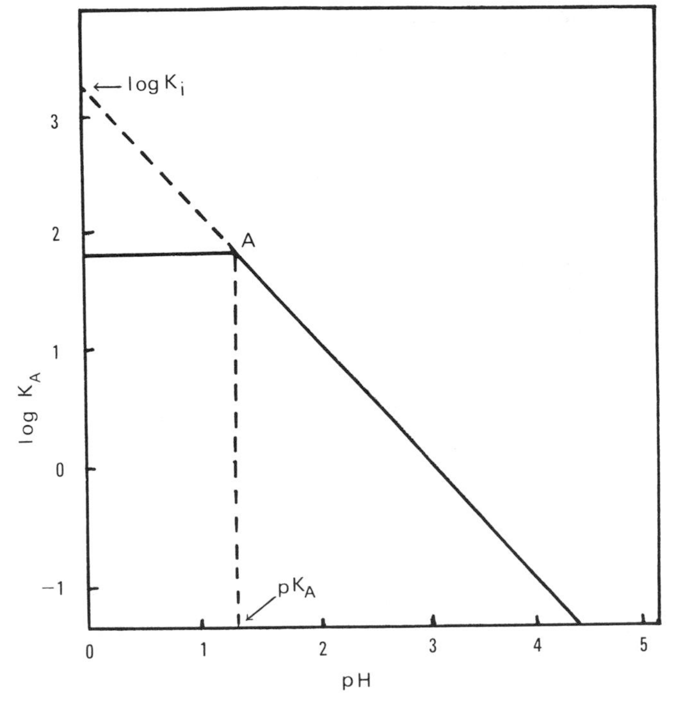

FIGURE 5.2 Variation of $M2B^+/M2BH$ as a function of pH in $HF:SbF_5$ (ref. 21).

However, when log R_A is plotted as a function of pH the authors noticed that in the case of M2BH below pH = 1.25 the ratio R_A becomes pH independent (Fig. 5.2). Thus, at higher acidity level another equilibrium must be taken into account, which is the protonation of the alkane

$$RH + H^+ \rightleftharpoons RH_2^+ \tag{4}$$

with

$$K_2 = \frac{[RH_2^+]}{[RH]\,[H^+]}$$

and the oxidation reaction becomes pH independent:

$$RH_2^+ \longrightarrow R^+ + H_2 \tag{5}$$

The acidity level at point A is representative of the basicity of the hydrocarbon if K_A is the acidity constant of the protonated alkane then $pH_A = pK_A$. On the other hand, as the redox Reactions 2 and 3 are combined in Reaction 1, the equilibrium constant K_1 is related to the standard potentials of the redox couples R^+/RH and H^+/H_2 according to the equation:

$$\frac{2.3\ RT}{2F} \log K_1 = E^0_{H^+/H_2} - E^0_{R^+/RH}$$

As K_1 can be determined experimentally and the oxidation potential of H^+/H_2 is known in the acidity range,[22] the oxidation potential $E_{R^+/RH}$ can be calculated. This allowed the authors to plot the B–D part of the potential-acidity diagram vs the H^+/H_2 system as shown in Figure 5.3 for isopentane (M_2BH). This diagram shows that oxidation of the alkane (M2BH) by H^+ gives the carbocation only at pH values below 5.7. In the stronger acids ($pH < pH_A$), it is the protonated alkane which is oxidized. At pH values higher than 5.7, oxidation of isopentane gives the alkane radical which dimerizes

$$\text{M2BH} \xrightarrow{-e^{\cdot}} \text{M2BH}^{\cdot +} \longrightarrow \text{M2B}^{\cdot} + \text{H}^+$$

$$2\text{M2B}^{\cdot} \longrightarrow (\text{M2B})_2$$

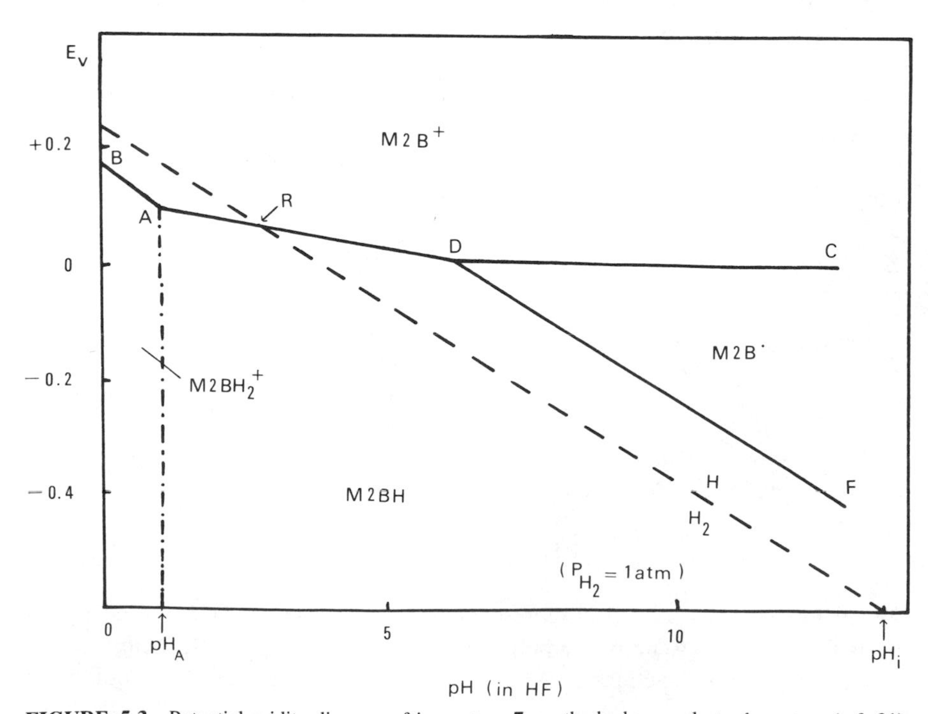

FIGURE 5.3 Potential-acidity diagram of isopentane **7** vs. the hydrogen electrode system (ref. 21).

or oxidized in a pH-independent process (DC);

$$M2B^{\cdot} \xrightarrow{-e^{\cdot}} M2B^{+}$$

An interesting point is the intersection point R at which crossover of the H^+/H_2 and the R^+/RH systems occurs. From this acidity level onwards, RH is oxidized spontaneously into R^+.

This potential acidity diagram (Pourbaix's type) has been determined for a large series of lower alkanes.[20] All of these results indicate two types of oxidation mechanism of the C—H bond: (1) oxidation of alkanes into carbenium ion at high acidity levels and (2) oxidation of alkanes into radicals at low acidity levels.

Using isopentane **7** as a reference alkane the authors calculated the Gibbs-free energy between the redox couples of various alkanes and the $M2BH/M2B^+$ couple. This leads to the standard potential of the alkane redox couples in HF (at $H_0 \sim -22.1$) (Table 5.1). The position of the redox couple R^+/RH vs the H^+/H_2 system leads to the determination of the oxidation pH or acidity level at which the Reaction (4) is quantitative.

These results are in agreement with the alkane behavior in superacid media and indicate the ease of oxidation of tertiary alkanes. However, high acidity levels are necessary for the oxidation of alkanes possessing only primary C—H bonds.

Once the alkane has been partly converted into the corresponding carbenium

TABLE 5.1 Standard Potential of Redox Couples of Alkanes in HF($H_0 \sim -22$) and Acidity Levels of Oxidation (pH_R in HF)

Alkane	$E_{R^+/RH}^0$	pH_R
C_1H	1.15	−34
C_2H	0.70	−17
C_3H	0.39	−4.3
n-C_4H	0.38	−5.6
i-C_4H	0.17	−2.0
n-C_5H	0.37	−5.2
i-C_5H	0.15	2.7
neo-C_5H	0.78	−2.1
n-C_6H	0.36	−5.0
2MPA[a]	0.14	3.0
3MBH[b]	0.16	2.4
22DMBH[c]	0.26	−1.5
23DMBH[d]	0.13	3.5

[a] 2-Methylpentane.
[b] 3-Methylbutane.
[c] 2,2-Dimethylbutane.
[d] 2,3-Dimethylbutane.

ion, the carbenium ion may undergo various reactions following intramolecular routes such as (1) skeletal rearrangement, (2) fragmentation, intermolecular routes such as (3), hydride transfer, (4) alkylation of another alkane molecule:

(1)

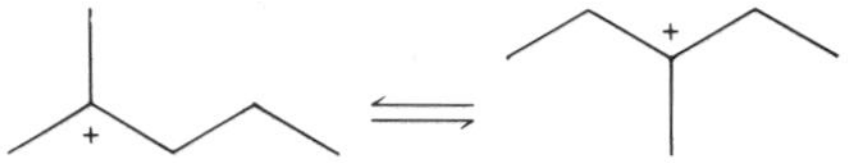

(2) 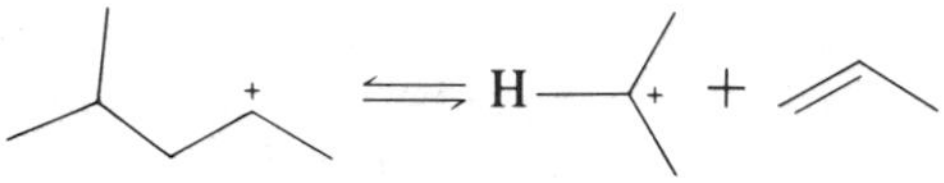 $\rightleftharpoons$ H— + +

(3) 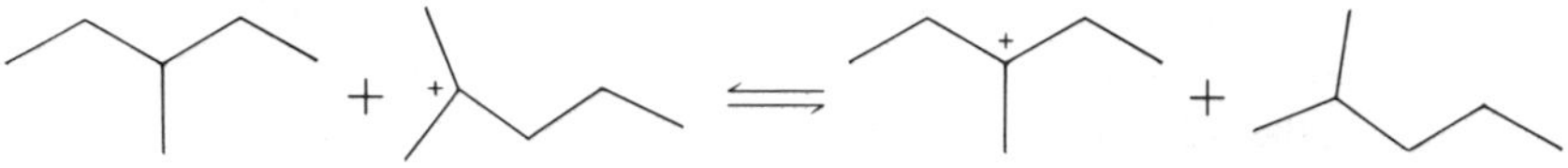 + $\rightleftharpoons$ +

(4) $CH_3^+ + CH_3—CH_3 \longrightarrow CH_3—CH_2—CH_3 + H^+$

The specificity of some of these reactions in superacid media will be discussed in the following sections.

5.1.3 Isomerization of Alkanes

Acid catalyzed isomerization of saturated hydrocarbons was first reported in 1933 by Nenitzescu and Dragan.[23] They found that when *n*-hexane was refluxed with aluminum chloride, it was converted into its branched isomers. This reaction is of major economic importance as the straight-chain $C_5—C_8$ alkanes are the main constituents of gasoline obtained by refining of the crude oil. Because the branched alkanes have a considerably higher octane number than their linear counterparts, the combustion properties of gasoline can be substantially improved by isomerization. Table 5.2 gives the octane number of a series of saturated and branched hydrocarbons.

The isomerization of *n*-butane **1** to isobutane **2** is of great importance because isobutane reacts under mild acidic conditions with olefins to give highly branched hydrocarbons in the gasoline range. A substantial number of investigations have been devoted to this isomerization reaction and a number of reviews are available.[24–26] The isomerization is an equilibrium reaction that can be catalyzed by various strong acids. In the industrial processes, aluminum chloride and chlorinated alumina are the most widely used catalysts. Whereas these catalysts become active only at temperatures above 80–100°C, superacids are capable of isomerizing alkanes at room temperature and below. The major advantage (besides energy saving) is that lower temperatures thermodynamically favor the most branched isomers (Table 5.3).

During the isomerization process of pentanes, hexanes, and heptanes, cracking of the hydrocarbon is an undesirable side reaction. The discovery that cracking can be substantially suppressed by hydrogen gas under pressure was of significant importance. In our present-day understanding, the effect of hydrogen is to quench

TABLE 5.2 Comparison of the Octane Numbers of Straight-Chain and Branched Paraffinic Hydrocarbons (Ref. 26)

Paraffinic hydrocarbons	Octane number
n-Butane	94
Isobutane	100
*n*Pentane	63
2-Methylbutane	90
2,2-Dimethylpropane (neopentane)	116
n-Hexane	32
2-Methylpentane	66
3-Methylpentane	75
2,2-Dimethylbutane (neohexane)	94.6
2,3-Dimethylbutane (bis-isopropyl)	95
n-Heptane	0
2-Methylhexane	45
3-Methylhexane	65
3-Ethylpentane	68
2,2-Dimethylpentane	80
2,3-Dimethylpentane	82
2,4-Dimethylpentane	80
3,3-Dimethylpentane	98
2,2,3-Trimethylbutane (triptane)	116
n-Octane	−19
2-Methylheptane	23.8
3-Methylheptane	35
4-Methylheptane	39
2,2,3-Trimethylpentane	105
2,2,4-Trimethylpentane	100
2,3,3-Trimethylpentane	99
2,3,4-Trimethylpentane	97

carbocationic sites through five coordinate carbocations to the related hydrocarbons, thus, decreasing the possibility of C—C bond cleavage reactions responsible for the acid-catalyzed cracking.

Isomerization of *n*-hexane **9** in superacid proceeds by the following three steps;

STEP 1. FORMATION OF THE CARBENIUM ION

9 ⇌ [H—C⁺—H five-coordinate carbonium ion] ⇌ [carbenium ion, +] + H_2

TABLE 5.3 Thermodynamic Isomerization Equilibria of Butanes, Pentanes, and Hexanes at Various Temperatures (Ref. 26).

	Temperature in °C			
Hydrocarbon	−6	38	66	93
Butanes				
Isobutane	85	75	65	57
n-Butane	15	25	35	43
Pentanes				
Isopentane	95	85	78	71
n-Pentane	5	15	22	29
2,2-Dimethylpropane (neopentane)	0	0	0	0
Hexanes				
2,2-Dimethylbutane (neohexane)	57	38	28	21
2,3-Dimethylbutane	11	10	9	9
2-Methylpentane	20	28	34	36
3-Methylpentane	8	13	15	17
n-Hexane	4	11	14	17

STEP 2. ISOMERIZATION OF THE CARBENIUM ION VIA HYDRIDE SHIFTS, ALKIDE SHIFTS, AND PROTONATED CYCLOPROPANE (FOR THE BRANCHING STEP)

$\rightleftharpoons$ H $\rightleftharpoons$ $\rightleftharpoons$ Isomers

STEP 3. HYDRIDE TRANSFER FROM THE ALKANE TO THE INCIPIENT CARBENIUM ION

iso-R^+ + $\rightleftharpoons$ + *iso*-RH

Whereas step 1 is stoichiometric, steps 2 and 3 form a catalytic cycle involving the continuous generation of carbenium ions via hydride transfer from a new hydrocarbon molecule (step 3) and isomerization of the corresponding carbenium ion (step 2). This catalytic cycle is controlled by two kinetic and two thermodynamic parameters that can help orient the isomer distribution depending on the reaction conditions. Step 2 is kinetically controlled by the relative rates of hydrogen shifts, alkyl shifts, and protonated cyclopropane formation and it is thermodynamically controlled by the relative stabilities of the secondary and tertiary ions. (This area has been thoroughly studied; see Chapter 3). Step 3, however, is kinetically controlled by the hydride transfer from excess of the starting hydrocarbon and by the relative thermodynamic stability of the various hydrocarbon isomers.

For these reasons, the outcome of the reaction will be very different depending on which thermodynamic or kinetic factor will be favored. In the presence of excess hydrocarbon in equilibrium with the catalytic phase and long contact times, the thermodynamic hydrocarbon isomer distribution is attained. However, in the presence of a large excess of acid, the product will reflect the thermodynamic stability of the intermediate carbenium ions (which, of course, is different from that of hydrocarbons) if rapid hydride transfer or quenching can be achieved. Torck and co-workers[16,27] have shown that the limiting step, in the isomerization of *n*-hexane **9** and *n*-pentane **6** with the $HF:SbF_5$ acid catalyst, is the hydride transfer with sufficient contact in a batch reactor, as indicated by the thermodynamic isomer distribution of C_6 isomers. Figure 5.4 shows the isomer distribution vs reaction time of *n*-hexane at 20°C.

It shows that 2-methylpentane (2MP), 3-methylpentane (3MP), and 2,3-dimethylbutane (2,3DMB) appear simultaneously and their concentration reaches a maximum at the same time. Their relative ratio stays the same at all times and is identical with the thermodynamically calculated one. 2,2-Dimethylbutane (2,2DMB), however, appears much more slowly. This is depicted in the following reaction scheme:

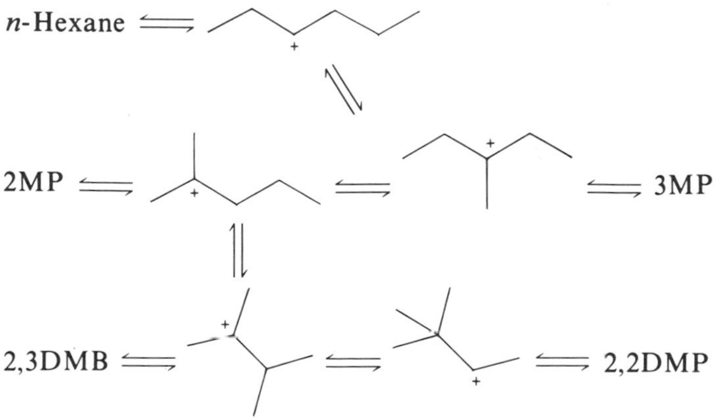

n-Hexane ⇌

2MP ⇌ ⇌ ⇌ 3MP

2,3DMB ⇌ ⇌ ⇌ 2,2DMP

The detailed kinetic study of this systems shows that after the initial period during which the catalytic phase is formed, the experimental rate constant of isomerization K_1 becomes

$$K_1 = K_{sp} \cdot [R^+] \cdot \beta \frac{Nc}{No}$$

in which β is the partition coefficient of the paraffin between the hydrocarbon and the catalytic phase, Nc is the number of moles in the catalytic phase, No is the total number of moles of paraffin at the start, $[R^+]$ is the concentration of carbenium ions, and K_{sp} is a temperature-dependent constant. Under external hydrogen pressure, the rate of isomerization slows down in agreement with the corresponding

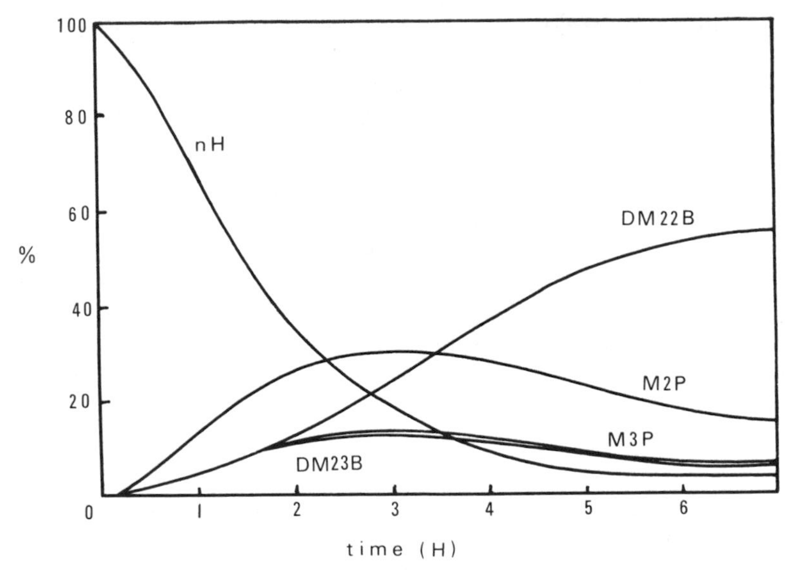

FIGURE 5.4 Distribution of *n*-hexane isomer vs. reaction time at 20° (ref. 16).

reduction of the carbenium ions into alkane (reverse step). This effect was found more pronounced for weaker superacids, SbF_5 content < 2 mol · liter^{-1}

The isomerization of *n*-pentane in superacids of the type $R_FSO_3H:SbF_5$ ($R_F = C_nF_{2n+1}$) has been investigated by Commeyras and co-workers.[28] The influence of parameters such as acidity (A), hydrocarbon concentration (B), nature of the perfluoroalkyl group (C), total pressure (D), hydrogen pressure (E), temperature (F), and agitation has been studied. Only A, C, E, and F have been found to have an influence on the isomerization reaction in accordance with such reactions in the $HF:SbF_5$ system.

In weaker superacids such as neat CF_3SO_3H, alkanes that have no tertiary hydrogen are isomerized only very slowly, as the acid is not strong enough to hydride abstract to form the initial carbocation. This lack of reactivity can be overcome by introducing initiator carbenium ions in the medium to start the catalytic process. For this purpose, alkenes may be added, which are directly converted into their corresponding carbenium ions by protonation, or alternatively the alkane may be electrochemically oxidized (anodic oxidation). Both methods are useful to initiate isomerization and cracking reactions. The latter method has been studied by Commeyras and co-workers[29] who came to the conclusion that it was much more favorable for side reactions such as condensation and cracking of alkanes than for simple isomerization reactions. Considering the low superacidity of the medium used, the result is in good agreement with the predictions of the acidity-potential diagrams described earlier.

$$RH \xrightarrow{-e^{\cdot}} RH^{\cdot+} \xrightarrow{-H^+} R^{\cdot} \xrightarrow{-e^{\cdot}} R^+$$

Recently, Olah[30] has developed a method wherein natural gas liquids containing saturated straight-chain hydrocarbons can be conveniently upgraded to highly branched hydrocarbons (gasoline upgrading) using $HF:BF_3$ catalyst. The addition of small amount of olefins, preferably butenes, helped the reaction rate. This can be readily explained by the formation of alkyl fluorides (HF addition to olefins), whereby an equilibrium concentration of cations is maintained in the system during the upgrading reaction. The gasoline upgrading process is also improved in the presence of hydrogen gas, which helps to suppress side reactions such as cracking and disproportionation and minimize the amount of hydrocarbon products entering the catalyst phase of the reaction mixture. The advantage of the above method is that the catalysts $HF:BF_3$ (being gases at ambient temperatures) can easily be recovered and recycled.

Olah has also found HSO_3F and related superacids as efficient catalysts for the isomerization of *n*-butane to isobutane.[31]

The difficulties encountered in handling liquid superacids and the need for product separation from catalyst in batch processes have stimulated research in the isomerization of alkanes over solid superacids. The isomerization of 2-methyl- and 3-methylpentane and 2,3-dimethylbutane, using SbF_5-intercalated graphite as a catalyst, has been studied in a continuous flow system.[32] The isomerization reaction carried out at −30°C, −17°C, and room temperature shows an unusually high activity of this catalyst. At −17°C conversions of over 50% with a selectivity over 99% in nonbranching isomerization could be achieved. For comparison, to obtain the same conversion with acidic zeolites, temperatures as high as 180°C are necessary. At room temperature, the activity of the catalyst is higher, but branching and cracking reactions compete. The study of the skeletal rearrangement of ^{13}C-labeled 2MP, 3MP, and 2,3DMB has been carried out over the same catalyst under similar conditions.[33] Despite the high conversion, very little label scrambling occurs. The isomerization process involves only intramolecular rearrangement of the hexyl ions and can be fully described in terms of 1,2 hydride, methide, or ethide shifts and protonated cyclopropane intermediates. The advantage of ^{13}C tracer technique is to differentiate between mono- and multimolecular processes and to remove the degeneracy of otherwise indistinguishable pathways. Combined with a flow system, it allows the investigation of the initial steps under kinetic control. For example, in the 2MP → 3MP isomerization process, it permits distinguishing between the methyl and the ethyl shift and estimating the relative rates of these two alkyde shifts.

When 3-methylpentane 3-^{13}C is isomerized to 2-methylpentane, the label distribution shows that the isomerization cannot be explained by a simple 1,2-methide shift:

4% 70% 15% 15%

30% of the 2-methylpentane has the ^{13}C label in a position that can be best explained by the ethyl shift. The recovered 3-methylpentane (96%) also shows a very large degree of internal ethyl shift.

96% recovered → 89% + 11%

The rate of ethyl shift in this reaction is three times as high as the apparent rate of conversion of 3-methyl- to 2-methylpentane. A simple calculation shows that the ethyl shift is about four times faster than the methyl shift in this isomerization process. The occurrence of a competitive ethyl shift in the 3-methyl-3-pentenium ion has been studied by Olah and co-workers under stable ion condition. More recently, Sommer and co-workers[34] have shown by using the deuterium-labeled 3-methyl pentenium ion that the ratio of ethyl:methyl shift was the same both in Magic Acid as well as in the presence of solid superacid. SbF_5-intercalated graphite has also been shown to exhibit a 99% selectivity in *n*-pentane skeletal isomerization at room temperature in a batch process[35] (48% conversion after 2 h). The selectivity in isomerization of hexane was much lower because of the competing cracking process described earlier. However, in the presence of HF, the selectivity of the catalyst substantially improves as shown by the skeletal isomerization of *n*-hexane (96% selectivity with 86% conversion after 2 h). The major drawback in the extended use of this catalyst system is its relatively rapid deactivation (studied by Heinerman and Gaaf[36]).

The isomerization of cycloalkanes over SbF_5-intercalated graphite can be achieved at room temperature without the usual ring opening and cracking reactions which occur at higher temperatures and lower acidities.[37] In the presence of excess hydrocarbon after several hours, the thermodynamic equilibrium is reached for the isomers. Interconversion between cyclohexane and methylcyclopentane yields the thermodynamic equilibrium mixture.

11 or **10** → **11** + **10**

The thermodynamic ratio for the neutral hydrocarbon isomerization is very different as compared with the isomerization of the corresponding ions. The large energy difference (> 10 kcal) between the secondary cyclohexyl cation **17** and the tertiary methylcyclopentyl ion **12** means that in the presence of excess superacid only the latter can be observed.[6]

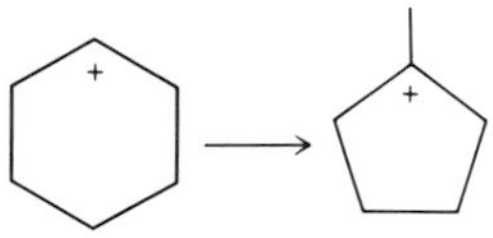

17 12

When *cis*- and *trans*-decalins **19** are treated with the same solid superacid at 0°C, the thermodynamic equilibrium is rapidly achieved.

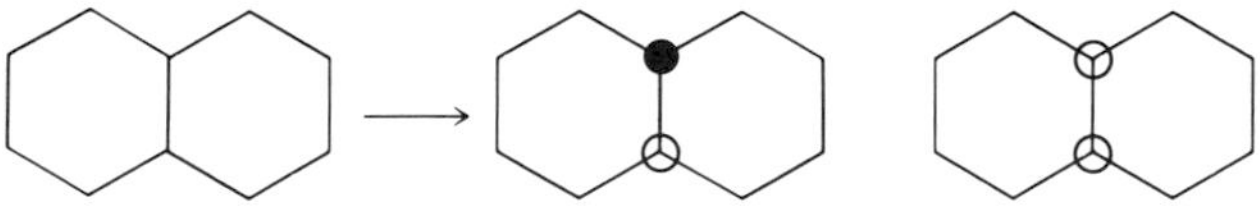

19 **19**-*trans* 98% **19**-*cis* 2%

In the reaction, decalin **19** serves as the solvent, the substrate, and the hydride donor. When the equilibrium is reached, the hydrocarbon can be separated from the catalyst by simple filtration. Perhydroindane **20** was also isomerized under mild condition at 0°C to the thermodynamic equilibrium mixture: 62% *trans*–38% *cis*.[38]

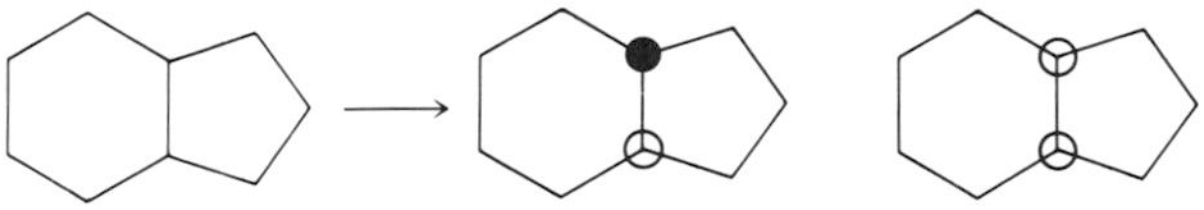

20 62%, **20**-*trans* 38%, **20**-*cis*

The potential of other solid superacid catalysts, such as Lewis acid-treated metal oxides for the skeletal isomerization of hydrocarbons, has been studied in a number of cases. The reaction of butane **1** with $SbF_5:SiO_2:TiO_2$ gave the highest conversion forming C_3, *iso*-C_4, *iso*-C_5, and traces of higher alkanes. $TiO_2:SbF_5$, on the other hand, gave the highest selectivity for skeletal isomerization of butane. With $SbF_5:Al_2O_3$, however, the conversions were very low (see Tables 5.4 and 5.5).[39,40]

Similarly, isomerization of pentane and 2-methylbutane over a number of SbF_5-treated metal oxides has been investigated. $TiO_2:ZrO_2:SbF_5$ system was the most reactive and at the maximum conversion, the selectivity for skeletal isomerization was found to be ca. 100%.[41] Table 5.5 summarizes the effect of temperature on desorption of SbF_5 out of the catalyst and shows that with increasing temperature the weight % of the retained Lewis acid decreases.

A comparison of the reactivity of SbF_5-treated metal oxides with that of HSO_3F-treated catalysts showed that the former is by far the better catalyst for reaction of

TABLE 5.4 Reaction of Butane and Isobutane over Solid Superacids at 20°C

Catalyst	Reactant	Time (h)	Distribution of Products (%)							
			Propane	Butane	Isobutane	Pentane	Isopentane	2,2-DMB[b]	Hexanes[c]	Heptanes
$SiO_2:Al_2O_3$	Butane	720	0	100	0	0	0	0	0	0
$SbF_5:SiO_2$		280	6.9	25.2	54.8	1.0	6.8	2.9	2.1	0
$SbF_5:TiO_2$		280	6.4	17.1	59.1	1.1	13.4	2.0	1.0	0
$SbF_5:Al_2O_3$		280	0.14	77.2	21.1	0.13	1.3	0.09	0	0
$SbF_5:TiO_2:SiO_2$		280	21.9	13.1	58.1	0.6	4.8	1.0	0.5	0
$SbF_5:SiO_2:Al_2O_3$		280	4.4	32.7	47.7	1.1	8.1	3.7	2.6	0
$SbF_5:SiO_2:Al_2O_3$		20	1.8	49.7	41.0	0.6	4.8	1.4	0.7	0
$SiO_2:Al_2O_3$	Isobutane	600	0	0	100	0	0	0	0	0
$SiO_2:Al_2O_3$		4[a]	0	0	99.9	0	<0.1	0	0	0
$TiO_2:SiO_2$		720	0	0	100	0	0	0	0	0
SbF_5		280	2.9	0.16	96.4	0	0.49	0	0	0
$SbF_5:SiO_2$		280	7.0	11.8	66.8	1.6	5.6	4.2	2.7	0.3
$SbF_5:Al_2O_3$		280	0.17	0.93	97.9	0.09	0.86	0	0	0
$SbF_5:SiO_2:Al_2O_3$		280	6.8	11.0	65.0	1.2	9.6	3.7	2.2	0.4
$SbF_5:SiO_2:Al_2O_3$		2	0.07	0.4	94.7	3.5	0.4	0.4	0.5	0

[a]Reacted at 200°C.

[b]2,2-Dimethylbutane.

[c]2,3-Dimethylbutane, 2-methylpentane, 3-methylpentane.

TABLE 5.5 Amount of SbF_5 Retained in Catalyst (wt%)

Catalyst[a]	SbF_5 Ads. Temp/°C	Evacuation Temperature			
		0°C	30°C	50°C	100°C
$SbF_5:SiO_2:Al_2O_3$	0	49.7	37.6	27.7	
	10	53.2	40.9	34.5	30.7
	30		34.1	32.2	
	100				17.2
$SbF_5:TiO_2:ZrO_2$	0	49.9	30.0	32.8	
	10		33.5	27.2	
	30			29.8	
	100				30.9
$SbF_5:SiO_2$	0	33.1	18.8	12.8	
	10			18.2	
	30			17.2	
$SbF_5:Al_2O_3$	0	37.1			
$SbF_5:TiO_2$	0	23.1			
$SbF_5:ZrO_2$	0	24.8			

[a]Adsorption-desorption cycle in SbF_5 treatment was repeated four times.

alkanes at room temperature, although the HSO_3F-treated catalyst showed some potential for isomerization of 1-butane.[42]

$SbF_5:SiO_2:Al_2O_3$ has been used to isomerize a series of alkanes at or below room temperature. Methylcyclopentane, cyclohexane, propane, butane, 2-methylpropane, and pentane all reacted at room temperature, whereas methane, ethane, and 2,2-dimethylpropane could not be activated.[42]

An ir study of $SbF_5:Al_2O_3$ after addition of pyridine to the catalyst shows an absorption at 1460 cm^{-1}, which is assigned to a pyridine coordinated with the Lewis acid site, and another at 1540 cm^{-1}, which is attributed to the pyridinium ion resulting from the protonation of pyridine by the Brønsted acid sites. When the catalyst is heated up to 300°C, the ir band of the pyridinium ion disappears, whereas the absorption for the Lewis acid is still present. The fact that the catalyst is still active for butane isomerization suggests that the Lewis acid sites are the active sites for the catalysis.[42]

Methane produced in the reaction of butane with $SbF_5:SiO_2:Al_2O_3$ did not contain any deuterium when the surface OH groups of the catalyst were replaced by OD. Furthermore, no hydrogen evolution could be detected in these reactions. Two alternative mechanisms have been suggested: (1) The reactions are initiated by hydride abstraction from the alkane by the Lewis acid to form a carbenium ion, and not by protonation of the C—C bond of butane. (2) Butane is protonated by the Brønsted acid to form a carbonium ion intermediate and the hydrogen formed is either used up to reduce SbF_5 or it loosely remains bound to the ion during the isomerization process.[42]

The isomerization of a large number of C_{10} hydrocarbons under strongly acidic conditions gives the unusually stable isomer adamantane **21.** The first such isomerization was reported by Schleyer in 1957.[43a] During a study of the facile aluminum chloride-catylized *endo*:*exo* isomerization of tetrahydrodicyclopentadiene **22,** difficulty was often encountered with a highly crystalline material that often clogged distillation heads. This crystalline material was found to be adamantane **21.** Adamantane **21** can be prepared from a variety of C_{10} precursors and involves a series of hydride and alkyde shifts (Scheme 1). The mechanism of the reaction has been reviewed in detail.[43b] Fluoroantimonic acid (HF:SbF_5) very effectively isomerizes tetrahydrodicyclopentadiene **22** into adamantane **21.**[43c]

Another example of superacid-catalyzed formation of an unusually stable highly

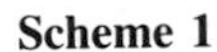

Scheme 1

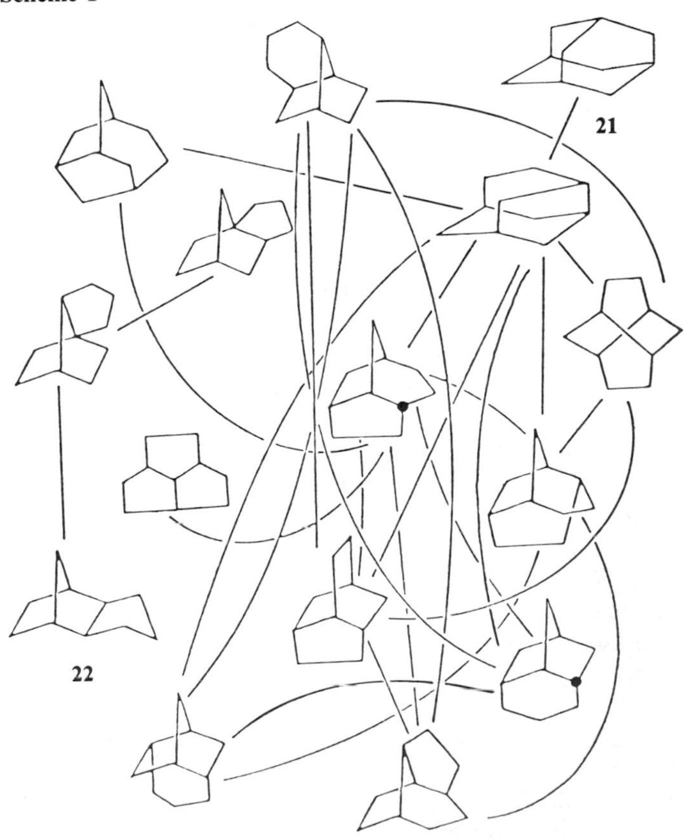

Rearrangement Map of C_{10} Hydrocarbons to Adamantane[43b]

symmetric hydrocarbon has been provided by Paquette[43d] in the synthesis of 1,16-dimethyl dodecahedrane **23.**

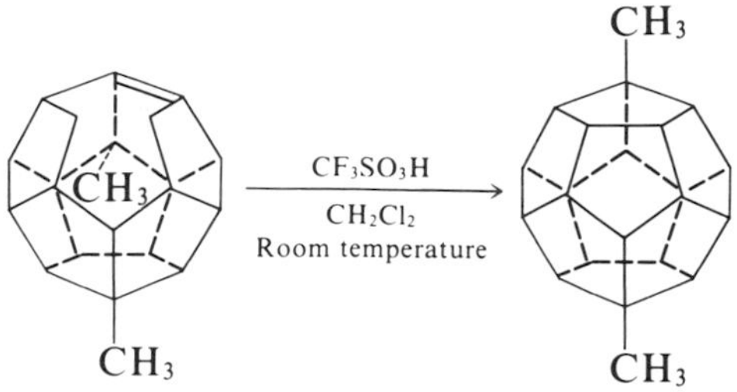
CH3
CF3SO3H
CH2Cl2
Room temperature
CH3
CH3

23

Two mechanistic schemes (**2** and **3**) via carbocationic intermediates have been proposed by Paquette.[43d]

Scheme 2

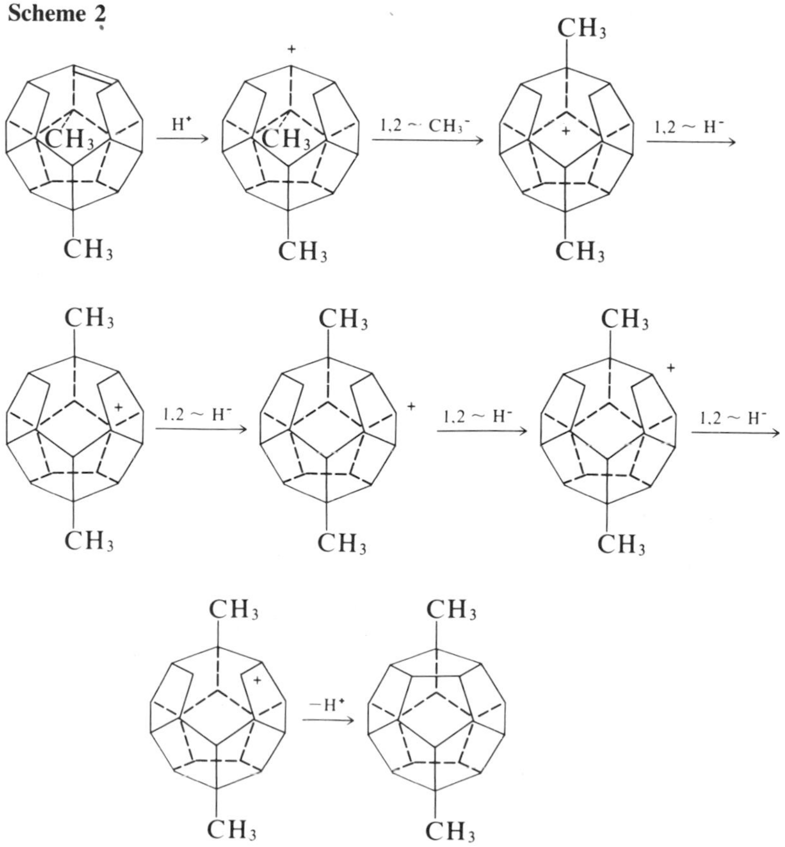
CH3
H+
1,2 ~ CH3-
1,2 ~ H-
1,2 ~ H-
1,2 ~ H-
1,2 ~ H-
−H+
CH3

23

Scheme 3

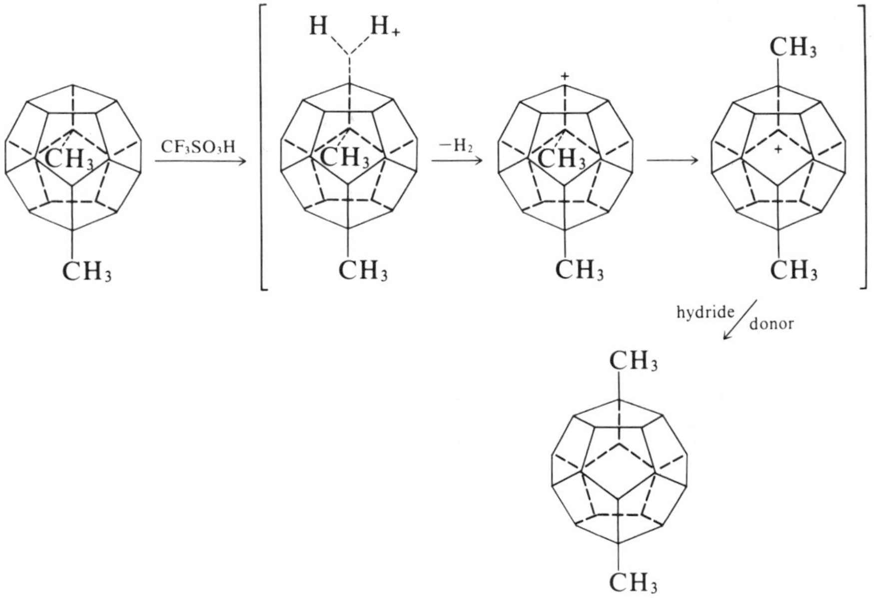

CF_3SO_3H $-H_2$ hydride donor

23

Scheme 2 involves Wagner-Meerwein shifts of the methyl group prior to cyclization followed by hydride shift to a number of cationic intermediates. The second scheme 3 depicts ring closure before methyl migration. The first step involves protolysis of the C—H bond next to the methyl-bearing carbon. The corresponding ion can then rearrange by a 1,2-methyl shift and yield 1,16-dimethyl-dodecahedrane **23** by hydride abstraction from a hydride donor.

5.1.4 Cleavage Reactions (β-Cleavage vs C—C Bond Protolysis)

The reduction in molecular weight of various fractions of crude oil is an important operation in petroleum chemistry. The process is called cracking. Catalytic cracking is usually achieved by passing the hydrocarbons over a metallic or acidic catalyst, such as crystalline zeolites at about 400–600°C. The molecular weight reduction involves carbocationic intermediates and the mechanism is based on the β-scission of carbenium ions

$$R{-}\overset{|}{\underset{|}{C}}{-}\overset{+}{\underset{|}{C}}{-}R' \longrightarrow R^+ + \;>C{=}C<^{R'}$$

The main goal of catalytic cracking is to upgrade higher boiling oils, which, through this process, yield lower hydrocarbons in the gasoline range.[44,45]

Historically, the first cracking catalyst used was aluminum trichloride. With the development of heterogeneous solid and supported acid catalysts, the use of $AlCl_3$ was soon superceded, since its activity was mainly due to the ability to bring about acid-catalyzed cleavage reactions.

The development of highly acidic superacid catalysts in the 1960s again focused attention on acid-catalyzed cracking reactions. $HSO_3F:SbF_5$, trade-named Magic Acid, derived its name (see Section 2.3.3) due to its remarkable ability to cleave higher-molecular-weight hydrocarbons, such as paraffin wax, to lower-molecular-weight components, preferentially C_4 and other branched isomers.

As a model for cracking of alkanes, the reaction of 2-methylpentane over SbF_5-intercalated graphite has been studied in a flow system, the hydrocarbon being diluted in a hydrogen stream.[32] A careful study of the product distribution vs time on stream showed that propane was the initial cracking product whereas isobutane and isopentane (as major cracking products) appear only later (Fig. 5.5).

This result can only be explained by the β-scission of the trivalent 2-methyl-4-pentenyl ion as the initial step in the cracking process. Based on this and on the product distribution vs time profile, a general Scheme 4 for the isomerization and cracking process of the methylpentanes has been proposed:[32]

The propene which is formed in the β-scission step never appears as a reaction product because it is alkylated immediately under the superacid condition by a C_6^+ carbenium ion, forming a C_9^+ carbocation which is easily cracked to form a C_4^+ or C_5^+ ion and the corresponding C_4 or C_5 alkene. The alkenes are further alkylated by a C_6^+ carbenium ion in a cyclic process of alkylation and cracking reactions. The C_4^+ or C_5^+ ions also give the corresponding alkanes (isobutane and isopentane) by hydride transfer from the starting methylpentane. This scheme, which occurs under superacidic conditions, is at variance with the scheme that was proposed for the cracking of C_6 alkanes under less acidic conditions.[29]

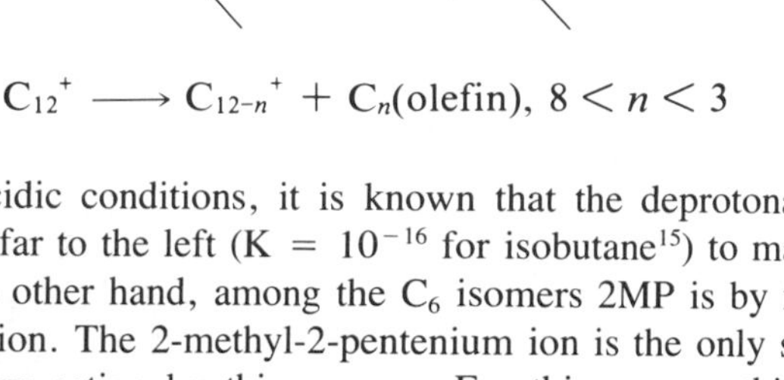

$$+\ H^+$$

$$\longrightarrow C_{12}^+$$

$$C_{12}^+ \longrightarrow C_{12-n}^+ + C_n(\text{olefin}),\ 8 < n < 3$$

Under superacidic conditions, it is known that the deprotonation (first type) equilibria lie too far to the left ($K = 10^{-16}$ for isobutane[15]) to make this pathway plausible. On the other hand, among the C_6 isomers 2MP is by far the easiest to cleave by β-scission. The 2-methyl-2-pentenium ion is the only species that does not give a primary cation by this process. For this reason, this ion is the key intermediate in the isomerization cracking reaction of C_6 alkanes.

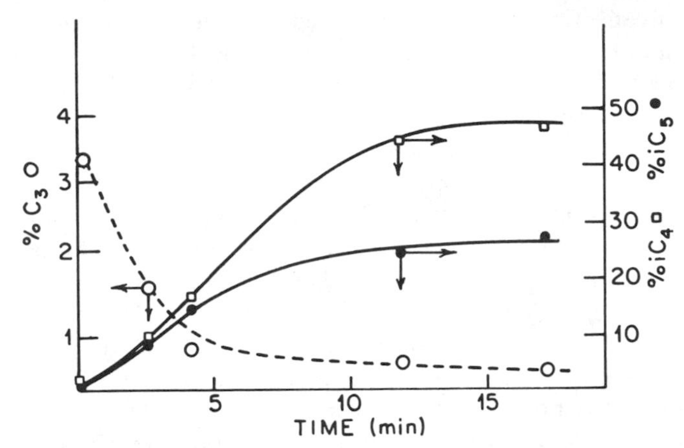

FIGURE 5.5 Distribution of the cracking products (mass%) of 2-methylpentane vs. time on stream at 20°C (ref. 32a).

Under superacid conditions, β-scission is not the only pathway by which hydrocarbons are cleaved. The C—C bond can also be cleaved by protolysis.

$$\rangle C{-}C\langle + H^+ \longrightarrow \left[\rangle C \overset{H}{\cdots} \overset{+}{} C\langle \right] \longrightarrow \overset{|}{C^+} + H{-}\overset{|}{\underset{|}{C}}{-}$$

The protolysis under superacid conditions has been studied independently by Olah et al.[46] and Brouwer et al.[4,47] The carbon–carbon cleavage in neopentane yielding methane and the *t*-butyl ion **1** occurs by a mechanism different from the usual β-scission of carbenium ions:

$$CH_3{-}\overset{CH_3}{\underset{CH_3}{C}}{-}CH_3 + H^+ \longrightarrow \left[CH_3{-}\overset{CH_3}{\underset{CH_3}{C}}\cdots\langle{}^{H}_{CH_3} \right]^+ \longrightarrow \underset{\mathbf{4}}{(CH_3)_3C^+} + CH_4$$

The protolysis occurs following the direct protonation of the σ-bond providing evidence for the σ-basicity of hydrocarbons. Under slightly different conditions, protolysis of a C—H occurs yielding rearranged *t*-pentyl ion **8** (*t*-amyl cation).

$$(CH_3)_3C{-}CH_3 + H^+ \longrightarrow \left[(CH_3)_3C{-}CH_2\cdots\langle{}^{H}_{H} \right]^+ \longrightarrow$$

$$\underset{\mathbf{8}}{CH_3{-}\overset{CH_3}{\underset{+}{C}}{-}CH_2CH_3} + H_2$$

In cycloalkanes, the C—C bond cleavage leads to ring opening.[6]

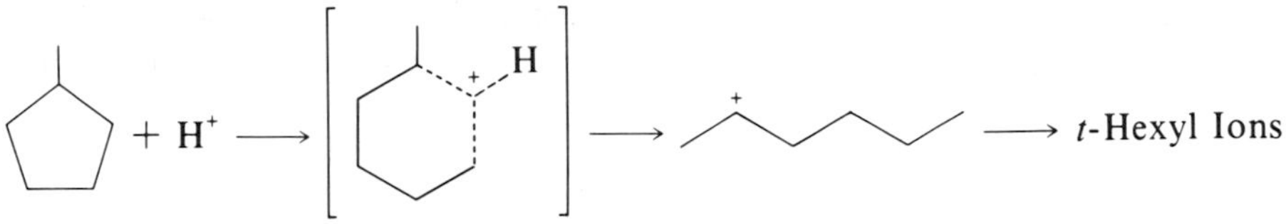

$$+ H^+ \longrightarrow [\ ] \longrightarrow \ \longrightarrow t\text{-Hexyl Ions}$$

This reaction is much faster than the carbon–carbon bond cleavage in neopentane, despite the initial formation of secondary carbenium ions. Norbornane is also cleaved in a fast reaction yielding substituted cyclopentyl ions. Thus, protonation of alkanes induces cleavage of the molecule by two competitive ways: (1) protolysis of a C—H

Scheme 4

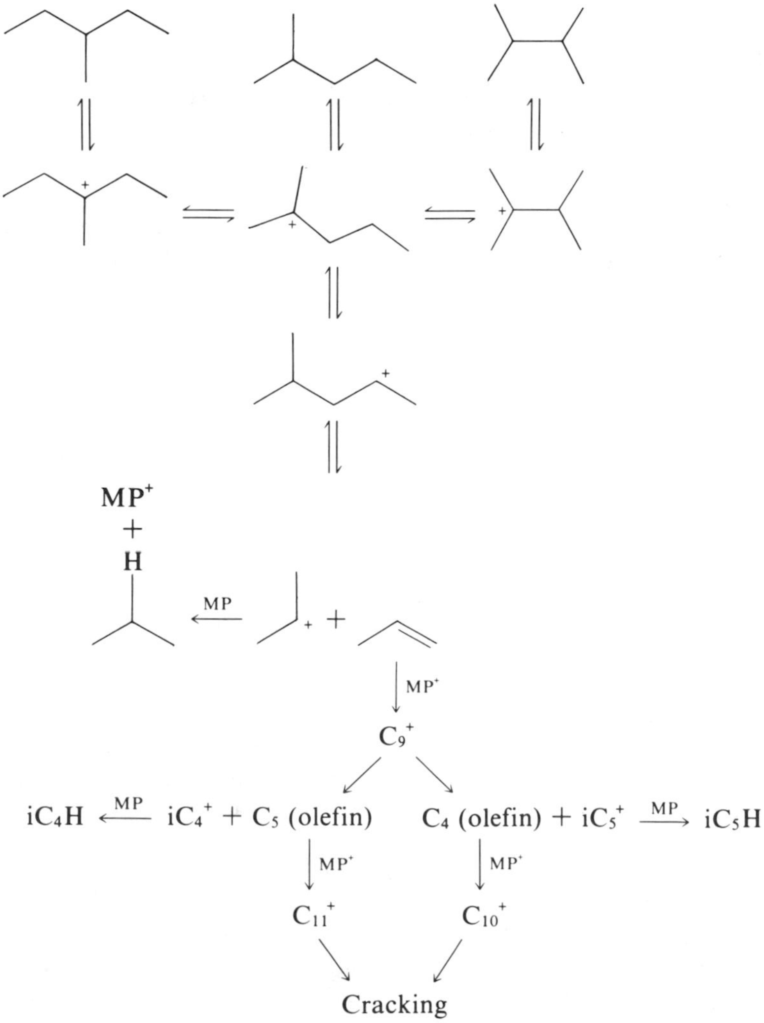

MP^+
+
H

MP

$\xrightarrow{MP^+}$

C_9^+

$iC_4H \xleftarrow{MP} iC_4^+ + C_5$ (olefin) $\quad C_4$ (olefin) $+ iC_5^+ \xrightarrow{MP} iC_5H$

$\downarrow MP^+ \quad \downarrow MP^+$

$C_{11}^+ \quad C_{10}^+$

Cracking

bond followed by β-scission of the carbenium ions and (2) direct protolysis of a C—C bond yielding a lower-molecular-weight alkane and a lower-molecular-weight carbenium ion. This reaction, which is of economic importance in the upgrading of higher boiling petroleum fractions to gasoline, has also been shown applicable to coal depolymerization and hydroliquefaction processes.[48] The cleavage of selected model compounds representing coal structural units in the presence of HF, BF_3, and under hydrogen pressure has been studied by Olah and co-workers.[48a] Bituminous coal (Illinois No. 6) could be pyridine-solubilized to the extent of 90% by treating it with superacidic $HF:BF_3$ catalyst in the presence of hydrogen gas at 105°C for 4 h. Under somewhat more elevated temperatures (150–170°C), cyclohexane extractibility of up to 22% and distillability of up to 28% is achieved. Addition of hydrogen donor solvents such as isopentane has been shown to improve the efficiency of coal conversion to cyclohexane-soluble products. The initial "depolymerization" of coal involves various protolytic cleavage reactions involving those of C—C bonds.

5.1.5 Alkylation of Alkanes and Oligocondensation of Lower Alkanes

The alkylation of alkanes by olefins, from a mechanistic point of view, must be considered as the alkylation by the carbenium ion formed by the protonation of the olefin. The well-known acid-catalyzed isobutane–isobutylene reaction demonstrates the mechanism rather well.

$$\text{isobutylene} + H^+ \rightleftharpoons t\text{-}C_4H_9^+$$

$$t\text{-}C_4H_9^+ + \text{isobutylene} \rightleftharpoons C_8H_{17}^+$$

$$C_8H_{17}^+ + H\text{-}C(CH_3)_3 \rightleftharpoons C_8H_{18} + t\text{-}C_4H_9^+$$

As is apparent in the last step, isobutane is not alkylated but transfers a hydride to the $C_8{}^+$ carbocation before being used up in the middle step as the electrophilic reagent (*t*-butyl cation **4**). The direct alkylation of isobutane by an incipient *t*-butyl cation would yield 2,2,3,3-tetramethylbutane **14**,[46] which, indeed was observed in small amounts in the reaction of *t*-butyl cation with isobutane under stable ion conditions at low temperatures (*vide infra*).

The alkylating ability of methyl and ethyl fluoride–antimony pentafluoride complexes have been investigated by Olah and his group[49,50] showing the extraordinary reactivity of these systems. Self-condensation was observed as well as alkane alkylation. When $CH_3F:SbF_5$ was reacted with excess of CH_3F at 0°C at first only an exchanging complex was observed in the ^{1}H-nmr spectrum. After 0.5 h, the starting material was converted into the *t*-butyl ion **4.** Similar reactions were ob-

$$FCH_3 + CH_3F \rightarrow SbF_5 \longrightarrow \left[FCH_2\text{---}\langle^{H}_{CH_3} \right]^+ SbF_6^- \longrightarrow FCH_2\text{—}CH_3 + H^+$$

$$FCH_2CH_3 + CH_3CH_2F \rightarrow SbF_5 \rightarrow\rightarrow FCH_2CH_2CH_2CH_3 \rightarrow\rightarrow \underset{\mathbf{4}}{(CH_3)_3C^+}$$

served with the ethyl fluoride–antimony pentafluoride complex. When the complex was treated with isobutane or isopentane direct alkylation products were observed.

$$CH_3\text{—}\overset{CH_3}{\underset{CH_3}{C}}\text{—}H + CH_3CH_2F \rightarrow SbF_5 \longrightarrow (CH_3)_3C\ CH_2CH_3$$

To improve the understanding of these alkane alkylation reactions, Olah and his group carried out experiments involving the alkylation of the lower alkanes by stable carbenium ions under controlled superacidic stable ion conditions.[46,51,52]

The σ-donor ability of the C—C and C—H bonds in alkanes was demonstrated from a variety of examples. The order of reactivity of single bonds was found to be: tertiary C—H > C—C > secondary C—H $\gg$ primary C—H, although various specific factors such as steric hindrance can influence the relative reactivities.

$$R\text{—}H + R'^+ \longrightarrow \left[R\text{---}\overset{H}{\vdots}\text{---}R' \right]^+ \longrightarrow R\text{—}R' + H^+$$

Typical alkylation reactions are those of propane, isobutane, and *n*-butane by the *t*-butyl or *sec*-butyl ion. These systems are somewhat interconvertible by competing hydride transfer and rearrangement of the carbenium ions. The reactions were carried out using alkyl carbenium ion hexafluoroantimonate salts prepared from the corresponding halides and antimony pentafluoride in sulfuryl chloride fluoride solution and treating them in the same solvent with alkanes. The reagents were mixed at -78°C warmed up to -20°C and quenched with ice water before analysis. The intermolecular hydride transfer between tertiary and secondary carbenium ions and alkanes is generally much faster than the alkylation reaction. Consequently the alkylation products are also those derived from the new alkanes and carbenium ions formed in the hydride transfer reaction.

Propylation of propane by the isopropyl cation **24,** for example, gives a significant amount (26% of the C_6 fraction) of the primary alkylation product:

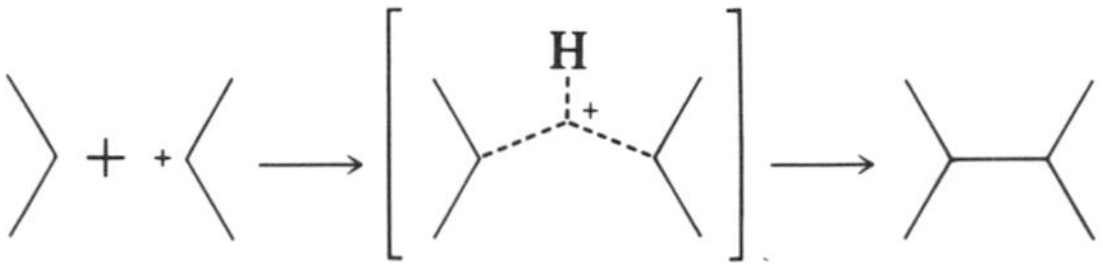

24

The C_6 isomer distribution, 2-methylpentane (28%), 3-methylpentane (14%), and *n*-hexane (32%) is very far from thermodynamic equilibrium, and the presence of these isomers indicates that not only isopropyl cation **24** but also *n*-propyl cation **25** are involved as intermediates (as shown by $^{13}C_2$—$^{13}C_1$ scrambling in the stable ion).[53]

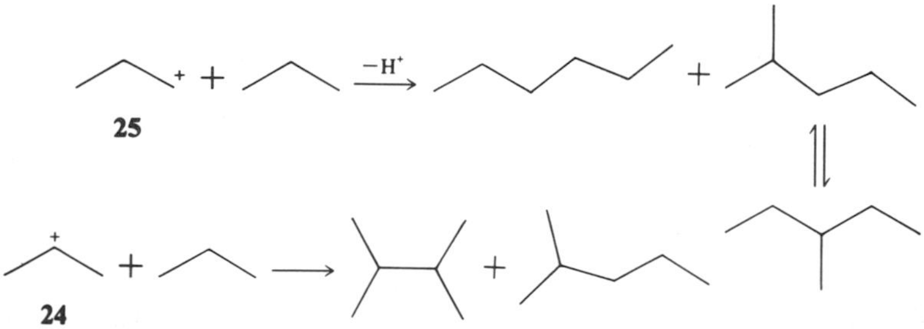

$-H^+$

25

24

The strong competition between alkylation and hydride transfer appears in the alkylation reaction of propane by butyl cations, or butanes by the propyl cation. The amount of C_7 alkylation products is rather low. This point is particularly emphasized in the reaction of propane by the *t*-butyl cation which yields only 10% of heptanes. In the interaction of propyl cation with isobutane the main reaction is hydride transfer from the isobutane to the propyl ion followed by alkylation of the propanc by thc propyl ions.

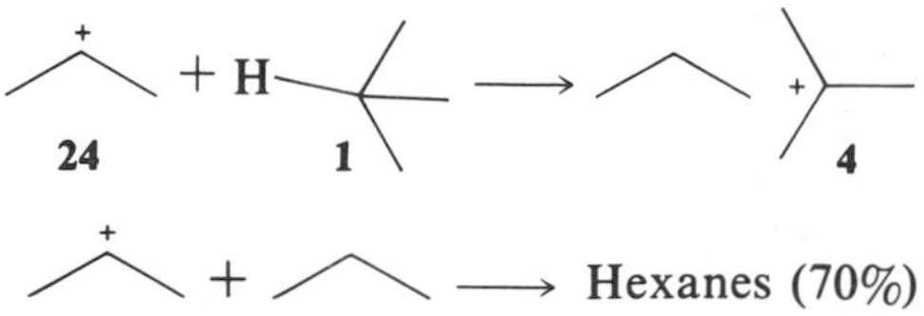

24 **1** **4**

⟶ Hexanes (70%)

Even the alkylation of isobutane by the *t*-butyl cation **4** despite the highly unfavorable steric interaction has been demonstrated[46] by the formation of small amounts of 2,2,3,3-tetramethylbutane **14.** This result also indicates that the related five-coordinate carbocationic transition state (or high lying intermediate) **26** of the degenerate isobutane-*t*-butyl cation hydride transfer reaction is not entirely linear, despite the highly crowded nature of the system.

4 1 26 14

2,2,3,3-Tetramethylbutane **14** was not formed when *n*-butane and *s*-butyl cations were reacted. The isomer distribution of the octane isomers for typical butyl cation–butane alkylations is shown in Table 5.6.

TABLE 5.6 Isomeric Octane Compositions Obtained in Typical Alkylations of Butanes with Butyl Cations

	C_4—C_4			
Octane	$(CH_3)_3CH$ $(CH_3)_3{}^+C$	$CH_3CH_2CH_2CH_3$ $(CH_3)_3{}^+C$	$CH_3CH_2CH_2CH_3$ $CH_3{}^+CH_2CHCH_3$	$(CH_3)_3CH$ $CH_3{}^+CHCH_2CH_3$
2,2,4-Trimethylpentane	18.0	4.0	3.8	8.5
2,2-Dimethylhexane			0.4	
2,2,3,3-Tetramethylbutane	1–2		Trace	1–2
2,5-Dimethylhexane	43.0	0.6	1.6	29.0
2,4-Dimethylhexane	7.6	Trace		6.6
2,2,3-Trimethylpentane	3.0	73.6	40.6	3.2
3,3-Dimethylhexane			12.3	7.1
2,3,4-Trimethylpentane	1.5	7.2	15.5	6.2
2,3,3-Trimethylpentane	3.6		3.8	8.8
2,3-Dimethylhexane	4.2	6.9		12.8
2-Methylheptane		Trace	10.3	6.7
3-Methylheptane	19.3	7.6	6.8	9.5
n-Octane	0.2		4.8	Trace

The protolytic condensation of methane in Magic Acid solution at 60°C is evidenced by the formation higher alkyl cations such as *t*-butyl and *t*-hexyl cations.

$$CH_4 \xrightarrow{H^+} \left[H_3C\cdots\langle^{H}_{H} \right]^+ \longrightarrow CH_3^+ + H_2$$

27

$$CH_3^+ + CH_4 \longrightarrow \left[H_3C\cdots H\cdots CH_3 \right]^+ \longrightarrow C_2H_5^+ + H_2$$

It is not necessary to assume a complete cleavage of $[CH_5]^+$ **27** to a free, energetically unfavorable methyl cation. The carbon–carbon bond formation can indeed be visualized as the C—H bond of methane reacting with the developing methyl cation.

$$CH_4 + \left[H_3C\cdots\langle^{H}_{H} \right]^+ \longrightarrow \left[H_3C\cdots H\cdots CH_3 \right]^+ + H_2$$

In order to overcome unfavorable thermodynamics hydrogen must be oxidatively removed (either by the superacid or added oxidant). Considering the abundance of methane in nature, the conversion of natural gas into branched liquid hydrocarbons in the gasoline range is of immense practical interest.

Alkylation of methane, ethane, propane, and *n*-butane by the ethyl cation generated via protonation of ethylene in superacid media has been studied by Siskin,[54] Sommer,[55] and Olah.[56] The difficulty lies in generating in a controlled way a very energetic primary carbenium ion in the presence of excess methane and at the same time avoiding oligocondensation of ethylene itself. Siskin carried out the reaction of methane–ethylene (86:14) gas mixture through a 10:1 HF:TaF_5 solution under pressure with strong mixing. Among the recovered reaction products 60% of C_3

$$CH_2{=}CH_2 \xrightarrow{HF} CH_3{-}CH_2^+ \xrightarrow{CH_4} \left[CH_3CH_2\cdots H\cdots CH_3 \right]^+ \xrightarrow{-H^+} CH_3CH_2CH_3$$

was found (propane and propylene). Propylene is formed when propane which is substantially a better hydride donor reacts with the ethyl cation:

$$CH_3CH_2CH_3 + CH_3CH_2^+ \longrightarrow CH_3{-}CH_3 + CH_3{-}\overset{+}{C}H{-}CH_3 \xrightarrow{-H^+} CH_3{-}CH{=}CH_2$$

Propane as a degradation product of polyethylene was ruled out because ethylene alone under the same conditions does not give any propane. Under similar conditions but under hydrogen pressure, polyethylene reacts quantitatively to form C_3 to C_6 alkanes, 85% of which are isobutane and isopentane. These results further substantiate the direct alkane–alkylation reaction and the intermediacy of the pentacoordinate carbonium ion. Siskin also found that when ethylene was allowed to react with ethane in a flow system, *n*-butane was obtained as the sole product, indicating that the ethyl cation is alkylating the primary C—H bond through a five-coordinate carbonium ion.

$$CH_2{=}CH_2 \xrightarrow{H^+} CH_3CH_2^+ \xrightarrow{CH_3{-}CH_3} \left[CH_3CH_2 \overset{H}{\cdots} CH_2CH_3 \right]^+$$

If the ethyl cation would have reacted with excess ethylene, primary 1-butyl cation would have been formed which irreversibly would have rearranged to the more stable *s*-butyl and subsequently *t*-butyl cation giving isobutane as the end product.

The yield of the alkene–alkane alkylation in homogeneous HF : TaF_5 depending on the alkene : alkane ratio has been investigated by Sommer and co-workers in a batch system with short reaction times.[55] The results support direct alkylation of methane, ethane, and propane by the ethyl cation and the product distribution depends on the alkene : alkane ratio (Fig. 5.6).

Despite the unfavorable experimental conditions in a batch system for kinetic controlled reactions, a selectivity of 80% in *n*-butane was achieved through ethylation of ethane. The results show, however, that to succeed in the direct alkylation the following conditions have to be met: (1) the olefin should be totally converted to the reactive cation (incomplete protonation favors the polymerization and cracking processes); this means the use of a large excess of acid and good mixing. (2) The alkylation product must be removed from the reaction mixture before it transfers a hydride to the reactive cation, in which case the reduction of the alkene is achieved. (3) The substrate to cation hydride transfer should not be easy; for this reason the reaction shows the best yield and selectivity when methane and ethane are used.

More recently, the direct ethylation of methane with ethylene has been investigated by Olah and his group[56] using ^{13}C-labeled methane (99.9% ^{13}C) over solid superacid catalysts such as TaF_5 : AlF_3, TaF_5, and SbF_5 : graphite. Product analyses by gas chromatography-mass spectrometry (GC-MS) are given in Table 5.7.

These results show a high selectivity in mono-labeled propane $^{13}CC_2H_8$ which

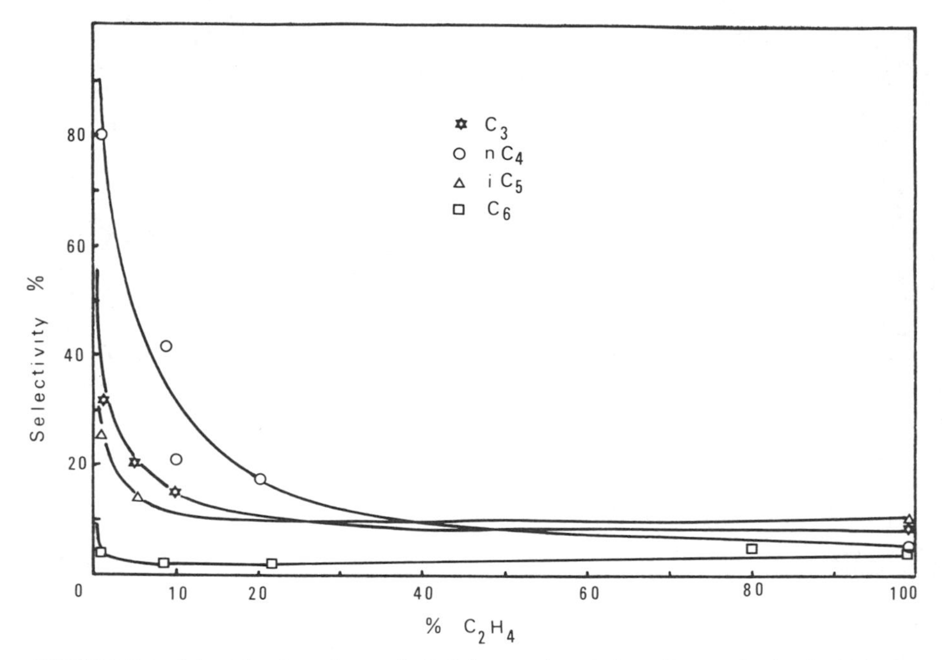

FIGURE 5.6 Selectivity dependence of alkylation on the ethylene-alkane ratio for the following reactions; ethylene + methane (✡), ethylene-ethane (○), ethylene + propane (△), ethylene + *n*-butane (□).

can only arise from direct electrophilic attack of the ethyl cation on methane via pentacoordinate carbonium ion.

$$CH_2{=}CH_2 \xrightarrow{H^+} CH_3{-}CH_2^+ \xrightarrow{^{13}CH_4} \left[CH_3{-}CH_2\cdots\overset{H}{\vdots}\cdots{}^{13}CH_3 \right]^+ \xrightarrow{-H^+} CH_3CH_2{}^{13}CH_3$$

An increase of the alkene : alkane ratio results in a significant decrease in single-labeled propane; ethylene polymerization-cracking and hydride transfer become the main reactions. This labeling experiment carried out under conditions where side reactions were negligible is indeed unequivocal proof for the direct alkylation of an alkane by a very reactive carbenium ion.

Polycondensation of alkanes over $HSO_3F:SbF_5$ has also been achieved by Roberts and Calihan.[57] Several low-molecular-weight alkanes such as methane, ethane, propane, butane, and isobutane were polymerized to highly branched oily oligomers with a molecular weight range from the molecular weight of monomers to around 700. These reactions again follow the same initial protolysis of the C—H or C—C bond which results in a very reactive carbenium ion. Similarly, the same workers[58] were also able to polycondense methane with small amount of olefins such as ethylene, propylene, butadiene, and styrene to yield oily polymethylene oligomer with a molecular weight ranging from 100 to 700.

TABLE 5.7 Ethylation of $^{13}CH_4$ with C_2H_4

Run	$^{13}CH_4:C_2H_4$	Catalyst[b]	Products Normalized (%)[c,d]				Label Content of C_3 Fraction (%)[e]	
			C_2H_6	C_3H_8	i-C_4H_{10}	C_2H_5F	$^{13}CC_2H_8$	C_3H_8
1	98.7:1.3	$TaF_5:AlF_3$	51.9	9.9	38.2		31	69
2	99.1:0.9	TaF_5		15.5	3.0	81.5	91	9
3	99.1:0.9	SbF_5:graphite	64.1	31.5		4.4	96	4

[a] All values reported are in mole percentage.

[b] Catalysts pretreated with HF for 30 s.

[c] Excluding methane.

[d] Trace amounts of $(CH_3)_2SiF_2$ detected in all runs are probably due to SiF_4 impurity (from HF) reacting with methane.

[e] Isobutane contained no ^{13}C label, thus it is derived from ethylene.

5.2 ALKYLATION OF AROMATIC HYDROCARBONS

Cumene is industrially produced by propylating benzene with propylene over supported acidic catalysts such as phosphoric acid. On the other hand, the largest-scale single industrial alkylation process, i.e., the ethylation of benzene with ethylene, is still carried out to a significant degree in the liquid phase using acid catalysts, since ethylene is less polar than propylene it requires more forcing conditions in the protolytic initiation step.

$$CH_2{=}CH_2 \xrightarrow{H^+} CH_3CH_2^+ \xrightarrow{C_6H_6} C_6H_5CH_2CH_3 + H^+$$

Consequently, stronger solid acids were needed to activate the ethylation reaction. The solid acids that were available earlier exhibited only limited acidity, which was sufficient to promote the propylation of benzene with propylene at reasonable temperatures and pressures, but it was not high enough to promote ethylation of benzene with ethylene under similar conditions.

The conventional resinsulfonic acids such as sulfonated polystyrenes (Dowex-50, Amberlite IR-112, and Permutit Q) are of moderate acidity with limited thermal stability. Therefore, they can be used only to catalyze alkylation of relatively reactive aromatic compounds (like phenol) with alkenes, alcohols, and alkyl halides. More recently Nafion-H has been found to be a suitable superacid catalyst in the 110–190°C temperature range (*vide infra*).

In 1976, Lalancette et al. studied the catalytic activity of graphite intercalated $AlCl_3$ and compared it with neat $AlCl_3$ in solution phase alkylations. Whereas the rate of alkylation slowed down using the intercalated catalyst, a higher selectivity toward monoalkylation was found (Table 5.8).[59]

The $AlCl_3$: graphite and graphite intercalates of related Lewis acid halides have been tested as solid catalysts for the gas-phase alkylation of aromatic hydrocarbons.[40] The catalytic activity of intercalation compounds of $AlCl_3$, $AlBr_3$, and $FeCl_3$ in graphite was measured toward two model reactions, the alkylation of benzene with ethylene and propylene and the transethylation of benzene with diethylbenzene.

$$C_6H_6 + RCH{=}CH_2 \longrightarrow C_6H_5CH(R)CH_3$$

$$R = H,\ CH_3$$

$$C_6H_4(C_2H_5)_2 + C_6H_6 \longrightarrow 2\ C_6H_5C_2H_5$$

TABLE 5.8 Electrophilic Substitutions Catalyzed by $AlCl_3$: Graphite

System	Substrate (S)	Reagent (R)	Temperature (°C)	Duration (h)	Molar Ratio (R/S)	Products		
							With $AlCl_3$	With $AlCl_3$: Graphite
Sealed tube	Benzene	Ethyl bromide	25	48	3	Benzene	2	29
						Monoethyl-	13	40
						Diethyl-	27	23
						Triethyl-	54	8
						Tetraethyl-	4	Traces
Sealed tube	Toluene	Ethyl bromide	25	24	3	Toluene	0	17
						Monoethyl-	Traces	52
						Diethyl-	41	31
						Triethyl-	60	Traces
Atmospheric	Toluene	Ethyl bromide	− 10	24	2	Toluene	5	5
						Monoethyl-	Traces	33
						Diethyl-	30	52
						Triethyl-	65	10
Sealed tube	Naphtalene	Ethyl bromide	25	44	3	Naphtalene	26	77
						Monoethyl-	39	21
						Diethyl- tetrahydro-	2	Traces
						Diethyl-	17	1
						Triethyl-	16	0.5
Sealed tube	Biphenyl	Ethyl bromide	25	48	3	Biphenyl	26	81
						Diethyl-	16	15
						Tetra–hexa ethyl-	33	4
						Octaethyl-	25	0

TABLE 5.8 (*Continued*)

System	Substrate (S)	Reagent (R)	Temperature (°C)	Duration (h)	Molar Ratio (R/S)	Products		
							With $AlCl_3$	With $AlCl_3$: Graphite
Atmospheric	Benzene	Ethylene	75	24	—	Benzene	0	3
						Monoethyl-	Traces	19
						Diethyl-	1.1	47
						Triethyl-	25	22
						Tetraethyl-	53	6
						Pentaethyl-	18	2
						Hexaethyl-	4	1
Atmospheric	Benzene	Propylene	60	0.5	—	Benzene	23	44
						Monoisopropyl-	56	54
						Diisopropyl-	18	2
						Triisopropyl-	3	Traces
Atmospheric	Toluene	Ethylene	80	1.45	—	Toluene	8	17
						Monoethyl-	49	51
						Diethyl-	40	26
Atmospheric	Toluene	Propylene	80	2.50	—	Toluene	Traces	4
						Monoisopropyl-	Traces	33
						Diisopropyl-	42	59
						Triisopropyl-	46	Traces
Atmospheric	Toluene	Isobutylene	80	1.15	—	Toluene	20	35
						Monoisobutyl-	66	63
						Diisobutyl-	13	1

The ethylation of benzene with ethylene proceeded to give high initial yield at temperatures as low as 125°C (Table 5.9) when the intercalated Lewis acid halide was $AlCl_3$. The initial yield is lower when $AlBr_3$ is used and very low when SbF_5 or $FeCl_3$ are intercalated in the graphite layers. Same results were observed when these catalysts were used to catalyze transethylation of benzene with diethylbenzene (Table 5.10).

Although in the case of relatively active catalysts ($AlCl_3$:graphite and $AlBr_3$:graphite) the initial conversions were high, their activity declined rapidly with the onstream time. The catalysts were totally deactivated after a period of 6–8 h. There are two possible reasons for the loss in catalytic activity of these systems.

1. Possible hydrolysis by traces of moisture present in the feed
2. Leaching of the metal halide from the graphite layers by the feed

Lalancette[60] indicated that intercalation process can be carried out in CCl_4 solution only if the Lewis acid halide is slightly soluble in the solvent. Therefore, $AlCl_3$ is probably eluted from the graphite by the feed. The rate of this process will increase with the increase in temperature, which increases the vapor pressure of the Lewis acid halide. Hence, there are several drawbacks in using these graphite intercalates.

On the other hand, the Nafion resin in its acidic form (Nafion-H) shows high activity in a variety of electrophilic reactions. Gas-phase alkylation of benzene with ethylene and propylene in a flow system proceeds at temperatures as low as 110°C over Nafion-H (Table 5.11).

$$C_6H_6 + RCH{=}CH_2 \xrightarrow[110\text{ to }190^\circ C]{\text{Nafion-H}} C_6H_5CH(R)CH_3$$

In the alkylation of ethylbenzene with ethylene with conventional acid catalysts under usual conditions *s*-butyl benzene is a by-product. *s*-Butylbenzene was detected when the reaction was carried out over catalysts such as supported phosphoric acid,[61] ferric phosphate,[61] or $AlCl_3$-NiO-SiO_2.[62] When Nafion-H or $AlCl_3$ are used, no such by-product is detected, probably due to fast dealkylation of *s*-butylbenzene under the more acidic conditions.

The high acidity of the Nafion-H catalyst is further demonstrated by its ability to promote both polyalkylation and isomerization. In reaction between benzene and ethylene at 190°C, 20% of the alkylated products are diethylbenzenes.[40] The isomer distribution of the diethylbenzenes is *o*-1%, *m*-75%, and *p*-34%. This composition is very close to the equilibrium composition of diethylbenzenes determined in solution chemistry with $AlCl_3$ catalyst and indicates that the reaction is thermodynamically controlled.

TABLE 5.9 Ethylation of Benzene with Ethylene over Graphite-Intercalated Metal Halides

Metal Halide	$AlCl_3$	$AlCl_3$	$AlCl_3$	$AlCl_3$	$AlBr_3$	SbF_5	$FeCl_3$
% Intercalated	16.6	16.6	16.6	28.4	4.5	26.1	50.5
(Temp, °C)	125	160	200	160	160	160	185
$[C_6H_6]/[C_2H_4]$ Ratio	3.3	3.4	3.2	3.5	3.3	3.6	3.2
Onstream Time (h)	% Conversion (Based on Ethene)						
1	61.7	41.7	15.1	60.4	31.7	4.8	
2	10.3	12.0	5.7	62.9	15.1		
3	5.4	7.3	2.4	43.8			
4	4.3	5.3	1.7	33.7			
5	2.9	5.7	1.9	26.5			
6	1.9	1.4	0.7	6.8			
7				2.6			
8				2.2			
9				1.7			
10				1.5			

TABLE 5.10 Transethylation of Benzene with Diethylbenzene over Graphite-Intercalated Metal Halides

Metal Halide	$AlCl_3$	$AlBr_3$	SbF_5
% Intercalated	28.4	4.5	26.1
(Temp, °C)	180	180	180
$[C_6H_6]/[C_6H_4Et_2]$ Ratio	4	4	4
Onstream Time (h)	% Conversion (Based on Diethylbenzene)		
1	45.0	66.2	2.4
2	70.4	51.8	
3	66.9	41.4	
4	70.3	3.1	
5	56.9		
6	16.2		
7	13.8		
8	15.2		
9	12.7		
10	10.4		

When olefins are used as alkylating agents, the catalytic activity of Nafion-H is slowly decreased, most probably due to some polymerization on the surface, which deactivates the catalytic sites. The activity decreases faster when more reactive branched alkenes are used.

The use of alcohols instead of olefins as the alkylating agents improves the lifetime of the catalyst. With alcohols, no ready polymerization takes place, since

TABLE 5.11 Alkylation of Benzene over Nafion-H Catalyst

Alkene	Temperature (°C)	$[C_6H_6]$/[Alkene]	Contact Time (s)	Alkene Conversion, (%)
Ethylene	110	4	7	10
	150	4	6	24
	180	4	6	36
	190	3.4	3.5	44
Propene	110	1.5	7	9
	150	1.5	6	16
	180	1.5	6	19
	180	3	4	21
	180	6	4	29

TABLE 5.12 Dehydration of Alcohols over Nafion-H

Alcohol	Temperature (°C)	Contact Time (s)	% Dehydration	Product	
				% Alkene	% Ether
i-PrOH	100	10	9		100
	130	9	28	45	55
	160	8	>97	100	
n-PrOH	130	4.5	8	47	53
	160	8	96	100	
t-BuOH	120	5	100	100	

water formed as by-product inhibits polymerization of any olefin formed (by dehydration) but does not affect the acidity of the catalyst at the reaction temperatures.

Alcohols can also undergo acid-catalyzed dehydration to give either the corresponding alkenes or the corresponding ethers. The product distribution of the dehydration of alcohols over Nafion-H catalyst shows temperature dependence (Table 5.12).[63] Alcohols are thus efficiently dehydrated in the gas phase over Nafion-H under relatively mild conditions with no evidence for any side reactions such as dehydrogenation or decomposition. At higher temperatures, olefin formation predominates.

Reaction of alcohols with benzene over Nafion-H catalysts gave the corresponding alkylbenzenes (Table 5.13). When propyl alcohol was the alkylating agent, no *n*-propylbenzene was detected, and the only product obtained was cumene.[40] This indicates the intermediacy of the isopropyl cation in the alkylation process.

The alkylation of aromatic hydrocarbons with methyl alcohol over Nafion-H catalysts, including the mechanistic aspects, has been studied in detail. The degree of conversion of methyl alcohol was much dependent on the nucleophilic reactivity of the aromatic hydrocarbon. For example, the reactivity of the isomeric xylenes was higher than that of toluene or benzene.

TABLE 5.13 Alkylation of Benzene with Alcohols over Nafion-H Catalyst

Alcohol	$[C_6H_6]/[ROH]$	Temperature (°C)	Contact Time (s)	% Alcohol Conversion
EtOH	2.6	180	9	3.5
	2.6	210	8	6
n-PrOH	0.85	110	10	0
	0.85	175	9	5
	2	175	9	17
i-PrOH	2	170	9	11
	2	210	8	16

A study of the temperature dependence of the reaction between toluene and methyl alcohol showed a substantial increase in xylene formation as the temperature was increased.[63] However, the overall yield of xylene at 209°C was lower than that expected from extrapolation of data obtained at lower temperatures. This is probably due to the increasing thermal instability of Nafion-H at temperatures around and above 200°C and hence, its reduced activity.

Yields and conversion of methyl alcohol were much higher when the aromatic substrate was phenol or anisole and their derivatives (Table 5.14).[64] In recent studies, various degrees of selectivity were reported in the gas-phase methylation of phenol toward the ortho substitution, using catalysts such as Al_2O_3,[65] TiO_2,[66] ZnO, Fe_2O_3,[67] ZnO · MO (M = Cu, Ba, Ca, Co, Mn, Mg, Ni),[68] MgO alone,[69] or mixed with oxides of Mn, Cu, Sn, Bi, Pb or Cr[70] at temperatures ranging from 250°C to 400°C. Unlike these examples, the selectivity toward ortho methylation when using Nafion-H as catalyst is somewhat lower,[64] probably due to the absence of basic sites on this solid catalyst.

In gas-phase methylation reactions over Nafion-H using methyl alcohol as the alkylation agent, the consumption of methyl alcohol was higher than that calculated by product analysis.[63,64] This is due to the formation of dimethyl ether as the by-product.

$$2CH_3OH \xrightarrow[150°C]{\text{Nafion-H}} CH_3OCH_3 + H_2O$$

Indeed, when neat methyl alcohol is passed over Nafion-H catalyst at temperatures over 150°C, dimethyl ether is the only product formed quantitatively with water as the by-product.[71]

Studies therefore were also carried out using dimethyl ether as the methylating agent over Nafion-H catalyst. The initial conversions were similar to those obtained with methyl alcohol as the methylating agent. However, the reactivity of the catalyst was found to decrease sharply with time probably due to the increased esterification of the acidic sites of the solid catalyst.[63] The reactivity of the catalyst could be, however, readily regenerated by passing steam over it at 185°C for 1 h. In general, it appears that the presence of water vapor (i.e., steam) in the systems catalyzed by Nafion-H, does not reduce the catalyst's activity. In cases where polymerization and other possible side reactions would lead to the deactivation of the catalyst, the presence of water helps to maintain the catalytic activity of Nafion-H catalyst. For example, when benzene was alkylated by dimethyl ether over Nafion-H, the use of benzene saturated with water slowed down the deactivation of the catalyst considerably.

Satisfactory results were obtained in the Nafion-H-catalyzed gas-phase alkylation of aromatic hydrocarbons with alkyl halides.[72]

$$ArH + RX \xrightarrow[140\text{–}200°C]{\text{Nafion-H}} ArR + HX$$

X = Halide

TABLE 5.14 Methylation of Phenol, Cresols, Anisole with Methyl Alcohol over Nafion-H Catalyst

	% Product Composition						
	Unreacted Starting Material	Anisole[a]	Methylanisoles	Dimethylanisoles	Phenol[a]	Cresols[a]	Xylenols
Phenol	37.3	37.2	9.7	1.0		10.4	4.4
Anisole	58.6		13.9	3.0	18.1	4.7	1.7
o-Cresol	51.4	0.1	23.4	4.7	0.4	0.6	19.4
m-Cresol	48.1	trace	26.0	5.8	0.2	1.0	18.9
p-Cresol	39.2	3.3	23.4	6.4	8.4	4.6	14.8

[a]Excluding starting material.

Alkyl halides are reactive Friedel–Crafts alkylating agents and give high conversions when alkylating benzene in the gas phase over Nafion-H catalyst. For example, in the alkylation of benzene with isopropyl chloride, conversions as high as 87% were achieved. Conversions, however, were temperature and contact time dependent (Table 5.15).[72]

The selectivity of the Nafion-H catalyst for monoalkylation has been found to be generally high. With a molar ratio of benzene:isopropyl chloride being 5:1, about 94% of the alkylate is monoalkylbenzene. This result is comparable to the highly selective monoalkylation reaction reported by Langlois et al.[73] They alkylated benzene with propene (5.2:1 m/m) over H_3PO_4-quartz catalyst at ~200°C and obtained cumene in 95% yield.

The results obtained in the gas-phase isopropylation of various aromatic hydrocarbons with isopropyl chloride over Nafion-H catalyst showed only a relatively small variation of reactivity in going from fluorobenzene to xylenes.[72] It has, therefore, been assumed that the reaction rate is controlled by the formation of a reactive electrophilic intermediate (possibly, protonated alkyl halide, or some form of incipient alkyl cation) rather than by σ-complex formation between the electrophile and the aromatic nucleus

$$R-\underset{H}{\overset{R'}{\underset{|}{\overset{|}{C}}}}-X \overset{H^+}{\rightleftharpoons} R-\underset{H}{\overset{R'}{\underset{|}{\overset{|}{C}}}}-\overset{+}{X}-H \rightleftharpoons R-\underset{H}{\overset{R'}{\underset{|}{\overset{|}{C^+}}}} + HX$$

The relatively minor differences observed in the degree of conversion in the reaction of various aromatic hydrocarbons with isopropyl chloride over Nafion-H are explained by the possibility of some dealkylation (i.e., the reverse reaction). Dealkylation reactions occur as a competitive process to the alkylation process. The more nucleophilic is the alkylated aromatic product, the higher the rate of dealkylation reaction.

Besides the advantage of their high reactivity toward alkylation reactions, primary and secondary alkyl halides show little tendency for dehydrohalogenation in Nafion-H-catalyzed gas-phase reactions.[72] Although a minor amount of olefin is reported to be formed, no polymer formation was observed on the catalyst. As a result, the catalytic activity of Nafion-H stays constant over prolonged onstream periods.[72]

The versatility of the catalytic activity of Nafion-H, is well demonstrated in the alkylation of aromatic hydrocarbons by carboxylic acid alkyl esters both in the gas-phase as well as in heterogeneous liquid-phase reactions.[74] Esters in the presence of conventional Lewis acid halide catalysts tend to give rise to acylation products along with the alkylation products.[75] The heterogeneous liquid-phase alkylation reactions have been generally carried out under reflux conditions. Two types of alkylating agent have been studied: (1) alkyl esters of carboxylic acids, preferentially those of oxalic acid and (2) alkyl chloroformates.

$$ArH + RCOOR' \xrightarrow{\text{Nafion-H}} ArR' + RCOOH$$

$$ArH + ClCOOR \xrightarrow{\text{Nafion-H}} ArR + HCl + CO_2$$

TABLE 5.15 Effect of Temperature and Contact Time on the Isopropylation of Benzene with Isopropyl Chloride over Nafion-H Catalyst

			Yield of Isopropylbenzene (%)					
Number	T (°C)	Feed Rate (ml/min)	Mono	Meta-di	Para-di	Tri	Ratio Mono/Di	Total Conversion Based on *i*-PrCl (%)
1	135	0.2	2.5	0.1	0.1	0.1	<12.5	3.0
2	150	0.2	6.4	0.4	0.3	0.1	<9.1	8.1
3	165	0.2	15.8	1.1	0.8	0.1	8.3	19.9
4	174	0.2	25.6	1.9	1.2	0.2	8.2	32.4
5	180	0.05	44.3	3.8	2.0	0.3	7.6	56.8
6	180	0.1	36.8	3.0	1.7	0.3	7.8	47.1
7	180	0.2	33.2	2.7	1.5	0.2	7.9	42.2
8	196	0.2	41.4	3.2	1.7	0.2	8.4	51.8
9	196	0.05	53.7	4.1	2.0	0.2	8.8	66.5
10	180	0.2[a]	63.0	5.5	2.4	0.3	8.0	78.7
11	180	0.1[a]	69.5	5.8	2.4	0.3	8.5	86.8
12	180	0.2[b]	55.0	2.5	1.3	0.0	14.5	62.6
13	180	0.2[c]	19.7	3.3	1.6	0.5	4.0	31.0

In all experiments, 1.00 g of Nafion-H was used as a catalyst and the ratio PhH:*i*-PrCl = 5:2 unless otherwise indicated.

[a] 2.00 g Nafion-H was used.

[b] PhH:*i*-PrCl = 5:1.

[c] PhH:*i*-PrCl = 5:4.

The advantage of alkyl chloroformates lies primarily in their volatile by-products (HCl and CO_2). Diethyl oxalate shows particularly good alkylating ability even under mild conditions (Table 5.16).

The gas-phase alkylation of toluene with alkyl chloroformate over Nafion-H was also reported to be an efficient one. It is interesting to compare the alkylating ability of methyl chloroformate with that of methyl alcohol on toluene under similar reaction conditions. For example a 59% conversion with methyl chloroformate was observed,[74] compared with 15% conversion with methyl alcohol.

The gas-phase alkylation of toluene with dimethyl and diethyl oxalate over Nafion-H was also reported.[74] The alkylating ability of diethyl oxalate is comparable with that of ethyl chloroformate. However, the alkylating ability of dimethyl oxalate is lower than that of methyl chloroformate.

When dialkylbenzenes are passed over Nafion-H at 160°C, both isomerization and disproportionation take place. Monoalkylbenzenes also disproportionate under these conditions.[75,76]

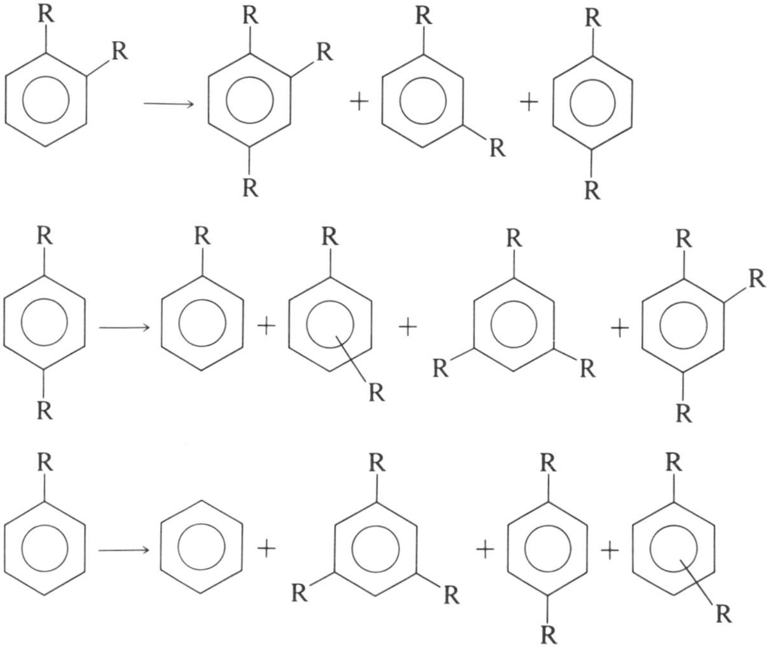

As expected, the aptitude for disproportionation of the aromatic compound depends upon the nature of the alkyl group, and the order of reactivity is isopropyl $>$ ethyl $>$ methyl. Due to their higher nucleophilicity, polyalkylbenzenes react faster than monoalkylbenzenes. This effect is pronounced in the case of methylbenzenes. Toluene itself shows little reactivity over Nafion-H at 193°C. Diethylbenzenes react much faster than dimethylbenzenes. The rate of conversion of diethylbenzenes over Nafion-H at 193°C is $\sim 5.10^{-5}$ mol $\cdot$ min^{-1} $\cdot$ g^{-1} of catalyst.[76] This is a low rate when compared with that using $AlCl_3$—HCl in the liquid phase at room temperature

TABLE 5.16 Nafion-H-Catalyzed Liquid-Phase Alkylation of Toluene with Esters and Haloesters[a]

Alkylating Agent	Temperature (°C)	% Conversion	% Isomer Distribution		
			ortho	meta	para
$MeOC(=O)Cl$	70–72	2	48	26	26
$EtOC(=O)Cl$	90	24	46	26	28
i-$PrOC(=O)—Cl$	110	80	42	21	37
CF_3CO_2Et	82	5.5	49	24	27
CCl_3CO_2Et	110	20	44	28	28
$(COOEt)_2$	110	50	48	24	28

[a]Reaction time, 12 h.

($\sim 10^{-4}$ mol · min^{-1} · g^{-1} catalyst).[77] However, one should bear in mind that Nafion-H is a truly insoluble heterogeneous catalyst, whereas in the case of $AlCl_3$—HCl a soluble complex is formed with the hydrocarbon and therefore the rates are not directly comparable. The equilibrium composition of the acid-catalyzed disproportionation of ethylbenzenes depends upon the nature of the catalyst.

The predominance of *meta*-diethylbenzene in the isomerization of diethylbenzenes is easily rationalized. Since isomerization reaction proceeds via arenium ion intermediates, the σ-complex derived from *meta*-diethylbenzenes is the most stable one. Moreover, *meta*-diethylbenzene is also the most basic of the isomeric diethylbenzenes.[78] Therefore, more acidic catalysts increase the amount of the meta-isomer at the expense of the para- (and ortho-) isomers, due to the increased stability of the substrate-catalyst complex.

Nafion-H appears to be a very useful catalyst for transalkylation reactions as indicated in these studies. Transalkylation of benzene with diethylbenzene, as well as with di-isopropylbenzene is efficiently catalyzed by Nafion-H in a flow system. The efficiency of the catalysts is, however, more limited when the transferring group is a methyl group.[75a] Beltrame and co-workers have also carried[75b] detailed mechanistic studies on the isomerization of xylenes over Nafion-H.

Nafion-H also very efficiently catalyzes the rearrangement of anisole, methylanisoles, and phenetole to ring-alkylated phenols and products of transalkylation when vapors of the alkyl aryl ethers are passed over it at temperatures higher than 160°C.[64,75] At these reaction temperatures, some of the starting alkyl phenyl ethers undergo cleavage of the alkyl group to give phenol.

In liquid phase alkylations besides conventional Friedel-Crafts systems, superacids which are capable of forming stable carbocations and onium ions have also found applications. Olah and his co-workers in their extensive studies have shown that alkyl halides readily ionize in SbF_5 to the corresponding alkylcarbenium hexafluroantimonates.[46] Tertiary and some secondary carbocations are remarkably stable in solutions of $SO_2:SbF_5$ and $SO_2ClF:SbF_5$, respectively. These systems were also found highly efficient aromatic alkylating agents[79a] (their ability to alkylate saturated aliphatic hydrocarbons was discussed in Section 5.1).

Methyl fluoride and ethyl fluoride form stable addition complexes with SbF_5 which are powerful methylating and ethylating agents, respectively.[49,50] In the study of alkyl halide–antimony pentafluoride systems, Olah and DeMember[79b] found that dialkylhalonium ions $R\overset{+}{X}R$ are formed when 2 mol (or excess) of an alkyl halide (except fluoride) are reacted with SbF_5 (see Chapter 4) in SO_2 or SO_2ClF solution. The alkylation of aromatic hydrocarbons, such as benzene, toluene, and ethylbenzene has been studied with highly electrophilic dimethyl- and diethylhalonium hexafluroantimonates **28** and **29** in SO_2ClF solution under superacidic conditions.[79c] Data for alkylations are summarized in Table 5.17.

$$ArH + R\overset{+}{X}R\ SbF_6^- \longrightarrow ArR + RX + HSbF_6$$

$$X = Cl, Br, I$$
$$\mathbf{28},\ R = CH_3$$
$$\mathbf{29},\ R = C_2H_5$$

Dimethylchloronium and bromonium ions give similar methylation results and are quite reactive even at temperatures as low as $-50°C$. The dimethyliodonium ion is less reactive and alkylate benzene and toluene in SO_2ClF solution only when allowed to react (if necessary under pressure) at or above 0°C.

The ethylation of toluene by diethylhalonium ions gives ethyltoluenes with ortho:para isomer ratios between 0.60 and 0.96. The ortho:para isomer ratios obtained for the alkylation of toluene in conventional Friedel–Crafts ethylations range from 1.17 to 1.84 (average ca. 1.60). Such differences are considered to be due to the steric ortho effect caused by diethylhalonium ions, and are in accordance with the most probable displacement reaction on the bulky diethylhalonium ions by the aromatic substrate. This can be envisioned to proceed through an S_N2-type transition state involving no free alkyl cations.

$$ArH + \underset{H}{\overset{CH_3\ \ H}{C}}{-}\overset{+}{X}{-}Et\ \ SbF_6^- \rightleftharpoons \left[HAr{\cdots}\underset{H}{\overset{CH_3\ \ H}{C}}{\cdots}X\,Et \right]^+ SbF_6^- \rightleftharpoons$$

$$\rightleftharpoons \left[Ar\genfrac{}{}{0pt}{}{CH_2CH_3}{H} \right]^+ SbF_6^- + CH_3CH_2X \longrightarrow ArCH_2CH_3 + HSbF_6$$

TABLE 5.17 Alkylation of Benzene, Toluene, and Ethylbenzene with Dimethyl and Diethylhalonium Fluoroantimonates in SO_2ClF[a]

Halonium Ion	Aromatic Substrate	Temperature (°C)	Time (min)	k_T/k_B	Isomer Distribution (%) Ortho	Meta	Para	Ortho/Para Ratio
$(CH_3)_2Cl^+$	Toluene	25	10		46.6	27.2	26.7	1.75
		0	1		51.8	16.2	32.1	1.61
		−50	5		52.3	15.7	32.0	1.63
		−50	2		58.6	13.0	28.4	2.06
$(CH_3)_2Br^+$	Toluene	−50	5		57.8	9.5	32.7	1.76
		−50	2		59.0	8.6	32.4	1.82
$(CH_3)_2I^+$	Toluene	25	10		46.2	15.6	38.1	1.27
		0	10		53.9	11.8	34.3	1.57
		−20	60			No reaction		
$(CH_3CH_2)_2Cl^+$	Benzene-Toluene	−78	1	4.9	33.4	28.1	38.5	0.86
		−78	2.5	4.8	31.4	24.6	44.0	0.72
	Ethylbenzene-Toluene	−78	2.5	1.10	31.9	19.3	48.8	0.65
$(CH_3CH_2)_2Br^+$	Benzene-Toluene	−78	1	4.0	38.7	19.3	42.0	0.92
		−78	5	4.5	36.0	18.2	45.8	0.78
	Ethylbenzene-Toluene	−78	5	1.2	32.8	14.5	52.8	0.62
					(25.2)	(19.4)	(55.4)	(0.45)
$(CH_3CH_2)_2I^+$	Benzene-Toluene	−45	5	4.1	44.0	10.2	45.8	0.96

[a]All data are average of at least three parallel experiments.

The alkylation data obtained from the reaction of dimethyl- and diethylhalonium ions provide evidence for direct alkylation of aromatics by dialkylhalonium ions. In addition, the data also indicate that dialkylhalonium ions are not necessarily involved as active alkylating agents in general Friedel–Crafts systems, although some of the reported anomalous alkylation results, particularly with alkyl iodides, could be attributed to reaction conditions favoring dialkylhalonium ion formation.

5.3 ACYLATION OF AROMATICS

Friedel–Crafts acylation of aromatics is of considerable practical value owing to the importance of aryl ketones and aldehydes as chemical intermediates. Whereas alkylation of aromatics with alkyl halides requires only catalytic amounts of catalysts, acylation to ketones generally necessitates equimolar or even some excess of the Friedel–Crafts catalysts. Usually one molar equivalent of catalyst combines with the acyl halide, giving a 1:1 addition compound, which then acts as the active acylating agent:

$$RCOX + AlX_3 \longrightarrow RCOX \longrightarrow AlCl_3 \rightleftharpoons RCO^+ \ AlCl_4^-$$

Evidence supporting the formation of 1:1 addition compounds is further substantiated by the actual isolation of stable acyl cation salts (Chapter III). It is, therefore, highly desirable to develop methods in which only a catalytic amount of Friedel–Crafts acid catalyst may be used for effective conversion.

Effenberger and Epple[80] showed that alkylbenzenes are efficiently acylated when ca. 1% triflic acid is added to the mixture (Table 5.18). It was shown that when other Brønsted or Lewis acids were used, the yield decreased drastically (Table 5.19). Perfluorobutanesulfonic acid was found similarly effective.[81]

$$\underset{X = Cl, OCOC_6H_5, OH}{R^1COX} + A\gamma H \xrightarrow[n = 1,4]{C_nF_{2n+1}SO_3H} A\gamma COR^1$$

Studies on mixed anhydrides of carboxylic acids and triflic acids have shown them to be extremely powerful acylating agents.[82] Similarly, higher perfluoroalkanesulfonic acids also form mixed anhydrides.[83]

Nafion-H has also been found to be an effective catalyst for the heterogenous acylation of aromatic hydrocarbons with aroyl chlorides and anhydrides.[84]

$$R\text{-}C_6H_4COCl + C_6H_5R' \longrightarrow R\text{-}C_6H_4\text{-}C(=O)\text{-}C_6H_4\text{-}R'$$

TABLE 5.18 Triflic Acid-Catalyzed Acylation of Aromatics with Acyl Chlorides (RCOCl)

Reactants		Conditions		Product	Yield	
(*1*), R	(*2*)	*T* (°C)	*t* (h)		(%)	*o*:*p*
C_6H_5	Benzene	80	8.5	Benzophenone	14	
C_6H_5	Chlorobenzene	132	5	2- and 4-Chlorobenzophenone	13	1:3
C_6H_5	Toluene	110	48	2- and 4-Methylbenzophenone	85	1:2
C_6H_5	*p*-Xylene	138	6	2,5-Dimethylbenzophenone	82	
p-NO_2-C_6H_4	Benzene	80	4	4-Nitrobenzophenone	82	
$(CH_3)_3C$	Anisole	154	12	*t*-Butyl *p*-methoxyphenyl ketone	54	
$(CH_3)_2CH$	Anisole	154	0.2	Isopropyl *p*-methoxyphenyl ketone	46	

The reported gas-phase acylations with Nafion-H catalyst were generally carried out at the boiling point of the hydrocarbon to be acylated. The yield of the aroylation reaction depends on the relative amount of the catalyst used. Optimum yields were obtained when 10–30% of Nafion-H was employed relative to the aroyl halide. Although this procedure allows very clean reactions with no complex formation and easy work-up procedures, it is presently limited to only aroylation. Attempted acetylation of aromatics with acetyl chloride under similar conditions led to thermal HCl elimination from the latter to form ketene and the products thereof. In the reaction of acetyl chloride by itself with Nafion-H, diketene was detected by ir and nmr spectroscopy.[84]

The reversibility of Friedel–Crafts acylation is only occasionally observed.[85] Schlosberg and Woodbury[86] have studied transacylation between tetramethylacetophenones and some arenes in superacidic HF:SbF_5 (5:1) and other strong superacid media. In fact, recently Keumi and co-workers[87] have been able to observe

TABLE 5.19 Catalytic Action of Brønsted and Lewis Acids in the Acylation of *p*-Xylene by Benzoyl Chloride

Catalyst	Amount (%)	Conditions		Yield (%)
		T (°C)	*t* (h)	
CF_3SO_3H	1	138	6	82
HSO_3F	1	138	6	20
p-CH_3-C_6H_4-SO_3H	1	138	6	31
H_2SO_4	1	138	6	28
$HClO_4$	1	138	6	14
CF_3COOH	2.6	138	10	21
$HPOF_2$	3.1	138	10	4
$AlCl_3$	2	138	15	26
$SnCl_4$	2	138	15	30

diprotonated acetylpentamethylbenzene intermediate **30** in $HSO_3F:SbF_5:SO_2ClF$ medium at low temperatures, which deacetylates to pentamethylbenzenium ion **31** at more elevated temperatures.

$$\mathbf{30} \xrightarrow{\Delta} \mathbf{31} + CH_3C\overset{+}{O}$$

Nafion-H has been found to be an efficient catalyst for the decarboxylation of aromatic carboxylic acids as well as deacetylation of aromatic ketones.[88]

5.4 CARBOXYLATION

The Koch–Haaf reaction[89] for the preparation of carboxylic acids from alkenes uses formic acid or carbon monoxide in strongly acidic solutions. The reaction between carbocations and carbon monoxide affording oxo-carbenium ions (acyl cations) is a key step in the Koch–Haaf reaction, and the topic has been reviewed by Hogeveen.[90] The original studies used sulfuric acid. Subsequent investigations found[91] the superacidic $HF:BF_3$ system very efficient, but due to the volatility of the catalyst system necessitate high pressure conditions.

More recently, the application of liquid superacids in the Koch–Haaf carboxylation has met with remarkable success. Triflic acid has been found to be by far superior to 95% H_2SO_4 for carboxylation of olefins, alcohols, and esters with carbon monoxide at atmospheric pressure. This is attributed to the high acidity of triflic acid and also the higher solubility of CO in this medium as compared with H_2SO_4. Moreover, triflic acid has the advantage that unlike H_2SO_4 it does not form electrophilic substitution products with aromatics and can be regenerated.[92]

Whereas the C_2—C_4 alcohols are not carboxylated under the usual Koch–Haaf conditions, carboxylation can be achieved in the $HF:SbF_5$ superacid system under extremely mild conditions.[93]

Using $HF:SbF_5$, Yoneda et al. have obtained dicarboxylic acids from diols by reaction with CO under mild conditions.[94] Some cyclization products were also obtained. The schemes 5 and 6 were suggested for the reaction.

The formation of C_6 or C_7 acids along with some ketones was reported in the reaction of isopentane, along with methylcyclopentane, and cyclohexane with CO in $HF:SbF_5$ at ambient temperatures and atmospheric pressure.[95]

Recently, Yoneda et al. have also found that other alkanes can also be carboxylated with CO in $HF:SbF_5$. Tertiary carbenium ions which are produced by protolysis of C—H bonds of branched alkanes in $HF:SbF_5$ undergo skeletal isomerization and diproportionation before reacting with CO.[96] It was also found that alkyl methyl ketones react with CO in the HF : SbF_5 superacid system to form oxocarboxylic acids after hydrolysis. Alkyl methyl ketones with a short alkyl chain (less than C_4) do not react under these conditions due to the proximity of the positive charge on the protonated ketone and the developing carbenium ion.[97]

It has been demonstrated by Olah et al. that α,β-unsaturated ketones are *O*-protonated in HF : SbF_5 forming hydroxyallylic cations which were directly observed

Scheme 5

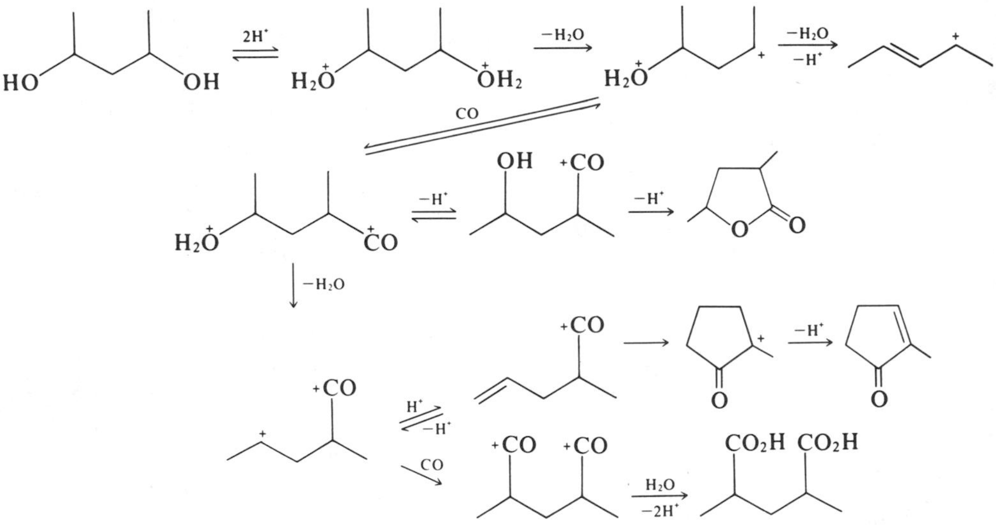

Scheme 6

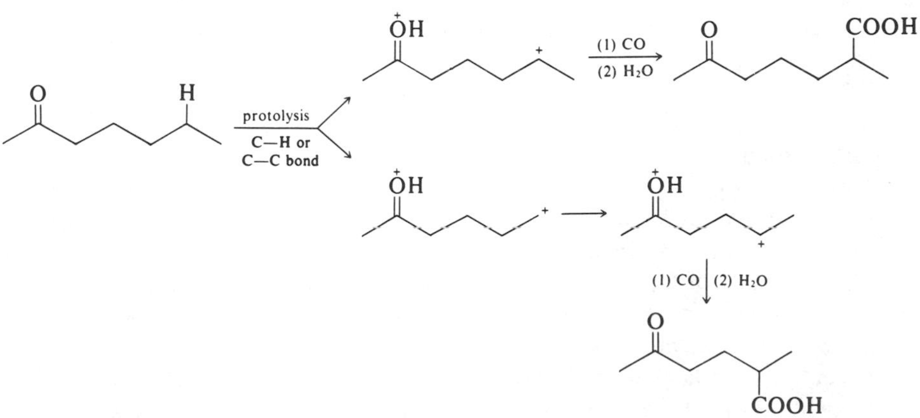

by nmr spectroscopy.[98] Jacquesy and Coustard have found indirect evidence for diprotonation of α,β-unsaturated ketones (enones) by trapping the dication with CO. The resulting acylium ion centers are then quenched with MeOH or benzene. An interesting synthetic method was therefore developed for carboxylation of bicyclic enones in superacid media at atmospheric pressure (Scheme 7).[99] When

Scheme 7

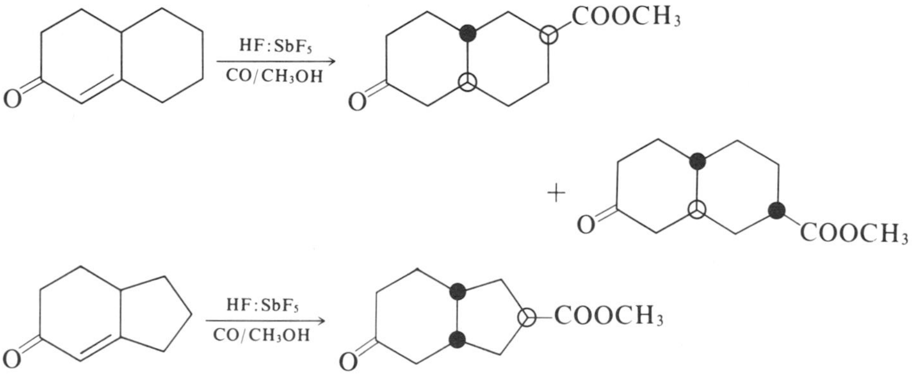

$HF:SbF_5$, CO/CH_3OH → $COOCH_3$ + $COOCH_3$

$HF:SbF_5$, CO/CH_3OH → $COOCH_3$

Scheme 8

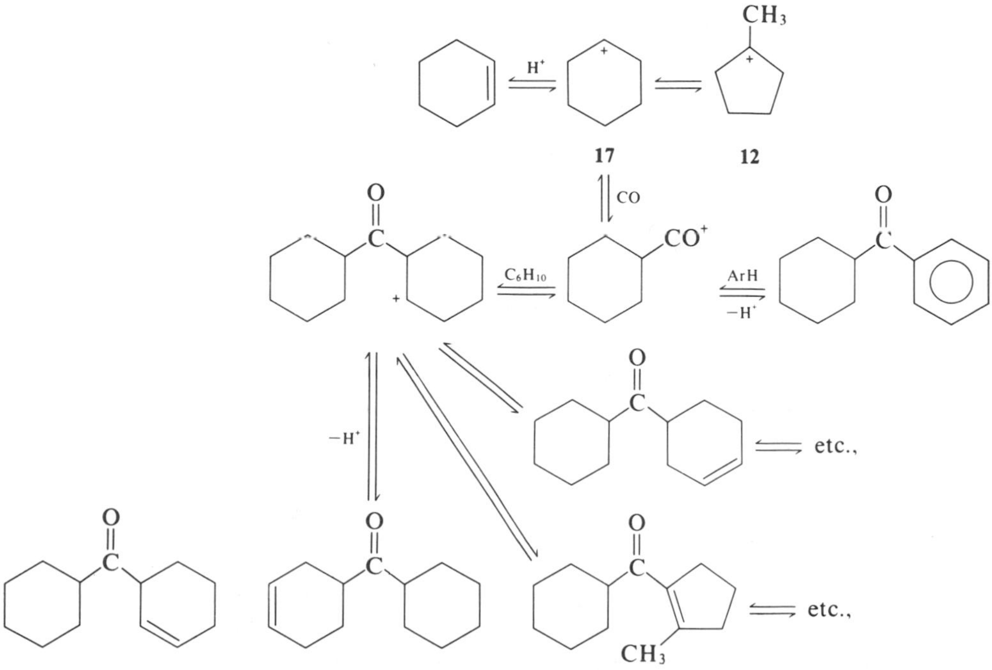

H^+ ⇌ 17 ⇌ 12 (CH_3)

CO; CO^+; C_6H_{10}; ArH, $-H^+$

$-H^+$

etc.,

etc.,

CH_3

cyclohexene is mixed with anhydrous triflic acid under a high pressure of carbon monoxide (120 atm) followed by the addition of benzene, cyclohexyl phenyl ketone and the isomeric cyclohexenyl cyclohexyl ketones are obtained with little isomerization of the initially formed cyclohexyl cation to methylcyclopentyl cation (Scheme 8).[100]

5.5 FORMYLATION

Aromatic formylation reactions are known to occur under Gatterman–Koch conditions using mixtures of CO + HCl + $AlCl_3$ and Cu_2Cl_2.[101–103] The use of superacidic HF:BF_3 and HF:SbF_5 as catalysts for aromatic formylation has been demonstrated.[104,105] Recent mechanistic studies by Olah and co-workers have shown that selectivity in formylation reactions strongly depends on the nature of the formylating agent.[106]

Among the most frequently used formylation methods, the Gatterman–Koch reaction shows the highest selectivity reflected both in the observed high $k_{toluene}:k_{benzene}$ rate ratios (155–860), as well as a high degree of para substitution (88.7–96%). Gross' formylation with dichloromethyl methyl ether[107] is somewhat less selective (k_T/k_B = 119; 60.4% para substitution), as is the Gatterman synthesis using $Zn(CN)_2$ and $AlCl_3$.[103] Friedel–Crafts-type formylation with formyl fluoride[108] gives a much lower selectivity (k_T/k_B = 34.6 and 53% para substitution) indicating that the HCOF:BF_3 system produces a more reactive electrophile (HCOF.BF_3 complex, but not necessarily a free formyl cation, HCO^+).

The lowest selectivity was observed in the case of HF:SbF_5-catalyzed formylation with CO in SO_2ClF solution at −95°C, which gave a k_T/k_B ratio of 1.6, and an isomer distribution of 45% *o*-, 2.7% *m*-, and 52.1% *p*-tolualdehydes. Under the superacidic conditions studied, CO is protonated to give the rapidly equilibrating (with the solvent acid system) protosolvated formyl cation, an obviously very reactive electrophilic reagent. When the reaction is carried out at 0°C using only excess aromatics as solvent, the selectivity becomes higher ($k_T/k_B \sim 25$) and giving an isomer distribution of 7.5% *o*-, 2.8% *m*-, and 89.8% *p*-tolualdehydes.

The formylation of hexadeuterobenzene, C_6D_6, with HCOF-BF_3 shows a kinetic hydrogen isotope effect of k_H/k_D = 2.68, based on the comparison of the reactivity of $C_6H_6/CH_3C_6H_5$ and $C_6D_6/CH_3C_6H_5$. This isotope effect is similar to that observed in Friedel-Crafts acetylation and propionylation reactions, and indicates that the proton elimination step is at least partially rate determining. The low substrate selectivity formylation with CO:HF:SbF_5 system, however, shows no primary isotope effect.

For nearly a century, Friedel–Crafts acylations were considered to give nearly exclusive para substitution of toluene. The reason accounting for this fact was considered to be steric. The present-day understanding of the mechanism of electrophilic aromatic substitution indicates that this is not necessarily the only reason. Para substitution is greatly favored if the transition state of highest energy is intermediate arenium ion (σ-complex) like, where a para methyl group is more stabilizing than an ortho (and much more than a meta). However, when the highest transition state is becoming increasingly "early" on the reaction path, the ratio of ortho:para substitution increases. Meta substitution always stays relatively low, generally less than 5–6% varying with the reactivity of the reagent within this limit. This substitution pattern is also observed in Friedel–Crafts-type formylation reactions. In these reactions, the involved substituting agents are obviously less space demanding than those of other acylation reactions. Steric effects consequently cannot be a significant factor affecting selectivity, which is primarily reflected in the changing ortho:para isomer ratio. The methyl group always remains a predominantly ortho:para directing, even in very low substrate selectivity reactions, and the meta isomer does not increase above 4%.

Regioselective formylation of toluene, *m*- and *p*-xylene, and mesitylene has been achieved by carbonylation in triflic acid at CO pressures of 90–125 atm. However, the use of six- to seven-fold excess of acid over arene is required to obtain high yields of the aldehydes.[100]

Attempts have been made[109] to observe the long-lived formyl cation under stable ion condition using ^{13}C-enriched carbon monoxide. However, even at very low temperatures proton exchange with the superacid solvent is fast on the nmr time scale.

5.6 THIO- AND DITHIOCARBOXYLATION

Aromatic carboxylic acid derivatives are generally prepared by Friedel–Crafts methods using phosgene, oxalyl chloride, or carbamoyl chlorides.[110] Carbon disulfide reacts with arenes in the presence of excess $AlCl_3$ catalyst to give dithiocarboxylic acids.[111,112] However, these reactions generally require at least a 2 mol excess of the strong Lewis acid catalyst and significant side reactions occur.

Olah and co-workers[113] have developed a mild method for the preparation of methyl and ethyl thio (dithio) benzoates by electrophilic substitution of aromatic hydrocarbons with *S*-methyl(*S*-ethyl) thiocarboxonium and dithiocarboxonium fluoroantimonates **32** and **33,** readily obtained by methylating (ethylating) carbonyl sulfide and carbon disulfide, respectively, with methyl(ethyl)fluoride–antimony pentafluoride complexes in SO_2 solution (Scheme 9).[49]

5.7 SULFONATION

Sulfonation of aromatic compounds is generally carried out with sulfuric acid, halosulfuric acids, or sulfur trioxide as reagents with or without solvent.[114] Friedel–Crafts catalysts such as aluminum chloride and boron trifluoride are effective catalysts in certain sulfonations with sulfuric acid and chlorosulfuric acid.

When SO_3 is used in fairly dilute solution, the attacking species is SO_3 itself. In concentrated sulfuric acid, however, the mechanism is more complex. Fuming sulfuric acid (in which the mole fraction of $SO_3 > 0.5$) is actually a mixture of SO_3 and ionized and nonionized monomers, dimers, trimers, and tetramers of H_2SO_4

Scheme 9

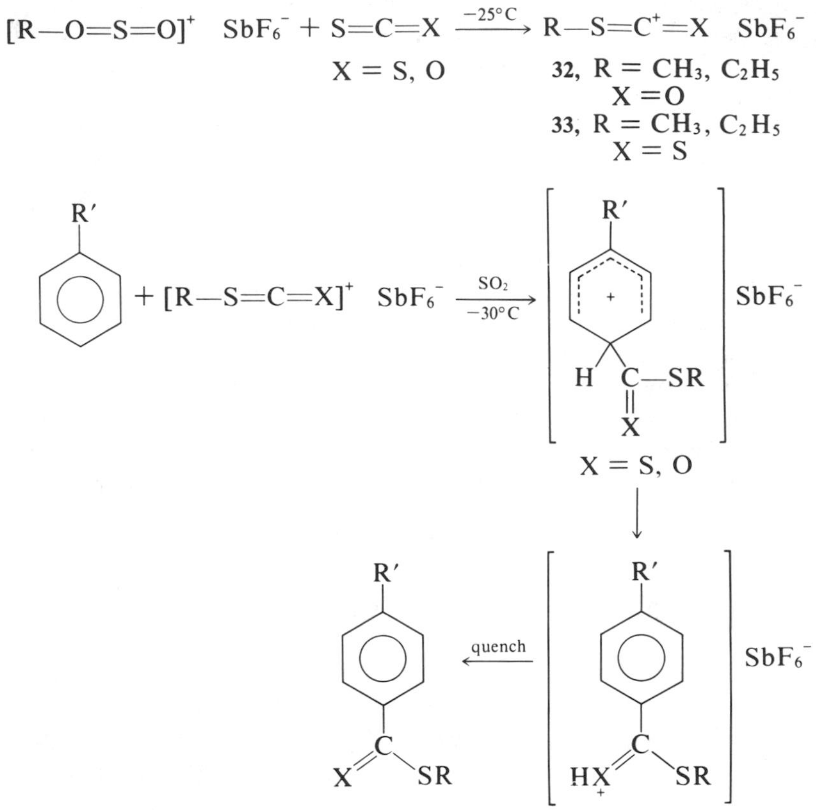

$$[R{-}O{=}S{=}O]^+ \; SbF_6^- + S{=}C{=}X \xrightarrow{-25°C} R{-}S{=}C^+{=}X \; SbF_6^-$$

$X = S, O$ — **32**, $R = CH_3, C_2H_5$, $X = O$; **33**, $R = CH_3, C_2H_5$, $X = S$

$$C_6H_5R' + [R{-}S{=}C{=}X]^+ \; SbF_6^- \xrightarrow[-30°C]{SO_2} [\text{arenium ion: } R'C_6H_5^+(H)(C(=X){-}SR)] \; SbF_6^-$$

$X = S, O$

$$\downarrow$$

$$R'C_6H_4C(=X)SR \xleftarrow{\text{quench}} [R'C_6H_4C(=\overset{+}{X}H)SR] \; SbF_6^-$$

(the latter three formed by dehydration). At higher water content, the tetramer and trimer disappear and the amount of dimer decreases. The reactive species in sulfuric acid thus depends on the amount of water in the acid and on the reactivity of the substrate. The reactive species in aqueous sulfuric acid are H_2SO_4 and $H_2S_2O_7$, the latter being more important at higher acid concentrations. In fuming sulfuric acid, $H_3S_2O_7^+$ and $H_2S_4O_{13}$ are also involved.[115]

Chlorosulfuric acid ($ClSO_3H$) reacts with aromatic hydrocarbons to give sulfonic acids, sulfonyl chlorides, and sulfones, the relative yields depending on the reaction conditions. The reaction with benzene with an equimolar amount of chlorosulfuric acid in sulfur dioxide as solvent at $-8°C$ yields mainly benzenesulfonic acid with only minor amounts of diphenyl sulfone.[116] However, with excess of the acid arylsulfonyl chloride is also formed.[117] Compared with $ClSO_3H$, HSO_3F is a poorer sulfonating agent and tends to give arylsulfonyl fluorides more easily; excess of halosulfuric acids give halosulfonation.

Nafion-H has also been used as a sulfonation catalyst.[118] When oleum and long-chain alkylbenzenes were separated from each other by a Nafion-H membrane, the membrane transported the sulfonating agent into the alkylbenzene at a rate convenient for dissipating the heat of the reaction. Reported yields of sulfonation products were 34% (4 h), 63% (6 h), and 86% after 22.5 h.

5.8 NITRATION

Conventional nitration of aromatic compounds is carried out using a mixture of nitric acid and sulfuric acid (mixed acid). There are, however, difficulties associated with the use of mixed acid.[119] In particular, the water formed as the reaction proceeds dilutes the acid and therefore reduces its strength. Also, the strong oxidizing ability of a mixed acid system makes it unsuitable to nitrate many acid-sensitive compounds. The disposal of the spent acid also poses a significant environmental problem. To overcome these difficulties of anhydrous Friedel–Crafts type nitrations catalyzed by strong acids, nitronium salts (such as $NO_2^+BF_4$, $NO_2^+PF_6$, $NO_2^+SbF_6^-$, etc.) were developed[120] which are extremely powerful nitrating agents.

Stable nitronium salts which are readily prepared from nitric acid (or nitrates) with HF and BF_3 (and other Lewis acids such as PF_5, SbF_5, etc.) nitrate[119] aromatics in organic solvents generally with close to quantitative yield. As HF and PF_5 (or BF_3) can be easily recovered and recycled, the method can be considered as a nitric acid nitration using a superacid catalyst.

$$HNO_3 + HF + 2BF_3 \rightleftharpoons NO_2^+BF_4^- + BF_3 \cdot H_2O$$

$$RONO_2 + HF + 2BF_3 \rightleftharpoons NO_2^+BF_4^- + BF_3 \cdot ROH$$

$$ArH + NO_2^+PF_6^- \ (BF_4^-) \longrightarrow ArNO_2 + HF + PF_5 \ (HF + BF_3)$$

The powerful nature of nitronium salts as nitrating agents is demonstrated in their ability to affect even trinitration[121] of benzene to trinitrobenzene **34.** Nitronium salts enable nitration of every conceivable aromatic substrate.

$NO_2^+BF_4^-$ in HSO_3F → 1,3,5-trinitrobenzene (O_2N, NO_2, NO_2)

34

The powerful nitronium ion salts are also capable of reacting with aliphatics. Electrophilic nitration of alkanes and cycloalkanes has been carried out with $NO_2^+PF_6^-$, NO_2^+, SbF_6^-, or $NO_2^+BF_4^-$ salts in CH_2Cl_2-tetramethylenesulfone or HSO_3F solution.[122] Table 5.20 lists some representative reactions of nitronium ion with various alkanes and cycloalkanes. The formation of nitroaliphatics indicates the insertion of nitronium ion into aliphatic σ-bonds involving two-electron, three-center-bonded, five-coordinate carbocations.

$$\text{Adamantane (1-H)} + NO_2^+ \rightleftharpoons \left[\text{1-adamantyl} \overset{+}{\underset{}{C}}(H)(NO_2) \right] \rightleftharpoons \text{1-nitroadamantane} + H^+$$

More selective nitronium ions such as *N*-nitropyridinium salts, which are readily prepared[123a,b] from the corresponding pyridine and nitronium salts, act as convenient transfer nitrating agents. Transfer nitrations are applicable to C-nitrations as well as to a variety of heteroatom nitrations. For example, they allow safe, acid-free preparation of alkyl nitrates and polynitrates from alcohols[123c] (polyols) in nearly quantitative yield.

$$1,3,5\text{-}(CH_3)_3C_6H_3 + \text{R-}C_5H_4\overset{+}{N}\text{-}NO_2\ PF_6^-(BF_4^-) \longrightarrow 1,3,5\text{-}(CH_3)_3C_6H_2NO_2 + \text{R-}C_5H_4\overset{+}{N}\text{-}H\ PF_6^-(BF_4^-)$$

$$\underset{OH}{CH_2}\text{—}\underset{OH}{CH}\text{—}\underset{OH}{CH_2} + 3\ \text{R-}C_5H_4\overset{+}{N}\text{-}NO_2\ (BF_4^-) \longrightarrow \underset{ONO_2}{CH_2}\text{—}\underset{ONO_2}{CH}\text{—}\underset{ONO_2}{CH_2} + 3\ \text{R-}C_5H_4\overset{+}{N}\text{-}H\ (BF_4^-)$$

$$R = H, CH_3, (CH_3)_2, (CH_3)_3$$

TABLE 5.20 Nitration and Nitrolysis of Alkanes and Cycloalkanes with $NO_2^+PF_6^-$

Hydrocarbon	Nitroalkane Products and Their Mol Ratio
Methane	CH_3NO_2
Ethane	$CH_3NO_2 > CH_3CH_2NO_2$, 2.9:1
Propane	$CH_3NO_2 > CH_3CH_2NO_2 > 2NO_2C_3H_7 > 1NO_2C_3H_7$, 2.8:1:0.5:0.1
Isobutane	t-$NO_2C_4H_9 > CH_3NO_2$, 3:1
n-Butane	$CH_3NO_2 > CH_3CH_2NO_2 > 2NO_2C_4H_9 \sim 1NO_2C_4H_9$, 5:4:1.5:1
Neopentane	$CH_3NO_2 > t$-$C_4H_9NO_2$, 3.3:1
Cyclohexane	Nitrocyclohexane
Adamantane	1-Nitroadamantane > 2-nitroadamantane, 17.5:1

Electrophilic nitration of olefins can also be carried out with nitronium salts in pyridinium polyhydrogen fluoride solution[124] (which also acts as solvent) to give high yields of nitrofluorinated alkanes. In the presence of added halide ions (iodide, bromide, chloride) the related haloalkanes are formed, and these can be dehydrohalogenated to nitroalkenes.

$NO_2^+BF_4^-$; pyridine $NH^+(HF)_nF^-$; F ; NO_2 ; $-HF$; NO_2 ; + ; NO_2

Until recently, the nitronium ion was recognized only as a nitrating agent. In recent work, it was found[125,126] to possess significant ambident reactivity and thus capable of acting as an oxidizing agent. Dialkyl- (aryl-) sulfides and selenides as well as trialkyl- (aryl-) phosphines, triarylarsines, and triarylstibines react with nitronium salts to give the corresponding oxides.

$$R{-}S{-}R \xrightarrow{NO_2^+} R{-}\overset{NO_2}{\underset{+}{S}}{-}R + R{-}\overset{ONO}{\underset{+}{S}}{-}R \longrightarrow R{-}\overset{O}{\overset{\|}{S}}{-}R + NO^+$$

$$R{-}Se{-}R \xrightarrow{NO_2^+} R{-}\overset{O}{\overset{\|}{Se}}{-}R + NO^+$$

$$R_3X \xrightarrow{NO_2^+} R_3X{=}O + NO^+$$

$$X = P, As, \text{ or } Sb$$

$$R = \text{alkyl, aryl}$$

Stable nitronium (NO_2^+) salts, particularly with PF_6^- and BF_4^- counterions, can act as mild, selective oxidative cleavage reagents[127] for a wide variety of functional groups. The following examples illustrate the utility of these methods.

$$Ar_2CHOMe \xrightarrow[(2)\ H_2O]{(1)\ NO_2^+X^-\ \text{or}\ NO^+X^-} Ar_2C{=}O$$

$$\text{cyclohexanone oxime } (C_6H_{10}{=}N{-}OH) \longrightarrow \text{cyclohexanone } (C_6H_{10}{=}O)$$

$$\text{cyclopentanone } N,N\text{-dimethylhydrazone } (C_5H_8{=}N{-}N(CH_3)_2) \longrightarrow \text{cyclopentanone } (C_5H_8{=}O)$$

$$\text{2-aryl-1,3-dithiolane } (Ar(H)C(SCH_2CH_2S)) \longrightarrow ArCHO$$

$$ArCH_2{-}O{-}C({=}O){-}R \longrightarrow RCOOH$$

$$R = \text{alkyl, aryl}$$

$$ArCH_2OH \longrightarrow ArCHO$$

The solid superacidic catalyst Nafion-H has also been found to catalyze effectively nitration reactions with various nitrating agents.[128] The nitrating agents employed were *n*-butyl nitrate, acetone cyanohydrin nitrate, and fuming nitric acid. In nitric acid nitrations, sulfuric acid can be substituted by Nafion-H and the water formed is azeotropically removed during the reaction (azeotropic nitration).[128]

Nitrobenzene was also prepared in a liquid flow system using a hollow tube of Nafion-H membrane. Benzene was passed over the outside of tube containing 70% nitric acid. The yield of nitrobenzene was 19%.[129]

The nitration of aromatic compounds has also been carried out with Nafion-H in the presence of mercuric nitrate as well as Hg(II)-impregnated Nafion-H catalyst.[130]

These studies indicate that Nafion-H:HNO_3 and Nafion-H:$(HNO_3)Hg^{2+}$ nitrate by different mechanisms. It was proposed that the mercury-containing catalyst operates in part by mercurating the arene followed by nitrodemercuration of the initial product.

Generally, nitration of aromatic compounds is considered to be an irreversible reaction. However, recently the reversibility of the reaction has been demonstrated.[131] 9-Nitroanthracene and pentamethylnitrobenzene transnitrate benzene, toluene, and mesitylene in the presence of $HF:SbF_5$ as well as Nafion-H.

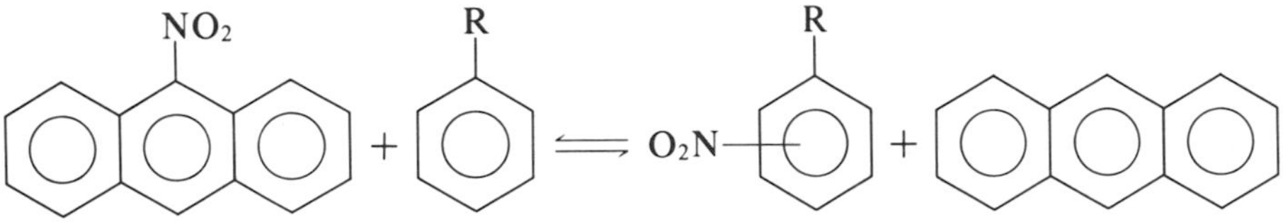

5.9 NITROSONIUM ION (NO^+)-INDUCED REACTIONS

Nitrosonium ion (NO^+) is the electrophilic species formed from nitrous acid media, which is responsible for such reactions as the diazotization of amines. Nitrosonium ion has been isolated as salts with a wide variety of counter ions such as BF_4^-, PF_6^-, SbF_6^-, etc.[132] The nitrosonium does not react with aromatics except in the case of activated systems (such as *N,N*-dimethylaniline or phenols). Frequently, they form colored π-complexes.[133,134] Nitrosonium is a powerful hydride abstracting agent. Cumene reacts with NO^+ to give various condensation products, which involves intermediate formation of cumyl cation.[135] Similarly, the nitrosonium ion is employed in the preparation of a variety of stabilized carbocations.[136] Some of the reactions of nitrosonium ion are depicted in Figure 5.7.[137]

The hydride abstraction reaction of NO^+ has been employed in a modified Ritter-type reaction[138] as well as in ionic fluorination[139] of bridgehead hydrocarbons.

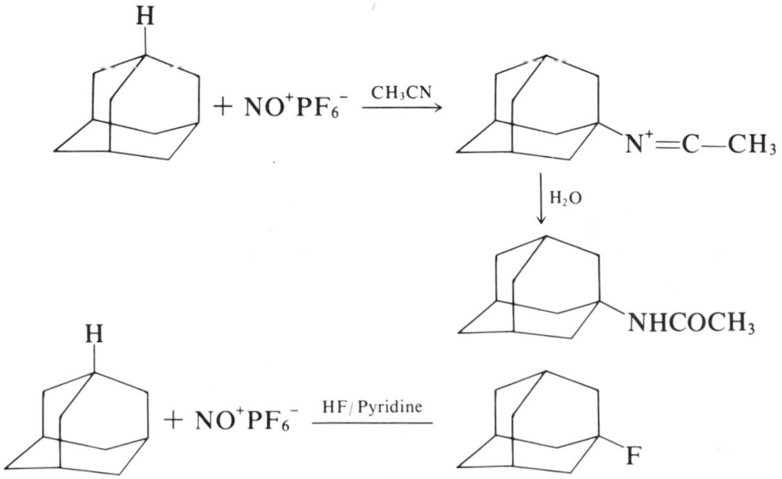

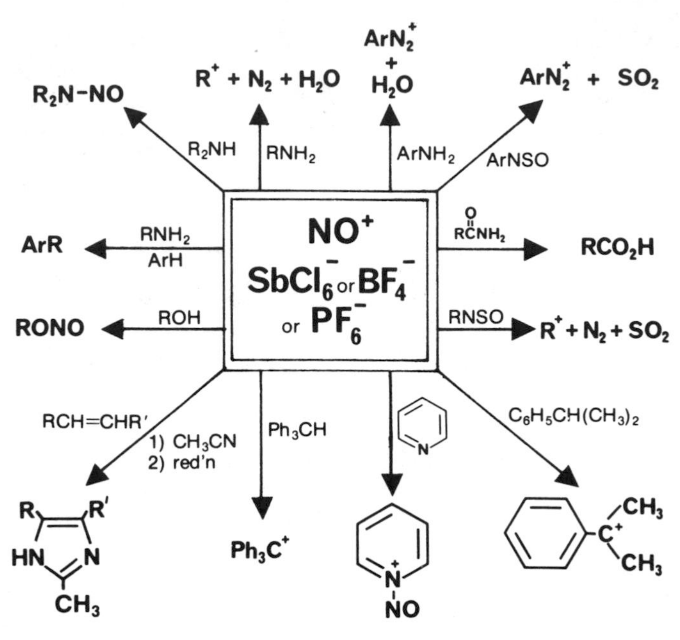

FIGURE 5.7 Some nitrosonium induced reactions.

Similarly, NO^+ is also capable of halogen abstraction from alkyl halides.[140,141] In the presence of suitable oxygen donor such as dimethylsulfoxide, nitrosonium ion can act as a nitrating agent.[125]

$$CH_3-\overset{+}{\underset{\substack{|\\CH_3}}{S}}-O^- + \overset{+}{N}O \rightleftharpoons CH_3-\overset{+}{\underset{\substack{|\\CH_3}}{S}}-ONO$$

$$\Big\Updownarrow$$

$$ArNO_2 \xleftarrow{ArH} CH_3-\overset{+}{\underset{\substack{|\\CH_3}}{S}}-NO_2$$

Nitrosonium ion can also act as a mild and selective oxidizing agent. It has been used to cleave oxidatively oximes, hydrazones, thioketals, ethers, etc., to their corresponding carbonyl compounds.[127]

5.10 HALOGENATION

The halogenation of saturated aliphatic hydrocarbons is usually achieved by free radical processes.[142] Ionic halogenation of alkanes has been reported under superacid catalysis. Olah and co-workers have carried[143] out chlorination and chlorolysis of alkanes in the presence of SbF_5, Al_2Cl_6, and $AgSbF_6$ catalysts. As a representative, the reaction of methane with $Cl_2:SbF_5$ is depicted below.

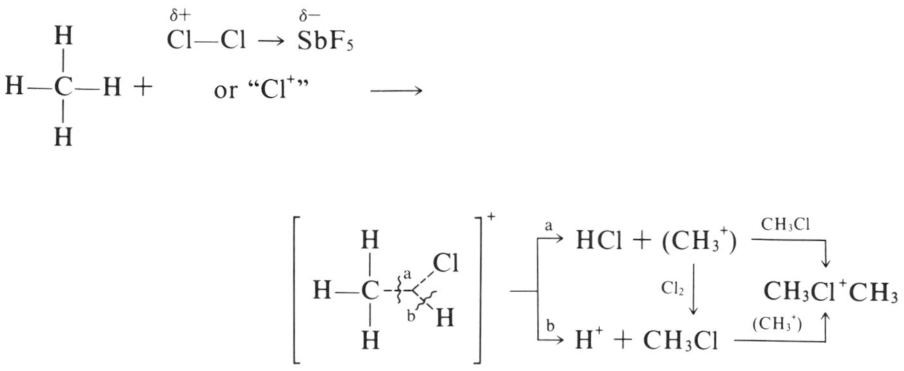

$$CH_4 + \overset{\delta+}{Cl}-Cl \rightarrow \overset{\delta-}{SbF_5} \text{ or } \text{“}Cl^+\text{”} \longrightarrow \left[H_3C\cdots\genfrac{}{}{0pt}{}{Cl}{H} \right]^+$$

$$\xrightarrow{a} HCl + (CH_3^+) \xrightarrow{CH_3Cl} CH_3Cl^+CH_3$$

$$\xrightarrow{b} H^+ + CH_3Cl \xrightarrow{(CH_3^+)} CH_3Cl^+CH_3$$

$$(CH_3^+) \xrightarrow{Cl_2} CH_3Cl$$

The results of $AgSbF_6$-catalyzed chlorinations are shown in Table 5.21. More recently, selective ionic chlorination of methane to methyl chloride has been achieved in the gas phase over solid superacid catalysts.[144]

Similarly, electrophilic bromination of alkanes has also been carried out.[145]

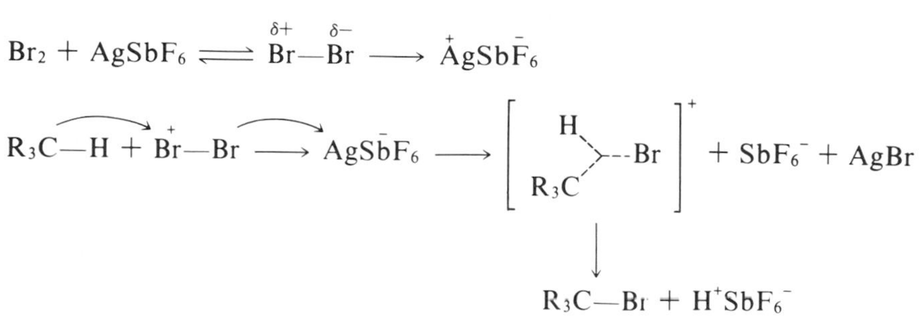

$$Br_2 + AgSbF_6 \rightleftharpoons \overset{\delta+}{Br}-\overset{\delta-}{Br} \longrightarrow \overset{+}{A}gSb\overset{-}{F_6}$$

$$R_3C-H + \overset{+}{Br}-Br \longrightarrow Ag\overset{-}{Sb}F_6 \longrightarrow \left[\genfrac{}{}{0pt}{}{H}{R_3C} \rangle\cdots Br \right]^+ + SbF_6^- + AgBr$$

$$\downarrow$$

$$R_3C-Br + H^+SbF_6^-$$

Even electrophilic fluorination of alkanes is possible. F_2 and fluoroxytrifluoromethane have been used to fluorinate tertiary centers in steroids and adamantanes by Barton and co-workers.[146] The strong influence of electron-withdrawing substituents on the substrates to the reaction rate as well as reaction selectivity in the presence of radical inhibitors seems to suggest the electrophilic nature of the reaction involving polarized, but not cationic fluorine species. Claims for the latter have been refuted.[147] Rozen and co-workers[148] have carried out direct electrophilic fluorination of hydrocarbons in the presence of chloroform. Fluorine appears to be strongly polarized in chloroform (hydrogen bonding with acidic proton of chloroform). However, so far no positively charged fluorine species is known in solution chemistry (i.e., fluoronium ions).

The halogenation of a wide variety of aromatic compounds proceeds readily in the presence of ferric chloride, aluminum chloride, and related Friedel-Crafts ca-

TABLE 5.21 $AgSbF_6$-Induced Chlorination of Alkanes with Chlorine in the Dark[a,b]

Alkane	Molar Ratio $Cl_2:AgSbF_6$:Alkane	Reaction Temperature (°C) and Time (min)	Reaction Products[c] (%)
Isobutane	5:1:10	−15,10	*t*-Butyl chloride (7.1)
		0,10	*t*-Butyl chloride (5.3)
Isopentane	5:1:10	−15,10	*t*-Pentyl chloride (0.7)
		0,10	*t*-Pentyl chloride (1.4)
		10,10	*t*-Pentyl chloride (5.0)
Cyclopropane	5:1:10	−15,10	*n*-Propyl chloride (40.0)
			Isopropyl chloride (31.5)
			1,3-Dichloropropane (1.2)
		−10,10	*n*-Propyl chloride (39.2)
			Isopropyl chloride (36.0)
			1,3-Dichloropropane (1.4)
Cyclopentane	5:1:10	−15,10	Cyclopentyl chloride (0.5)
			1,2-Dichlorocyclopentane (1.8)
		−15,20	Cyclopentyl chloride (0.3)
			1,2-Dichlorocyclopentane (1.9)

		−15,60	Cyclopentyl chloride (0.5)
			1,2-Dichlorocyclopentane (7.1)
		25,180	Cyclopentyl chloride (0.3)
			1,2-Dichlorocyclopentane (3.0)
Cyclohexane	5:1:10	−15,10	Cyclohexyl chloride (5.1)
		−15,60	Cyclohexyl chloride (3.3)
Norbornane	5:1:10	25,15	*exo*-2-Chloronorbornane (66.6)
			endo-2-Chloronorbornane (15.8)
			7-Chloronorbornane (7.6)
		25,30	2-*exo*-Chloronorbornane (69.5)
			2-*endo*-Chloronorbornane (18.3)
			7-Chloronorbornane (6.0)
Adamantane	5:1:10	−15,5	1-Chloroadamantane (62.4)
			2-Chloroadamantane (1.2)
			1-Hydroxyadamantane (36.3)

[a] In CH_2Cl_2.
[b] Analysis by gas-liquid chromatography.
[c] Yields of the reaction products based on the amount of $AgSbF_6$.

talysts. Halogenating agents generally used are elementary chlorine, bromine, or iodine and interhalogen compounds (such as iodine monochloride, bromine monochloride, etc.). These reactions were reviewed[149] and are outside the scope of the present discussion.

Generally, electrophilic halogenation of phenols leads to corresponding ortho:para substituted products. The synthesis of meta substituted products is considered difficult.[120]

The protonation of phenols and alkyl phenyl ethers in superacids has been extensively investigated.[98] Both oxygen protonation as well as ring protonation occurs. In general, oxygen protonation is observed for unsubstituted phenols and unsubstituted alkyl phenyl ethers, usually at low temperatures. Substitution of electron-releasing groups in the aromatic ring causes ring carbon protonation to predominate.

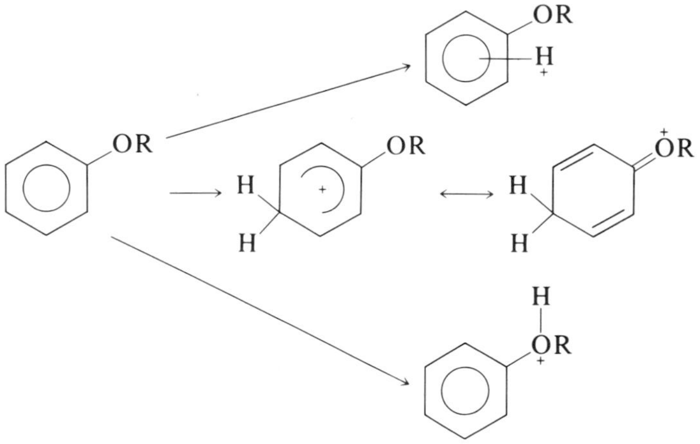

Recently, Jacquesy and co-workers[150] have succeeded in preparing *m*-bromophenols from phenols using Br_2:HF:SbF_5 system.

The *o*-protonated phenol or alkylphenyl ether, which is in equilibrium with the neutral precursor, reacts with the reactive Br^+ in the HF:SbF_5 medium leading to only meta-brominated phenols. The ring protonated phenols are unreactive toward electrophilic bromine in the superacid medium. More recently, Jacquesy et al.[151] have also studied the mechanism of isomerization of *o*- and *p*-bromophenols to metabromophenols using HF:SbF_5 and CF_3SO_3H superacid systems.

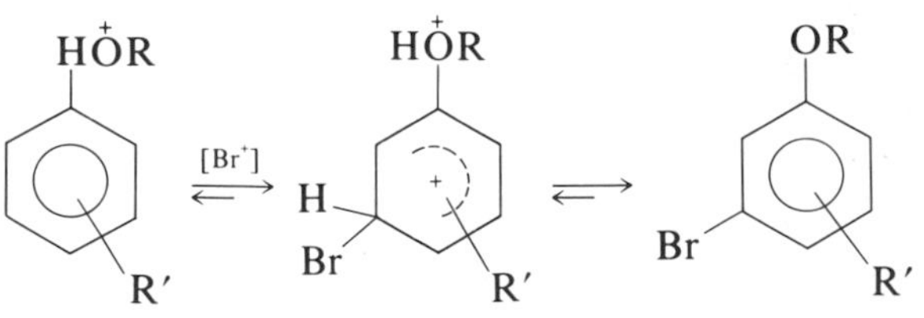

5.11 AMINATION

Electrophilic amination of aromatics is not a widely used reaction. However, attempts have been made with reagents such as hydroxyl ammonium salts,[152] hydroxylamine-*o*-sulfonic acid,[153] hydrazoic acid,[154] as well as organic azides,[155] mostly under Lewis acid-catalyzed conditions (and also with thermal initiation[156] or photolysis).[157] Kovacic and co-workers have found that haloamines[158] can be used as reagents for aromatic amination giving preferential meta subtitution in presence of large excess of $AlCl_3$ catalyst. They have also investigated[159] the aromatic amination reaction with hydrazoic acid catalyzed by $AlCl_3$ or H_2SO_4, but the reaction condition necessitate a 2:1 ratio of catalyst to azide at reflux temperature, giving only modest yields.

Recently Olah and co-workers have carried out a comprehensive investigation of aminodiazonium ions under superacid conditions using ^{13}C and ^{15}N-nmr spectroscopic methods[160] (see Chapter 4). They also studied the electrophilic amination ability of these aminodiazonium ions. It has been found that NaN_3 (or trimethylsilyl azide) reacts with $AlCl_3$ and dry HCl in situ to form aminodiazonium tetrachloroaluminate **35**,[161,162] which reacts with a variety of arenes to give the corresponding aromatic amines. Best results for aromatic aminations were obtained when an excess of the aromatic substrate itself was used as the reaction medium. The results are summarized in Table 5.22.

$$NaN_3 + AlCl_3 \longrightarrow AlCl_2N_3 + NaCl$$

$$AlCl_2N_3 + 2HCl \longrightarrow \underset{\mathbf{35}}{H_2N{-}\overset{+}{N}{\equiv}N\ AlCl_4^-}$$

$$35 + ArH \longrightarrow ArNH_2 \cdot HCl + N_2 + AlCl_3$$

5.12 OXYFUNCTIONALIZATION

Converting alkanes and aromatics in a controlled way into their oxygenated compounds is of substantial interest. The discovery and development of superacidic systems and weakly nucleophilic solvent such as $HSO_3F:SbF_5:SO_2$, $HSO_3F:SbF_5:SO_2ClF$, and $HF:SbF_5:SO_2ClF$ has enabled the preparation and study of a variety of carbocations (see Chapter 3). In connection with these studies, it was also found that electrophilic oxygenation of alkanes with ozone (O_3) and hydrogen peroxide (H_2O_2) takes place readily in the presence of superacids under typical electrophilic conditions.[163] The reactions giving oxyfunctionalized products of alkanes can be explained as proceeding via initial electrophilic attack by protonated ozone i.e. ($\overset{+}{O}_3H$) or the hydrogen peroxonium ion $H_3O_2^+$, respectively, on the σ-bonds of alkanes through pentacoordinated carbonium ions.

TABLE 5.22 Yield and Isomer Distribution of Aromatic Amines Obtained by Amination of Aromatic Substrates with Aminodiazonium Tetrachloroaluminate

		Isomer Distribution (%)		
	Yield (%)	Ortho	Meta	Para
C_6H_6	63.3			
CH_3–C_6H_5	72.6	47.3	13.8	38.9
$C_6H_3(CH_3)_3$ (1,3,5-)	77.8			
$C_6H_2(CH_3)_4$ (1,2,3,4-)	39.0			
C_6H_5–Cl	25.1	28.5	15.3	56.2
H_3C–C_6H_4–CH_3 (p-)	69.8			
C_6H_5–OCH_3	48.7	74.4	4	21.6
C_6H_5–NO_2	1.5			

5.12.1 Oxygenation with Hydrogen Peroxide

Hydrogen peroxide (H_2O_2) in superacid media is protonated to hydrogen peroxonium ion ($H_3O_2^+$). Recently, Christe et al.[164] have reported characterization and even isolation of several peroxonium salts. The ^{17}O-nmr spectrum of $H_3O_2^+$ has also been obtained.[165]

The hydrogen peroxonium ion may be considered as an incipient OH^+ ion capable of electrophilic hydroxylation of single (σ) bonds in alkanes, and thus be able to effect reactions similar to such previously described electrophilic reactions as protolysis, alkylation, chlorination (chlorolysis), and nitration (nitrolysis).

The reaction of branched-chain alkanes with hydrogen peroxide in Magic Acid–

TABLE 5.23 Products of the Reaction of Branched-Chain Alkanes with H_2O_2 in $HSO_3F:SbF_5:SO_2ClF$ Solution

Alkane	Alkane [mmol]	H_2O_2 [mmol]	Temperature [b] [°C]	Major Products
C—C(C)(H)—C	2	2	−78 → −20	$(CH_3)_2C{=}\overset{+}{O}CH_3$
	2	4	−78 → −20	$(CH_3)_2C{=}\overset{+}{O}CH_3$
	2	6	−78 → −20	$(CH_3)_2C{=}\overset{+}{O}CH_3$
	2	6	+20	$(CH_3)_2C{=}\overset{+}{O}CH_3$ (trace), DAP[a] (25%), CH_3OH (50%), $CH_3CO{—}O{—}CH_3$ (25%)
C—C—C(C)(H)—C	2	3	−78 → −20	$(CH_3)_2C{=}\overset{+}{O}C_2H_5$, $C_2H_5(CH_3)_2C^+$
	2	6	−78 → −20	$(CH_3)_2C{=}\overset{+}{O}C_2H_5$
C—C(C)(C)—C(C)(H)—C	2	4	−78	$(CH_3)_2C{=}\overset{+}{O}CH_3$ (50%), $(CH_3)_2C{=}\overset{+}{O}H$ (50%)
	2	6	−40	$(CH_3)_2C{=}\overset{+}{O}CH_3$ (50%), DAP (50%)

[a] DAP, Dimeric acetone peroxide.

SO_2ClF solution has been carried out with various ratios of alkane and hydrogen peroxide, and at different temperatures.[166] Some of the results are summarized in Table 5.23.

As neither hydrogen peroxide nor Magic Acid–SO_2ClF alone led to any reaction under the conditions employed, the reaction must be considered to proceed via electrophilic hydroxylation. Protonated hydrogen peroxide inserts into the C—H bond of the alkane. A typical reaction path is as depicted in Scheme 10 for isobutane.

The reaction proceeds via a pentacoordinate hydroxycarbonium ion transition state, which cleaves to either *t*-butyl alcohol or the *t*-butyl cation. Since 1 mol of isobutane requires 2 mol of hydrogen peroxide to complete the reaction, one can conclude that intermediate alcohol or carbocation reacts with excess hydrogen peroxide, giving *t*-butyl hydroperoxide. The superacid-induced rearrangement and cleavage of the hydroperoxide results in very rapid formation of the dimethylmethylcarboxonium ion, which upon hydrolysis gives acetone and methyl alcohol.

When the reaction was carried out at room temperature, by means of passing isobutane into a solution of Magic Acid and excess hydrogen peroxide, the formation of methyl alcohol, methyl acetate, and some dimethylmethylcarboxonium ion together with dimeric acetone peroxide was observed. These results clearly show that

Scheme 10

$$HOOH \overset{H^+}{\rightleftharpoons} H\overset{H}{\underset{+}{O}}OH \equiv (\overset{+}{O}HH_2O)$$

$$(CH_3)_3C{-}H \xrightarrow[-78^\circ C]{H_3O_2^+} \left[(CH_3)_3C\cdots\begin{matrix} H \\ OH \end{matrix} \right]^+ \xrightarrow{-H^+} (CH_3)_3C{-}OH$$

$$\left[(CH_3)_3C\cdots\begin{matrix} H \\ OH \end{matrix} \right]^+ \xrightarrow{-H_2O} (CH_3)_3C^+ \qquad (CH_3)_3C{-}OH \xrightarrow[-H_2O]{H^+} (CH_3)_3C^+$$

$$\text{Acetone} + \text{methanol} \xleftarrow{H_2O} \begin{matrix} H_3C \\ H_3C \end{matrix}\!\!>C{=}\overset{+}{O}{-}CH_3 \xleftarrow{-H_2O} (CH_3)_3C{-}O^+{-}OH \xleftarrow{H_2O_2} (CH_3)_3C^+$$

the products observed can be rationalized as arising from hydrolysis of the carboxonium ion and from Baeyer-Villiger oxidation of acetone.

The mechanism was substantiated by independent treatment of alkane hydroperoxides with Magic Acid.[167] Similarly, Baeyer-Villiger oxidation of several ketones in the presence of H_2O_2 and superacids gave similar product compositions.[168]

Under the same reaction conditions as employed for branched-chain alkanes, straight-chain alkanes such as ethane, propane, butane, and even methane gave related oxygenation products.[166]

Methane, when reacted with hydrogen peroxide-Magic Acid above 0°C, gave mainly methyl alcohol. A similar result was obtained with hydrogen peroxide–HSO_3F at 60°C. Ethane with hydrogen peroxide–Magic Acid at -40°C gave ethyl alcohol. The reaction of propane with hydrogen peroxide takes place more easily than that of methane or ethane and yields isopropyl alcohol as the initial oxidation product. On raising the temperature, isopropyl alcohol gave acetone, which underwent further oxidation with hydrogen peroxide, giving dimeric acetone peroxide, methyl acetate, methyl alcohol, and acetic acid.

Although there have been reports of the direct, one-step hydroxylation of aromatic compounds with peracids in the presence of acid catalysts, monohydroxylated products, i.e., phenols, have generally been obtained in only low yields.[168] Although moderate to good yields of phenols, based on the amount of hydrogen peroxide used, were reported for the $AlCl_3$-catalyzed reaction of simple aromatics with hydrogen peroxide, a 10-fold excess of the aromatics was used over hydrogen peroxide.[168k] The conversion of the aromatics thus was low, probably due to the fact that introduction of an OH group into the aromatic ring markedly increases its reactivity and thus tends to promote further side reactions.[169]

It is well recognized that phenols are completely protonated in superacidic solutions.[98] This raised the possibility that protonated phenols, once formed in these media, might resist further electrophilic attack. Electrophilic hydroxylations of aromatics with hydrogen peroxide (98%) in superacidic media has been achieved by Olah and Ohnishi[170] in Magic Acid which allows clean, high-yield preparation of monohydroxylated products. Benzene, alkylbenzenes, and halobenzenes are efficiently hydroxylated at low temperatures (-75°C). The obtained yields and isomer distributions are shown in Table 5.24. Subsequently, Olah and co-workers found that benzene and alkylbenzenes are smoothly hydroxylated using 30% H_2O_2 in $HF:BF_3$ acid system at low temperatures.[171] The above method is particularly attractive because the acid system is recoverable and recyclable.

All these reactions involve hydroxylation of aromatics by hydrogen peroxonium ion.

$$\text{R-C}_6\text{H}_5 + H_3O_2^+ \xrightarrow{-H_2O} [\text{R-C}_6\text{H}_5(\text{OH})(\text{H})]^+ \longrightarrow \text{R-C}_6\text{H}_4\text{-OH}$$

Olah has also hydroxylated[172] naphthalene to the corresponding naphthols. Depending upon the acid strength of the medium employed preferentially, α- or β-naphthol is formed.

Jacquesy and co-workers[173a,b] have developed a new route to resorcinols by reacting alkyl-substituted phenols or their ethers with H_2O_2 in $HF:SbF_5$. The products observed are formed by the reaction of $H_3O_2^+$ either on neutral substrate or on the corresponding *O*-protonated ions. When the C-protonated form is highly stabilized by alkyl substituents, no hydroxylation occurs. Higher phenol ethers (R = Et, *n* = Pr, *i*-Pr) isomerize[173c] and are then hydroxylated to dialkyl resorcinols. Even hydroxylations of aromatic aldehydes and ketones in $HF:SbF_5$ medium has been achieved with H_2O_2.[174a] No Baeyer-Villiger oxidation products were detected in these reactions. The procedure has also been adopted for the hydroxylation of aromatic amines.[174b]

It is noteworthy to point out that polyhydroxylation in most of the cases is practically suppressed under the reaction conditions because phenols are totally *O*-protonated in the superacid media and are thus deactivated against further electrophilic attack or secondary oxidation. However, recently it has been shown by Jacquesy et al.[174c] that hydroxylation of α- and β-naphthols can be achieved in $H_2O_2:HF:SbF_5$ under certain conditions. The hydroxylation occurs selectively on the nonphenolic ring.

5.12.2 Oxygenation with Ozone

Ozone can be depicted as the resonance hybrid of cannonical structures **36a–d**[175] and one might expect ozone to react as a 1,3 dipole, an electrophile, or a nucleophile.

TABLE 5.24 Yields and Isomer Distributions of the Hydroxylation of Aromatics[a]

Starting Aromatic Substrate	% Isomer Distribution[b]				% Yield[c]
Benzene					67
Fluorobenzene	24 (2)	3 (3)	73 (4)		82
Chlorobenzene	28 (2)	7 (3)	65 (4)		53
Toluene	71 (2)	6 (3)	23 (4)		67
Ethylbenzene	68 (2)	6 (3)	26 (4)		70
s-Butylbenzene	49 (2)	11 (3)	40 (4)		55
Isobutylbenzene	65 (2)	7 (3)	28 (4)		83
n-Amylbenzene	64 (2)	7 (3)	29 (4)		67
o-Xylene	12 (2,6)	59 (2,3)	29 (3,4)		63
m-Xylene	16 (2,6)	2 (2,5)	82 (2,4)	1 (2,3)	73
p-Xylene	64 (2,5)	36 (2,4)			65
1,2,3-Trimethylbenzene	3 (2,3,6)	91 (2,3,4)	6 (3,4,5)		43
1,2,4-Trimethylbenzene	9 (2,4,6)	30 (2,3,6)	61 (2,3,5 + 3,4,6)		57
1,3,5-Trimethylbenzene	100 (2,4,6)				57

[a]In $HSO_3F:SO_2ClF$ solution at dry-ice temperature.

[b]Based on chromatographic analysis of quenched phenolic products. Parentheses show position of substituent(s).

[c]Based on aromatics used.

The electrophilic nature of ozone has been recognized in its reactions towards alkenes, alkynes, arenes, amines, sulfides, phosphines, etc.[176–180] Reactions of ozones as a nucleophile, however, are less well documented.[181]

$$\bar{O}-\overset{+}{O}=O \longleftrightarrow O=\overset{+}{O}-\bar{O} \longleftrightarrow \overset{+}{O}-O-\bar{O} \longleftrightarrow \bar{O}-O-\overset{+}{O}$$

36 a b c d

It was shown in the reaction of ozone with carbenium ions[182] that the initial attack as expected is alkylation of ozone giving rise to intermediate trioxide, which then undergoes carbon to oxygen alkyl group migration with simultaneous cleavage of a molecule of oxygen, similar to the acid-catalyzed rearrangement of hydroperoxides to carboxonium ions (Hock reaction).

$$R^2-C(R^1)(R^3)X \xrightarrow{SbF_5:SO_2ClF} R^2-C^+(R^1)(R^3) \xrightarrow{O_3} \left[R^2-C(R^1)(R^3)-O-O-O\right]^+$$

X = F, Cl, OH

$$\xrightarrow{-O_2} \left[R^2-C(R^1)(R^3)-O^+\right] \longrightarrow R^2R^3C=\overset{+}{O}-R^1$$

When a stream of oxygen containing 15% ozone was passed through a solution of isobutane in $HSO_3F:SbF_5:SO_2ClF$ solution held at −78°C, the colorless solution immediately turned brown in color. ^1H- and ^{13}C-nmr spectra of the resultant solution were consistent with formation of the dimethylmethylcarboxonium ion in 45% yield together with trace amounts of acetylium ion (CH_3CO^+).[183]

Similar treatment of isopentane, 2,3-dimethylbutane, and 2,2,3-trimethylbutane resulted in formation of related carboxonium ions as the major product (Table 5.25).

For the reaction of ozone with alkanes under superacid conditions, two mechanistic pathways could be considered. The first possible pathway is the formation of an alkylcarbenium ion via protolysis of the alkane prior to quenching of the ion by ozone, as shown in Scheme **11.** Alkylcarbenium ions may also be generated via initial oxidation of the alkane to an alcohol followed by protonation and ionization. There have already been a number of reports of ozone reacting with alkanes to give alcohols and ketones.[184–186] In both cases, intermediary alkylcarbenium ions would then undergo nucleophilic reaction with ozone as described earlier.

The products obtained from isobutane and isoalkanes (Table 5.25) are in accord with the above-discussed mechanism. However, the relative rate of formation of the dimethylmethylcarboxonium ion from isobutane is considerably faster than that of the *t*-butyl cation from isobutane in the absence of ozone under the same conditions.[182] Indeed, a solution of isobutane in excess Magic Acid-SO_2ClF solution

Scheme 11

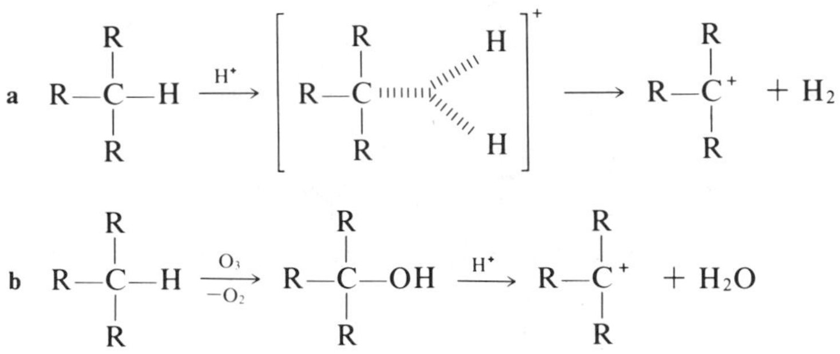

$$\text{a}\quad R_3C{-}H \xrightarrow{H^+} \left[R_3C\cdots\begin{matrix}H\\H\end{matrix} \right]^+ \longrightarrow R_3C^+ + H_2$$

$$\text{b}\quad R_3C{-}H \xrightarrow[-O_2]{O_3} R_3C{-}OH \xrightarrow{H^+} R_3C^+ + H_2O$$

TABLE 5.25 Products of the Reaction of Branched Alkanes with Ozone in Magic Acid–SO_2 ClF at −78°C (Ref. 182).

Alkane	Products
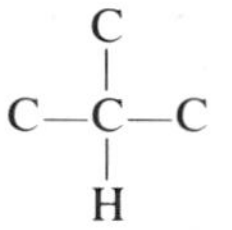 $(C)_3CH$ (C—C(C)(H)—C)	$(CH_3)_2C{=}\overset{+}{O}CH_3$
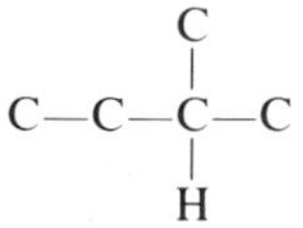 C—C—C(C)(H)—C	$(CH_3)_2C{=}\overset{+}{O}C_2H_5$
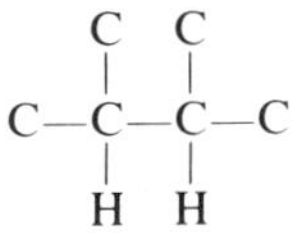 C—C(C)(H)—C(C)(H)—C	$(CH_3)_2C{=}\overset{+}{O}CH(CH_3)_2$ (60%), $(CH_3)_2C{=}\overset{+}{O}H$ (40%)
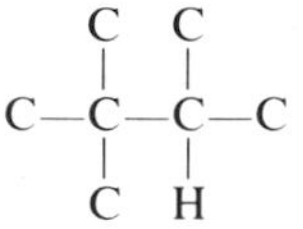 C—C(C)(C)—C(C)(H)—C	$(CH_3)_2C{=}\overset{+}{O}CH_3$ (50%), $(CH_3)_2C{=}\overset{+}{O}H$ (50%)

showed only trace amounts of the *t*-butyl cation after standing for 5 h at −78°C. Passage of a stream of oxygen gas through the solution for 10 times longer a period than in the ozonization experiments showed no effect. It was only when ozone was introduced into the system, that rapid reaction took place.

On the other hand, *t*-butyl alcohol itself in Magic Acid–SO_2ClF solution gave the *t*-butyl cation readily and quantitatively, even at −78°C. In the presence of ozone, however, under the same conditions it gave the dimethylmethylcarboxonium ion.

Although isobutane does not give any oxidation products in the absence of Magic Acid under the same low-temperature ozonization conditions, it was not possible for the authors[182] to determine whether formation of intermediate oxidation products, such as alcohols, plays any role in the ozonization of alkanes in Magic Acid. There is no experimental evidence for reactions proceeding via the intermediacy of carbenium ions; whether the initial oxidation step of alkanes to alcohols is important. This oxidation, indeed, was found to be extremely slow in the acidic media studied.

The most probable reaction path postulated for these reactions is electrophilic attack by protonated ozone on the alkanes, resulting in oxygen insertion into the involved σ-bonds and cleavage of H_2O_2 from the pentacoordinated trioxide insertion transition state giving a highly reactive oxenium ion intermediate which immediately rearranges to the corresponding carboxonium ion, which can be hydrolyzed to ketone and alcoholic products.

$$^{-}O{-}\overset{+}{O}{=}O + H^+ \longrightarrow HO{-}\overset{+}{O}{=}O \longleftrightarrow HO{-}O{-}\overset{+}{O}$$

$$R_3C{-}H + \overset{+}{O}{-}O{-}OH \longrightarrow \left[R_3C\cdots\begin{matrix} O{-}O{-}OH \\ H \end{matrix} \right]^+$$

$$\xrightarrow{-H^+} \left[R_3C{-}\overset{+}{O}{-} \right] \longrightarrow R_2C{=}\overset{+}{O}R + H_2O_2$$

Since ozone has a strong 1,3-dipole,[175] or at least has a strong polarizability (even if a singlet biradical structure is also feasible), it is expected to be readily protonated in superacids, in an analogous manner to its alkylation by alkylcarbenium ions. Protonated ozone O_3H^+, once formed, should have a much higher affinity (i.e., be a more powerful electrophile) for σ-donor single bonds in alkanes than neutral ozone.

Attempts to directly observe[182] protonated ozone by ^{1}H-nmr spectroscopy, were inconclusive because of probable fast hydrogen exchange with the acid system (may be through diprotonated $O_3H_2^{2+}$) and also the difficulty in differentiating between shifts of HO_3^+ and H_3O^+.

Straight-chain alkanes also efficiently react with ozone in Magic Acid at $-78°C$ in SO_2ClF solution. Ethane gave protonated acetaldehyde as the major reaction product together with some acetyleum ion. Reaction of methane, however, is rather complex and involves oxidative oligocondensation to *t*-butyl cation which reacts with ozone to give methylated acetone (Scheme 12).

Similar reactions have been investigated[163,182] with a wide variety of alkanes. Cycloalkanes in particular give cyclic carboxonium ions along with protonated ketones. The reaction of cyclopentane in Scheme 13.

Scheme 12

$$H_3C{-}CH_2{-}H \xrightarrow{O_3H^+} \left[H_3C{-}CH_2 \cdots \begin{matrix} H \\ O_3H \end{matrix} \right]^+ \longrightarrow \longrightarrow \begin{matrix} H_3C \\ H \end{matrix} C{=}\overset{+}{O}{-}H$$

$$H_3C{-}CH_2{-}H \xrightarrow{O_3H^+} \left[\begin{matrix} O_3H \\ H_3C \cdots \quad \cdots CH_3 \end{matrix} \right]^+ \longrightarrow \longrightarrow CH_3CO^+$$

$$CH_4 \xrightarrow{O_3H^+} \left[H_3C \cdots \begin{matrix} H \\ O_3H \end{matrix} \right]^+ \xrightarrow{CH_4} CH_3{-}\overset{+}{C}H_2 \xrightarrow[-H_2]{CH_4} CH_3{-}\overset{+}{C}H{-}CH_3 \xrightarrow[-H_2]{CH_4} CH_3{-}\underset{+}{\overset{CH_3}{C}}{-}CH_3$$

$$CH_3{-}\overset{+}{C}H{-}CH_3 \xrightarrow{O_3} \begin{matrix} H_3C \\ H_3C \end{matrix} C{=}\overset{+}{O}{-}H$$

$$CH_3{-}\underset{+}{\overset{CH_3}{C}}{-}CH_3 \xrightarrow{O_3} \begin{matrix} H_3C \\ H_3C \end{matrix} C{=}\overset{+}{O}{-}CH_3$$

Even electrophilic oxygenation of functionalized compounds has been achieved.[163] Alcohols are oxidized to the corresponding keto-alcohols. Ketones and aldehydes are oxidized to their corresponding dicarbonyl compounds (Scheme 14).

In the case of carbonyl compounds, the C—H bond located farther than γ-position seems to react with ozone in the presence of Magic Acid. It appears that the strong electron-withdrawing effect of protonated carbonyl group is sufficient to inhibit reaction of these C—H bonds (in α-, β-, and γ-positions). Jacquesy and co-workers[187] have recently shown that protonated ozone in $HF:SbF_5$ reacts with 3-keto steroids bearing various substituents such as carbonyl, hydroxy, and acetoxy groups at the 17 positions. In situ diprotonation of the substrate in the superacid medium directs the electrophilic attack of ozone to the B- or C-ring methylenes. The position of oxidation depends both on the steric hindrance of the corresponding axial C—H bond and the remoteness of the positive charges initially present (diprotonated species) in the molecule. For example, 3,17-diketosteroids only lead to the formation of the corresponding 3,6,7- and 3,7,17-triketones, whereas oxidation

Scheme 13

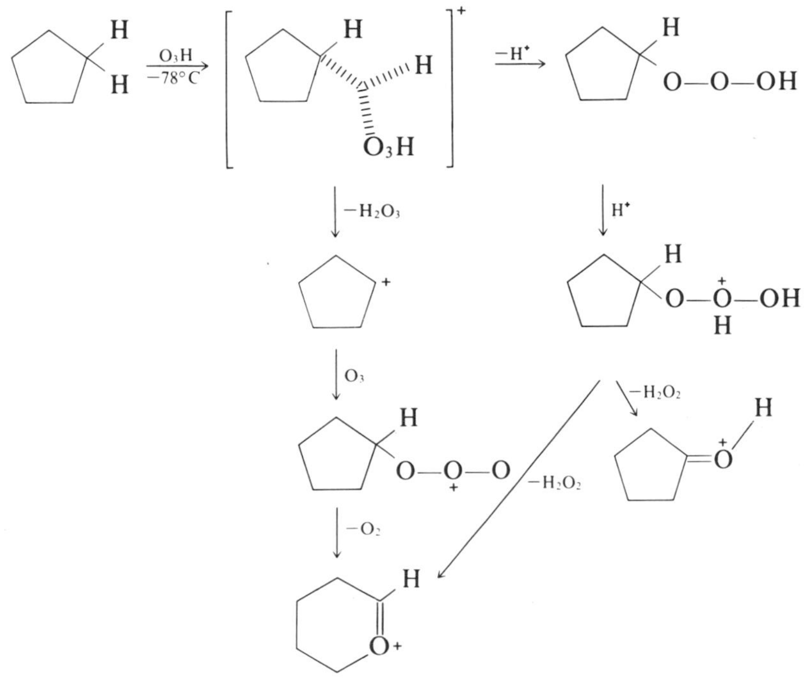

H
H
$\xrightarrow[-78^\circ C]{O_3H^+}$
[H, H, O_3H]$^+$
$\xrightarrow{-H^+}$
H
O—O—OH

$\downarrow -H_2O_3$

(cyclopentyl cation) +

$\downarrow O_3$

H
O—$\overset{}{O}^+$—O

$\downarrow -O_2$

H
O^+

$\downarrow H^+$

H
O—$\overset{+}{O}$—OH
H

$-H_2O_2$

H
$\overset{+}{O}$

$-H_2O_2$

at the 11 or 12 position is also observed when an OH or OAc group is present at the 17*b* position. In all cases, the 6- to 7-keto ratio appears to be higher for substrates

Scheme 14

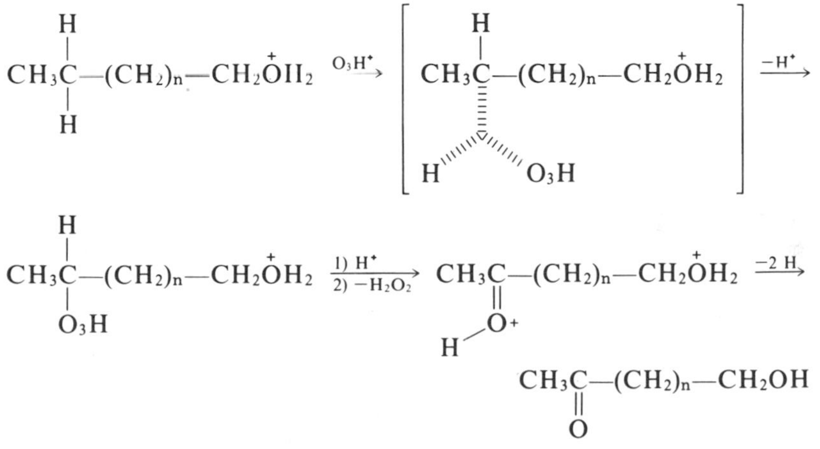

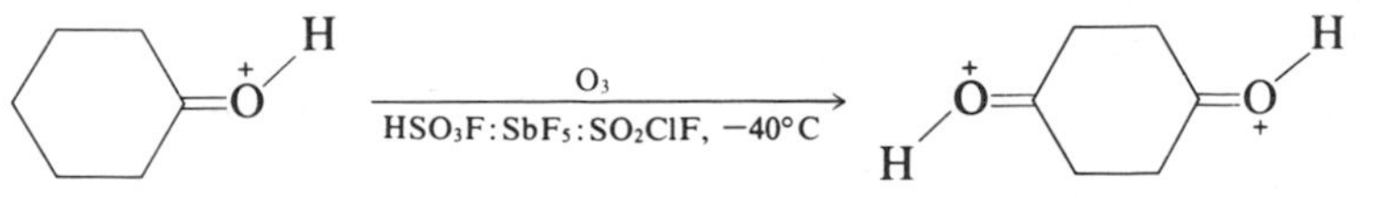

CH_3CH_2—$(CH_2)_n$—$CH_2\overset{+}{O}H_2$ $\xrightarrow{O_3H^+}$ [$CH_3\overset{H}{C}$—$(CH_2)_n$—$CH_2\overset{+}{O}H_2$ with H, O_3H] $\xrightarrow{-H^+}$

$CH_3\underset{O_3H}{\overset{H}{C}}$—$(CH_2)_n$—$CH_2\overset{+}{O}H_2$ $\xrightarrow[2)\ -H_2O_2]{1)\ H^+}$ $CH_3\underset{\|\ O^+-H}{C}$—$(CH_2)_n$—$CH_2\overset{+}{O}H_2$ $\xrightarrow{-2\ H^+}$

$CH_3\underset{\|\ O}{C}$—$(CH_2)_n$—CH_2OH

(cyclohexanone)=$\overset{+}{O}$—H $\xrightarrow[HSO_3F:SbF_5:SO_2ClF,\ -40^\circ C]{O_3}$ H—$\overset{+}{O}$=(cyclohexane)=$\overset{+}{O}$—H having a *cis* A/B ring junction as compared with those with a *trans* one. Low reactivity of the 11 position, especially for substrates with a *cis* A/B ring junction was observed. Oxidation of the tertiary carbon atoms, however, was not observed in any of these systems (Scheme 15).

Scheme 15

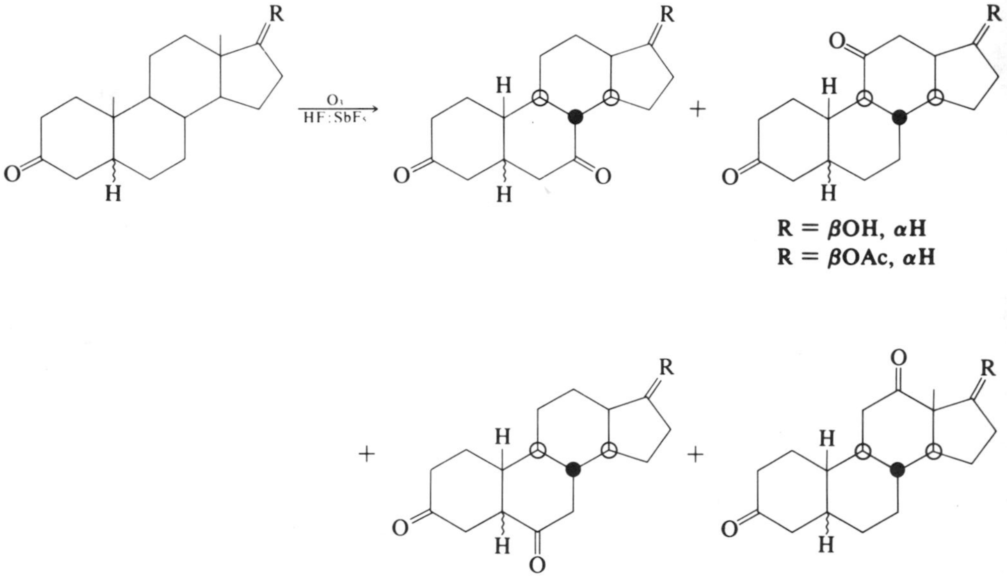

O_3, HF:SbF_5

R = βOH, αH
R = βOAc, αH

5.13 POLYMERIZATION

The key initiation step in cationic polymerization of alkenes is the formation of a carbocationic intermediate, which can then react with excess monomer to start propagation. The kinetics and mechanisms of cationic polymerization and polycondensation have been studied extensively.[188,189] Kennedy and Marechal have pointed out that only cations of moderate reactivity are useful initiators, since stable ions such as arenium ions were found to be unreactive for olefin polymerization. On the other hand, energetic alkyl cations such as $CH_3CH_2^+$ were too reactive and gave side products.[188b]

It has been shown by Olah et al.[189,190] that cationic polymerization of alkenes can be initiated by stable alkyl or acyl cations as well as nitronium ion salts.

Trivalent carbenium ions play a key role, not only in the acid-catalyzed polymerization of alkenes, but also in the polycondensation of arenes (π-bonded monomers) as well as in the cationic polymerization of ethers, sulfides, and nitrogen compounds (nonbonded electron-pair donor monomers). On the other hand, pentacoordinated carbonium ions play the key role in the electrophilic reactions of σ-bonds (single bonds), including the oligocondensation of alkanes and the cocondensation of alkanes and alkenes (Section 5.1.5).

$$RF + BF_3 \longrightarrow R^+BF_4^- + C_2H_5CH{=}CH_2 \longrightarrow C_2H_5\overset{+}{C}HCH_2R\ \overset{-}{B}F_4$$

$$\xrightarrow{n(C_2H_5CH=CH_2)} C_2H_5\underset{\displaystyle CH_2R}{CH}{-}[CH_2\overset{\displaystyle C_2H_5}{CH}{-}]_{n-1}{-}CH_2\overset{+}{C}HC_2H_5\ \overset{-}{B}F_4$$

$$\longrightarrow C_2H_5\underset{\displaystyle CH_2R}{CH}{-}[{-}CH_2\overset{\displaystyle C_2H_5}{CH}{-}]_{n-1}{-}CH{=}CH{-}C_2H_5 + HF + BF_3$$

In general, the cationic polymerization of olefins should be considered as a typical example of the general carbocationic reactivity in electrophilic reactions, and all other suggested mechanisms can be looked upon as only differing in the nature of the initial electrophiles, always leading to the key trivalent alkyl cation which then initiates the polymerization reaction.

$$(CH_3)_2C{=}CH_2 \xrightarrow{E^+} (CH_3)_2{-}\overset{+}{C}{-}CH_2{-}E \xrightarrow{\text{Olefin}} \text{Polymer}$$

The solid superacid Nafion-H is also a good polymerization catalyst. 2-Methylpropene has been polymerized with Nafion-H. At 145°C only oligomers (dimers to tetramers) were obtained; decreasing the temperature increased the molecular weight of the oligomers.[191,192] The oligomerization reaction has also been studied for higher alkenes. For example, in the liquid phase, the addition of 1% by weight Nafion-H to 1-decene at 150°C gave, after 5 h, a 65% conversion to oligomeric products, of which 55% consisted of the trimer.[193] Under similar conditions, 5-decene was converted to a 4:1 mixture of dimers and trimers, which was hydrogenated to a lubricating-type oil.[194]

Nafion-H has also been used as a catalyst for the oligomerization of styrene. The reaction was studied by Higashimura and co-workers.[195,196]

In the patent literature, there are several reports of the cationic polymerization of tetrahydrofuran (THF) with Nafion-H. In most cases, small amounts of acetic anhydride were added so the initial polymer had a terminal acetate group that could be hydrolyzed to the free hydroxyl. THF has been homopolymerized,[197,198] and copolymerized with 1,4-butanediol,[199] ethylene oxide,[200,202] or dioxolane.[203]

5.14 MISCELLANEOUS REACTIONS

Acid-catalyzed rearrangements of natural products, particularly terpenes, have been studied extensively.[204–210] This research has unravelled numerous unique rearrangements that are both synthetically and mechanistically significant. Moreover, the recognition that cationic cyclizations play a key role in biogenesis[210–214] of isoprenoids (terpenes) has provided incentive to mimic in vitro[215–219] many of these rearrangements by generating the appropriate carbocations.

The literature[204–210] abounds with examples of rearrangement of a wide variety of terpenoids with varying acid catalysts. One of the most significant area is the synthesis of triterpenoids, i.e., steroids by van Tamelen,[219] Johnson,[217] and others. However, most of these rearrangements have been studied under mild conditions, in nucleophilic media with relatively weak acids. Under these conditions of low acidity, only a very minute amount of the reactant exists as carbocation at any time because the electron deficient intermediate reacts immediately with available nucleophiles, both external and internal via a number of competitive, irreversible reactions. In other words, only short lived, transient intermediates are involved. Consequently quite often, therefore, complex mixtures of products are formed in the rearrangements during mild acid catalysis, as the products obtained are generally comparable in energy content.

The use of carbocation-stabilizing superacids like HSO_3F, $SbF_5:HSO_3F$, CF_3SO_3H, $HF:SbF_5$, $HF:BF_3$, etc. to alter the normal course of acid-catalyzed rearrangements has offered unique possibilities, since the stable carbocation would have time to explore many internal escape paths (through a probably shallow potential energy surface), but would not convert to neutral products. In recent years, considerable progress has been made in this area especially by the pioneering work of Jacquesy and co-workers.

5.14.1 Rearrangements and Cyclization Reactions

As early as 1893, it was discovered[220] that camphor **37** rearranges to 3,4-dimethylacetophenone **38** in conc. H_2SO_4 and the mechanism of the reaction has been elucidated by Roberts[221] and Rodig.[222]

H_2SO_4

37 **38**

However, recently Jacquesy and co-workers[223] have shown that in superacidic $HF:SbF_5$ medium, the reaction takes a different course. Camphor gives a mixture of three ketones. Sorensen and co-workers[224,225] have studied the fate of observable Camphene hydrocation prepared from isoborneol **39,** camphene **40,** or tricyclene

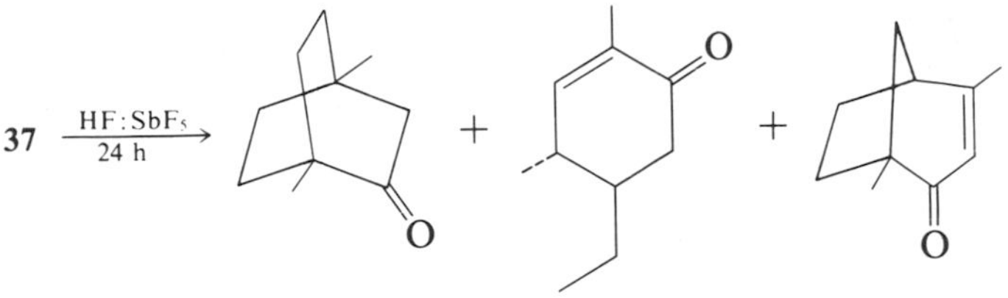

37 $\xrightarrow[24\ h]{HF:SbF_5}$ + +

O O O

41 in HSO_3F acid medium. The intermediate cycloalkenyl cation can also be prepared by the protonation of α-terpineol, sabinene, and β-pinene.

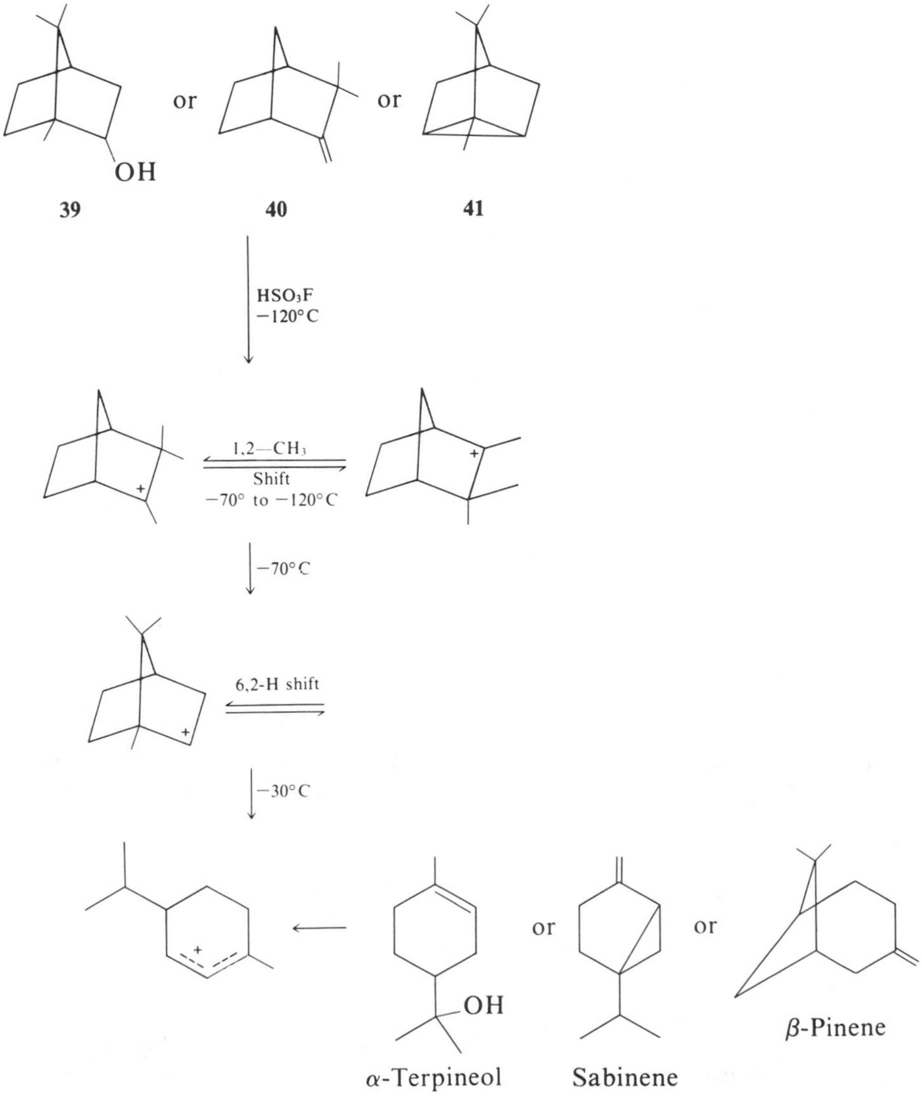

OH or or

39 **40** **41**

HSO_3F
−120°C

1,2—CH_3
Shift
−70° to −120°C

−70°C

6,2-H shift

−30°C

OH or or

α-Terpineol Sabinene β-Pinene

The fascinating rearrangement[226] of acyclic monoterpenes geraniol **42** and nerol **43** to a stable observable oxonium ion in $HSO_3F:SO_2:CS_2$ at $-78°C$ represents yet another dramatic example of altered course of terpene rearrangement in superacid medium. Careful quenching of this ion led to the isolation of 3β,6α,6aα-trimethyl-*cis*-perhydrocyclopenta[b]furan **44** in excellent yield.

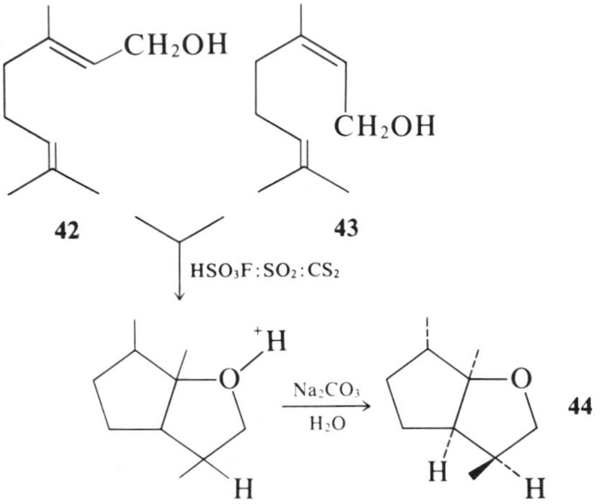

CH_2OH CH_2OH

42 **43**

$HSO_3F:SO_2:CS_2$

^+H O H $\xrightarrow[H_2O]{Na_2CO_3}$ O H H **44**

Farnum and co-workers[227] have shown that longifolene **45** and isolongifolene **46** in HSO_3F media give a mixture of cyclohexenyl cations at different temperatures. Quenching these cations provides some unusual sesquiterpene like C_{15}-hexahydronaphthalenes (Scheme 16).

The rearrangement of some resin acids **47, 48,** and **49** in superacidic HSO_3F and $ClSO_3H$ media has also been studied.[228] Jacquesy et al.[229,230] have developed a novel isomerization of pregnan-3,20-diones **50** to mixture of isomers that also contains 13-α isomers. The reaction is proposed to occur through the cleavage of C_{13}-C_{17} bond (Scheme 17).

Similarly, $HF:SbF_5$-induced isomerization of androsta 4,6-diene-3,17 dione **51** has been studied in detail,[231] which led to a new entry into the 9-methylestrane series **52.** Also, methods have been developed for the synthesis of isoestrane derivatives[232] and other methyl-substituted estrane dione derivatives of unnatural configurations.[233] $HF:SbF_5$ superacid medium is also capable of demethylating aromatic ethers. This reaction has been successfully employed in the synthesis of 11-deoxyanthracyclines **53** (Scheme 18).[234]

Jacquesy and Jouannetaud in a series of papers have demonstrated the efficacy of aromatic cyclization reactions in $HF:SbF_5$ medium. 1,3-bis(methoxyphenyl) propanes and other substituted phenyl propanes **54**-R give tricyclic spiro enones and substituted indenes in the superacid medium (Scheme 19).[235]

Monocyclic phenols and their methyl ethers react with benzene in $HF:SbF_5$ medium to provide 4,4-disubstituted cyclohexenones **55.**[236,237] *p*-Methylanisole gives three products that also include an interesting tricyclic ketone **56** (Scheme 20).

Scheme 16

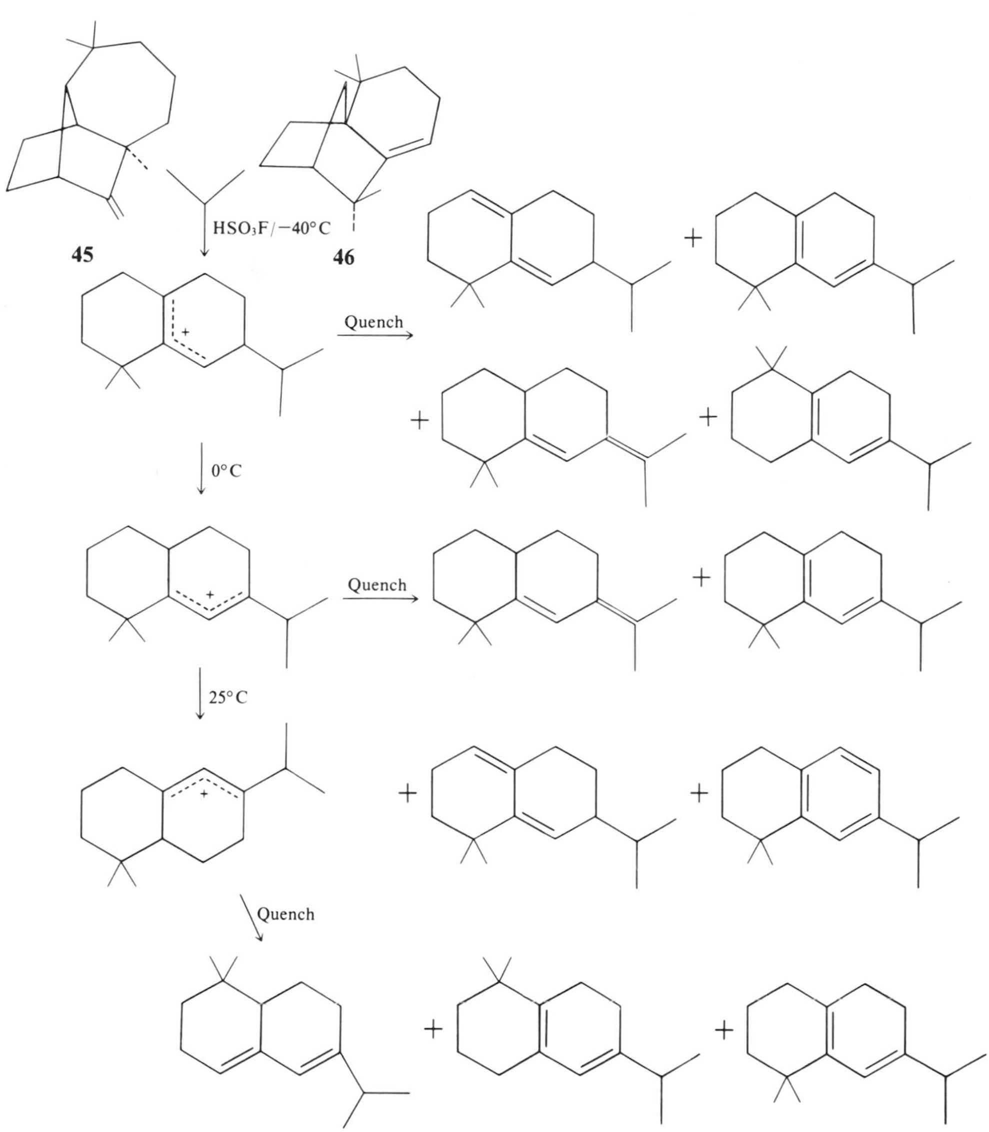

45

46

$HSO_3F/-40°C$

Quench

0°C

Quench

25°C

Quench

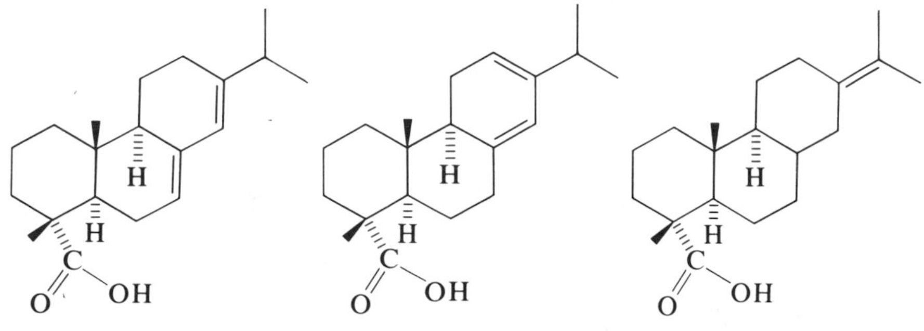

H

H

C

O

OH

47

H

H

C

O

OH

48

H

H

C

O

OH

49

Scheme 17

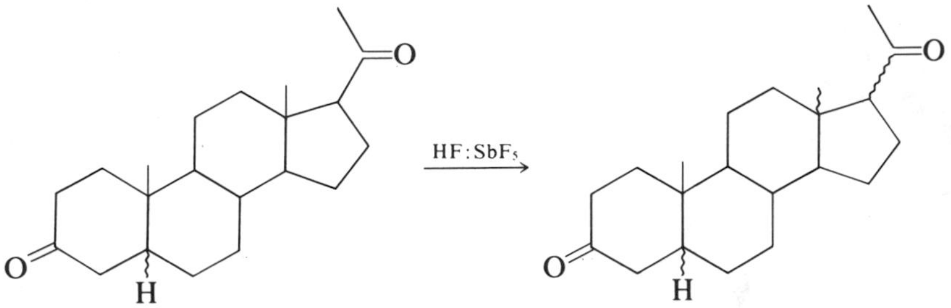

HF:SbF$_5$

50 a; 5 α H
b; 5 β H

Mixture of four isomers

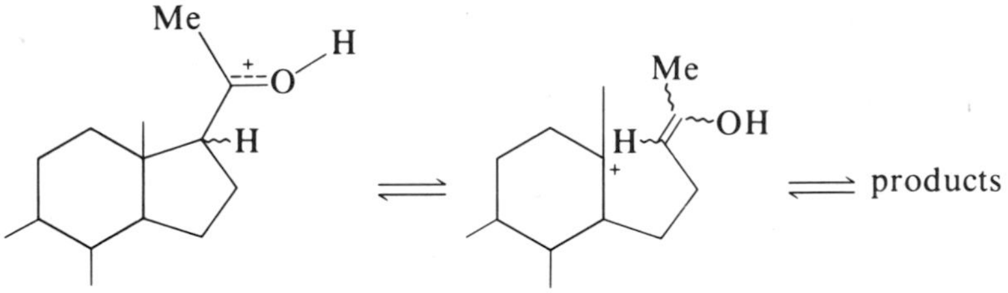

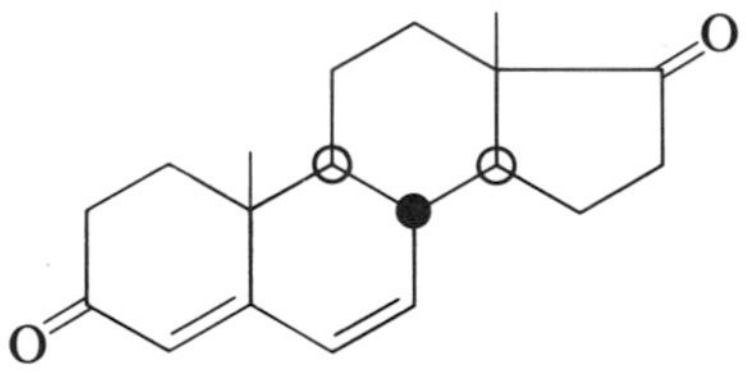

Me H H OH products

51

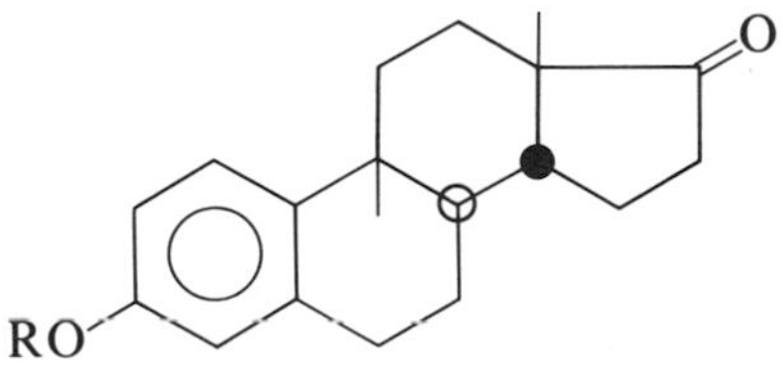

RO

52

Scheme 18

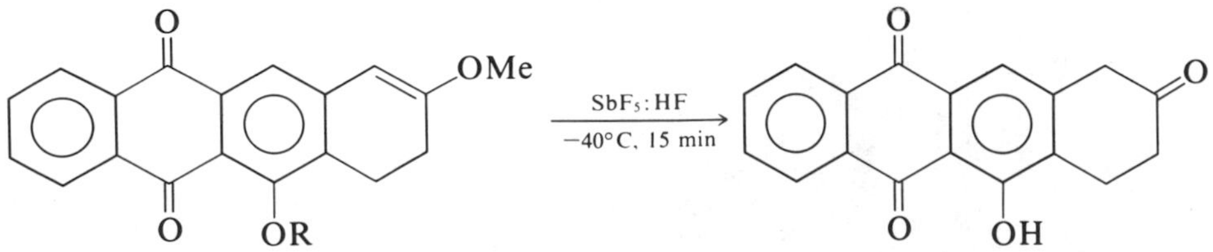

OMe

SbF_5:HF
$-40°C$, 15 min

O OR

R = H; R = Et

O OH

53

Scheme 19

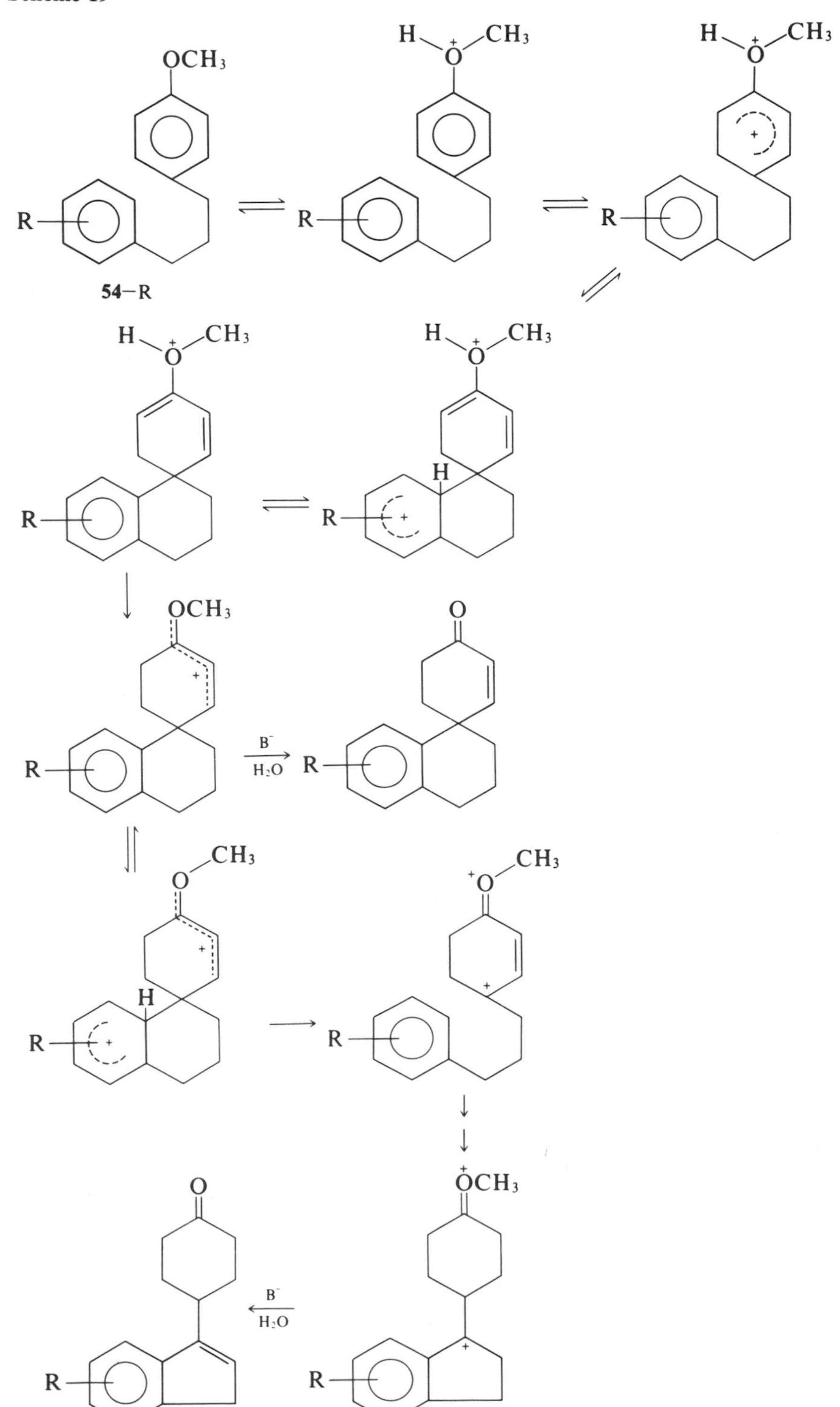

Scheme 20

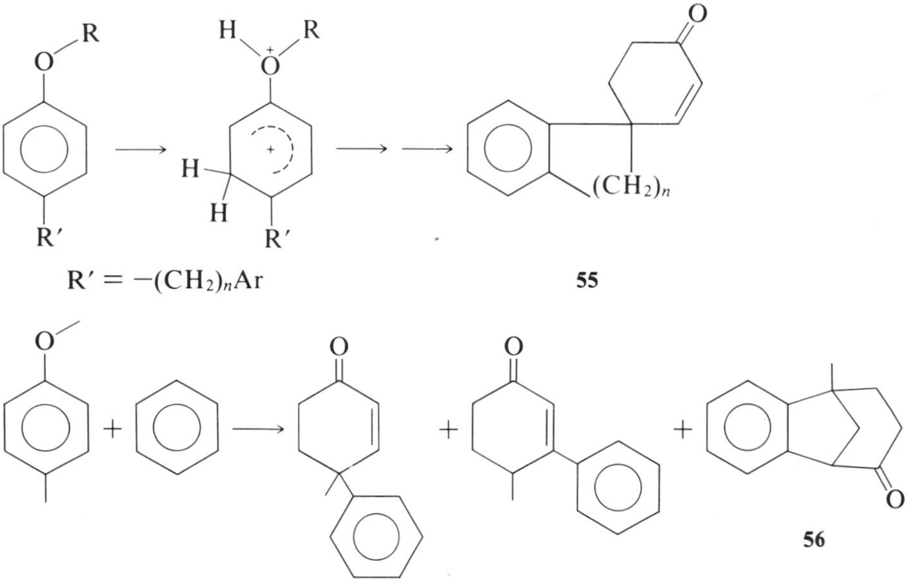

R′ = −$(CH_2)_n$Ar

55

56

By utilizing the above methods, a convenient route to the tetracyclic ketone **58,** a derivative of the benzobicyclo[3.2.1]heptane skeleton, has been developed[238] using 2′-methoxy,5′-methyl 1,3-diphenyl propane **57** (Scheme 21).

5.14.2 Phenol–Dienone Rearrangements

Conversion of phenol to a dienone and vice versa is an important transformation and is widely used in organic natural product synthesis. Jacquesy and co-workers have demonstrated that this rearrangement occurs very efficiently in superacid solutions. Treatment of estrone derivatives **59**-R in HF:SbF_5 followed by aqueous bicarbonate work up lead to estra-4,9-dien-3,7-dione **60** (Scheme 22). The reaction also occurred in HSO_3F:SbF_5 medium. The intermediate tricationic species and their isomers of the above rearrangement process have been characterized by ^{1}H-nmr spectroscopy.[240] As models, Jacquesy et al.[241] have also investigated the rearrangement of simple bicyclic phenols and phenolic ethers in HF:SbF_5 or HSO_3F:SbF_5 medium (Scheme 23).

Similarly, many A-norsteroids **61**-R have been subjected to phenol–dienone rearrangement in HF:SbF_5 medium. ^{1}H-nmr spectroscopic studies at low temperature confirms the formation of *O*-protonated intermediates which subsequently rearrange to diprotonated precursors of the dienones **62**-R (Scheme 24).[242]

Androsta-1,4,6-triene-3,17-dione **63** when treated with ordinary acid catalysts such as acetic anhydride/*p*-toluenesulfonic acid gives the meta phenolic product **64.** However, under HF:SbF_5 catalysts at −50°C the phenolic product **65** is obtained in 75% yield (Scheme 25).[243] The mechanism of the above discussed phenol-dienone and dienone–phenol rearrangement has been investigated in detail.[244,245]

Scheme 21

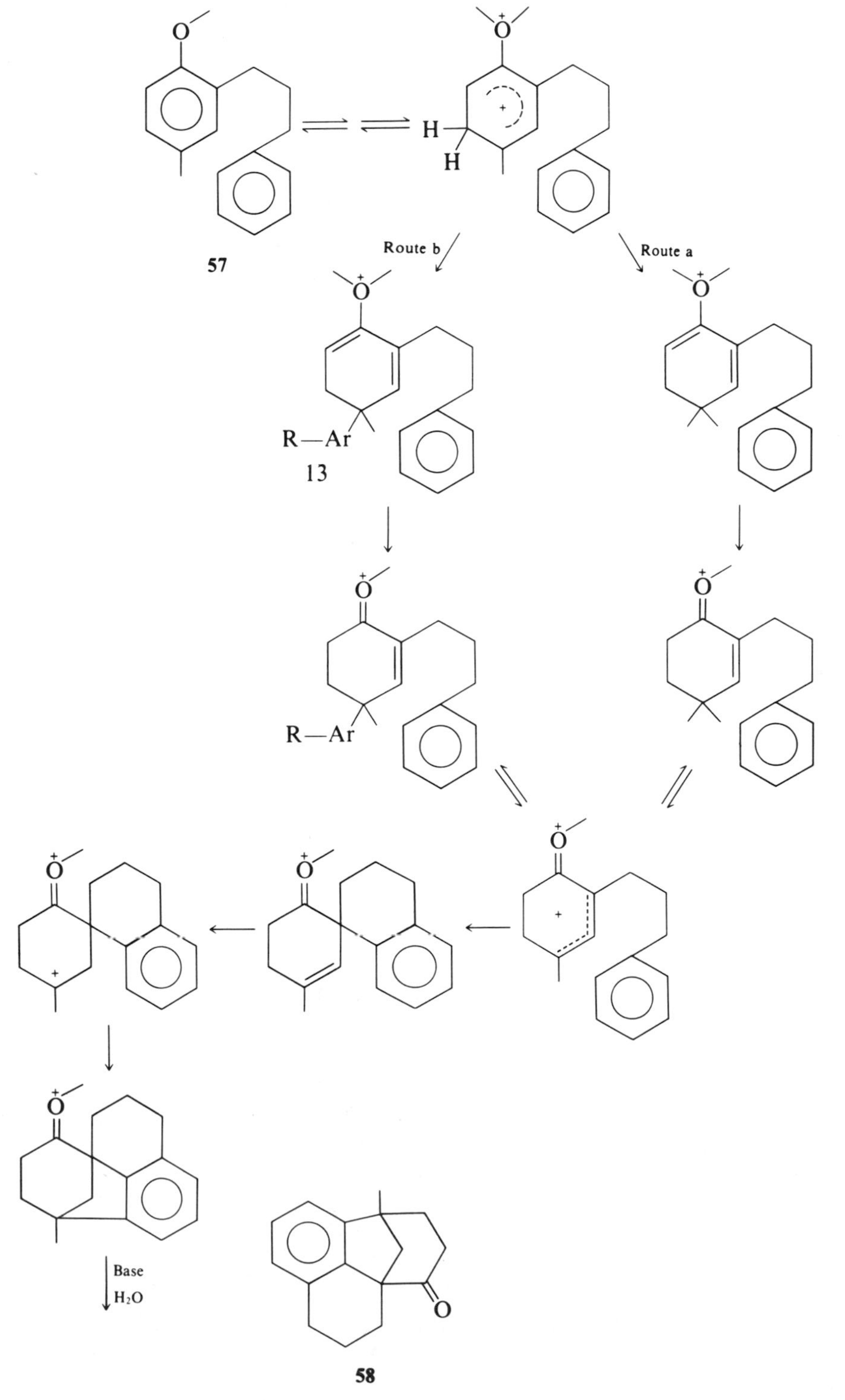

57

Route b

Route a

R—Ar

13

R—Ar

Base

H_2O

58

Scheme 22

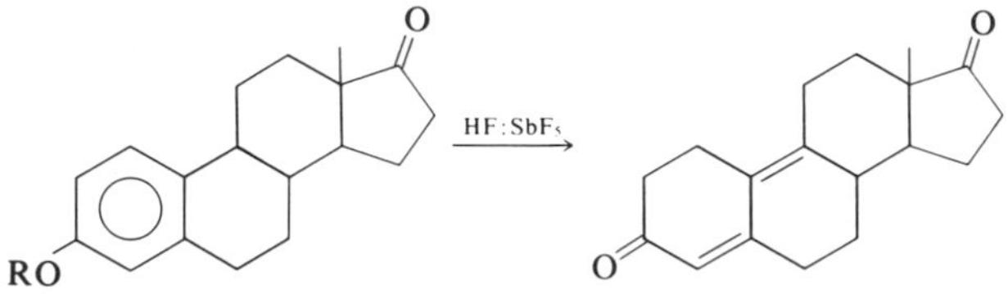

HF:SbF5

59-R, R = H, CH_3, Ac

60

Scheme 23

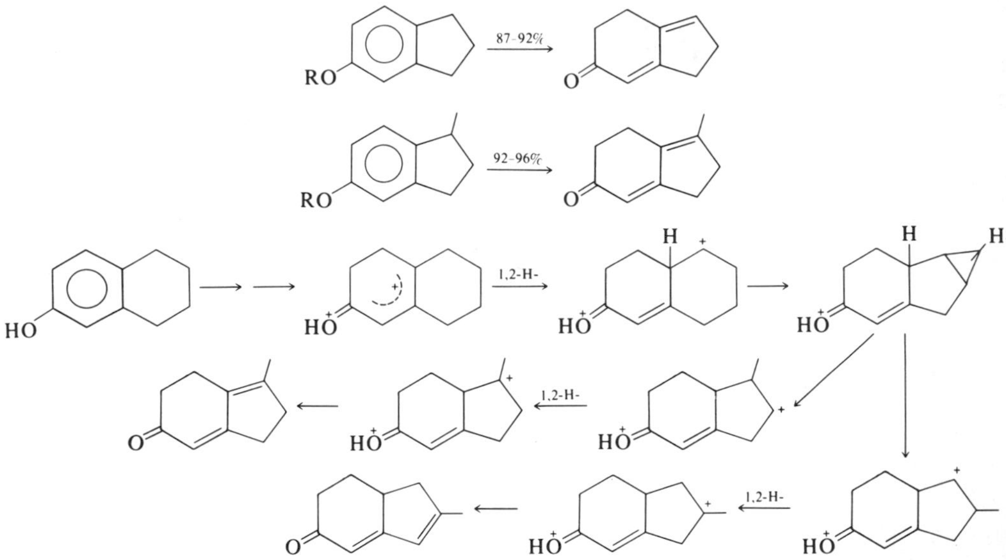

87–92%

92–96%

1,2-H-

5.14.3 Ionic Hydrogenation

In catalytic hydrogenations, usually hydrogen is activated (generally by noble metal catalysts). In contrast, in ionic hydrogenations an acid catalyst activates the hydrocarbon-forming carbocationic sites which then are quenched by hydrogen to reduced products. A significant number of ionic hydrogenations were studied and a review of earlier work is available.[246a] Hydride transfer in superacidic media is a very versatile reaction, especially in isomerization and cracking of hydrocarbons (Sections 5.1.3 and 5.1.4). The reaction has been extensively employed in liquifaction of coal using tetralin, methylcyclopentane, etc., as hydride donors in the presence of strong Lewis acids.[246b] In fact, Olah and co-workers[48] have found that coal can be depolymerized in the presence of HF:BF_3 under hydrogen pressure at modest temperatures (100–170°C). Addition of isopentane to this reaction sub-

Scheme 24

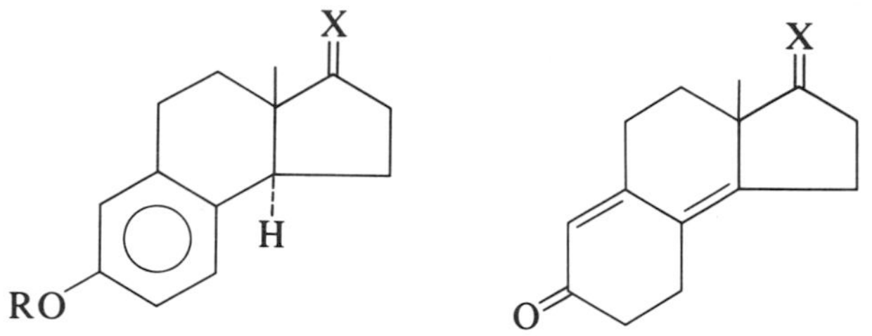

61-R	62-R
R = CH_3; X = 0	X = 0
R = H; X = 0	
R = CH_3; X = βOH, αH	X = βOH, αH
R = CH_3; X = βOCOH, αH	X = βOCOH, αH
R = CH_3; X = $\beta OCOCH_3$, αH	X = $\beta OCOCH_3$, αH
R = CH_3; X = $\beta COCH_3$, αH	X = $\beta COCH_3$, αH

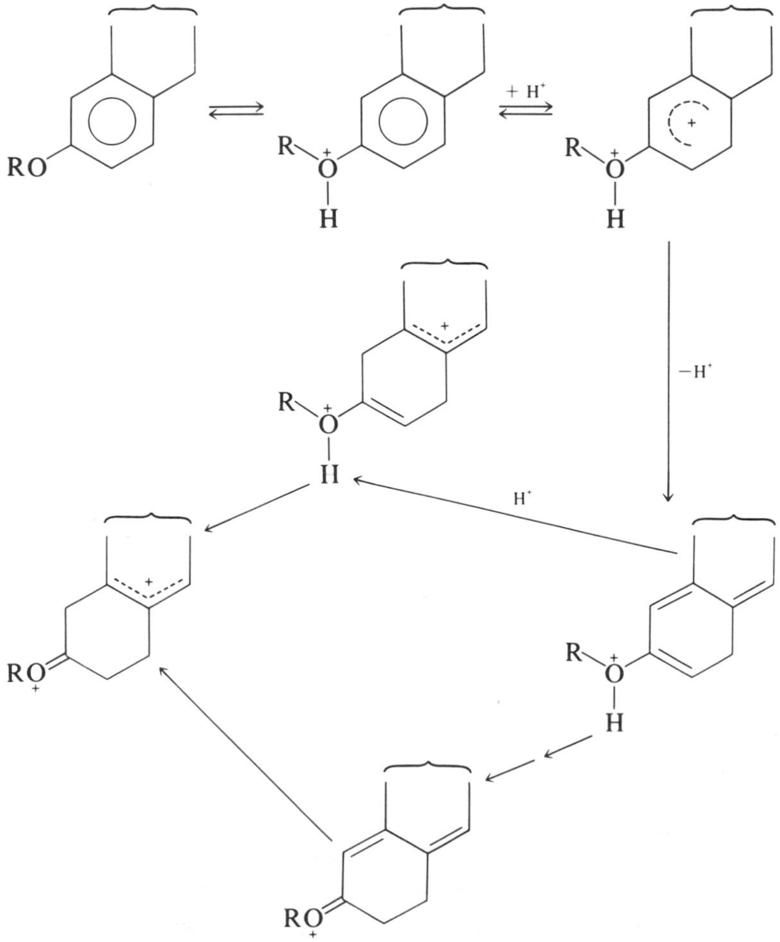

Scheme 25

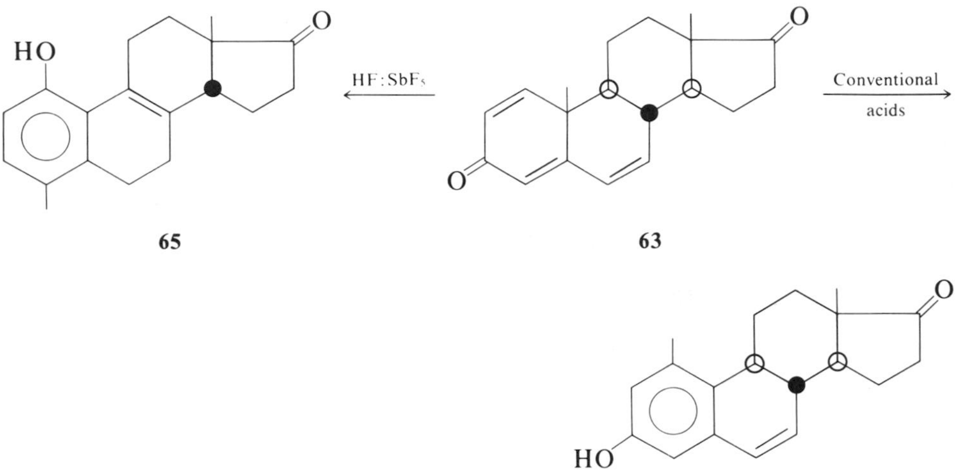

65 ← $HF:SbF_5$ — 63 — Conventional acids → 64

stantially improved depolymerization (*vide supra*). Similar reactions have been studied using $HF:TaF_5$ as catalysts.[246c]

The hydride transfer reaction catalyzed by strong acids has also been successfully adapted in the natural product chemistry. Jacquesy and co-workers[247] have found that protonated dienone and enones of steroid nucleus can be conveniently reduced in $HF:SbF_5$ medium under hydrogen pressure. The reaction has also been carried out with added hydrocarbons as hydride donors.[248,249] The following mechanism has been proposed.

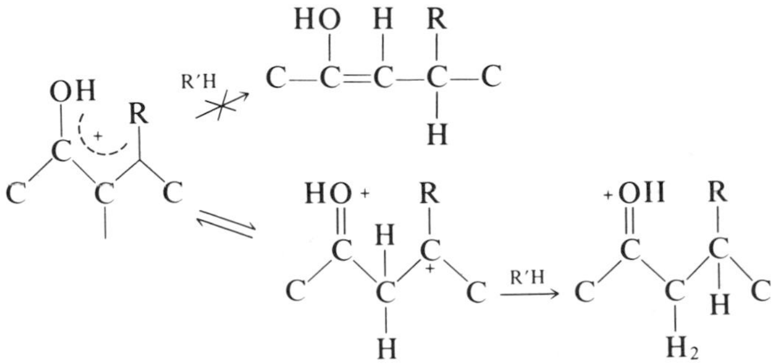

As model studies, several bicyclic compounds have also been reduced using methylcyclopentane in $HF:SbF_5$.[250] Hydroxytetralin **66** undergoes phenol-dienone rearrangement before reduction. Estrone and an unsaturated estrane have been reduced in $HF:SbF_5$ medium under hydrogen pressure yielding an anthrasteroid, whose structure and absolute configuration have been determined.[251] Recently, during the reductive isomerization of 7-β-methyl-14-isoestr-4-ene-3,17-dione **67** in

$HF:SbF_5$/methylcyclopentane at 0°C, it was found[252] that a 1,3-hydride shift occurs followed by kinetically controlled hydride transfer. The mechanism of the reaction

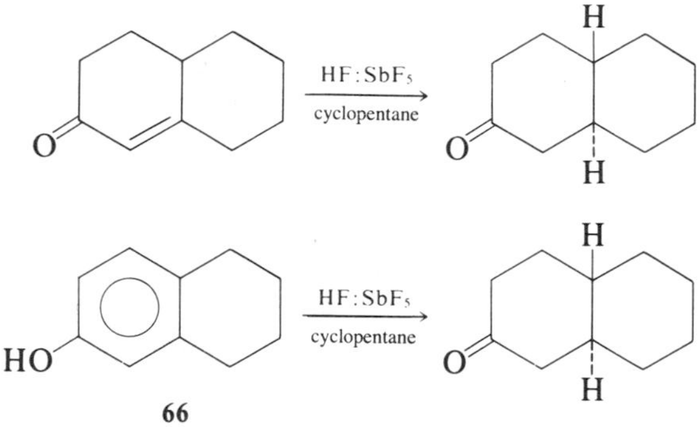

HF:SbF₅ cyclopentane

HF:SbF₅ cyclopentane

66

was confirmed by employing a deuterated hydride donor (cyclohexane-d_{12}) as well as a specifically deuterium-labeled starting steroid.

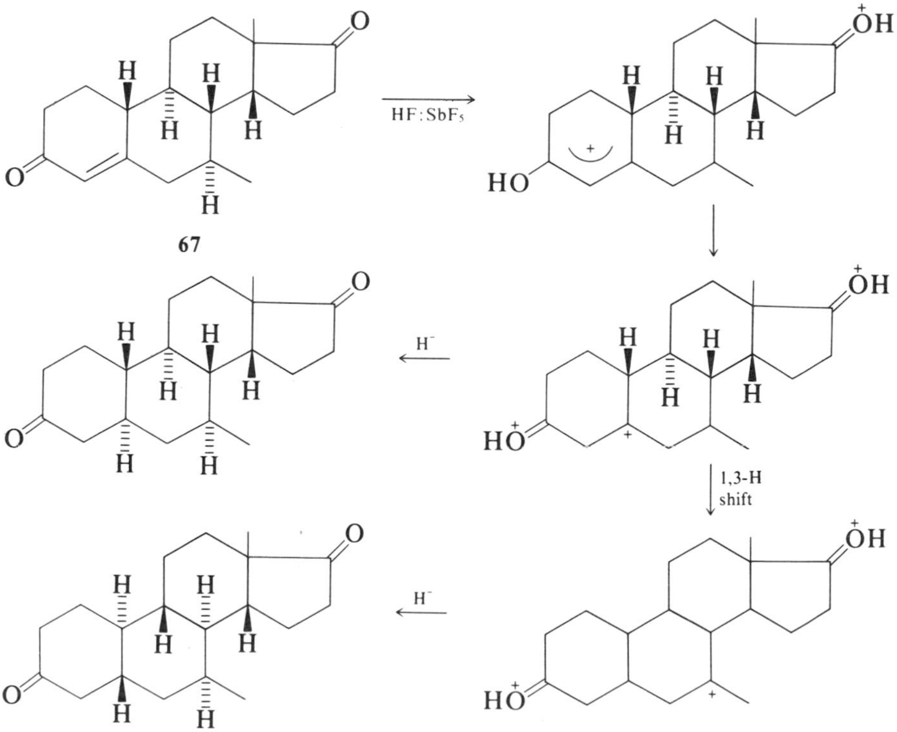

HF:SbF₅

67

1,3-H shift

The ionic transfer hydrogenation in superacidic media has also been studied in the case of bicyclic[4.n.0]enones both under kinetic[253] and thermodynamic control.[254] Kinetically controlled hydride transfer between the hydrocarbon hydride source and the substrate often involves skeletal rearrangements. This has been recently demonstrated in the reduction of androsta-4,6-diene-3,17-dione **51**.[255]

REFERENCES

1. F. Asinger, *Paraffins, Chemistry and Technology,* Pergamon Press, New York, 1965.
2. H. S. Bloch, H. Pines, L. Schmerling, *J. Am. Chem. Soc. 68,* 153 (1946).
3a. G. A. Olah, J. Lukas, *J. Am. Chem. Soc. 89,* 2227 (1967).
3b. G. A. Olah, J. Lukas, *Ibid. 89,* 4739 (1967).
4a. A. F. Bickel, G. J. Gaasbeek, H. Hogeveen, J. M. Oelderick, J. C. Platteuw, *Chem. Commun.* 634 (1967).
4b. H. Hogeveen, A. F. Bickel, *Ibid.* 635 (1967).
5. G. K. S. Prakash, A. Husain, G. A. Olah, *Angew. Chem. 95,* 51 (1983).
6. G. A. Olah, J. Lukas, *J. Am. Chem. Soc. 90,* 933 (1968).
7. G. A. Olah, R. H. Schlosberg, *J. Am. Chem. Soc. 90,* 2726 (1968).
8. H. Hogeveen, C. J. Gaasbeek, *Recl. Trav. Chim. Pays-Bas. 87,* 319 (1968).
9. G. A. Olah, G. Klopman, R. H. Schlosberg, *J. Am. Chem. Soc. 91,* 3261 (1969).
10a. G. A. Olah, J. Shen, R. H. Schlosberg, *J. Am. Chem. Soc. 92,* 3831 (1970).
10b. T. Oka, *Phys. Rev. Lett. 43,* 531 (1980).
11. H. Hogeveen, A. F. Bickel, *Recl. Trav. Chim. Pays-Bas. 86,* 1313 (1967).
12. H. Pines, N. E. Hoffman, in *Friedel–Crafts and Related Reactions,* Vol. II, G. A. Olah Ed., Wiley-Interscience, New York, 1964, p. 1216.
13. G. A. Olah, Y. Halpern, J. Shen, Y. K. Mo, *J. Am. Chem. Soc. 93,* 1251 (1971).
14. J. W. Otvos, D. P. Stevenson, C. D. Wagner, O. Beeck, *J. Am. Chem. Soc. 73,* 5741 (1951).
15. H. Hoffman, C. J. Gaasbeek, A. F. Bickel, *Recl. Trav. Chim. Pays-Bas. 88,* 703 (1969).
16. R. Bonnifay, B. Torck, J. M. Hellin, *Bull. Soc. Chim. Fr.* 808 (1977).
17. J. Lucas, P. A. Kramer, A. P. Kouwenhoven, *Recl. Trav. Chim. Pays-Bas. 92,* 44 (1973).
18. J. W. Larsen, *J. Am. Chem. Soc. 99,* 4379 (1977).
19. J. Bertram, J. P. Coleman, M. Fleischmann, D. Pletcher, *J. Chem. Soc. Perk. II,* 374 (1973).
20. For a review, see P. L. Fabre, J. Devynck, B. Tremillon, *Chem. Rev. 82,* 591 (1982).
21. J. Devynck, P. L. Fabre, A. Ben Hadid, B. Tremillon, *J. Chem. Res(s).* 200 (1979).
22. J. Devynck, A. Ben Hadid, P. L. Fabre, *J. Inorg. Nucl. Chem. 41,* 1159 (1979).
23. C. D. Nenitzescu, A. Dragan, *Ber. Dtsch, Chem. Ges. 66,* 1892 (1933).
24. H. Pines, N. E. Joffman, *Friedel–Crafts and Related Reactions,* Vol. II, G. A. Olah, Ed., Wiley-Interscience, New York, 1964, p. 1211.
25. C. D. Nenitzescu, in *Carbonium Ions,* Vol. II, G. A. Olah, P. v. R. Schleyer, Eds., Wiley-Interscience, New York, 1970, p. 490.
26. F. Asinger, in *Paraffins,* Pergamon Press, New York, 1968, p. 695.
27a. R. Bonnifay, B. Torck, J. M. Hellin, *Bull. Soc. Chim. Fr.,* 1057 (1977).
27b. R. Bonnifay, B. Torck, J. M. Hellin, *Ibid.,* 36 (1978).
28. D. Brunel, J. Itier, A. Commeyras, R. Phan Tan Luu., D. Mathieu, *Bull. Soc. Chim. Fr. II,* 249 and 257 (1979).
29a. A. Germain, P. Ortega, A. Commeyras, *Nouv. J. Chim. 3,* 415 (1979).
29b. H. Choucroun, A. Germain, D. Brunel, A. Commeyras, *Ibid. 5,* 39 (1980).
29c. H. Choucroun, A. Germain, D. Brunel, A. Commeyras, *Ibid. 7,* 83 (1982).
30. G. A. Olah, U.S. Patent application filed.
31. G. A. Olah, U.S. Patent application filed.
32a. F. Le Normand, F. Fajula, F. Gault, J. Sommer, *Nouv. J. Chim. 6,* 411 (1982).
32b. F. Le Normand, F. Fajula, J. Sommer, *Ibid., 6,* 291 (1982).

33. F. Le Normand, F. Fajula, F. Gault, J. Sommer, *Nouv. J. Chim. 6,* 417 (1982).
34. R. Jost, K. Laali, J. Sommer, *Nouv. J. Chim. 7,* 79 (1983).
35. N. Yoneda, T. Fukuhara, T. Abe, A. Suzuki, *Chem. Lett.* 1485 (1981).
36. J. J. L. Heinerman, J. Gaaf, *J. Mol. Cat. 11,* 215 (1981).
37. K. Laali, M. Muller, J. Sommer, *Chem. Commun.* 1088 (1980).
38. N. L. Allinger, J. L. Coke, *J. Am. Chem. Soc., 82,* 2553 (1960).
39. G. A. Olah, J. Kaspi, J. Bukala, *J. Org. Chem. 42,* 4187 (1977).
40. G. A. Olah, U.S. Patent, 4,116,880 (1978).
41. G. A. Olah, J. Kaspi, *J. Org. Chem. 42,* 3046 (1977).
42. G. A. Olah, J. R. DeMember, J. Shen, *J. Am. Chem. Soc. 95,* 4952 (1973).
43a. P. v. R. Schleyer, *J. Am. Chem. Soc. 79,* 3292 (1957).
43b. R. C. Fort, Jr., in *Adamantane, the Chemistry of Diamond Molecules,* P. Gassman, Ed., Marcel Dekker, New York, 1976.
43c. G. A. Olah, J. A. Olah, *Synthesis,* 488 (1973).
43d. L. Paquette, A. Balogh, *J. Am. Chem. Soc. 104,* 774 (1982).
44. G. C. Schuit, H. Hoog, J. Verhuis, *Recl. Trav. Chim. Pays-Bas. 59,* 793 (1940); Brit. Pat. 535054 (1941).
45. H. Pines, U.S. Pat. 2405516194.6 (H. Pines to Universal Oil); *Chem. Abst. 41,* 474 (1947).
46. For a review, see G. A. Olah, *Carbocations and Electrophilic Reactions,* Verlag Chemie, John Wiley & Sons, New York, 1974.
47. For a review, see D. M. Brouwer, H. Hogeveen, *Progr. Phys. Org. Chem., 9,* 179 (1972).
48a. G. A. Olah, M. Bruce, E. H. Edelson, A. Husain, *Fuel 63,* 1130 (1984).
48b. G. A. Olah, A. Husain, *Fuel 63,* 1427 (1984).
48c. G. A. Olah, M. Bruce, E. H. Edelson, A. Husain, *Fuel 63,* 1432 (1984).
49. G. A. Olah, J. R. DeMember, R. H. Schlosberg, *J. Am. Chem. Soc. 91,* 2112 (1969).
50. G. A. Olah, J. R. DeMember, R. H. Schlosberg, Y. Halpern, *J. Am. Chem. Soc. 94,* 156 (1972).
51. G. A. Olah, Y. K. Mo, J. A. Olah, *J. Am. Chem. Soc. 95,* 4939 (1973).
52. G. A. Olah, J. R. DeMember, J. Shen, *J. Am. Chem. Soc. 95,* 4952 (1973).
53. G. A. Olah, A. M. White, *J. Am. Chem. Soc. 91,* 5801 (1969).
54. M. Siskin, *J. Am. Chem. Soc. 98,* 5413 (1976).
55. J. Sommer, M. Muller, K. Laali, *Nouv. J. Chim. 6,* 3 (1982).
56. G. A. Olah, J. D. Felberg, K. Lammertsma, *J. Am. Chem. Soc. 105,* 6529 (1983).
57. D. T. Roberts, Jr., L. E. Calihan, *J. Macromol. Sci (Chem).* A7(8), 1629 (1973).
58. D. T. Roberts, Jr., L. E. Calihan, *J. Macromol. Sci (Chem).* A7(8), 1641 (1973).
59. J. M. Lalancette, M. J. Faurrier-Breault, R. Thiffault, *Can. J. Chem. 52,* 589 (1974).
60. J. M. Lalancette, L. Rey, J. Lafontaine, *Can. J. Chem. 54,* 2505 (1976).
61. W. M. Kurts, J. E. Nickels, J. J. McGovern, B. B. Corsa, *J. Org. Chem. 16,* 699 (1951).
62. G. C. Bailey, J. A. Reid, U.S. Patent 2,519,099.
63. J. Kaspi, D. D. Montgomery, G. A. Olah, *J. Org. Chem. 43,* 3147 (1978).
64. J. Kaspi, G. A. Olah, *J. Org. Chem. 43,* 3142 (1978).
65a. S. V. Kannan, C. N. Pillai, *Ind. J. Chem. 8,* 1144 (1970).
65b. Y. Ogata, K. Sakanishi, H. Hosai, *Kogyo Kagaku Zasshi 72,* 1102 (1969); *Chem. Abs. 71,* 80842 (1969).
65c. H. L. Schlichting, A. D. Barbopoulos, W. H. Prahl, U.S. Patent 3,426,358 (1969).
66. J. A. Sharp, R. E. Dean, British Patent 1,125,087 (1968); *Chem. Abs. 70,* 37452 (1969).

67a. T. Kotanigawa, *Bull. Chem. Soc. Jpn. 47,* 950 2466 (1974).

67b. T. Wada, *Japan Kokai,* 75, 76, 033 (1975); *Chem. Abs. 83,* 178557 (1975).

68. T. Kotamigawa, H. Yamamoto, K. Shimokawa, Y. Yoshiba, *Bull. Chem. Soc. Jpn. 44,* 1961 (1971).

69. Y. Iohikawa, Y. Yamanaka, O. Kobayashi, *Japan Kokai,* 73 99 129 (1973); *Chem. Abs. 80,* 95477 (1974); *Japan Kokai,* 73 96 529 (1973); *Chem. Abs. 80,* 95481 (1974); *Japan Kokai,* 73 99 128 (1973); *Chem. Abs. 80,* 95486 (1974).

70. W. Funakashi, T. Urasaki, I. Oda, T. Shima, *Japan Kokai,* 73 97 825 (1973); *Chem. Abs. 80,* 59675 (1974); *Japan Kokai,* 74 07 235 (1974); *Chem. Abs. 80,* 120534 (1974); *Japan Kokai,* 74 13 128 (1974); *Chem. Abs. 80,* 120532 (1974); *Japan Kokai,* 74 14 432 (1974); *Chem. Abs. 80,* 120527 (1974); *Japan Kokai,* 74 18 834 (1974); *Chem. Abs. 81,* 3586 (1974).

71. G. A. Olah, D. Meidar, unpublished results.

72. G. A. Olah, D. Meidar, *Nouv. J. Chim. 3,* 269 (1979).

73. G. E. Langlois, U.S. Patent 2,713,600 (July 19, 1955); *Chem. Abs. 50,* 5738 (1956).

74. G. A. Olah, D. Meidar, R. Malhotra, J. A. Olah, S. C. Narang, *J. Catalysis 61,* 96 (1980).

75a. G. A. Olah, J. Kaspi, *Nouv. J. Chim. 2,* 581 (1978).

75b. P. Beltrame, P. L. Beltrame, P. Carniti, M. Magnoni, *Gazz. Chim. Ital. 108,* 651 (1978).

75c. P. L. Beltrame, P. Carniti, G. Nespoli, *Ind. Eng. Chem. Prod. Res. Dev. 19,* 205 (1980).

76. G. A. Olah, J. Kaspi, *Nouv. J. Chim. 2,* 585 (1978).

77. G. A. Olah, M. W. Meyer, N. A. Overchuk, *J. Org. Chem. 29,* 2313 (1964).

78a. D. A. McCaulay, in *Friedel-Crafts and Related Reactions,* Vol. 2, G. A. Olah, Ed., Wiley-Interscience, New York, 1964, Chap. 24.

78b. H. Matsumoto, H. Take, Y. Yoneda, *J. Catalysis, 11,* 211 (1968); *Ibid. 19,* 113 (1970).

79a. G. A. Olah, unpublished results.

79b. G. A. Olah, J. R. DeMember, *Ibid. 91,* 2113 (1969); *92,* 718, 2562 (1970).

79c. G. A. Olah, J. R. DeMember, Y. K. Mo, J. J. Svoboda, P. Schilling, J. A. Olah, *Ibid. 96,* 884 (1974).

80. F. Effenberger, E. Sohn, G. Epple, *Angew. Chem. Int. Ed. Engl. 11,* 300 (1972).

81. F. Effenberger, E. Sohn, G. Epple, *Chem. Ber. 116,* 1195 (1983).

82. R. D. Howells, J. D. McCown, *Chem. Rev. 77,* 69 (1977).

83. T. R. Forbus, Jr., J. C. Martin, *J. Org. Chem. 44,* 313 (1979).

84. G. A. Olah, R. Malhotra, S. C. Narang, J. A. Olah, *Synthesis,* 672 (1978).

85. G. Baddeley, A. G. Pendleton, *J. Chem. Soc.* 807 (1952); W. M. Schubert, H. K. Latomette, *J. Am. Chem. Soc. 74,* 1829 (1952); I. Agranat, Y. S. Shih, Y. Bentar, *J. Am. Chem. Soc. 96,* 1259 (1974); I. Agranat, Y. Bentar, Y. S. Shih, *J. Am. Chem. Soc. 99,* 7068 (1977); A. D. Andreou, R. V. Bulbulian, P. H. Gore, *J. Chem. Res.* 225 (1980); A. D. Andreou, R. V. Bulbulian, P. H. Gore, D. F. C. Morris, E. L. Short, *J. Chem. Soc. Perkin II,* 830 (1981).

86. R. H. Schlosberg, R. P. Woodbury, *J. Org. Chem. 37,* 2627 (1972).

87. T. Keumi, T. Morita, T. Shimada, M. Ikeda, N. Teshima, H. Kitajima, G. K. S. Prakash, *J. Am. Chem. Soc.* submitted.

88. G. A. Olah, K. Laali, A. K. Mehrotra, *J. Org. Chem. 48,* 3360 (1983).

89. G. A. Olah, J. A. Olah, *Friedel-Crafts and Related Reactions,* Vol. 3, Part 2, Wiley-Interscience, New York, 1964, pp. 1272.

90. H. Hogeveen, *Adv. Phys. Org. Chem. 10,* 29 (1973).

91. W. F. Gresham, G. E. Tabet, U.S. Pat. 2,485,237 (1946).

92. B. L. Booth, T. A. El-Fekky, *J. Chem. Soc. Perk. I,* 2441 (1979).

93. Y. Takahashi, N. Tamita, N. Yoneda, A. Susuki, *Chem. Lett.* 997 (1977).

94. N. Yoneda, Y. Takahashi, Y. Sakai, A. Susuki, *Chem. Lett.* 1151 (1978).
95. R. Paatz, G. Weisberger, *Chem. Ber. 100,* 984 (1967).
96. N. Yoneda, T. Fukuhara, Y. Takahashi, A. Susuki, *Chem. Lett.* 17 (1983).
97. N. Yoneda, H. Sato, T. Fukuhara, Y. Takahashi, A. Susuki, *Chem. Lett.* 19 (1983).
98. G. A. Olah, A. M. White, D. M. O'Brien, *Chem. Rev. 70,* 591 (1970).
99. J. M. Coustard, J. C. Jacquesy, *J. Chem. Res(S),* 280 (1977).
100. B. L. Booth, T. A. El-Fekky, G. F. M. Noori, *J. Chem. Soc. Perk. I,* 181 (1980).
101. L. Gatterman, J. A. Koch, *Chem. Ber. 30,* 1622 (1897).
102. N. N. Crounse, *Org. Reactions 5,* 290 (1949).
103. G. A. Olah, S. J. Kuhn, *Friedel-Crafts and Related Reactions,* Vol. 3, Part 2, Wiley-Interscience, New York, 1964, p. 1153.
104. S. Fujiyama, M. Takagawa, S. Kajiyama, G. P. 2, 425591 (1974).
105. J. M. Delderick, A. Kwantes, B. P. 1,123,966 (1968).
106. G. A. Olah, F. Pelizza, S. Kobayashi, J. A. Olah, *J. Am. Chem. Soc. 98,* 296 (1976).
107. A. Rieche, H. Gross, E. Hoft, *Chem. Ber. 93,* 88 (1960).
108. G. A. Olah, A. Kuhn, *J. Am. Chem. Soc. 82,* 2380 (1961).
109. G. A. Olah, G. K. S. Prakash, unpublished results.
110. G. A. Olah, J. A. Olah in *Friedel-Crafts and Related Reactions,* Vol. 3, Part 2, G. A. Olah Ed., Wiley-Interscience, New York, 1964, p. 1257.
111a. H. Jorg, *Chem. Ber. 60,* 1466 (1972).
111b. A. Treibs, R. Friess, *Justus. Liebigs Ann. Chem. 737,* 173 (1970).
111c. P. D. George, *J. Org. Chem. 26,* 4235 (1961).
112. P. D. George, *J. Org. Chem. 26,* 4235 (1961).
113. G. A. Olah, M. R. Bruce, F. L. Clouet, *J. Org. Chem. 46,* 438 (1981).
114. H. Cerfontain, in *Mechanistic Aspects in Aromatic Sulfonation and Desulfonations,* Wiley-Interscience, New York, 1968.
115. A. Koeberg-Telder, H. Cerfontain, *Recl. Trav. Chim. Pays-Bas. 90,* 193 (1971).
116. L. Leirson, R. W. Bost, R. LeBaron, *Ind. Eng. Chem. 40,* 508 (1948).
117. L. I. Levina, S. N. Patrakova, D. A. Patruskev, *J. Gen. Chem. U.S.S.R. 28,* 2464 (1958).
118. R. J. Vaughn, U.S. Pat. 4308215; *Chem. Abs. 96,* 87450a (1982).
119a. For a current review see G. A. Olah, S. C. Narang, J. A. Olah, K. Lammertsma, *Proc. Natl. Acad. Sci. USA, 79,* 4487 (1982).
119b. K. Schofield, *Aromatic Nitration,* Cambridge Univ. Press, New York, 1971.
120. G. A. Olah, S. J. Kuhn, A. Mlinko, *J. Chem. Soc. 80,* 5871 (1958).
121. G. A. Olah, H. C. Lin, *Synthesis,* 444 (1974).
122. G. A. Olah, H. C. Lin, *J. Am. Chem. Soc. 93,* 1259 (1971).
123a. G. A. Olah, J. A. Olah, N. A. Overchuk, *J. Org. Chem. 30,* 3373 (1965).
123b. C. A. Cupas, R. L. Pearson, *J. Am. Chem. Soc. 90,* 4742 (1968).
123c. G. A. Olah, S. C. Narang, R. L. Pearson, C. A. Cupas, *Synthesis,* 452 (1978).
124. G. A. Olah, M. Nojima, *Synthesis,* 785 (1973).
125. G. A. Olah, B. G. B. Gupta, S. C. Narang, *J. Am. Chem. Soc. 101,* 5317 (1979).
126a. G. A. Olah, B. G. B. Gupta, A. G. Luna, S. C. Narang, *J. Org. Chem. 48,* 1760 (1983).
126b. G. A. Olah, B. G. B. Gupta, *J. Org. Chem. 48,* 3585 (1983).
127a. G. A. Olah, T. L. Ho, *Synthesis,* 609 (1976).
127b. *Ibid.* 610 (1976).
127c. *J. Org. Chem. 42,* 3097 (1977).

127d. T. L. Ho, G. A. Olah, *Synthesis,* 418 (1977).
127e. G. A. Olah, S. C. Narang, G. F. Salem, B. G. B. Gupta, *Ibid.* 273 (1979).
128. G. A. Olah, R. Malhotra, S. C. Narang, *J. Org. Chem. 43,* 4628 (1978).
129. R. J. Vaughn, Australian Pat. 506423, *Chem. Abs. 93,* 46166 (1980).
130. G. A. Olah, V. V. Krishnamurthy, S. C. Narang, *J. Org. Chem. 47,* 596 (1982).
131. G. A. Olah, S. C. Narang, R. Malhotra, J. A. Olah, *J. Am. Chem. Soc. 101,* 1805 (1979).
132. D. Cook in *Friedel-Crafts and Related Reactions,* Vol. I, G. A. Olah Ed., Wiley-Interscience, New York, 1963, p. 769.
133. C. K. Ingold, *Structure and Mechanism in Organic Chemistry,* Cornell University Press, Ithaca, N.Y., 1953.
134. J. Allan, J. Podstata, D. Snobl, J. Jarkovsky, *Tetrahedron Lett. 40,* 3565 (1965).
135. G. A. Olah, N. Friedman, *J. Am. Chem. Soc. 88,* 5330 (1966).
136. G. A. Olah, G. Salem, J. S. Staral, T. L. Ho, *J. Org. Chem. 43,* 173 (1978).
137. G. A. Olah, *Acc. Chem. Res. 13,* 330 (1980).
138. G. A. Olah, B. G. B. Gupta, *J. Org. Chem. 45,* 3532 (1980).
139. G. A. Olah, J. Shih, B. P. Singh, B. G. B. Gupta, *J. Org. Chem. 48,* 3356 (1983).
140. G. A. Olah, B. G. B. Gupta, S. C. Narang, *Synthesis,* 274 (1979).
141. G. A. Olah, J. Shih, B. P. Singh, B. G. B. Gupta, *Synthesis,* 713 (1983).
142. M. L. Poutsma, *Methods in Free-Radical Chemistry,* Vol. II, E. S. Husyer Ed., Marcel Dekker, New York, 1969.
143a. G. A. Olah, Y. K. Mo, *J. Am. Chem. Soc. 94,* 6864 (1972).
143b. G. A. Olah, R. Renner, P. Schilling, Y. K. Mo, *J. Am. Chem. Soc. 95,* 7686 (1973).
144. G. A. Olah, U.S. Patent, applied.
145. G. A. Olah, P. Schilling, *J. Am. Chem. Soc. 95,* 7680 (1973).
146. D. Alker, D. H. R. Barton, R. H. Hesse, J. L. James, R. E. Markwell, M. M. Pechet, S. Rozen, T. Takeshita, H. T. Toh, *Nouv. J. Chim. 4,* 239 (1980).
147. K. O. Christe, *J. Fluo. Chem.* 519 (1983); K. O. Christe, *J. Fluo. Chem.* 269 (1984).
148. C. Gal, S. Rozen, *Tetrahedron Lett.* 449 (1984).
149. H. P. Braendlin, E. T. McBeein, in *Friedel-Crafts and Related Reactions,* Vol. 3, G. A. Olah Ed., Part 2, p. 1517, Wiley-Interscience, New York, 1964.
150a. J. C. Jacquesy, M. P. Jouannetaud, S. Makani, *Chem. Commun.* 110 (1980).
150b. J. C. Jacquesy, M. P. Jouannetaud, S. Makani, *Nouv. J. Chim. 4,* 747 (1980).
151. J. C. Jacquesy, M. P. Jouannetaud, *Tetrahedron Lett.* 1673 (1982).
152a. C. Graebe, *Ber. Dtsch. Chem. Ges. 34,* 1778 (1901).
152b. G. F. C. R. Jaubert, *Hebd. Seances Acad. Sci. 132,* 841 (1901).
152c. J. F. de Turski, German Patent 287 756, 1914.
152d. British Patent 626 661, 1949.
152e. U.S. Patent 2,585,355, 1952.
152f. A. Bertho, *Ber. Dtsch. Chem. Ges. 59,* 589 (1926).
153a. R. N. Keller, P. A. S. Smith, *J. Am. Chem. Soc. 66,* 1122, 1944.
153b. *Ibid. 68,* 899 (1946).
153c. P. A. S. Smith, Ph.D Thesis, University of Michigan, Ann Arbor, 1944.
154a. K. F. Schmidt, *Ber. Dtsch. Chem. Ges. 57,* 704 (1924).
154b. K. F. Schmidt, *Ibid. 58,* 2413 (1925).
154c. K. F. Schmidt, P. U. S. Zutavern, U.S. Patent 1,637,661 (1927).
154d. G. M. Hoop, J. M. Tedder, *J. Chem. Soc.* 4685 (1961).

155a. W. Borche, H. Hahn, *Ber. Dtsch. Chem. Ges. 82,* 260 (1949).
155b. E. Bamberger, *Liebigs Ann. Chem. 424,* 233 (1921).
155c. E. Bamberger, J. Brun, *Helv. Chim. Acta 6,* 935 (1923).
155d. C. L. Arcus, J. V. Evans, *J. Chem. Soc.* 789 (1958).
156a. P. A. S. Smith, J. H. Hall, *J. Am. Chem. Soc. 84,* 480 (1962).
156b. G. Smolinsky, *Ibid. 83,* 2489 (1961).
156c. T. Curtius, *J. Prakt. Chem. 125,* 303 (1930).
156d. J. F. Heacock, M. T. Edmison, *J. Am. Chem. Soc. 82,* 3460 (1960).
156e. J. F. Tilney-Bassett, *J. Chem. Soc. 2517* (1962).
156f. J. E. Leffler, Y. Tsumo, *J. Org. Chem. 28,* 902 (1963).
156g. T. Curtius, A. Bertho, *Ber. Dtsch. Chem. Ges. 59,* 565 (1926).
157a. D. H. R. Barton, L. R. Morgan, Jr., *J. Chem. Soc.* 623 (1962).
157b. K. Stewart, *Trans. Faraday Soc. 41,* 663 (1945).
158a. P. Kocacic, J. A. Levinsky, C. T. Goralski, *J. Am. Chem. Soc. 88,* 100 (1966).
158b. J. W. Strand, P. Kovacic, *Ibid. 95,* 2977 (1973).
159. P. Kovacic, R. L. Russell, R. P. Bennett, *J. Am. Chem. Soc. 86,* 1588 (1964).
160. A. L. Mertens, K. Lammertsma, M. Arvanaghi, G. A. Olah, *J. Am. Chem. Soc. 105,* 5657 (1983).
161. E. Wiberg, H. C. Michaud, *Naturforsch. B, 9B,* 495 (1954).
162. N. Wiberg, W. C. Joo, K. H. Schmid, *Anorg. Allg. Chem.* 394, 197 (1972).
163. G. A. Olah, D. G. Parker, N. Yoneda, *Angew. Chem. Int. Ed. Engl. 17,* 909 (1978).
164. K. O. Christe, W. W. Wilson, E. C. Curtis, *Inorg. Chem. 18,* 2578 (1976).
165. G. A. Olah, A. L. Berrier, G. K. S. Prakash, *J. Am. Chem. Soc. 104,* 2373 (1982).
166. G. A. Olah, N. Yoneda, D. G. Parker, *J. Am. Chem. Soc. 99,* 483 (1977).
167. G. A. Olah, D. G. Parker, N. Yoneda, F. Pelizza, *J. Am. Chem. Soc. 98,* 2245 (1976).
168a. D. H. Derbyshire, W. A. Waters, *Nature (London) 164,* 401 (1950).
168b. R. D. Chambers, P. Goggin, W. K. R. Musgrave, *J. Chem. Soc.* 1804 (1959).
168c. J. D. McClure, P. H. Williams, *J. Org. Chem. 27,* 24 (1962).
168d. J. D. McClure, P. H. Williams, *Ibid. 27,* 627 (1962).
168e. C. A. Buehler, H. Hart, *J. Am. Chem. Soc. 85,* 2117 (1963).
168f. A. J. Davidson, R. O. C. Norman, *J. Chem. Soc.* 5404 (1964).
168g. H. Hart, C. A. Buehler, *J. Org. Chem. 29,* 2397 (1964).
168h. H. Hart, C. A. Buehler, A. J. Waring, *Adv. Chem. Ser. 51,* 1 (1965).
168i. S. Hashimoto, W. Koike, *Bull. Chem. Soc. Jpn. 43,* 293 (1970).
168j. J. A. Vesely, L. Schmerling, *J. Org. Chem. 35,* 4028 (1970).
168k. M. K. Kurz, G. J. Johnson, *Ibid. 36,* 3184 (1971).
169. R. O. C. Norman, R. Taylor, *Electrophilic Substitution in Benzenoid Compounds,* Elsevier, Amsterdam, 1965.
170. G. A. Olah, R. Ohnishi, *J. Org. Chem. 43,* 865 (1978).
171. G. A. Olah, A. P. Fung, T. Keumi, *J. Org. Chem. 46,* 4305 (1981).
172. G. A. Olah, unpublished results.
173a. J. P. Gesson, J. C. Jacquesy, M. P. Jouannetaud, *Chem. Commun.* 1128 (1980).
173b. J. P. Gesson, J. C. Jacquesy, M. P. Jouannetaud, *Nouv. J. Chim. 6,* 477 (1982).
173c. J. P. Gesson, L. D. Giusto, J. C. Jacquesy, *Tetrahedron 34,* 1715 (1978).
174a. J. P. Gesson, J. C. Jacquesy, M. P. Jouannetaud, G. Morelett, *Tetrahedron Lett.* 3095 (1983).

174b. J. C. Jacquesy et al. unpublished results.

174c. J. C. Jacquesy, M. P. Jouannetaud, G. Morelett, *Ibid.* 3099 (1983).

175. R. Trambarulo, S. N. Ghosh, C. A. Barrus, W. Gordy, *J. Chem. Phys. 24,* 851 (1953).

176. P. D. Bartlett, M. Stiles, *J. Am. Chem. Soc. 77,* 2806 (1955).

177. J. P. Wibault, E. L. J. Sixma, L. W. E. Kampschidt, H. Boer, *Recl. Trav. Chim. Pays-Bas 69,* 1355 (1950).

178. J. P. Wibault, E. L. J. Sixma, *Recl. Trav. Chim. Pays-Bas 71,* 76 (1951).

179. P. S. Bailey, *Chem. Rev. 58,* 925 (1958).

180. L. Horner, H. Schaefer, W. Ludwig, *Chem. Ber. 91,* 75 (1958).

181. P. S. Bailey, J. W. Ward, R. E. Hornish, F. E. Potts, *Adv. Chem. Ser. 112,* 1 (1972).

182. G. A. Olah, N. Yoneda, D. G. Parker, *J. Am. Chem. Soc. 98,* 5261 (1976).

183. Further oxidation products, i.e., acetylium ion and CO_2, were reported to be observed in a number of the reactions studied. Such secondary oxidations are not induced by ozone.

184. T. M. Hellman, G. A. Hamilton, *J. Am. Chem. Soc. 96,* 1530 (1974).

185. M. C. Whiting, A. J. N. Bolt, J. H. Parrish, *Adv. Chem. Ser. 77,* 4 (1968).

186. D. O. Williamson, R. J. Cvetanovic, *J. Am. Chem. Soc. 92,* 2949 (1970).

187a. J. C. Jacquesy, R. Jacquesy, L. Lamande, C. Narbonne, J. F. Patoiseau, Y. Vidal, *Nouv. J. Chim. 6,* 589 (1982).

187b. J. C. Jaquesy, J. F. Patoiseau, *Tetrahedron Lett.* 1499 (1977).

188a. J. P. Kennedy, in *Polymer Chemistry of Synthetic Elastomers,* J. P. Kennedy, E. G. M. Tornauist, Eds., Wiley-Interscience, New York, 1968.

188b. J. P. Kennedy, E. Marechal, *Carbocationic Polymerization,* Wiley-Interscience, New York, 1982.

189. G. A. Olah, H. W. Quinn, S. J. Kuhn, *J. Am. Chem. Soc. 82,* 426 (1960).

190. G. A. Olah, *Macromol. Chem. 175,* 1039 (1974).

191. G. Pruckmayr, U.K. Pat. 2,025,955; *Chem. Abs. 93,* 47469V (1980).

192. G. A. Pruckmayr, U.S. Pat. 4,153,786; *Chem. Abs. 91,* p58077a (1979).

193. Sanyo Chemical Industries, Ltd., Japanese Patent 8 226 626; *Chem. Abs. 97,* 72986b (1982).

194. E. T. Marquis, L. W. Watts, W. H. Braker, J. W. Aden, Belgian Patent 891 533; *Chem. Abs. 97,* 185195m (1982).

195. H. Hasegawa, T. Higashimura, *Polymer J. 11,* 737 (1979).

196. T. Higashimura, H. Nishii, *J. Polym. Sci. Polym. Chem., 15,* 329 (1977).

197. G. E. Heinsohn et al., *Ger. Offen.* 2,709,280, *Chem. Abs., 87,* P202353.

198. G. E. Heinsohn, I. M. Robinson, G. Pruckmeyr, W. W. Gilbert, U.S. Pat 4,163,115; *Chem. Abs., 91,* P158335b (1979).

199. E. I. duPont de Nemours & Co., Belg. Pat. 868,381, *Chem. Abs. 90,* p. 204877j (1979).

200. G. Pruckmayr, R. H. Weir, U.S. 4,120,903, *Chem. Abs. 90,* P24000S (1979).

201. G. Pruckmayr, U.S. 4,139,567, *Chem. Abs. 90,* 186,474u (1979).

202. E. I. duPont de Nemours & Co. Fr. Demande, 2429234; *Chem. Abs. 92,* P199059 (1980).

203. W. R. Cares, U.S. Pat. 4,065,512, *Chem. Abs., 89,* P6805a (1978).

204. J. Simonsen, *The Terpenes,* Vols. 1 and 2 (1947, 1949); J. Simonsen, D. H. R. Barton, Vol. 3 (1949); J. Simonsen, W. C. J. Ross, Vols. 4 and 5 (1957); Cambridge University Press, Cambridge (1947–57).

205. J. King, P. de Mayo, in *Molecular Rearrangements,* P. de Mayo, Ed., Vol. 2, Interscience, New York, 1964.

206. G. Ourisson, *Proc. Chem. Soc.* 274 (1964).

207. T. S. Santhanakrishnan, *J. Sci. Ind. Res. 12,* 631 (1965).

208. W. G. Dauben, *J. Agr. Food Chem. 22,* 156 (1974).
209. T. S. Sorensen, *Acc. Chem. Res. 9,* 257 (1976).
210. J. K. Sutherland, *Chem. Soc. Rev.* 265 (1980).
211. L. Ruzicka, A. Eschenmoser, H. Heusser, *Experimentia 9,* 357 (1953); L. Ruzicka, *Proc. Chem. Soc.* 341 (1959).
212. J. B. Hendrickson, *Tetrahedron 7,* 83 (1959).
213. J. H. Richards, J. B. Hendrickson, *The Biosynthesis of Steroids, Terpenes and Acetogenins,* Benjamin, New York, 1964, p. 173–288.
214. W. Parker, J. S. Roberts, R. Ramage, *Quart. Rev. (London) 21,* 331 (1967).
215. E. E. van Tamelen, *Fortschr. Chem. Org. Naturstoffe 19,* 245 (1961); *Acc. Chem. Res. 1,* 111 (1968).
216. W. S. Johnson, *Acc. Chem. Res. 1,* 1 (1968).
217. W. S. Johnson, *Angew. Chem. Int. Ed. Engl. 15,* 9 (1976).
218. S. Dev, *Acc. Chem. Res. 14,* 82 (1981).
219. E. E. van Tamelen, *Acc. Chem. Res. 8,* 152 (1975).
220. H. E. Armstrong, F. S. Kipping, *J. Chem. Soc. (London) 63,* 75 (1893).
221. R. P. Lutz, J. d. Roberts, *J. Am. Chem. Soc. 84,* 3715 (1962).
222. O. R. Rodig, R. J. Sysko, *J. Am. Chem. Soc. 94,* 6475 (1972).
223. J. C. Jacquesy, R. Jacquesy, J. F. Patoiseau, *Tetrahedron Lett. 32,* 1699 (1976).
224. E. Huang, K. Ranganayakalu, T. S. Sorensen, *J. Am. Chem. Soc. 94,* 1780 (1972).
225. R. P. Haseltine, T. S. Sorensen, *Can. J. Chem. 53,* 1067 (1975).
226. D. V. Banthorpe, P. A. Boullier, W. D. Fordham, *J. Chem. Soc. Perkin I,* 1637 (1974).
227a. D. G. Garnum, G. Mehta, *Chem. Commun.* 1643 (1968).
227b. D. G. Farnum, R. A. Mader, G. Mehta, *J. Am. Chem. Soc. 95,* 8692 (1973).
228. G. Mehta, S. K. Kapoor, *Tetrahedron Lett.* 2385 (1973).
229. J. C. Jacquesy, R. Jacquesy, J. F. Patoiseau, *Chem. Commun. 20,* 785 (1973).
230. J. C. Jacquesy, R. Jacquesy, J. F. Patoiseau, *Bull. Soc. Chim. Fr. II,* 1959 (1974).
231. R. Jacquesy, H. L. Ung, *Tetrahedron, 32,* 1375 (1976).
232. J. C. Jacquesy, R. Jacquesy, C. Narbonne, *Bull. Soc. Chim. Fr. II,* 1240 (1976).
233. R. Jacquesy, C. Narbonne, *Bull. Soc. Chim. Fr. II,* 163 (1978).
234. J. P. Gesson, J. C. Jacquesy, M. Mondon, *Tetrahedron Lett.* 3351 (1980).
235. J. C. Jacquesy, M. P. Jouannetaud, *Bull. Soc. Chim. Fr. II,* 202 (1978).
236. J. C. Jacquesy, M. P. Jouannetaud, *Bull. Soc. Chim. Fr. II,* 267 (1980).
237. J. C. Jacquesy, M. P. Jouannetaud, *Bull. Soc. Chim. Fr. II,* 295 (1980).
238. J. P. Gesson, J. C. Jacquesy, M. P. Jouannetaud, *Bull. Soc. Chim. Fr. II,* 305 (1980).
239. J. P. Gesson, J. C. Jacquesy, R. Jacquesy, *Tetrahedron Lett.* 4733 (1971).
240a. J. P. Gesson, J. C. Jacquesy, R. Jacquesy, *Bull. Soc. Chim. Fr.* 1433 (1973).
240b. J. M. Coustard, J. P. Gesson, J. C. Jacquesy, *Tetrahedron Lett.* 4932 (1972).
241a. J. M. Coustard, J. C. Jacquesy, *Tetrahedron Lett.* 1341 (1972).
241b. J. M. Coustard, J. C. Jacquesy, *Bull. Soc. Chim. Fr.* 2098 (1973).
242. J. P. Gesson, J. C. Jacquesy, *Tetrahedron* 3631 (1973).
243. J. C. Jacquesy, R. Jacquesy, U. H. Ly, *Tetrahedron Lett.* 2199 (1974).
244. J. P. Gesson, J. C. Jacquesy, R. Jacquesy, G. Joly, *Bull. Soc. Chim. Fr.* 1179 (1976).
245. R. Jacquesy, H. L. Ung, *Tetrahedron 33,* 2543 (1977).
246a. D. N. Kurchanov, Z. N. Parnes, M. I. Kalintin, N. M. Lojm, *Ionic Hydrogenation,* Izdatelstvo Chimia, Moscow, 1979, chapter 9, p. 167 (in Russian).

246b. D. D. Whitehurst, T. O. Mitchell, M. Farcasiu, in *Coal Liquefaction,* Academic Press, New York, 1980.

246c. J. Wristers, *J. Am. Chem. Soc., 97,* 4312 (1975).

247. J. C. Jacquesy, R. Jacquesy, G. Joly, *Tetrahedron Lett.* 4433 (1974).

248. J. C. Jacquesy, R. Jacquesy, G. Joly, *Bull. Soc. Chim. Fr.* 2283 (1975).

249. J. C. Jacquesy, R. Jacquesy, G. Joly, *Bull. Soc. Chim. Fr.* 2289 (1975).

250. J. M. Coustard, M. H. Douteau, J. C. Jacquesy, R. Jacquesy, *Tetrahedron Lett.* 2029 (1975).

251. J. C. Jacquesy, R. Jacquesy, G. Joly, *Tetrahedron* 2237 (1975).

252. R. Jacquesy, C. Narbonne, *Chem. Commun.* 765 (1979).

253. J. M. Coustard, M. H. Douteau, R. Jacquesy, P. Longevialle, D. Zimmermann, *J. Chem. Res(S),* 16 (1978).

254. J. M. Coustard, M. H. Douteau, R. Jacquesy, *J. Chem. Res(S),* 18 (1979).

255. R. Jacquesy, C. Narbonne, H. L. Ung, *J. Chem. Res(S),* 288 (1979).

OUTLOOK

Acids have played a fundamental role in the entire history of chemistry. Mineral acids such as sulfuric and perchloric acid were until recently considered as the strongest acids. In the last two decades, emergence of acid systems up to 10^{15} times stronger than 100% sulfuric acid has opened up a whole new area of chemistry. These superacids, reviewed in the present monograph in a systematic way, range from liquid systems such as antimony pentafluoride-fluorosulfuric acid (Magic Acid), antimony pentafluoride–hydrogen fluoride, antimony pentafluoride–trifluoromethanesulfonic acid, and related acids containing arsenic, tantalum, and niobium pentafluorides and the like, to solid superacids such as polymeric perfluorinated resinsulfonic acids (Nafion-H), immobilized or intercalated antimony pentafluoride and related Lewis acid-based systems. The exceedingly high acidity of these systems covering a broad range, allows many weak bases to undergo acid-catalyzed reactions, which otherwise would not take place using conventional acid catalysts. Many hydrocarbon conversions, including those of methane and alkanes are some of the examples of applications catalyzed by superacids that in the future are expected to gain increasing importance. Usually acid-catalyzed reactions, such as isomerization of alkanes, carried out with mineral acids or conventional Friedel–Crafts systems necessitate relatively high temperatures. However, superacids promote such reactions at much lower temperatures, thus allowing favorable isomer distribution of branched hydrocarbons (increasing the octane number significantly).

A whole array of potential weakly basic reagents, such as CO_2, CO, O_3, O_2(singlet), N_2O, Cl_2, Br_2, and the like are starting to show electrophilic reactivity under superacid-catalyzed conditions. These reactions could lead to novel and practical applications and some of them are already emerging. Ionic reactions generally show much higher selectivity than their free radical counterparts. The ability of superacidic catalysts to cleave carbon–carbon bonds with ease has resulted in novel hydrocarbon cracking applications, particularly in systems where metal-based catalysts are susceptible to deactivation (poisoning) by impurities such as sulfur and nitrogen compounds. This is of particular interest in processing of heavy oils, tar sands, oil shale, and in coal conversion–liquefaction processes.

Superacid-catalyzed reactions and ionic reagents (obtained under superacidic, stable ion conditions) are gaining increasing significance in synthetic and natural product chemistry and this trend is expected to continue and expand. Some special applications, such as use of intercalated antimony or arsenic pentafluoride/graphite (or polyacetylene) as highly conducting (even superconducting) materials has attracted and will continue to attract considerable attention.

Future development of new and improved superacidic systems, particularly allowing long onstream time in catalytic applications without deactivation and ease of regeneration is of particular interest. Applications of superacids are foreseen to expand in catalysis, synthetic chemistry, as well as in preparation and study of reactive ionic intermediates.

INDEX